Human Geography

Issues for the 21st Century

PEARSON EDUCATION

We work with leading authors to develop the
strongest educational materials in Geography,
bringing cutting-edge thinking and best learning
practice to a global market.

Under a range of well-known imprints, including
Prentice Hall, we craft high quality
print and electronic publications which help
readers to understand and apply their content,
whether studying or at work.

To find out more about the complete range of our
publishing please visit us on the World Wide Web at:
www.pearsoneduc.com

Human Geography

Issues for the 21st Century

Edited by

Peter Daniels School of Geography and Environmental Sciences, The University of Birmingham

Michael Bradshaw Department of Geography, The University of Leicester

Denis Shaw School of Geography and Environmental Sciences, The University of Birmingham

James Sidaway School of Geography and Environmental Sciences, The University of Birmingham

Prentice Hall

An imprint of **Pearson Education**

Harlow, England · London · New York · Reading, Massachusetts · San Francisco · Toronto · Don Mills, Ontario · Sydney
Tokyo · Singapore · Hong Kong · Seoul · Taipei · Cape Town · Madrid · Mexico City · Amsterdam · Munich · Paris · Milan

Pearson Education Limited
Edinburgh Gate
Harlow
Essex CM20 2JE
England

and Associated Companies throughout the world

Visit us on the World Wide Web at:
http://www.pearsoneduc.com

First published 2001

ISBN 0 582 36799 9

British Library Cataloguing-in-Publication Data
A catalogue record for this book is available from the British Library

Library of Congress Cataloging-in-Publication Data
A catalog record for this book is available from the Library of Congress

10 9 8 7 6 5 4 3 2 1
05 04 03 02 01

Typeset by 3 in 10.5 pt Bembo
Printed and bound in Great Britain by Ashford Colour Press Ltd., Gosport

CONTENTS

PLATES

FIGURES

TABLES

CONTRIBUTORS

Ian Bowler Professor of Human Geography, Department of Geography, University of Leicester. Agricultural geography; rural development; food policy.

Michael Bradshaw Professor of Human Geography, Department of Geography, University of Leicester. Political and economic geography; economic transformation in the post-socialist states and regional change in the Russian Federation.

John Bryson Senior Lecturer in Economic Geography, School of Geography and Environmental Sciences, University of Birmingham. Economic geography; knowledge economies.

Tony Champion Professor of Geography, Department of Geography, University of Newcastle-upon-Tyne. Population distribution trends; migration and residential mobility; housing and land use planning.

Allan Cochrane Professor of Public Policy, Faculty of Social Sciences, The Open University. Urban studies, particularly urban policy; the local state and restructuring.

Philip Crang Reader in Geography, Department of Geography, Royal Holloway, University of London. Cultural and economic geography, especially commodity culture.

Peter W. Daniels Professor of Geography, School of Geography and Environmental Sciences, University of Birmingham. Geography of service industries; trade, foreign direct investment and internationalization of services.

Nick Henry Reader in Urban and Regional Development, Centre for Urban and Regional Development Studies, The University of Newcastle upon Tyne. Economic geography; regional development.

Brian Ilbery Professor of Geography, Division of Geography, Coventry University. Agricultural change and policy.

Murray Low Lecturer in Geography, Department of Geography, University of Reading. Political geography; urban politics; social theory.

Cheryl McEwan   Lecturer in Geography, School of Geography and Environmental Sciences, University of Birmingham. Feminist and cultural geographies; multiculturalism; citizenship and difference in South Africa.

Jane Pollard   Lecturer in Geography, School of Geography and Environmental Sciences, University of Birmingham. Financial services; regional economic development.

Marcus Power   Lecturer in Geography, School of Geography, University of Leeds. Geographies of development; the politics of ethnic and gender identities in Africa; globalization and geopolitics.

Denis J.B. Shaw   Reader in Russian Geography, School of Geography and Environmental Sciences, University of Birmingham. Historical geography of Russia; history of geographical thought.

James D. Sidaway   Lecturer in Geography, School of Geography and Environmental Sciences, University of Birmingham. Transformations in Southern Africa; regional communities; Portuguese and Spanish geopolitical discourses; the sociology of geographical knowledges.

Terry R. Slater   Reader in Historical Geography, School of Geography and Environmental Sciences, University of Birmingham. Historical geography; urban morphology; medieval towns.

Editing a book is a learning experience. If nothing else, we have discovered that even amongst just four human geographers there is considerable diversity in our academic interests and motivations, our approaches and priorities for understanding the world around us, and our philosophical and methodological approaches towards the study of human geography. The importance each of us attaches to the significance of key concepts and processes such as capitalism, globalization, technological change, resource management, culture and identity, or geopolitics is also equally varied. For those familiar with human geographers this perhaps comes as no surprise! Yet from such diversity emerges unity. It is built around a shared belief that at the beginning of the twenty-first century many of the issues outlined in this book remain dependent on the ways in which human activities are shaped by space, place and time. Other disciplines such as economics, sociology or political studies are, in our opinion rather belatedly, acknowledging this. Therefore the challenge for human geographers at the beginning of the new millennium is not just how to use their skills and insights to address some of the issues identified in this book, but to take the lead in providing explanations or at least posing different questions.

An academic rationale for a book (see Chapter 1) counts for little unless it is matched by a will to transform some initial and sketchy ideas into something more tangible. A very early source of encouragement was Matthew Smith, Social Sciences Editor at Pearson Education. He used some market analysis undertaken in 1996 to try to convince us that we should consider compiling a proposal for an introductory human geography textbook suitable for adoption as a course text in the UK. Ever since some exploratory discussions at the Institute of British Geographers' Conference in Exeter (1997) about the potential for such a project and how it might be delivered, Matthew has been unswerving in his commitment to making it happen. We hope that the faith he has consistently shown in all those involved will have proved worthwhile. This external influence coincided with some internal discussions amongst the human geographers in the School of Geography and Environmental Sciences at the University of Birmingham about the structure and contents of its first-year human geography modules. The approach that we were contemplating, elements of which we were already teaching, coincided with the views that came through via colleagues in the market analysis undertaken in the UK and in North America. We were already faced with large numbers of first-year students and the prospect of future increases. Taking account of the legal restrictions on producing course packs and the intense competition amongst our students for the learning resources in the library, there seemed to be a case for an introductory course text.

Having identified the opportunity, there was still a requirement to assemble the team of editors and authors who would subscribe to the book proposal to be considered by Pearson Education. It was hoped that it could be an exclusively 'Birmingham' project but it proved difficult to match the proposed themes in the book with those colleagues available and willing to be involved. We are extremely grateful to all our colleagues in the School who enthusiastically embraced the idea from the start as well as the one or two 'reluctants' who needed to be persuaded to come on board. We hope that they feel that their involvement has proved worthwhile. It did not prove difficult to identify the authors to cover the topics that could not be covered internally and we were delighted by their positive and enthusiastic responses to our invitations to join the team. The response to the book proposal from external reviewers was helpful and many of their comments and suggestions have been incorporated in the completed work.

The discharge of our responsibilities as editors has undoubtedly been helped by the willingness of all the authors in the team to keep to deadlines or to respond promptly to the inevitable queries and clarifications that arise as a book makes the transition from first draft chapters to page proofs. This has been especially important during the later stages of conversion from the final manuscript to finished product when the time schedule has been particularly tight. How did editors and publishers manage before the days of e-mail and the facsimile machine? A sense of common purpose has been fostered by a number of half-day meetings in Birmingham hosted by Matthew Smith and arranged with the help of Claire Clark at the School of Geography and Environmental Sciences. These meetings have facilitated the sharing of ideas, information about progress, opportunities to comment on the design and format of the book, and discussions about the design and structure of the companion web-site. Such is the pace of development in learning support and the expectations of course instructors that the latter was not an anticipated requirement when we accepted the original contract. This is probably the first introductory human geography textbook put together by UK-based human geographers to incorporate dedicated web support and we are grateful to all the authors for agreeing to undertake some of the additional tasks involved, both now and in the future. A glossary and a bibliography are more 'standard' features and we would like to thank Jon Oldfield for the prompt and efficient manner with which he liaised with the editors to compile the former. He also checked the entries in the consolidated bibliography at the end of the book.

We would like to acknowledge the invaluable contribution of several individuals whose specialist knowledge and skills have been vital, especially following delivery of the final manuscript to Pearson Education. Carol Gardiner, who channelled a steady stream of queries through the editors to authors, undertook copy editing with tact and patience; this ensured that there was adequate time to research and return the information requested. One of the more demanding tasks was undertaken by the picture researcher, Faith Perkins, who had to use suggestions/ideas at various levels of detail from the authors to prepare a portfolio of plates and slides from which those included in the book could be selected. She not only produced a wide selection to an extremely demanding schedule but also found time to provide an excellent lunch during 'selection day' at her home in Newbury. While some of the roughs of the figures in the book were capably prepared by Ann Anckorn and Kevin Burkhill in the Drawing Office in the School of Geography and Environmental Sciences, all the final artwork was undertaken by David Hoxley who often liaised patiently and constructively with individual authors about their detailed requirements. Suki Cheyne came to a half-day meeting in Birmingham to present her plans for the dedicated web page for the book and provoked a wide-ranging discussion about aspects of the content that has hopefully benefited the finished product.

Claire Brewer, Senior Editor in the Production Department at Pearson Education deserves our particular thanks for the cheerful and ever helpful way in which she has acted as the 'go between' dealing with the numerous queries and problems that inevitably arise when working on a book of this kind. She always kept us fully informed about the production schedule and has given us deadlines for the completion of various tasks that suggested some empathy for the numerous demands on the editors' time, not least during the May/June examination period! There are no doubt numerous other 'unseen' individuals at Pearson Education to whom we owe our thanks, not least the book design team and those involved with the promotion and marketing of the book.

Of course, it is on the basis of the decisions made by the editors (with guidance and advice from Pearson Education) that the book will ultimately be judged. Throughout its gestation we have held regular meetings to pool ideas, to debate first drafts, to sort out the division of labour between us, to address and argue about problems or to agree priorities and deadlines for specific tasks. During all of this we have certainly learned something about each other; our strengths and weaknesses and how to work as an editorial team. It is pleasing to report that we are still smiling and talking to each other!

Peter Daniels
Michael Bradshaw
Denis Shaw
James Sidaway

Birmingham, July 2000

PUBLISHER'S ACKNOWLEDGEMENTS

We are grateful to the following for permission to reproduce copyright material:
Figures 1.2(a) and 1.2(b) from Sheratt, A. (ed.) (1980) *The Cambridge Encyclopedia of Archaeology*, reprinted by permission of Cambridge University Press; Figure 1.3 from Whittlesey, D. 'Major agricultural regions of the earth' *Annals of the Association of American Geographers* 2b (1936), copyright Association of American Geographers, reproduced by permission of Blackwell Publishers; Figure 1.5 from Salway, P. (ed.) (1981) *Roman Britain (Oxford History of England)*, © Oxford University Press 1981, reproduced by permission of Oxford University Press; Figures 1.6, 1.7 and 3.2 from *Times Atlas of World History* (1989), reprinted by permission of HarperCollins Publishers; Figure 2.1 from Gutkind, E. A. (1964) *Urban Development in Central Europe*, reproduced by permission of Macmillan Ltd; Figure 26(a) from *Black's General Atlas*, reproduced by permission of A & C Black (Publishers) Ltd; Figure 2.7 from Giles and Goodall (1992) *Yorkshire Textile Mills 1770–1930* (pub HMSO), © Crown Copyright, NMR, reproduced by permission of National Monuments Record; Figure 2.9 reprinted by permission of Popperfoto; Figure 2.10 from Kotler, M. (1990) 'The morphological evolution of a nineteenth-century city centre; Lodz, Poland, 1825–1973' in Slater, T. R. (ed.) *The Built Form of Western Cities* (pub Leicester University Press), copyright Continuum, reproduced by permission of Cassell; Figure 3.1 from Mitchell, J. B. (1962) *Great Britain: Geographical Essay*, by permission of Cambridge University Press and the author; Figure 5.1 from Veldman, J. (1984) 'Proposal for a theoretical basis for the human geography of rural areas' in Groenendijk, C. G. and Thissen, F. (eds) *The Changing Countryside* (pub Geo Books), reprinted by permission of Elsevier; Figure 5.2 from Cloke, P. (1985) 'Counterurbanisation: a rural perspective' *Geography* 70 and Figure 5.3 from Ogden, P. E. (1985) 'Counterurbanisation in France: the results of the 1982 population census' *Geography* 70, reprinted by permission of The Geographical Association; Figures 8.1 and 8.3 from Rees, J. (1985) *Natural Resources: Allocation, Economics and Policy*, Second Edition (pub Routledge), Figure 8.2 and Table 8.1 from Rees, J. (1991) 'Resources and environment: scarcity and sustainability' in Bennett, R. and Estall, R. (eds) *Global Change and Challenge: Geography in the 1990s* (pub Routledge), reprinted by permission of Taylor & Francis Books Ltd; Figure 8.4 from Jones, G. and Hollier, G. (1997) *Resources, Society and Environmental Management*, reprinted by permission of Paul Chapman Publishing and the authors; Figure 8.11 from Odell, P. R. (1989) 'Draining the world of energy' in Johnston,

R. J. and Taylor, P. J. (eds) *A World in Crisis?*, Second Edition, reprinted by permission of Blackwell Publishers; Figure 9.1 reprinted by permission of R. J. Smedley, Wensleydale Dairy Products, Aardman Animations Ltd and J. M. Nuttall; Figure 9.2(a) from Ilbery, B. and Kneafsey, M. 'Producer constructors of quality in regional speciality food production: a case study from South West England' *Journal of Rural Studies* 16, 2000, reprinted by permission of Elsevier Science Ltd; Figure 9.2(b) from Ilbery, B. and Kneafsey, M. 'Registering regional speciality food and drink products in the UK: the case of PDOs and PGIs' *Area* 32 (4) 2000, pp. 317–326 and Figure 15.4 from Mackinder, H. 'The geographical pivot of history' *The Geographical Journal* 23, 1904, p. 121, reprinted by permission of the Royal Geographical Society (with The Institute of British Geographers); Figure 10.1 from *The World Bank, Uganda and the Pearl of Africa* (pub 1998), reproduced by permission of Channel 4 and International Broadcasting Trust; Figures 10.2 and 10.4 from *World Bank Atlas 1998*, reprinted by permission of World Bank; Figure 10.3 from 'The World Bank and the State: a Recipe for Change?' by Nicholas Hildyard, Bretton Woods Project, reprinted by permission of the author; Figure 11.3 copyright © 1997 Larry Landweber and the Internet Society, reprinted with permission; Figure 12.2 from Knox, P. et al. (1988) *The US: A Contemporary Human Geography* (pub Longman), reprinted by permission of Pearson Education Ltd; Figure 12.3 from Bloomfield, G. (1981) 'The changing spatial organization of the multinational corporations in the world automotive industry' in Hamilton, F. E. and Hinge, G. J. (eds) *Spatial Analysis, Industry and the Industrial Environment Volume II: International Industrial Systems*, © John Wiley & Sons Ltd, reproduced with permission; Figure 12.4 from Harryson, S. (ed.) (1998) *Japanese Technology and Innovation Management*, reprinted by permission of Edward Elgar Publishing Ltd; Figures 13.1 and 13.2 from Laulajainen, R. (1998) *Financial Geography*, reproduced by permission of the author; Figure 15.1, from the *Minneapolis Journal*, 1912, reprinted by permission of the Star Tribune; Figure 15.2 from Hecht, S. and Cockburn, A. (1989) *The Fate of the Forest*, reprinted by permission of Verso; Figure 15.5 from O'Laughlin, J. and Van der Wusten, H. (1990) 'Political geography the pan regions' *Geographical Review* 80, reprinted by permission of the American Geographical Society; Figure 16.1 from Chailand, G. (1993) 'Introduction' in Chailand, G. (ed.) *A People Without a Country: The Kurds and Kurdistan*, Second Edition, reprinted by permission of Zed Books Ltd; Figure 17.2 from Nash, G. B., Jeffrey, J. R. et al. (1990) *The American People: Creating a Nation and a Society*, Second Edition and Table 5.2 from Joseph, A. and Phillips, D. (1984) *Accessibility and Utilization: Geographical Perspectives on Health Care Delivery*, reprinted by permission of Addison Wesley Longman; Table 1.1 from Harris D. R. (ed.) (1996) *The Origins and Spread of Agriculture and Pastoralism in Eurasia*, reproduced by permission of UCL Press; Table 2.3 from Moore, R. I. (ed.) (1981) *The Hamlyn Historical Atlas*, copyright George Philip Ltd, reprinted by permission of Octopus Publishing Group; Table 4.1 from *Philip's Atlas of the World. Comprehensive Edition*, Eighth Edition (pub George Philip, in association with the RGS and IBG), copyright George Philip Ltd, reprinted by permission of Octopus Publishing Group; Table 5.1 from Siddiquee, N. A. (1996) 'Health for all by 2000? An assessment of the delivery of health care services in rural Bangladesh' first published in *South Asia* Vol. XIX, No. 1 (1996), pp. 49–70, reprinted by permission of the South Asian Studies Association; BP Amoco plc for Tables 8.2, 8.3, 8.5, 8.6 and 8.7 from *BP Amoco Statistical Review of World Energy 2000*, reprinted by permission of BP Amoco; Tables 8.8 and 8.9 from the OECD/IEA website www.iea.org/stats/files/keystats/stats98.htm, © OECD/IEA, reprinted by permission of the Organisation for

Economic Co-operation and Development and the International Energy Agency; Table 9.2 from Ilbery, B. and Kneafsey, M. 'Product and place: promoting quality products and services in the lagging rural regions of the European Union' *European Urban and Regional Studies* 5, 1998, reprinted by permission of Sage Publications Ltd and the authors; Table 11.13 from UNCTAD (1999) *United Nations Conference on Trade and Development World Investment Report 1999: Foreign Direct Investment and the Challenge of Development*, reprinted by permission of the United Nations Publications Board; Box 11.4, Figure from *Key Indicators of the Labour Market (KILM) 1999*, page 477, Figure 18a, copyright © International Labour Organization 1999, reprinted by permission of the ILO; Plates 1.1, 1.10, 3.8, 6.2, 12.4, 13.3, 15.6, 17.3 and 17.6 reprinted by permission of Topham Picturepoint; Plate 1.2 reprinted by permission of B & C Alexander; Plates 1.3, 3.10, 4.5, 4.8, 4.10, 4.12, 6.4, 7.4, 7.5, 8.7, 10.5, 11.10, 12.1, 13.2, 15.7, 1.2, and 16.4 reprinted by permission of Format Photographers; Plate 1.4 reprinted by permission of Michael Holford and C. M. Dixon; Plate 1.5 reprinted by permission of Aerofilms; Plates 1.6, 3.2, 3.3, 11.9, 17.1 and 17.2 reprinted by permission of Mary Evans Picture Library; Plate 1.7 reprinted by permission of South American Pictures; Plates 1.8, 4.9, 5.1, 5.3 and 11.8 reprinted by permission of Geophotos; Plates 3.1, 3.4, 4.1, 4.2, 4.3, 4.4, 4.6, 4.7, 12.5, 13.1, 15.4 and 16.8 reprinted by permission of Corbis UK; Plate 3.5 reprinted by permission of the Imperial War Museum; Plates 3.6, 3.11, 4.15, 5.4, 7.1, 7.6, 7.7, 8.2, 8.6, 8.8, 10.4, 12.3, 14.1, 14.6, 14.7, 15.1, 15.2 and 17.8 reprinted by permission of Popperfoto; Plates 3.9 and 14.3 reprinted by permission of Vintage Ad Gallery; Plates 4.11 and 10.1 reprinted by permission of Images of India Picture Agency; Plates 4.13, 5.5, 7.2, 7.3, 8.1 and 14.5 reprinted by permission of Panos Pictures; Plates 4.14, 7.8, 11.3 and 16.7 reprinted by permission of Camera Press; Plate 5.2 reprinted by permission of Magnum Photos; Plates 6.1, 6.5, 15.5 and 16.1 by permission of Trip: Plate 6.3 reprinted by permission of Sky B; Plate 6.6 by permission of J. Allan Cash; Plates 8.3, 17.4, 17.5 and 17.7 by permission of Associated Press; Plates 8.4, 11.2 and 11.4 by permission of Quadrant Picture Library; Plates 8.5 and 11.6 by permission of Science Photograph Library; Plates 8.9 and 10.2 by permission of Still Pictures; Plate 9.2 reprinted by permission of R. J. Smedley, Wensleydale Dairy Products, Aardman Animations and J. M. Nuttall; Plate 10.3 by permission of Oxfam; Plates 11.1, 11.5, 14.8, 14.10, 16.3 and 16.9 reprinted by permission of Alan Thomas; Plate 11.7 reprinted by permission of Microsoft; Plate 12.2 reprinted by permission of Birmingham City Council Department of Leisure and Community Services; Plate 12.6 reprinted by permission of Air Flight Service Mapping Photographers; Plate 12.7 reprinted by permission of Terence Soames Photography; Plates 14.2 and 14.3 reprinted by permission of The Advertising Archives; Plate 14.4 reprinted by permission of Redferns Ebet Roberts; Plate 14.9 reprinted permission of Bluewater Complex Kent; Plate 14.11 reprinted by permission of Ministry of Sound; Plate 15.3 from *The World of General Haushofer: Geopolitics in Action* by Andreas Dorpalen, © 1942 by Andreas Dorpalen, reprinted by permission of Henry Holt and Company, LLC; Plate 16.5 reprinted by permission of Robert Harding Picture Library; and Plate 16.6 reprinted by permission of John Frost Historical Newspaper Collection.

The Hereford Mappa Mundi is reproduced by permission of the Dean and Chapter of Hereford Cathedral Mappa Mundi Trust. Plates 3.7, 9.1, 9.2, 13.1 and 15.3 were supplied by the authors.

The extract from the poem 'Slough' by John Betjeman is reproduced by permission of John Murray (Publishers) Ltd; the extract from 'We and They' by Rudyard Kipling is reproduced by permission of A. P. Watt Ltd on behalf of The National Trust for Places of Historical Interest or Natural Beauty.

While every effort has been made to trace the owners of copyright material, in a few cases this has proved impossible and we take this opportunity to offer our apologies to any copyright holders whose rights we have unwittingly infringed.

A Companion Web Site accompanies *Human Geography: Issues for the 21st Century* edited by Daniels, Bradshaw, Shaw and Sidaway

Visit the *Introduction to Geographical Information Systems* Companion Web Site at *www.booksites.net/daniels* to find valuable teaching and learning material including:

For Students:
- Study material designed to help you improve your results
- Extensive links to valuable resources on the web
- Essay questions to help you check your understanding
- Annual updates of developments in the field of Human Geography
- Online Glossary and Bibliography
- Search for specific information on the site

For Lecturers:
- A secure, password protected site with teaching material
- Powerpoints of key artwork from the book, to assist in lecturing
- Answers to the essay questions found in the student area of the web site
- A syllabus manager that will build and host a course web page

PETER DANIELS•MICHAEL
BRADSHAW•DENIS SHAW•
JAMES SIDAWAY

INTRODUCTION

Geography:
a discipline for the twenty-first century

Human geography underwent numerous changes during the twentieth century. In its pursuit of explanations for the forces that shaped human society it tended to overstate and then to overlook the importance of diversity and difference. Model building and the application of quantitative analysis reinforced the latter tendency, at least until the last quarter of the century. Since then, human geographers have more consciously embraced diversity and difference amongst economies, societies and cultures. This is the source of some of the major challenges that face human society at the turn of the century.

In the 1970s the so-called 'crisis of relevance' in human geography kindled an interest in the underlying economic causes of inequality in capitalist societies. These structural approaches were later criticized for being too deterministic and for placing too much emphasis on the economic as the explanation for inequality. Political, social and cultural considerations were therefore increasingly incorporated in both contemporary and historical geographical studies. While human geographers remained within their sub-disciplines such as urban, political, historical or cultural geography, the boundaries between them started to blur. This was encouraged by 'a recent movement in philosophy, arts and social sciences characterized by scepticism towards the grand claims ... of the modern era, and their privileged vantage point, stressing in its place an openness to a range of voices in social enquiry' (Johnston, Gregory and Smith, *Dictionary of Human Geography*, 1994: 4). We refer to this movement as postmodernism. This enabled the breaking down of boundaries between sub-disciplines and the bringing together of urban, historical, economic and other geographies. Regional geography, that had long been out of favour since the mid-twentieth century, attracted renewed interest in response to a desire to develop a more integrated approach to geographical explanation. For some, outside the discipline, the end of the Cold War in the late 1980s brought the 'end of history'. For others, the development of information and telecommunications technology and the globalization of the economy were creating a 'borderless world'; some were even proclaiming the 'end of geography'.

On the contrary, geography still matters. In fact we would argue that, given the challenges of the contemporary world, it matters even more. As we shall see, globalization is a geographically uneven and contested process. It is a symptom of an increasingly competitive and interconnected world that has the potential to impose social, economic and cultural uniformity. Yet, a sense of place remains as critical as ever to maintaining identity and a successful role in the world. While the enhanced mobility of capital is generating increased inequality between the rich and the poor, the diversity of, and difference between, places is now used to attract investment and stake a claim to resources.

For the most part, the evolution of human geography has been driven by the acquisition of ideas, concepts and theories from other disciplines. The challenge for human geographers has been how to apply these in a geographical context. Thus, economic geography, for example, took ideas from economics and business studies and applied them to help explain how economic activity is distributed across space. The geography of economic activities was the outcome of the process. The job of the geographer was to map and describe that pattern of economic activity. This relegated the economic geographer (and geographers more generally) to the role of supplier of information for others to decipher.

Recently, however, the boot has shifted to the other foot. The advent of globalization and postmodernism has encouraged other social scientists to embrace geography. Of course, geographers have always claimed that *geography matters* for all the social sciences. For some reason, this is only acknowledged more widely when internationally renowned economists, political scientists and sociologists actually say that it must be included in their analyses and interpretations. But as geographers we should beware. They choose their own 'take' on geography, one that is often antiquated and simplistic – reinventing industrial location models in the name of 'geographical economics', for example. Nevertheless, geographers can take some comfort from the fact that other disciplines are at last recognizing the *spatiality of human society*.

Put simply, the fact that social and economic processes take place across space matters. Human geography is not just about *describing* the spatial manifestations of economy and society; it is about *explaining* how space helps shape economies, societies and social processes. Thus, geography is not a passive outcome; it is a critical component of social and economic processes. More than that, geography in its broader definition provides a unique interface between the human and the natural worlds. Without being too pompous, we would argue that geography is *the* discipline for the twenty-first century, because many of the challenges that face humankind are at that interface between human societies and natural environments.

Partly because of the emergence of feminist geographies, human geographers have become more aware that knowledge is socially constructed. The way you see the world is a function of who you are. Most views of the world, be they academic, scientific, artistic, political, economic or spiritual, are masculine constructions. Equally, human geographers have come to realize that much of the knowledge and understanding that they claim to be universal is in fact 'Eurocentric'. Thus, 1 January 2000 was not the beginning of a new century or a new millennium for many of the world's populations; yet, an instant measured on the time scale of a particular religion which, until recently, dominated Western society, became a global event. Similarly, that Eurocentrism has given a very 'white' view of a multi-ethnic world. This means that geographers must now confront the ways in which their own assumptions and value judgments shape the way they look at the world. They must accept that *all* descriptions of the world are culturally determined, often politically motivated, and always contested.

As social scientists, few human geographers now actually believe that there is a straight-forward objective reality around us that we are all seeking to describe and explain. For example, the fact that this book has been written by a group of white, predominantly male, geographers whose lives and careers have largely been spent in the United Kingdom, and mostly outside London, has influenced the way it has been written. A straw poll would probably quickly reveal that there are differences between the authors, and amongst the editors, as to the contents of the book, or how it should have been written. What you see is the result of compromise. The same book written by a group of geographers in Australia, Russia or South Africa would undoubtedly be different. At the same time, each individual reader of this book will interpret its contents in different ways. The position that you occupy in your society and where you are on the globe will shape the way that you look at the world. Given the anticipated market, the majority of the readers of this book will probably reside in the so-called 'developed world'. For all readers the most important geographical fact shaping your life-chances is where you were born. You may think that being at a university or college places you in a privileged position in society; in fact being born in the 'developed world' puts you from birth in a relatively privileged position amongst humankind. So, think about your position and privilege; realize that this book presents the views of human geographers working at a particular place and at a particular time. We have set out to challenge the reader to think about the ways that human geography interprets the major social, cultural, economic and related issues that face the world at the turn of the century. It is for you to use this book to inform your own insights and opinions as geographers; for some it could be part of a life-long engagement with human geography.

Approach of the book

Writing an introductory textbook is a challenging task. No matter what its contents, or how they are organized, it will not satisfy everyone. It not only has to engage the attention and interest of the students who will learn with the book, it also has to satisfy the instructors who will teach with it. Thus, as practising human geographers we are caught on the horns of a dilemma: how to make the text as accessible and as user-friendly as possible to students, while not appearing to be too naïve to the instructor assigning it. We also have to acknowledge that each geography programme is different, at least in detail if not overall objective, and that a textbook to fulfil every need would be very large indeed! Broadly speaking, however, most first-year geography programmes will have a course that offers an introduction to human geography. Such a course is nearly always compulsory. It is often supported with a 'world regional geography' course or a course devoted to topical 'issues in human geography'. Increasingly (though it was always the case in North America) these courses are required to meet the needs of non-geography students who are taking geography as an option or elective. This poses a major challenge for the instructor confronted by a class comprising 'wanna-be' human geographers, physical geographers who might rather be elsewhere, and non-geographers who are there out of interest or requirement. Such a diverse audience also poses a challenge to the textbook writer.

If you browse the shelves in your university or college library, you will come across a variety of human geography textbooks. Broadly speaking these are of three types: systematic texts organized by major sub-discipline in human geography, topical texts organized by key topics or issues, and world regional texts using major regions to introduce human geography. It will also be evident that these books are clearly aimed at different markets or audiences. Assigning a course text is almost a matter of fact in the US (students are aghast if there is no 'course text'), but elsewhere, certainly in the UK, assigned texts are a relatively recent phenomenon. Increasing student numbers, reduced library budgets, and the resultant pressure on library resources mean that an assigned text is seen as a necessary *starting point* for an introductory course in human geography. With students' budgets also becoming stretched, there is also a good case for purchasing a textbook that will cover more than one course and that will be of lasting use in later years.

There is no doubt that introductory textbooks do not always travel well. This is because they are often designed for a particular educational system; but more than that they also reflect a specific version of what is important in the world and what constitutes human geography. In addressing this problem, we have sought to produce a textbook that offers an *issue-oriented* introduction to human geography. The major sections focus on important issues facing the world at the beginning of the twenty-first century, but they also relate to major sub-disciplines in human geography. No introductory text can hope to be comprehensive. We believe that the issues we have selected are important and will continue to be so for some time. At the same time, they also reflect the interests and expertise of the authors and our sense of what captures the enthusiasm and imagination of the students we have taught over the years. The individual chapters bring together ideas from a variety of fields in human geography. They also incorporate coverage of the globe, drawing ideas from a variety of places to reflect both diversity and difference as well as the increasingly interdependent nature of the world in which we live. While this integrated approach may not sit well with the traditional ways of producing introductory human geography textbooks, we believe that it acknowledges the complexity of human society as well as the practice of contemporary human geography.

Structure of the book

This book is divided into five sections, supported by an Introduction and a brief Conclusion. Section 1 provides an essential historical context for the four issue-oriented sections that follow. While it is natural for us to proclaim that geography matters, it is also the case that history matters. A cursory examination of an atlas printed at the turn of the last century will show you that over the past century the world has changed a great deal. Many of the contemporary issues discussed in this book, such as the geographies of development and of inequality, owe much to the way the world looked in 1900. This is not historical determinism, rather an acknowledgment that history is critical to an understanding of contemporary human geography. However, if we cast our minds even further back into the records of antiquity we are reminded that societies are transitory, nothing lasts forever. Great civilizations have come and gone and this should teach us that there is nothing permanent about the current world order. Section 1 highlights some of these issues and then focuses on the emergence of capitalism and its relationship with the making of the twentieth-century world.

The remaining four sections comprise groups of chapters that focus on a related set of issues. It makes sense to work with Section 1 first; after that the sections can be utilized in whatever order you see fit. Each section starts with a brief summary of the important issues covered by each of the chapters and there is logic to the order in which the chapters are presented within each section. Section 2 focuses on social and cultural issues within an urban and rural context and at a global scale. Section 3 examines the inter-relationship between population change, resource production and consumption, and global economic development. Section 4 examines the globalization of economic activity, the transition of production from Fordism to post-Fordism, the emergence of a global financial system and means of exchange, and the importance of consumption to an understanding of the geography of economy. Section 5 considers a variety of political geographies at differing scales from the global, to the local, to the individual. In the Conclusion we reflect on the challenges facing human geography at the turn of the century (for most) and the new millennium (for many).

How to use the book

Each chapter is designed to introduce a set of specific issues and particular fields in human geography. It begins with a list of topics covered. The text is broken into manageable sections with supporting materials; important concepts and specific examples are presented in boxes. Each chapter concludes with a list of learning outcomes. These provide a means of assessing if you have understood the content of the chapter. To assist you in reading beyond the text, each chapter has a set of recommended readings that you should be able to find in your library. Where relevant, we have also included a list of web-sites which contain information you may find useful in your own research and writing. In addition, at the back of the book there is a glossary of important terms and an extensive bibliography. It is not normal to include such a lengthy bibliography in an introductory text; however, we want this book to remain an essential reference tool for the duration of your geography programme. Each chapter contains references to the bibliography giving you access to the specialist reading that is essential when you wish to research or otherwise follow through on specific topics in detail. One way of using this book is to read the chapters to support your lectures and seminars, the recommended readings to prepare for examinations, and the references, web-sites and bibliography to research and write essays or term papers. In other word, reading this textbook is the beginning of your introduction to human geography, not the end!

Learning resources

As we noted above, each chapter provides a set of recommended readings and, where relevant, useful web-sites. This chapter is no exception. Here we consider other resources that provide an introduction to human geography. The readings focus on other introductory texts and the study skills necessary to conduct your own research. The web-site guide lists places to visit to find out more about geography as a discipline and as a profession.

Further reading

This first set of readings provides a sample of other introductory textbooks that you can use to supplement your reading. Individual chapters from them are cited elsewhere as references in this book, but they are also worth browsing to get a sense of the different ways in which human geography is presented.

Cloke, P., Crang, P. and **Goodwin, M.** (eds) (1999) *Introducing Human Geographies*, Arnold, London. An innovative text comprised of 34 short chapters that reflect the diversity of approaches in contemporary human geography. Designed more for the UK market, but well worth delving into wherever you may be.

Knox, P. and **Marston, S.A.** (1998) *Human Geography: Places and Regions in Global Context*, Prentice Hall, Upper Saddle River, New Jersey. A comprehensive text written for the US market with lots of colour pictures and diagrams. Probably the best US-style textbook on the market at present.

The Shape of the World: Explanations in Human Geography A set of texts published in 1995 by the UK's Open University to support their distance learning courses. They fall somewhere between an introductory text and a more specialist book, but are well worth consulting and have been used widely in the production of this text. Complete references can be found in the bibliography.

> Volume 1: *Geographical Worlds*, edited by John Allen and Doreen Massey.
> Volume 2: *A Shrinking World? Global Unevenness and Inequality*, edited by John Allen and Chris Hamnett.
> Volume 3: *An Overcrowded World? Population, Resources and Environment*, edited by Philip Sarre and John Blunden.
> Volume 4: *A Place in the World? Places, Culture and Globalization*, edited by Doreen Massey and Pat Jess.
> Volume 5: *A Global World? Re-ordering Political Space*, edited by James Anderson, Chris Brook and Allan Cochrane.

Study aids

This section is dedicated to publications that will assist you in conducting geographical research. The list is comprised of dictionaries, readers (which bring together a collection of journal articles, book chapters and commissioned essays to survey a particular field) and books on methodology and theory in human geography.

Agnew, J., Livingstone, D. and **Rogers, A.** (eds) (1996) *Human Geography: An Essential Anthology*, Blackwell, Oxford. A collection of classic geographical writings and a sober reminder that there is very little that is new.

Bryson, J., Henry, N., Keeble, D. and **Martin, R.** (eds) (1999) *The Economic Geography Reader*, Wiley, Chichester. We include this as an example of the recent trends toward producing readers that bring together readings in a specific area of human geography. A number of publishers are now producing readers and they are a useful way of getting access to journal materials that your library may not hold.

Daniels, S. and **Lee, R.** (eds) (1996) *Exploring Human Geography: A Reader*, Arnold, London. The individual articles were not intended for a student audience, but the collection provides a survey of recent critical human geography.

Flowerdew, R. and **Martin, D.** (1997) *Methods in Human Geography: A Guide for Students Doing a Research Project*, Longman, Harlow. An indispensable guide to a wide variety of the research techniques employed by contemporary human geography.

Johnston, R.J. (1997) *Geography and Geographers: Anglo-American Human Geography since 1945*, 5th edition, Arnold, London. Provides an accessible survey of the major trends in human geography since 1945 in the English-speaking world. Includes an interesting discussion of geography as a discipline that might help you to understand what academic geographers do.

Johnston, R.J., Gregory, D., Pratt, G. and **Watts, M.** (eds) (2000) *The Dictionary of Human Geography*, 4th edition, Blackwell, Oxford. A fully revised and updated edition 960 pages in length. An indispensable reference tool that can double as a computer stand or door stop!

Livingstone, D. (1992) *The Geographical Tradition*, Blackwell, Oxford. A study of the history of human geography.

Rogers, A., Viles, H. and **Goudie, A.** (eds) (1992) *The Student's Companion to Geography*, Blackwell, Oxford. Now rather dated, but still serves as a useful introduction; the various sections contain information on the fields of geography, methods and data sources.

Women and Geography Study Group (1997) *Feminist Geographies: Explorations in Diversity and Difference*, Longman, Harlow. A good way to begin to understand how feminist geographies have reshaped the study of human geography.

Useful web-sites

The Internet contains a wealth of information that is of relevance to an introductory course on human geography. It is impossible to cover all of the potential sources here. This section provides access to the sites of some of the major geographical societies in the UK and North America. All of these sites provide links to societies and organizations across the globe, as well as links to geography-related sites. In addition, many of these sites now carry links to electronic versions of their journals and other publications as well as information and links to resources such as maps, statistics, atlases or photographic collections. Your own department may have a web-site; visit it and find out what the staff who teach you are up to. Visit departments across the world and discover how geography is taught in different countries. Increasingly, academic publishers are also posting electronic versions of the journals that they produce. Thus, the Internet provides you with your own virtual guide to human geography.

www.amergeog.org The web-site of the American Geographical Society. The Society was established in 1851 and is the oldest professional geographical organization in the United States. The society publishes the *Geographical Review* and *Focus*.

www.aag.org The web-site of the Association of American Geographers. Founded in

1904, the AAG is the primary academic geography organization in the United States. This site provides a wealth of information on the Association's activities, as well as access to its two scholarly journals: the *Annals of the Association of American Geographers* and the *Professional Geographer*, both of which are available online.

www.unwindsor.ca/cag This site is the home of the Canadian Association of Geographers and contains information on the Association's activities and publications, which include the *Canadian Geographer*. Its front page has numerous links to other geographical societies throughout the world.

www.geography.org.uk The web-site of the Geographical Association, which was founded in 1893 and is the national subject teaching organization for all geographers in the UK. The Association publishes three journals: *Primary Geographer*, *Teaching Geography* and *Geography*.

www.igu.net.org The web-site of the International Geographical Union, the international umbrella organization for academic geography. This web-site contains a comprehensive list of geography departments throughout the world.

www.nationalgeographic.org The public face of geography in the English-speaking world. Home to the *National Geographic* magazine, the site also provides access to the maps and photographs that appear in its magazine as well as a searchable index.

www.rgs.org The home of the Royal Geographical Society (with the Institute of British Geographers). The RGS was founded in 1830 and the RGS-IBG is the primary academic organization in UK geography. The web-site provides information on the societies as well as electronic access to its three learned journals: *Area*, the *Geographical Journal* and *Transactions*.

www.utexas.edu/depts/grg/virtdept/resources/contents.htm This is not a learned society, but the home of the University of Texas's virtual department's guide to Internet Resource for Geographers. A one-stop-shop for information on the web, the guide provides information on journals, professional organizations, academic departments and much more.

The world before globalization:
changing scales of experience

If present-day globalization means anything, it refers to a world marked by increasing intercommunication and interaction at a global scale. The corollary is that people's lives in the past were not as immediately affected by what happens elsewhere in the world, they were more ignorant of such happenings, and they had fewer possibilities to travel and to communicate across great distances. Changing scales of experience imply the growing significance of larger scales of activity and communication to human life as time has gone on. However, it would be a mistake to imagine that linkages over large distances are an entirely recent phenomenon, even at the global scale. This section shows how centuries and even millennia ago people, ideas and phenomena travelled vast distances, though usually taking much longer to do so than would be the case today. Geographical diffusion has been an important engine in social change, and change has been the norm rather than the exception throughout human history.

The present-day world has been moulded and is being remoulded by the forces of capitalism (which is defined in Chapter 2). As modern capitalism spread across the globe it inevitably changed the societies which preceded it. But the pre-capitalist societies, while adapting to capitalism, did not necessarily lose all their distinctive features, and capitalism itself was changed in the process of interacting with such societies. The fact that human societies in the modern world differ among themselves in all kinds of ways is only partly the result of the way global capitalism works and of the ways in which societies are now adapting to the various forces affecting them. Societies also differ because of their past histories, histories which stretch back into pre-capitalist times. The world into which modern capitalism spread was itself already enormously varied and had been changing in innumerable ways over thousands of years.

Chapter 1 of this section considers what the world was like before capitalism and what if any generalizations can be made about this long period of human history. Chapter 2 analyses the rise and spread of modern capitalism down to the end of the nineteenth century. Its history during this period is intimately linked to the rise of the European powers and to the worldwide spread of their influence. This, of course, is not to deny that capitalism probably sprouted, and then died, in other societies in other periods, or that it might have gained worldwide significance on the basis of a non-European core had circumstances been different. During the twentieth century, which is the subject of Chapter 3, capitalism has become a truly global phenomenon and has radically changed in the process. This chapter thus provides a more immediate background for the sections that follow.

In summary, this part of the book emphasizes the variety and complexity of the world before globalization. And it demonstrates that only by knowing something about that world can globalization itself be understood.

CHAPTER **1**

DENIS SHAW

Pre-capitalist worlds

Topics covered

- Why geographers need to know about the past

- Ways of classifying past and present societies

- Bands, tribes, chiefdoms, regulated states, market-based states

- Hunting and gathering

- The invention and spread of agriculture

- The rise of cities, states and civilizations

- Medieval feudalism

- Problems of studying past societies

Much of this book is about how capitalism has spread across the globe and even now is remoulding human societies in new and unexpected ways. This chapter is about what the world was like before capitalism. In other words it is about how human beings have lived during most of their existence on this planet.

Capitalism has changed the world profoundly and often very quickly. In the first half of the twentieth century it was still possible for geographers and others to travel to regions where its influence was minimal and where people still lived in pre-capitalist societies. Today this is much less true. Even so the geographer can still find numerous areas where the way of life has many pre-capitalist traits, though perhaps increasingly influenced by the modern world. Let us consider some examples.

About 2 per cent of the population of Australia is composed of Aborigines whose ancestors first settled the continent some 50,000 years before white Europeans (Broome 1994). Today their way of life varies between that of suburban professionals, and remote outstations and homelands where people live partly off the land in traditional style. When

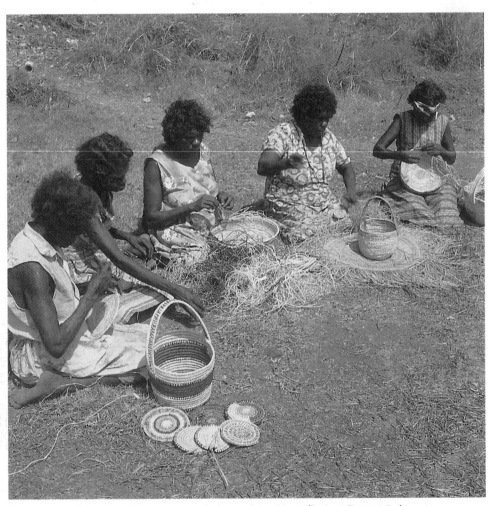

Plate 1.1   Aboriginal-Australian women producing tourist artifacts. (Topham Picturepoint)

the British established their first penal colony on Australian soil in 1788, more Aborigines lived on the continent than today, and they were divided into some 500 groups displaying a wide variety of languages, cultures, economies and technologies. All were hunters and gatherers leading a semi-nomadic existence and using some similar tools, like stone core hammers, knives, scrapers and axe heads as well as wooden implements like spears, digging sticks and vessels. However, there was also a variety in technology and economy that reflected the diverse environments in which Aboriginal peoples lived. Whether living on coasts, riverbanks, in woodland or desert regions, they were efficient and skilled in exploiting the flora and fauna of their surroundings. Aboriginal peoples who hunt and fish today are often using skills honed by their ancestors over thousands of years. Their languages and cultures similarly reflect the experiences of many hundreds of generations.

The Inuit peoples living in the Arctic, mainly coastal, territories of Greenland, northern Canada, Alaska and the extreme eastern tip of Russia, are another group who in the past lived almost entirely by hunting and gathering (Sugden 1982). Hunting and gathering still play an important role in many of the remoter communities. Like the Australian Aborigines, the Inuit showed a remarkable ability to exploit the resources of what in their case is an extremely harsh environment. Because of their environment, Inuit communities demonstrated less economic variation than the Aborigines but they were far from uniform. There were marked seasonal movements depending on the character of the local resources as groups moved between exploiting the fish and mammals of the open ocean, hunting seals on the ice, and seeking for caribou, musk ox and similar fauna inland. The rich Inuit culture is also far from uniform but by no means as varied as the Aboriginal (there are only two interrelated languages, for example). Considering the vast geographical spaces that the Inuit occupy, the latter fact is remarkable and may reflect a relative ease

Plate 1.2   An Inuit hunter sits on a modern skidoo. (B&C Alexander)

of communication by sledge and boat as well as the frequency of migrations. Like the Aborigines, the Inuit have had to adapt in the recent period to an intrusive, mainly white culture, but are still sufficiently conscious of their distinctive histories and life-styles to have a strong sense of identity.

The Australian Aborigines and the Inuit are but two of many human groups existing either now or in the recent past who have lived by hunting and gathering (see Figure 1.1). The present-day world also contains other groups whose way of life to greater or lesser extent reflects pre-capitalist characteristics. Pastoral nomads, for example, live mainly by raising and herding domestic animals (cattle, sheep, goats, camels, yak, reindeer and others), that provide them with food, clothing and other necessities. Pastoralists are particularly found in marginal (semi-arid, sub-arctic, sub-alpine) lands in parts of Eurasia and Africa. Cases in point are some of the peoples (for example, Nentsy, Evenki, Chukchi) who live in northern Russia, especially in the northern coniferous forests and tundra (Vitebsky 1996). Many of these peoples have traditionally lived by herding reindeer (though hunting and freshwater fishing are also important to them). Because reindeer are very demanding on the environment, the herders are traditionally nomadic, following set routes in accordance with seasonal changes. In the twentieth century the northern peoples were subjected to considerable intrusions on to their territories, especially in the wake of the Soviet government's determination to exploit the region for its minerals and other resources. The environmental consequences were very serious. There were also ideologically inspired attempts to force the northern peoples to abandon their nomadic ways.

Plate 1.3   Pastoral farmers with their cattle, West Africa. (Format Photographers/Maggie Murray)

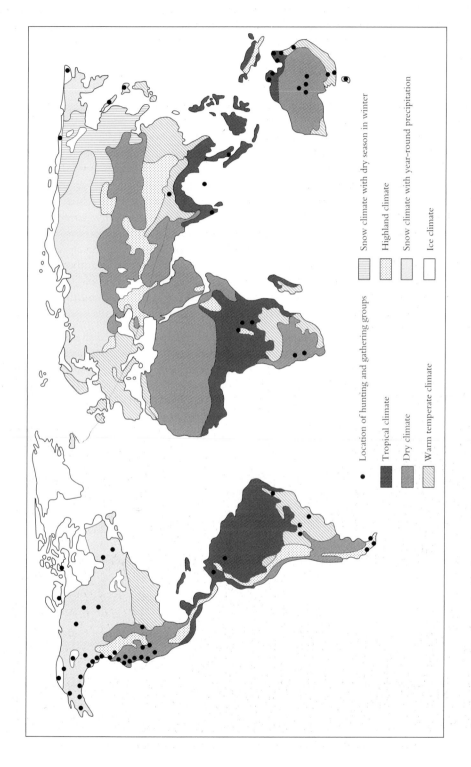

Figure 1.1 World distribution of hunter gatherers today and in the recent past, with indication of their habitats (based on a classification of climate).
Source: Adapted from Simmons, 1996: 48

Like the other cases discussed already, therefore, the northern minorities have had to make adjustments in their traditional ways, but reindeer herding is still vital to many of them.

As a final example of peoples whose way of life has yet to be completely changed by capitalism, one might mention the many peasant farmers still to be found in parts of sub-Saharan Africa, southern Asia and elsewhere. They still grow much of their own food and market relationships have yet to become of central importance (Wolf 1966). To the extent that such people eat and otherwise depend on what they grow and raise rather than on what they can sell or earn outside the farm, their way of life is similar to that of peasants over the centuries. Of course nowadays the number of peasants is declining (relatively at least) and they are increasingly exposed to the influences of the outside world. But again they remind us of a world before capitalism.

The point, then, is that today's world still contains many societies whose ways of life differ from those lived by most of this book's readers and recall earlier periods in human history. In order to understand those ways of life and the world as it is now, the geographer must know something about the world before capitalism. Learning about the world before capitalism helps us to see how capitalism has changed the world and the plusses and minuses of that process.

A number of arguments can be advanced to suggest why it is important to know something about the world before capitalism:

- Understanding the past helps geographers and others to understand themselves and their societies. Societies and individuals are products of the past, not just of the present. The present cannot be understood in ignorance of the past. Studying the past provides answers not merely to questions about how things were but also about how things are now (and, more tentatively, about how things will be).
- An example of the latter is the character of the physical environment in which people now live. Human beings have lived on this planet for so many thousands of years that over vast areas the physical environment has been profoundly modified. The world in which we now live is the product of thousands of generations of human activity.
- A further important point is that Aborigines, Inuit and other **indigenous** peoples (those peoples native to a particular territory which was later colonized, particularly by Europeans) have recently been asserting their rights, for example, by claiming rights to local resources long ago usurped by outsiders and by trying to protect their distinctive cultures. They feel they are the victims of the past. By studying that past and trying to understand the variety of cultures and ways of life which exist in the world and how they came to be, geographers are more likely to understand and respect such feelings.

Thus, while human geography is primarily about the present, it cannot afford to ignore the past. Some of the above points will be illustrated in the following pages.

1.1   Making sense of the past

Despite the spread of capitalism and globalization, the world of today remains immensely varied and complex. The same is true of the world in the past. The question for the geographer is how to make sense of this complex past; how to make it amenable to geographical analysis and understanding.

Different scholars have tried different ways of answering this question, but all suffer from shortcomings. For example, the economist W.W. Rostow, in his famous theory of economic growth typical of capitalist development, simply described the pre-industrial period as 'traditional society' (Rostow 1971: 4–6) (see Box 2.1). Yet the fact is that pre-industrial human societies ranged from the smallest communities of hunters and gatherers to societies as sophisticated and geographically extended as the Roman Empire, ancient China and feudal Europe. In as much as such societies existed in the past, and not today, it is not possible to observe them directly. But the more scholars learn about them, the more they realize how unlike one another these early societies were.

One reason why Rostow oversimplified the pre-industrial past was that he was not much interested in it. He was really interested in modernity, as exemplified by the United States and its Western allies. Furthermore his theory can be described as 'unilinear evolutionary' in as much as it suggested that all societies should pass through a series of set stages of economic development before finally arriving at the age of 'high mass consumption'. The United States (where Rostow lived) and other 'developed' countries have already reached this age. Evolutionary models have certainly long been a popular way whereby Western societies have tried to make sense of the past. One of the attractions of such models has no doubt been their tendency to suggest that the West is the most 'developed' and thus most 'progressive' part of the world and, by implication, the best. But quite apart from the questionable assumption that all societies are seeking to imitate the West, this raises important issues about the meaning of 'progress' and 'development', two terms that certainly carry very positive connotations in the West. For example, while those parts of the world usually deemed most developed and wealthiest certainly use most energy and enjoy access to more material goods, they also make huge demands on the environment, perhaps ultimately to everyone's undoing. Thus what is progress in this context? Similarly, it has often been noted that the price people tend to pay for more 'development' and wealth is that they have to work harder and for more hours. Unless hard work is regarded as a virtue in itself, this again raises questions about the nature of progress. The point here is not to disparage all forms of evolutionary theory or idea (for example, to deny that human societies have, by and large, become more complex and spatially extended through time: see Dodgshon 1987). Rather, it is to suggest that what is most recent or new or complex is not necessarily best. Thus, in what follows, words like 'modern' do not imply 'better', nor do 'ancient' or 'primitive' imply 'worse'.

An alternative method of trying to make sense of the past, and one which does not have the unilinear evolutionary structure of, say, Rostow's theory, is to classify human societies into a series of types. Classification can be described as a way of simplifying a complex world by grouping phenomena together which are regarded as having some common feature or property, particularly where the latter is deemed especially significant. Thus Marxists have commonly grouped societies in accordance with their prevailing mode of production, or in other words with the way in which material production is related to social structure (Hindess and Hirst 1975). As a geographer, primarily interested in spatial structure, Robert Sack classified societies in accordance with their use of territoriality (Sack 1986). The point is that there is no right or wrong mode of classification. It all depends on what the purpose of the classification is. Most social theorists have wanted to claim that their mode of classifying societies is particularly significant for under-

standing how and why societies differ. The problem is that there is a large measure of disagreement about what the best mode is, and all have their shortcomings.

1.2   A classification of human societies

As an example of the kinds of difficulties that face any attempt to classify past and present
human societies, this chapter will focus on one mode of classification that has had widespread appeal not only for human geographers but also for anthropologists and other
scholars. This is the mode of classifying human societies into bands, tribes, chiefdoms and
states (regulated and market-based) (see Bobek 1962; Service 1971; Dodgshon 1987).
Boxes 1.1 to 1.5 describe, in simplified form, the major characteristics of each type of
human society according to this classification. It will quickly be seen that the classification
describes a range of societies from the simplest and most primitive to the capitalist societies
of today. What it does not suggest is the circumstances under which one kind of society
may change into, or be succeeded by another.

One reason why this mode of classification has had such appeal is because it suggests
that over the course of human history larger and more complex societies have tended to
appear. It also relates size and complexity to the way societies occupy space and their
relationships to their physical environment. However, it is worth reiterating that this is
only *one* way of classifying human societies. Indeed some scholars reject the whole idea of
classifying human societies because they feel it leads to misapprehensions about how
societies are formed and survive, and even to dangerous notions of unilinear evolution. It
thus needs to be remembered that this scheme is suggestive only, and does not necessarily
describe the way any given society has actually changed through time. In particular, it is
not stipulating that societies must develop through a series of set stages, or that one type
of society necessarily leads to the next.

Having indicated in outline the main features of the classification scheme, subsequent
sections will spell out in more detail the social processes that might have helped to produce, change and perhaps destroy different kinds of human society. In so doing some of
the limitations of the classification system used will become apparent. Before proceeding,
however, it is worth making one important point. Our knowledge of **prehistoric** societies (societies which have left no written records) derives mainly from archaeological
data, or material left by such societies and the traces of their environmental impacts.
Methods for analysing such data have become ever more sophisticated, but the data themselves obviously become scarcer the older they are. Moreover, there are many kinds of
questions, for example about the way early people viewed the world around them, which
cannot be answered directly from the archaeological evidence. New discoveries are always
liable to change our understanding of past societies, and especially of the earliest ones. Our
knowledge of the latter societies, and of the dates attaching to them, must therefore be
regarded as especially tentative.

1.3   Hunting and gathering

The time when the first human-type species (hominids) first appeared on earth is very
uncertain, but a date of two and a half million years ago is often given. About 800,000

Plate 1.4 Artifacts from early human sites: flints (Michael Holford), bone harpoons and a human skull. (C.M. Dixon)

years ago is when *Homo erectus*, forerunner of the modern human species, is believed to have become dominant, and the modern species (*Homo sapiens sapiens*) at least 40,000 years ago (but see Foley 1995, for a contrary view). Since agriculture appears to have arisen less than 10,000 years ago, hunting and gathering in bands have been the basic occupations for much of humanity's existence (Box 1.1).

As noted already, only a limited amount is known about early human societies. Much has to be inferred, for example from hunter gatherers who are still in existence or who existed until quite recently. Needless to say, such inference can be dangerous because modern hunter gatherers have probably been influenced by other, more 'advanced' societies that now share their world. Their way of life is thus unlikely to be identical with those who existed in the remote past.

The distribution of hunter gatherers in the recent past shows that they have existed in

Box 1.1   Bands

- Bands are societies of hunter gatherers. Most of human history has been lived in this way.
- Bands live by hunting wild animals and gathering food from the surrounding flora.
- Because of the pressures they place on the surrounding environment, bands are almost always nomadic.
- Bands are the smallest and simplest of human groups.
- Bands rarely exceed 500 people.
- For most of the time band members generally live in smaller groups of 25–100 people, which allows for efficient exploitation of the environment.
- Bands usually have a minimum of social differentiation and functional specialization.

every major climatic zone and have practised enormously varied economies, depending on the nature and potentials of the local environment (Figure 1.1). This fact alone suggests that a simplistic evolutionary model of human societies will fail to do justice to their complex histories. Some generalizations do seem possible, however – for example, that in recent times groups living in the highest latitudes (above 60 degrees) have relied largely on hunting, and those below 39 degrees on plant collecting. Fishing seems to have played an important role for many groups living in intermediate latitudes (Lee 1968).

Hunter gatherers have lived in this world for such a long period that today's environment is the way it is partly as a result of their activities. For example, any notion that early human beings lived in complete harmony with their environments or always had 'sustainable' economies is simply untrue. Thus hunter gatherers are credited with the extinction of many animals. An outstanding case is North America where perhaps two thirds of the large mammal fauna living there just before the arrival of the ancestors of modern native Americans (via the Bering land bridge, possibly about 12,000 years ago) subsequently disappeared. The most likely explanation is the effects of hunting. The discovery of fire, its use by hominids long predating the appearance of *Homo sapiens sapiens*, was also an important instrument whereby human beings changed their environments. Some scholars have argued that the present-day appearance of such major biomes as the African savannas and the prairies of North America is the result of the human use of fire over thousands of years (Simmons 1996). More recently, contacts between hunter gatherers and agricultural and industrial societies frequently had even more far-reaching environmental effects.

Scholars have wondered about processes whereby bands evolved into tribes, or hunter gathering economies into agricultural ones (Harris 1996). It has been pointed out, for example, that where the natural environment was especially favourable, the packing of bands often became denser leading to less mobility and perhaps to semi-permanent settlement and closer interaction between bands. There may also have been a much greater degree of manipulation of the environment and its resources than might be expected in a pure hunter gatherer economy. Altogether the life of bands was much more variable and a good deal less stable than the above classification scheme suggests. The boundary between band and tribe is thus blurred.

1.4   Human settlement and agriculture

Just as the boundary line between band and tribe may not always be easy to define in practice, the same can be said of the boundary between tribe and chiefdom. Some scholars, for example, doubt whether the category tribe (Box 1.2) is particularly helpful, implying as it does the lack of a hierarchical social structure (Friedman and Rowlands 1977). Such scholars tend to believe that the appearance of agriculture will have sparked off competition for access to the best land, or at least rules whereby such land was allocated and inherited, and that some people and groups will inevitably have lost out. They therefore see a social hierarchy (characteristic of chiefdoms and states – see Boxes 1.3 and 1.4) beginning to emerge even as agriculture and the process of permanent settlement began.

Be that as it may, it is now widely accepted that the traditional picture of agriculture

Box 1.2   Tribes

- Tribes appeared with the invention and spread of agriculture.
- Agriculture usually demands considerable investment of human effort into a relatively small area. This reduces the propensity to migrate (but pastoral nomads are an exception).
- The appearance of agriculture usually allows more people to live in a smaller space. Agricultural societies thus tend to be bigger than bands, with higher population densities.
- Settlements tend to become more fixed than in bands.
- The greater spatial fixity of society encourages intermarriage and the development of extended family networks. Tribes thus develop a sense of kinship and of common descent.
- Socially, tribes are relatively egalitarian, at least as between kinship groups.

Box 1.3   Chiefdoms

- Chiefdoms arose with the emergence of 'ranked and stratified' societies.
- 'Ranked' societies are societies where groups and individuals have, on a relatively permanent basis, different degrees of status and power.
- 'Stratified' societies are societies where groups and individuals have, on a relatively permanent basis, different degrees of material wealth.
- A chiefdom implies the presence of a permanent ruling group and/or individual, though kinship remains the chief bond between ruler and subject.
- Chiefdoms imply a greater degree of centralization and control within society than in a tribe (shown, for example, in the imposition of taxes and tribute on the ordinary subjects). Spatially this might be reflected in the greater importance accorded to a central settlement (proto-city) where the ruler resides.
- The first chiefdoms began to appear about 3000 BC in Europe, but earlier elsewhere.

> **Box 1.4   Regulated (pre-modern) states**
>
> - Whereas the chiefdom is organized around the principle of kinship, the state is organized on the basis of territory (viz. one is subject to the state if one resides in the territory of that state).
> - States, whilst also based on social inequality, tend to be larger and administratively more complex than chiefdoms. This implies greater functional specialization and greater probability of the rise of urban forms.
> - Pre-modern states existed in a world before modern capitalism. The market was not central to their functioning and was often controlled ('regulated') in various ways. Thus economic relationships were generally subordinated to political, social and religious considerations.
> - Other forms of social relationship, for instance landholding, also often tended to be controlled (regulated) rather than being determined by the market.

> **Box 1.5   Market-based states**
>
> - Market-based states are modern states whose development is closely linked to that of capitalism and thus of the world economy.
> - The market is of key importance to their functioning.

being invented in a small number of 'hearths' and then spreading across the globe is far too simple. Bands of hunters and gatherers are known to develop an often intimate knowledge of their local environments, and no doubt human beings understood much about the factors influencing plant and animal development long before they began to practise full-blown agriculture. David Harris, for example, talks of a 'continuum of people–plant interaction', starting at the simplest level with wild plant food procurement or foraging and ending with cultivation and full-blown agriculture (Table 1.1). By the latter stage, the domestication of plants and animals had led to the modification of species which could then no longer survive in the wild (Harris 1989; 1996a,b). The important point is that Harris is not propounding an inevitable evolutionary path that is applicable to all societies. What he is saying is that the development of agriculture was a long drawn-out affair and that there must have been much trial and error before it finally emerged in some places in a fully recognizable form. Some prehistoric northern Australian Aborigines, for example, knew about agriculture, but never adopted it.

Something can, however, be said about when and perhaps why agriculture appeared. The archaeological evidence provides most information about where and when the latter stages of Harris's 'continuum' emerged. For example, it is possible to detect and date the remains of domesticated varieties of plants and animals. According to Simmons (1996: 93) there were probably three foci for the initial surges of domestication: around 7000 BC (to use the Christian chronology) in south-west Asia, around 6000 BC in south-east Asia, and around 5000 BC in Meso-America. Different species of plants were associated with each,

Table 1.1   An evolutionary classification of systems of (above) plant and (below) animal exploitation

Wild plant-food procurement	Plant-food production		Crop-production dominant
	Wild plant-food production dominant		
Gathering and collecting	Cultivation with small-scale clearance of vegetation and minimal tillage	Cultivation with larger-scale land clearance and systematic tillage	Agriculture based largely or exclusively on the cultivation of domesticated plants

Decreasing dependence on wild plants ⟶

Increasing dependence on domesticated plants ⟶

Time ⟶

Predation	Protection	Domestication
Generalized hunting (and scavenging)	Taming	Livestock raising by settled agriculturalists
Specialized hunting (and scavenging)	Protective herding	Transhumance
Fishing	Free-range management	Nomadic pastoralism

Decreasing dependence on wild animals ⟶

Increasing dependence on domesticated animals ⟶

Time ⟶

Source: After Harris (1996a: 4).

for example: wheat, barley and oats in south-west Asia; rice in south-east Asia; and maize, squash and gourds in Meso-America. Domesticated animals also began to appear about the same time (though the dog, domesticated from the wolf, appeared before agriculture). Sheep, goats, pigs and cattle are all associated with south-west Asia (Simmons 1996: 87–134) though pigs may also have been domesticated independently in eastern Asia.

As to why agriculture was adopted, scholars differ. The simple answer, that it was more efficient than hunting and gathering, simply will not do in naturally productive environments where it clearly involved much greater effort relative to the return achieved. Some scholars favour population pressure. Others, however, argue that this view is over-deterministic and point out that even hunter gatherers had the means of controlling population growth. Such scholars tend to favour more complex explanations that embrace cultural preferences and choices as well as environmental and population pressures (Maisels 1993: 25–31).

Just as the factors that led up to the appearance and spread of agriculture are by no means straightforward, the same is also true of its further development. No simple evo-lutionary model, such as the one proposed by Esther Boserup involving an increasing intensification of agricultural land use through time, can account for how agriculture has changed and developed throughout the world (Boserup 1965). For example, it is no longer generally accepted, as it once was, that extensive forms of agriculture like shifting agriculture usually preceded farming in permanent fields. Britain may be unusual in Europe in having undergone a slow transition to permanent forms of agriculture (Dodgshon 1987: 66–70). Similarly pastoral nomadism is no longer regarded as more primitive than cultivation. Indeed, many scholars have argued that pastoralists cannot exist in isolation from cultivators and therefore cannot have preceded them. Throughout the world, traditional agriculturalists showed a remarkable ability to adapt their practices to local environmental conditions. Eventually they largely displaced the hunter gatherers (Figure 1.2) and gave rise to an enormous variety of agricultural systems (Figure 1.3).

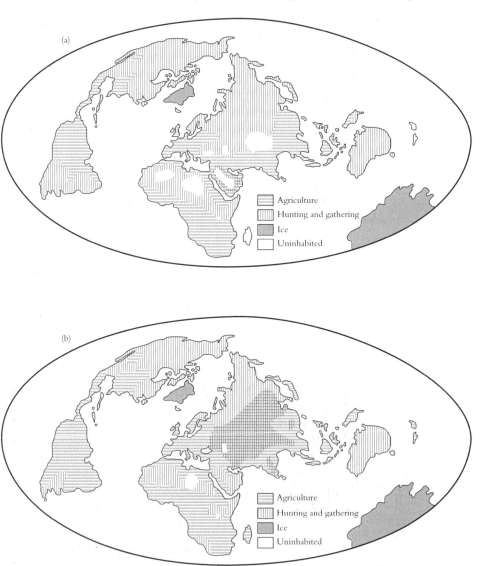

Figure 1.2   World distribution of agriculture and hunter gathering about AD 1(a) and AD 1500(b).
Source: Sherratt, 1980: 97, 117

Figure 1.3 Agricultural regions of the world. *Source:* After Whittlesey, 1936

Flat polar quartic
equal-area projection

Equator

1 Nomadic herding
2 Livestock
3 Primitive subsistence agriculture
4a Intensive subsistence, wet rice dominant
4b Intensive subsistence, wet rice *not* dominant
5 Plantations and small farms
6 Mediterranean agriculture
7 Commercial grain farming
8 Crop and livestock farming
9 Commercial dairy farming
10 Commercial gardening and fruit
11 Little or no agriculture

0   2500 miles
0   4000 km

Agriculture's effects on the landscape were much greater than the effects of hunting and gathering. It changed the vegetation cover of wide areas as forests and grasslands disappeared under fields, practices like terracing and irrigation were introduced, and the effects on soil composition abetted erosion in some places. The grazing of animals and other forms of resource use frequently changed the species composition of forests, grasslands and other areas. Agricultural systems, crops and domesticated animals spread way beyond their initial locations. An outstanding example is the cultivation of rice, which may have started in east Asia around 5000 BC, if not before, and spread as far as Egypt and Sicily by Roman times. Such developments long preceded the European overseas expansion beginning in the fifteenth century AD which was to have even more far-reaching consequences for both the geography of agriculture and for the environment (see Chapter 2).

There is no doubt that the spread of agriculture (whether as a result of migration or through cultural diffusion) had far-reaching effects on human society, encouraging permanent settlement and the emergence of tribal systems (though Crone, 1986, questions the nature of the link between agriculture and tribes). Tribal members, living permanently side by side, almost inevitably developed similar cultural traits such as common languages. As suggested in Box 1.2, tribes also typically develop myths of common ancestry (indeed, this seems to be part of their definition). Scholars have long pondered whether such myths might sometimes have a basis in biological fact, and whether early tribes might have formed the basis for some of the ethnic divisions apparent today. For example, some scholars now believe that the present-day geography of language groups (for example, that of the Indo-European languages, which are spread across much of Europe and parts of south Asia) can be partly explained by the prehistoric migrations of interrelated peoples speaking similar languages. They also spread agricultural practices into these regions. Such scholars see a correspondence between the present-day geography of languages and the geography of human genetic patterns which science is now beginning to reveal for Europe and other areas (Renfrew 1996). Other scholars, however, are more cautious. It has been argued, for example, that different human groups, settling in the same region, may gradually have adopted similar languages and cultural patterns, thus slowly fusing into one people or ethnic group even though they were not biologically interrelated. Thus, according to this argument, the Slavs of eastern Europe might not be descendants of one original tribe, but of different groups who gradually adopted the same culture (Dolukhanov 1996). The question, then, is how far tribal (and nowadays ethnic) differences are purely cultural and how far biological. It is a question to which there is no sure answer. What can be said is that the migration of peoples, which certainly played a big role in history even after the adoption of permanent settlement and agriculture, must often have led to the intermixing of peoples and to frequent redefinitions of tribal and ethnic divisions.

1.5   Cities and civilization

The invention of agriculture was only one of the events that moved human societies on a road that ultimately led to the kinds of societies which predominate in the world today. Others included the appearance of cities, states and civilizations. Cities and states usually go together. Maisels, for example, argues that the functional specialization associated with cities allowed rulers to rule more effectively and also put a greater social distance between

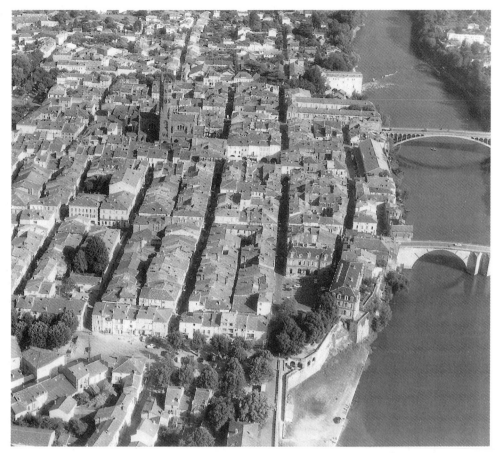

Plate 1.5 A medieval walled town, Villeneuve-sur-Lot, showing evidence of planning. (Aerofilms)

them and their subjects (Maisels 1993: 302). The gathering of specialists around the person of a state's ruler fostered urban life, commerce and the many other things (such as writing, technology, religion, and speculative thought) with which is associated the word 'civilization'. Indeed, the very words 'city' and 'civilization' have a common Latin root (*civitas*, meaning 'state').

As is the case with the appearance of agriculture, the processes that eventually gave rise to cities and civilizations are far from straightforward. It is clear that the development of a social hierarchy was a necessary precursor, but what produced that? Some scholars have argued that competition for land in the face of population pressure, inequalities fostered by inheritance rules, and the emergence of powerful leaders who organized inter-tribal exchanges, acted as 'tribal bankers' in times of food shortage, or proved their military prowess, are all possible explanations. But while these and other factors may have produced **chiefdoms** (see Box 1.3), something more seems to be required to explain the rise of cities, states and civilizations. As Maisels writes, chiefs enjoy only 'hegemony' over their peoples, and are bound by rules of kinship which imply mutual obligations. Chiefdoms were notoriously unstable. Kings and other state rulers, by contrast, exercise 'sovereignty' (ultimate power) over a **territory** and its peoples (Maisels 1993: 199).

Scholars are generally agreed that cities and states required, in the first place, a mini-mum threshold population density, which only settled agriculture would normally permit. Only settled agriculture can produce the reliable food surpluses needed to feed the special-ists and non-agriculturalists who lived in the first cities. But equally something else was required to persuade the agriculturalists to yield up their surpluses to the cities. Trade was one possibility, but probably not the main one in what, after all, were still relatively simple societies. A much more likely candidate was the political power and influence exercised by incipient rulers. There is no agreed theory that explains how such power arose (Dodgshon 1987: 139–45). Some have argued that the origins of the state lie in the ability of an individual or group to monopolize control over some function that is of great value to the community as a whole. Examples might include trade and exchange (including the storage of needed goods), administration, religion or the engineering of irrigation works. Others prefer to see the explanation in the social divisions produced by such factors as conquest by outsiders, or struggles between groups over limited resources. The problems with all such monocausal explanations of the rise of the state, however, is to show how the kind of social inequality that might be produced by such processes then transforms itself into the stable and legitimate power which a ruling group exercises in the state. Other scholars, therefore, have proposed that states came into being as a result of differ-ent processes working together in a contingent way rather than being caused determinis-tically by a single overriding process (Maisels 1993: 199–220).

Whatever may be the explanation for the initial development of states and associated cities, they gave rise to major changes in human societies (see Box 1.4). Rather than being organized around the principle of kinship like the tribe and the chiefdom, for example, states were organized on the territorial principle. This means that the rulers of states exer-cised their rule over defined territories and their inhabitants, no matter who the latter hap-pened to be. This proved an extremely powerful way of organizing human societies. States thus appeared in different parts of the globe. The first organized states are believed to have developed in Sumer in present-day Iraq around 3000 BC. According to Maisels, these took the form of city-states. In China the first state, associated with the Shang dynasty, arose about 1500 BC, and by about 1000 BC the Maya civilization had appeared in Meso-America. In Europe the first states are commonly taken to be the Greek city-states which began to develop in eastern Greece and the neighbouring islands about the eighth century BC. Whether the state arose independently in different places across the globe, or whether the idea of the state was somehow diffused from one or two initial foci, is uncertain.

Early states took many different forms. All faced difficulty in enforcing and maintain-ing their territoriality in the context of poor communications, and different states approached this problem in different ways. The Greek city-states, for example, typically occupied only small areas and restricted their population sizes. When population in any one state exceeded a certain threshold, that state would found a self-governing colony elsewhere on the sea coast. Travel by sea was generally easier than overland travel and thus settlements could keep in touch. In this way the Greeks eventually colonized much of the Mediterranean coast (Figure 1.4). The huge Roman Empire, by contrast, could not rely solely on sea travel. This was a much more centralized polity and only by having superb military organization, an elaborate system of roads and fortifications, good administration, and a state-controlled religion centred on the person of the emperor, could it survive for several centuries (Figure 1.5 and Box 1.6). Another huge empire was the Mongol one that controlled much of Asia and a good slice of central and eastern Europe in the thirteenth

THE CANOE IN A BREEZE.

Plate 1.6 A nineteenth century Polynesian canoe. Travel by water was the most efficient form in pre-industrial times, allowing vast distances to be covered in remarkably short periods of time. (Mary Evans Picture Library)

Figure 1.4 The Greek colonization of the ancient Mediterranean. *Source*: After Pounds, 1947: 49

Figure 1.5   Roman Britain. *Source*: © Oxford University Press 1981. Redrawn from map VI from *Roman Britain* (*Oxford History of England*) edited by Peter Salway (1981) by permission of Oxford University Press

> **Box 1.6** The Roman Empire at its zenith (first to fourth centuries AD)
>
> The Roman Empire was one of the greatest achievements of state-building in the pre-modern period and is testimony to the way in which huge distances could be crossed and enormous territories controlled in spite of the absence of modern industrial technologies. At its greatest extent the empire's east–west axis stretched from the Caucasus to Cape Finisterre in north-west Spain, a distance of about 2,800 miles. North–south it reached from Hadrian's Wall, near the present-day Anglo-Scottish border, to the fringes of the Sahara in North Africa, some 1,600 miles. The empire was characterized by two major languages (Latin and Greek), low tariff barriers, a common currency, a common code of laws, and the basis of a common system of education and culture for the Romanized elite. Rome was a truly 'universal state' which lasted in its mature form for half a millennium, and in its eastern (Byzantine) manifestation until the capture of Constantinople by the Turks in 1453.
>
> Like all early states, Rome had problems in enforcing its territoriality – the sheer scale of the empire meant that there were always difficulties in maintaining control over its far-flung provinces. That the empire was able to endure for so long (albeit with many vicissitudes) was a tribute to its highly developed capacity for organization and an astuteness in gaining and retaining local loyalties. The acquisition of new territories proceeded by a mixture of conquest, colonization by Roman and Latin colonists, the establishment of vassal kingdoms and powers, alliances (often forced), and other means. The Roman army, reformed by the Emperor Augustus after 13 BC, was a masterpiece of military organization superior to any of Rome's enemies in this period. The seas and waterways were patrolled by the navy, protecting merchant shipping. Towns and military bases were interconnected by a network of fine roads and an official transport system (the *cursus publicus*) with inns and posting stations at regular intervals speeded official communications. There were even roadbooks (*itineraria*) for the guidance of the many travellers, including some tourists. Fortified lines, of which Hadrian's Wall is the most celebrated, protected the most vulnerable frontiers. But the empire was not held together only by military force. It was also by judicious extensions of Roman citizenship and other privileges, and by fostering helpful religious practices like emperor worship, that local loyalties were ensured.
>
> The empire's economic and cultural achievements were many. Long-distance trade in such items as grain, metals and luxury goods was important, for example, though there were many hindrances to commerce and most people depended on local agriculture and manufacture. It must also be remembered that there was a heavy dependence on slavery. In the end the Romans could never be assured of political stability, even in their heartlands. Reliance on the army contributed to the empire's eventual undoing – generals and their armies competed for political power, the economy was undermined by overtaxation and other factors, and growing military weakness invited rebellion, and invasion by 'barbarians'.

and fourteenth centuries AD. The Mongols were pastoral nomads whose formidable warrior skills were based on their horsemanship and their ability to outmanoeuvre their enemies. But nomadic empires had a tendency to contract as quickly as they arose. It was difficult for the emperor to maintain control over his swift-moving armies of mounted warriors. Finally, medieval Europe's answer to the problem of territoriality was to decentralize political power through kings to territorial lords and minor nobles (see Box 1.7).

Box 1.7   European feudalism

- Feudal society was hierarchical – ranked and stratified.

- The majority of the population were subsistence farmers, engaged in extensive forms of agriculture (for example, by cultivating strips in open fields shared with others, raising some livestock), and exploiting local resources like pastureland, woodland and fish.

- Many of the rural population were obligated to their lords and landholders, for example not to move away from their villages, to work the lord's land, to pay the tribute and taxes the lord and the state required, to serve the lord militarily, and to perform other services.

- Lords held their land conditionally from their overlord and ultimately from the king or other ruler. Land could not easily be sold. Lords exercised jurisdiction over those living on their estates.

- Lords owed their overlords (and they in turn their overlords and so on) allegiance or fealty (hence the term feudalism), for example in supporting them with their military forces and performing other services as required.

- Feudal states were quite decentralized, parcelled out into lordships and they in turn into jurisdictions of various kinds.

- Towns were generally small and few and far between. Most were centres of trade and crafts. Their residents (merchants, burgesses) often enjoyed freedoms and privileges ('liberties') denied to most rural dwellers.

- Merchants and itinerant traders joined together networks of local and long-distance trade, linking local fairs and markets with towns, ports and so on. Craft production and trade were generally subject to a variety of restrictions and controls, and often played only a minor role in the lives of rural dwellers.

- The medieval world view was strongly influenced by the Church, whose priests and officials were among the few educated people and were universally present.

In return for their rights to their land and other privileges, lesser nobles owed their lords a duty of military support in time of need, which meant having to appear with an army of knights, retainers and other soldiers when ordered to do so. The lords in turn owed loyalty (or fealty, hence the term 'feudalism') to their overlords, and so on in a hierarchy which ended with kings and other rulers. Medieval European states were thus quite decentralized and unstable, as regional lords and their underlings were often tempted to rebel (Figure 1.6). Only after centuries of struggle between rulers, lords and ordinary subjects did the European states of today finally appear.

A modern geographer, looking at maps of past states, might be surprised at how large some of them were, given their many problems. They are a lesson in how human ingenuity can cope with geographical difficulty. But it is important to remember that before modern communications the geographical constraints to state power were formidable. Pre-modern states and empires were inherently unstable, and the bigger they were, the more unstable they tended to be. Only modern communications have allowed states to overcome this problem to some extent, but even then imperfectly, as future chapters will show.

Figure 1.6 France towards the end of the feudal period. showing the political fragmentation typical of many parts of Europe prior to the appearance of modern states. *Source*: Based on *Times Atlas of World History*, 1989: 15.1

1.6 Pre-capitalist societies

It should be clear by now that pre-capitalist societies were by no means simple. They differed among themselves in all kinds of ways. At the same time, when compared with present-day capitalist societies, they appear to have a number of distinguishing features. The final section will consider some of the more obvious of these features.

Geographers and others have sometimes been tempted to think that past societies, lacking modern communications, were essentially small in scale and localized, with life based around the community. It is hoped that enough has been said already to suggest how over-simplified this view is. Huge empires had to be administered and defended, great cities like classical Rome had to be fed. Classical Rome imported some 17 million bushels

Plate 1.7 Teotihuacán, near Mexico City, probably the first city in the Western Hemisphere. This great city flourished between the first and eighth centuries AD, was linked to the Mayan culture of Yucatan and Guatemala, and influenced the later culture of the Aztecs. (South American Pictures/Tony Morrison)

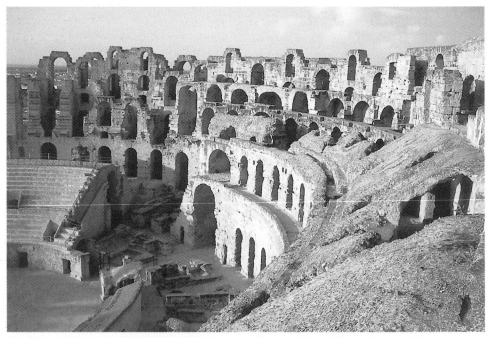

Plate 1.8 A Roman amphitheatre (El Jem), an important feature of large Roman cities, testifying to the wealth and sophistication of the empire. (Geophotos/Dr Tony Waltham)

of wheat each year from Egypt, North Africa and Sicily (Simmons 1996: 109). In the Roman Empire and other societies, high-value goods often travelled much greater distances. Even the Inuit, living the lives of hunters and gatherers, sometimes travelled hundreds of miles each year on hunting trips, and their geographical knowledge frequently covered an even wider span. Enough has been said above about migrations and the diffusion of artefacts and practices across great distances to suggest how space could be overcome even if individuals were immobile, at least by today's standards.

There was, however, no easily avoiding the problems of communication in premodern societies. Most people in agricultural societies lived lives which were bound to their villages and the surrounding regions. Their work consisted essentially in winning the means of subsistence for themselves and their families from their environment, as well as paying the taxes and meeting the other demands which were made upon them. From the time of the appearance of states down until the twentieth century the lives of most people were lived in the context of such **peasant communities**. As mentioned earlier, such communities are still characteristic of parts of the world today, though increasingly eroded by commercialism and growing contacts with modernity. Certain scholars, impressed by the many different forms that peasant societies have taken across the world and yet also by the things they seem to have in common, have tried to construct a specialized branch of social science out of their study (Shanin 1971).

Scholars researching traditional peasant societies have frequently pointed to the many cultural differences between such rural dwellers and the cities which ruled over them and which they might occasionally visit. A case in point is religion. Religion has been a fundamental factor in social organization and in outlook in virtually every human society down to the twentieth century. Early states were generally associated with an official religion which legitimized the established order and which was often practised in cities but which was rarely fully understood by the peasants, most of whom were probably illiterate. Even where the peasants officially followed the same religion as the elite, as in medieval Europe, they almost always interpreted its teachings in their own ways, mixing them with their own superstitions and 'pagan' beliefs. The life and outlook of the ruler and the elite, if not of all cities, were entirely alien to them. And equally *their* way of life was alien to the cities.

For this reason, it would be completely misleading to think of the subjects of precapitalist states as 'citizens' in the modern sense. Traditional rulers knew little of the countryside, where most of their subjects lived, and cared even less (except, perhaps, where they had landed estates, and even then the running of the estate could be left to officials). What mattered to rulers was law and order, and extracting the taxes and tribute the state needed to maintain itself in existence. The idea that rulers should care for the welfare of their ordinary subjects, let alone consult them about their policies, is very modern indeed.

It is difficult for most people living in a world dominated by capitalist relationships to conceive of societies where this was not the case. Before capitalism, most people were engaged in subsistence activities: hunting, gathering, farming, fishing or whatever. They might have to pay taxes or tribute, and they might trade some of what they gathered or produced, but market trade, generally organized by a small minority of merchants and traders, was often quite marginal to them. Moreover, trade, or more properly exchange, frequently took different forms from the market-type exchange which is so familiar today (Polanyi 1968; Dodgshon 1987: 193–224). For example, in many primitive societies

Plate 1.9   The celebrated Mappa Mundi or world map, dating from the fourteenth century AD, in Hereford Cathedral, England. Constructed to show Jerusalem in the centre of the world, this map typifies the medieval Christian view of a world explicable only in religious terms and symbols. (By permission of the Dean and Chapter of Hereford Cathedral Mappa Mundi Trust)

exchanges often took the form of ceremonial gift-giving and the mutual exchange of goods and tokens (**reciprocity**). The point was to maintain alliances and social relationships, much as today gifts are exchanged within the family and among friends. Bargaining and profit-making have little place in this kind of activity. In chiefdoms and early states, the levying of taxes and tribute by rulers and lords was also usual, the goods or money so

Plate 1.10 Part of the Terracotta Army, China. This remarkable archaeological site, believed to date to the end of the third century BC, consists of some 6,000 life-size statues of soldiers and horses. It probably forms part of the burial site of the Emperor Shi Huangdi and reflects the size and wealth of the Chinese empire over which he ruled. (Topham Picturepoint/Associated Press)

accumulated being used for a variety of purposes (**redistribution**). Where market trade did exist in early societies (which was quite often the case), it was frequently hedged about by laws and restrictions of various kinds. There might be laws against the taking of interest on loans, for example, regulations on price, or restrictions on where trade might take place or who might engage in it. Only with the rise of modern capitalism did the market become central to the way societies functioned.

It is very difficult to generalize about pre-capitalist societies and almost any generalization is open to objection. Box 1.7 describes the main features of the feudal society that existed in medieval Europe, the society that was the seedbed for the development of world capitalism (see Chapter 2). Even here, however, one must be careful – the features described would be more or less true, depending on the date and the place being considered.

1.7 Conclusion

Capitalism was preceded by millennia of human existence on the earth during which human societies took many different forms and helped mould the world that can be seen today. This chapter has focused on one way of trying to make sense of that past world,

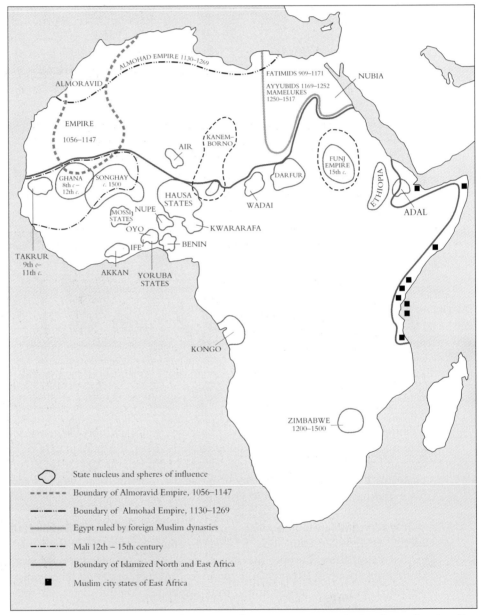

Figure 1.7 The great empires of Africa (AD 900–1500) before European colonization. *Source:* Based on *Times Atlas of World History*, 1989: 136–7

but also pointed to some of the difficulties and controversies which surround any attempt to do so. It is clear that human societies have on the whole become bigger and more complex through time, and their environmental impacts greater. What is less clear is the exact nature of the processes which have produced change, and who have been the winners and losers. Change has evidently been the norm rather than the exception, and this became particularly so with the rise of capitalism, the subject of the next chapter.

Learning outcomes

By the end of this chapter, you should begin to appreciate that:

- The surviving evidence necessarily limits knowledge of past societies, since they cannot be observed directly. In general, the further back in time past societies existed, the more uncertain knowledge about them becomes.

- People in the past lived very different lives from those of most people alive today. But it is important not to oversimplify or overgeneralize about past societies that varied enormously in their structures and ways of life and in their degree of complexity.

- Classifying societies into types is a way of making sense of the past and of trying to understand the geographical characteristics of varied societies. But all classifications have their shortcomings.

- Change has always been a characteristic of human societies. Societies have never been static, or completely isolated from the outside world. Societies have always changed their physical environments.

- The greater part of human history has been spent in hunter gatherer communities.

- Overcoming the friction of space was a severe problem in pre-capitalist societies. But it is important not to exaggerate its effects – people often travelled, and often migrated, long distances even in 'primitive' societies, ideas and artefacts travelled vast distances, and continental empires were successfully established.

- The landscapes that can be seen today have been profoundly influenced by past societies, including hunter gatherer ones. Our world is the product of generations of human activity.

- We must not assume that people in the past had the same values and outlook as we do. We should not assume that our ways of doing things are necessarily better (or indeed worse) than past ways, or than the ways of other peoples alive in the world today.

- Change and progress are two different things. Social change almost always means winners and losers. The human record suggests that the latter have often been the majority.

Further reading

Crone, P. (1989) *Pre-Industrial Societies*, Blackwell, Oxford. An excellent introduction to the character of pre-industrial societies 'of the civilized kind', i.e. from the invention of cities and civilization. Written for undergraduates.

Dodgshon, R.A. (1987) *The European Past: Social Evolution and Spatial Order*, Macmillan, Basingstoke. An original examination of European prehistory and history, taken from a geographical perspective. Based on the social classification system high-lighted in this chapter. Very detailed, but repays careful study.

Dodgshon, R.A. and **Butlin, R.A.** (eds) (1990)   *An Historical Geography of England and Wales*, 2nd edition, Academic Press, London. An historical geography of the region from prehistoric times to 1939. Chapters 4 and 5 illustrate the issue of feudalism.

Hodges, R. (1988)   *Primitive and Peasant Markets*, Blackwell, Oxford. A lucid introduction for undergraduates of the intricacies of exchange in pre-capitalist societies and of the development of markets. Contains a considerable amount of geographical material.

Past Worlds. The Times Atlas of Archaeology (1988)   Times Books, London. A lavishly illustrated survey of archaeology. Students need to remember that subsequent research has thrown doubt on some of the earliest dates.

Simmons, I.G. (1996)   *Changing the Face of the Earth: Culture, Environment, History*, 2nd edition, Blackwell, Oxford. A history of the impact of humankind on the natural environment from the earliest times to the present. Contains excellent chapters on pre-capitalist societies, especially from the environmental point of view, but tends to be a little technical in places.

The Times Atlas of World History (1989)   3rd edition, Guild Publishing, London. A highly acclaimed survey of world history beginning with human origins. This is in fact more a history book with maps than a true atlas and fails to provide detailed maps despite its large format. But it is lavishly illustrated and contains a wealth of factual and explanatory material.

CHAPTER 2

TERRY SLATER

The rise and spread of capitalism

Between 1500 and 1900 there was a fundamental change in the way an increasingly large part of the world was organized. Beginning in particular parts of England, and spreading to other parts of Europe, the capitalist economic system was to change, or influence substantially, not only the economies, but also the political, social, and cultural dimensions of newly powerful nation-states. By 1900, capitalism was the dominant socio-economic system over a large part of the world thanks to the colonial empires of those nation-states.

2.1   What is capitalism?

A number of writers have provided theoretical frameworks for understanding the workings and ramifications of the capitalist system. The first was devised by the Scottish philosopher, historian, and father of classical economics, Adam Smith, in his book *An Inquiry into the Nature and Causes of the Wealth of Nations* (1776). Like all theoretical models, Smith's is a simplification of reality but it introduced a terminology and series of conceptualizations which are still familiar to us today (Table 2.1). Smith assumed the model was driven by people's selfish desires for gain and self-interest. Thus, production takes place to generate profit; surplus profits are accumulated as capital; and the basic rule of the system is 'accumulate or perish'. Integral to the system is the determination of prices which, in a free market, said Smith, are determined by the supply of, and demand for, the factors of production. In order to maximize their profits, industrialists will always seek to minimize their costs of production, including the wages that they pay their workforce, so as to out-compete other producers by having lower prices. One way of doing this was through the **division of labour**, dividing manufacturing tasks into simple, repetitive operations which could be performed by unskilled, and therefore cheap, labour (see Chapter 12). Another important way in which manufacturers could seek **competitive advantage** over their fellows was to be first in the use of new technology. The significance of the invention and adoption of new ways of doing things, whether through science and technology, or ideas and organization, has often been crucial to firms, industries and regions in getting ahead of competitors.

2.1.1   Cyclical characteristics of economic development

Smith went on to model the workings of a national economy over a year (macroeconomics) to elucidate the way in which capitalism works in a series of interconnected cycles. Smith's model simplified the economy into only two sectors: agriculture and manufacturing. Services are considered only briefly; they were classified as 'unproductive', but useful, and therefore having value (Skinner 1970). The cyclical characteristics of capitalism have been studied subsequently by other economists, most notably the Russian

Table 2.1   Smith's micro-economic model

Factors of production	Society	Monetary return
Land	Proprietors	Rent
Labour	Labourers	Wages
Capital	Industrialists	Profit

Table 2.2   Logistic and Kondratieff cycles

Cycle	Period	Growth phase	Stagnation phase
Logistic II	c. 1450–c. 1750	c. 1450–c. 1600	c. 1600–c. 1750
KI	1770/80–1845/50	1770–1815	1815–1845/50
KII	1845/50–1890/96	1845/50–1875	1875–1890/96

Kondratieff (1925), whose name is now used for the roughly fifty-year long-wave cycles of boom and depression which have characterized the capitalist world since the mid-eighteenth century (Table 2.2).

In seeking the explanation for these cycles of growth and stagnation, Schumpeter (1939) argued that technical innovation leading to the development of new industries was the key to understanding the growth phase of the cycle. Because technical innovation is spatially uneven, then so too has been the geography of economic development under capitalism. The question then arises as to why the geography of innovation is uneven. Recent research has suggested that the socio-institutional structures of regions of innovation, or of those lacking innovation, are the key to explanation; in other words whether educational, governmental and social organizations encourage or discourage enterprise and change. Others have criticized the technological determinism of these theories (Mahon 1987).

A powerful, though almost certainly false, model of change and transformation within the capitalist system is the 'stages of development' model through which all capitalist societies, regions or states are proposed to pass. The best-known of these is Rostow's five-stage model of British economic history (1975) (Box 2.1). The danger of such stage

---

Box 2.1   **Rostow's 'stages of development' model**

Stage One: Traditional society

limited technology and little social change

Stage Two: Preconditions for take-off

commercialization of agriculture and growth of extractive industries; investment in better transport; growth of social elite

Stage Three: Take-off to industrialization

development of manufacturing; modernization of social/political systems

Stage Four: Drive to maturity

development of wider industrial and commercial base; at least 10 per cent of national income invested in industry

Stage Five: High mass consumption

exploitation of comparative advantage in international trade models is that they encourage the belief that any society can move forward through the stages from 'traditional society' to 'high mass-consumption'.

It is important to understand that parts of the capitalist system that Smith first theorized in the eighteenth century had been in existence for many hundreds of years: the division of labour for some tasks within family groups, for example, and the use of money to facilitate exchange. However, most scholars agree that the capitalist system began to cohere into an integrated whole in a particular place (England), at a particular time (the seventeenth century), and that it then took another long period (until the mid-nineteenth century) before it had matured in the states of north-west Europe and in North America (Green and Sutcliffe 1987: 6–7) (Box 2.2).

Box 2.2

Characteristics of states with a mature capitalist economy and society

- The majority of the population in these countries were wage-labourers forming a 'working class' who sold their labour power for wages in cash or kind so as to purchase food and other commodities to survive.
- The majority of these wage-labourers were male and worked outside the home; females increasingly 'worked' in the home, both in the domestic care of their families and by 'selling' surplus labour time for minimal wages.
- Wage-labourers were increasingly closely supervised by managers of the production processes.
- The means of production of wage-labourers was owned by capitalists whose aim was to make a profit on their investment.
- There was an increasing disparity between the 'profit' of capitalists and the 'wages' of workers with an increasing propensity for conflict between the two.
- The vast majority of goods and services, including fixed property like land, were distributed through monetary exchange.

2.2   Other perspectives, other stories

2.2.1 Marxism

The viewpoints of orthodox economics are not the only interpretation of the transformation of large parts of the world over the past three hundred years. Karl Marx (1867) took a very different perspective in his analysis of capitalism in the mid-nineteenth century, *Das Kapital* (see Box 3.2). He proposed that profit arises out of the way in which capitalists (the bourgeoisie) dominate labour (the workers) in an unequal class-based relationship. Capitalists, as a class, he said, can only reproduce themselves by accumulating profit, and continuing to expand the basis of profit, at the expense of labour by, for example, increasing the length of the working day or reorganizing the work process to increase productivity. This is because capitalists are in mutual competition. If they do not exploit labour, they perish. This unequal class relationship was perceived by Marx to lead inevitably to class conflict. In his later writing he sketched out ways in which labour could gain control of the means of production and thereby 'throw off their chains'. This was to have dramatic long-term consequences for the socio-political organization of the world through most of the twentieth century between the Russian Revolution and the upheavals of 1989 (see Chapter 3). We need to note, however, that Marx says little of a third class-based group which became increasingly significant as the nineteenth century progressed, the middle classes of professionals, administrators and educators. Marx's writing is also dominated by an historical, but not a geographical, perspective: it traces class relationships in time but not in space. Marxist interpretations of capitalist development have therefore always had difficulty in conceptualizing uneven spatial development. The geographer David Harvey has perhaps done most to begin to close this gap with his studies of the working of capitalism in the production of urban spaces, notably of the transformation of Paris in the third quarter of the nineteenth century (Harvey 1985).

2.2.2 World systems theory

Another way of conceptualizing the nature of these changes is to be found in the writings of Immanuel Wallerstein (1979), whose 'world-systems theory' provides a threefold cat-egorization of historical socio-economic systems (Box 2.3). Wallerstein proceeded to delimit the spatial characteristics of the capitalist world economy into core, periphery and semi-periphery. These terms are not used in the everyday sense but signify areas in which particular processes operate. For Wallerstein, core processes are those characterized by relatively high wages, advanced technology and diversified production; periphery pro-cesses are characterized by low wages, simple technology and limited production. Between the two is the semi-periphery: areas that exploit the periphery but which are exploited by the core, so that they exhibit a mixture of both core and peripheral processes

Box 2.3	Wallerstein's 'world-systems' theory

Mini-systems

These are based upon reciprocal-lineage modes of production where small kin-based groups in local territories dominate. They tend to develop, dominate and dissolve over short time periods (see Chapter 1).

World empires

These are based upon redistributive-tributary modes of production where large numbers of agri-cultural producers generate a surplus sufficient to allow for artisans and a military-religious-bureaucratic ruling class (see Chapter 1). Wallerstein suggests there have been many such world empires from the Roman Empire, to feudal Europe, and the Chinese dynasties.

The world-economy

This is the entity based upon the capitalist mode of production based on profitability. The European capitalist world-economy is the only one that has survived the vulnerable first stages of develop-ment and has been able to expand to eliminate other mini-systems and world empires (Taylor, 1989).

(Taylor 1989: 16–17). In the period between the seventeenth and the end of the nineteenth century some regions of the world were in the core (north-west Europe, for example) or periphery (central Africa) throughout the period. Other regions moved from the semi-periphery, into either the core (the southern states of the USA, for example) or the periphery (the states emerging from the ruins of the Ottoman/Turkish empire). Wallerstein's theory has been very influential in explanations of the growth of globalization, but it is an explanation only at the 'structural' level. It says little about the complexities of the social and economic networks that enable the system to work, or of the resistance by groups or individuals trying to change it (Ogborn 1999).

2.3   The transition from feudalism to capitalism

The transformation of the European economy into a capitalist one was a long-drawn-out process and historians continue to argue the precise causes of the changes that took place. Writers using an avowedly Marxist frame of reference (Dobb 1946; Kaye 1984) have made many of the most significant contributions in this debate. To them, the development of a class of wage labourers is therefore crucial, together with the assumption of political power by the new class of capitalists. Geographers have been particularly concerned to trace the outworking of these historical processes of social and economic transformation in the particular space economies of local regions (Gregory 1982; Langton 1984).

The disintegration of feudalism (see Box 1.7), with its carefully controlled market system, was considerably speeded by the after-effects of the Black Death in mid-fourteenth century Europe. The European population was reduced by between one third and a half, and towns were especially hard hit. The resulting labour shortage meant that enterprise and new ways of doing things were more likely to be rewarded, since the feudal elite required more revenue to maintain their power. At the same time, the slackening of social and cultural controls meant that new ways of thinking could flourish, especially in the growing towns.

By the later fifteenth century larger European towns were dominated by what has come to be called **merchant capitalism**. Merchants were both the providers of capital and principal traders in a regionally-specialized, complex, Europe-wide trading nexus based not on luxury goods as in the medieval period, but on bulky staples like grain and timber, and on an increasing array of manufactured products. Towns which flourished economically in fifteenth- and sixteenth-century Europe were those whose inhabitants were able to copy and manufacture imported products more cheaply than the exporting region, and develop a constant stream of new, innovative, marketable products (Knox and Agnew 1994: 154–61) (Box 2.4). For most commentators, the key question then becomes: why did these processes of transformation coalesce first in sixteenth- and seventeenth-century England?

We can begin to answer this question by suggesting that, since land was one of the key factors of production, critical in the sixteenth century was the enormous transfer of land from conservative ecclesiastical ownership to secular ownership. This was the consequence of the dissolution of the monasteries by Henry VIII in the later 1530s. Even the monastic buildings could be adapted by capitalist manufacturers, as the Wiltshire clothier William Stumpe showed at Malmesbury Abbey where he installed 300 weavers in the 1540s (Chandler 1993: 487–9). Land was also being transformed in the sixteenth and

> ### Box 2.4 Characteristics of the period of merchant capitalism in Europe
>
> - Increasing numbers of people sold their labour for money wages. They ceased to work on their own land.
>
> - Increased numbers of wage-labourers led to increased consumer demand for food, clothing and household goods; in many places there was a notable rise in living standards.
>
> - More producers of both agricultural and craft-manufactured products began to accumulate capital as they produced for this growing market. These prosperous yeoman farmers and manufacturers were the foundation of a new class of capitalists.
>
> - The removal of feudal market restrictions and controls by guilds on production saw the accumulation of capital by merchants who began to invest in the reorganization of production on a capitalist basis.
>
> - Technical innovations transformed industries; the most significant was probably the development of the printing press by Gutenberg, in Mainz, enabling knowledge to be diffused cheaply and rapidly.
>
> - The rediscovery of Classical knowledge led to new ways of seeing the world and, consequently, to its rebuilding to reflect these new images, especially in towns.
>
> - The more individualistic and humanist perspective brought about by the Renaissance led to new ways of thinking about people's relationship with God, which was ultimately to split the Western Christian Church asunder and societies with it.
>
> - Ultimately, too, these changes led to political revolution whereby the new capitalist class became dominant in the governance of nation-states.

seventeenth centuries through the process of enclosure. This enabled livestock to be raised more efficiently for the urban meat markets and experimentation in new agricultural techniques to be undertaken by yeoman farmers (Butlin 1993: 178–9).

In the seventeenth century it was the social, religious and political transformation of England that was critical according to Dobb (1946). The struggles between Crown and Parliament during the Civil War and Commonwealth (1642–60) are seen as a conflict between landowners and capitalist yeomen and manufacturers. Though this struggle was continued through the Restoration (1660) and Glorious Revolution (1688), by the end of the seventeenth century the capitalist bourgeoisie were politically predominant and were able to transform the state to their own advantage.

Elsewhere in Europe landowners remained pre-eminent over a predominantly peasant agricultural workforce and this period is known as the 'Age of Absolutism'. Landowners, especially the rulers of small states, amassed enormous wealth from their control of land. This wealth was expended on increasingly spectacular landscapes of display: huge palaces, filled with works of art and every luxury, set in carefully designed, and often very intricate, landscapes reliant on vast inputs of labour for their maintenance (Figure 2.1). King Louis XIV of France had set the model at his palace of Versailles, outside Paris, using the resources of a much larger state. Such conspicuous displays of wealth and privilege by the few led, ultimately, to a much bloodier revolution than in England. The French Revolution saw

Figure 2.1   The palace, gardens, park and town of Karlsruhe, Germany, founded in 1715 by the Margrave Karl Wilhelm of Baden. *Source*: Gutkind, 1964: 301

the slaughter of not simply the royal house, but of the landowners, intellectuals and the bourgeoisie, since this was the first of the workers' revolutions, with its cry of '*Liberté, egalité, fraternité*'. However, it led rapidly to the totalitarian militarism of Napoleon.

2.4   An expanding world

As the European economic and cultural world was gradually transformed, European travellers began to voyage beyond the shores of Europe. They travelled overland into Asia, drawn both by a thirst for greater knowledge but, more significantly, by the rewards which could be reaped from direct exploitation of scarce commodities for European consumption. These included spices, sugar, silk, muslins, porcelain and the like which had previously been traded through the eastern Mediterranean. In the late fifteenth century, however, this region was coming under Turkish control. The Portuguese, funded by Italian bankers, were the first to navigate round Africa into the Indian Ocean, with its Arab-dominated sea-based trading networks, whilst soon after Christopher Columbus, again funded by Italian (Genoese) merchant capital, sailed westwards across the Atlantic (in 1492) to open the Americas to European exploitation.

For Spain and Portugal, this is the heroic 'Age of the Navigators', as they moved rapidly to take control of these new maritime trade routes looking out to the Atlantic. The merchants of Lisbon and Seville grew rich on this trade. For the rest of the world it marks the beginning of the European colonial empires which were to dominate and control almost every aspect of life for the next four hundred years. The Portuguese established a far-flung network of fortified port bases around Africa and the Indian Ocean through which trade in gold, spices and textiles was conducted (Figure 2.2). Equally significant was their use of African slaves to produce sugar in plantation farms in their Brazilian territory. The Spanish founded a more militaristic, oppressive and exploitative empire in central and south America which saw the decimation of local populations through warfare and the catastrophic effects of European diseases, especially smallpox and measles. It also brought

Figure 2.2 The Portuguese fortress and trading station of Sofala, in modern Mozambique, in 1558. For a time it was the administrative centre for all Portuguese trading with India. *Source*: Carneiro, 1990

enormous inflows of gold and silver bullion into Spain and, ultimately, onwards into the European economy. Blaut (1993) sees this as the event which more than any other kick-started the European capitalist world system.

Fairly quickly, control of the products of this new maritime world economy moved from south-west to north-west Europe. The Dutch began to compete successfully with the Portuguese in the Indian Ocean and the centres of European capital moved from Lisbon and Seville first to Antwerp by the mid-sixteenth century, and then to Amsterdam at the end of the century. The merchants of these ports successfully integrated the Atlantic trade with that flowing from the Baltic and from central Europe via the Rhine valley. It was at Antwerp that the first stock exchange was established enabling capital raised in one business to fund investment in another. Amsterdam's rise was based on having greater military and naval power than its neighbour and using it to enforce the privileges of a distinctive trading monopoly: the state-licensed joint-stock charter company (the Dutch East India Company) (Dodgshon 1998: 74–83).

At the same time, French and English fishermen had followed John Cabot's exploratory voyage across the North Atlantic to begin to exploit the rich fishing grounds off Newfoundland and, subsequently, to trade with native Americans for fur pelts. From these small beginnings French and English merchant influence began to expand through the seventeenth century in the Indian Ocean (using similar company trading monopolies to the Dutch); in the Caribbean, where sugar plantations and slave labour were copied from the Portuguese; and in North America. It superseded the Dutch in the early eighteenth century through superiority of arms, but it was not until the later eighteenth century that the British emerged dominant in the developing world economy. Consequently the primate city of the European capitalist system moved once more, from Amsterdam to London (Dodgshon 1998).

The colonialist trading systems that developed in the period up to 1770 have often been described in simple dualist, oppositional, terms (core–periphery; dominant–subordinate; metropolitan–colonial, for example). Blaut (1993) is especially critical of the Eurocentric character of this type of argument. More recently, a more complex, multi-layered interpretation has been posited. We need to note, first, that different parts of the developing world economy operated in somewhat different ways, were based on different product interactions between raw materials, manufacture and consumption, involved complex transportation flows, and required different politico-military frameworks to make them function properly. Meinig's (1986) model of the Anglo-French mid-eighteenth-century North Atlantic economy uncovers some of these complexities in terms of the commercial, political and social systems that were required for the geography of colonial capitalism and Ogborn (1999) has added details of networks of individuals and patterns of resistance (Figure 2.3).

2.4.1 Colonial commerce

In Meinig's North Atlantic commercial system, London and Paris acted as the source of finance, commercial intelligence and marketing (London's Royal Exchange was founded in 1566). The expanding industries of particular regions of western Europe were growing, in part because they supplied a developing colonial market with manufactured goods such as textiles, tools, armaments and household necessities. These goods were assembled in Atlantic ports such as Bristol, Liverpool, Bordeaux and Le Havre. After crossing the

Figure 2.3 The eighteenth-century North Atlantic trading system between Britain, Africa and North and Central America. *Source:* Bayley, 1989: 46–7

Atlantic they were stored, disassembled and distributed from colonial ports such as New York, Charleston and New Orleans (Figures 2.4, 2.5) by major traders. They sold on to local traders towards the colonial frontier. At the frontier a barter system of exchange was as likely to be found as a money economy. None the less, fur, fish, timber and agricultural

PLAN OF
NEW ORLEANS
THE CAPITAL OF
LOUISIANA;
with the Disposition of
its Quarters and Canals
as they have been traced
by Mr de la Tour
in the Year 1720.

REFERENCES

A. *House of the Intendant*
B. *Convent of the Capuchin Fryars*
C. *The Jail*
D. *Guard Room*
E. *Hospital and Convent of the Ursulines*
F. *Guard House of the Inhabitants*
G. *Place where the Windmill Stood*

British Fathoms

RIVER ST LOUIS

Called by the Natives

MISSISSIPI and MESHASSEPI

Figure 2.4   Plan of the French colonial entrepôt city of New Orleans at the head of the Mississippi delta in 1720, published in London in 1759 by Thomas Jefferys.

Figure 2.5   Madame John's Legacy, c. 1788. One of the oldest houses in New Orleans, it was built during the Spanish colonial period (1765–1803). Commercial functions occupy the brick-built ground floor, residence the balcony floor.

produce flowed in the opposite direction for storage and transhipment back to Europe (Figure 2.3).

In North America, tobacco, rice, sugar and cotton were the major products. All were produced on the plantation system using slave labour. By 1750, some 50–60,000 people were being forcibly transported across the Atlantic each year from all parts of the western African seaboard. Many died en route or shortly afterwards but, altogether, some 3.8 million Africans had been transported by 1750, just over half to Latin America, the remainder to English, French and Dutch North America and the West Indies. These staggering figures meant that African Americans were easily the largest new culture group to be established in both North and South America in the colonial period (Meinig 1986: 226–31). By 1750, the trade in slaves was dominated by large European companies using specially constructed and fitted ships, but many smaller traders from the Atlantic ports would sail first to African trading stations to exchange manufactured products for slaves. The slaves would then be taken to the West Indies or Charleston, with ships returning to Liverpool or Bristol with a cargo of sugar, tobacco or rice, and sometimes slaves too. Most were destined for domestic service in aristocratic households (Figure 2.3).

2.4.2 Colonial society

The slave trade was perhaps the most significant aspect of the cultural transformation effected by the eighteenth-century capitalist world economy in the Americas and, it should not be forgotten, in Africa. One of the distinctive features of this transformation in the Americas was the 'othering' of Africans by Europeans. They were regarded as in every way inferior to Europeans; they were legally and socially defined as different; and they therefore lived their lives separately from white settlers. Where Africans were in the majority, as they were in the West Indies, and in the Carolinas and Louisiana, this enabled distinctive African-American cultures to develop within the confines of political and economic enslavement.

There were, of course, other trans-Atlantic migrations throughout the time period covered in this chapter. The earliest included those seeking to escape religious persecution in Europe, or to establish religious utopias in the new continent. The Puritan 'Founding Fathers' of New England were in this category and, right through to the Mormon migration in the early nineteenth century, such groups continued to leave Europe's shores for what was perceived to be a better new world. Religious faith also inspired another group of settlers who came to convert the native people of the Americas to Christianity. The competition in Europe between Catholic and Protestant versions of the faith – between Reformation and Counter-Reformation – was exported to the colonies since both native Americans and African slaves were perceived as being in need of conversion. Consequently, Jesuit monasteries or Franciscan friaries often stand at the heart of Spanish colonial towns and cities, whilst their dedicatory saints gave name to them: San Francisco, Los Angeles, San Diego and San Antonio, for example (Conzen 1990).

By the beginning of the eighteenth century, significant migrations of Scots, Irish, French, Germans, Dutch, Swiss and Moravians had taken place, but the dominant migrant group were the English. Between 1700 and 1775 the population of the eastern colonies grew tenfold to some 2.5 million and, by 1820, the population of the United States surpassed that of Britain (Lemon 1987: 121). Pennsylvania was the colony both with the most rapid growth and with the greatest variety of European settlers. Once the initial phase of settlement was passed, these new Americans were a people of agricultural plenty, and of good health and longevity. They were also overwhelmingly rural; only 5 per cent of them lived in towns. The losers were native Americans who were dispossessed of their lands east of the Appalachians and often taken into slavery.

2.4.3 Colonial politics

There were major differences in the way European states governed their colonial empires. The Spanish developed a highly centralized system in which all aspects of policy were laid down in Madrid. The *Laws of the Indies*, which laid down in meticulous detail how colonization should take place and how new towns should be planned, are symptomatic of this centralization (Nostrand 1987) (Box 2.5). The French governed through a similar centralized structure of military, administrative and ecclesiastical strands of authority linked to government departments and the Crown in Paris. The English, in contrast, allowed each colony to be virtually self-governing and the links to London were many and various. English colonial plantation owners and merchants were not averse to threatening a governor if their capitalist trading relations were threatened. It should occasion no surprise that it was the English colonies that eventually rebelled and fought for their independence. Overlapping these colonial territories were the older territorial patterns of native groupings, those of the Iroquois and Creeks remaining well into the eighteenth century (Meinig 1986: 262).

> **Box 2.5**
>
> ## Summary of the *Laws of the Indies* by Philip II of Spain, 1573
>
> **Laws 1–14**
>
> Establish that exploration cannot take place without royal permission; that governors should learn all they can about their territory; they should consult with and negotiate with local 'Indians'. 'Discoverers' should take possession of land and should name rivers, hills and settlements.
>
> **Laws 15–31**
>
> Lay down that Spaniards should treat 'Indians' in a friendly manner; they should help priests in their work of conversion.
>
> **Laws 32–42**
>
> Concern the type of region that should be explored, pacified, settled and brought under Spanish mandate; the siting of towns in such regions, and the establishment of local government.
>
> **Laws 43–109**
>
> Establish the legal and taxation regimes for new colonies and lay down that each town should have at least thirty families. The ideal size of house plots, farms, herds, commons is established.
>
> **Laws 110–128**
>
> Provide details of the plans of new towns, including the size of the plaza, the orientation and breadth of the streets, the location of churches and public buildings, and the allocation of house plots by lot.
>
> **Laws 129–135**
>
> Describe provision for common land and farms and the character of domestic buildings.
>
> **Laws 136–148**
>
> Lay down relationships with local 'Indians' and the way in which they should be converted to Christianity so that they 'can live civilly'. (Source: Crouch *et al.* 1982: 6–19)

The Revolutionary Wars of the 1770s, marking the foundation of an independent United States, provide a convenient termination for the first phase of the colonialist expansion of the core European states into the periphery of the Americas. Bergeson and Schoenberg (1980) have shown how this first cycle of the colonial expansion of the capitalist system coincided with the expansion and contraction of the second logistic cycle (the longest amplitude economic cycle: Table 2.2) (Knox and Agnew 1994: 9).

2.5 Imperialism and racism

A second phase of colonial expansion, through the nineteenth century, was geographically distinct in that it was focused on Asia and, especially after 1880, on Africa. It is also marked by the development of a much more virulent racism in the 'othering' of all non-Europeans. The consolidation of European states in the nineteenth century, especially of Germany and Italy, and the increasing economic competition they offered to the eighteenth-century leaders of the world economy, Britain and France, led them to seek new opportunities through the colonial exploitation of hitherto economically peripheral lands. Britain seized the Dutch Cape Colony at the tip of southern Africa in 1806 to prevent its use by Napoleon's navy. By 1820, 4,000 British settler graziers were in conflict with African pastoralists, despite the fact that the low wage-labour of those same Africans was essential to maintain the farms (Lester 1999). At much the same time, the first British farmers were beginning to establish themselves in Australia as its period as a convict colony began to draw to a close. Over extensive parts of the Indian sub-continent the collapse of the Mughal Empire saw its replacement with a variety of British administrations backed by British soldiers. After the 1857–9 Indian Mutiny, India became a Crown Colony and British administration, schools and economic investments followed.

Meanwhile, in Britain itself, Christian-inspired reformers succeeded in bringing an end to the slave trade, and then to slavery. Sadly, the invective of both populist and bourgeois writing opposed to abolition characterized Africans, Australian Aborigines, Indians and native Americans as not only unchristian and therefore 'uncivilized' but as less than human; inferior to Europeans in every way; both savage and yet child-like, and therefore needing to be controlled and disciplined. By the middle of the nineteenth century these attitudes were widespread amongst all classes in Britain and were fuelled further by popular antagonism to events such as the Indian Mutiny (1857), the Zulu rebellion (1879), and the Maori wars (1860s). By the later nineteenth century, Darwin's ideas on evolution were being misused to give a scientific veneer of respectability to this 'othering' of non-Europeans, to the extent that Australian Aborigines were being shot as 'vermin' rather than seen as fellow humans.

The mid-nineteenth century marked both the deliverance of one continent from colonialism to self-government, as South America broke free from Spanish and Portuguese control, and the colonizing of another, as European powers raced to divide Africa amongst themselves. The interior of Africa remained largely unknown to Europeans until the 1860s except for Christian missionaries. But by the end of the century, borders had been delineated and fought over, natural resources were being rapaciously exploited, and the polities, cultures and economies of native peoples had been disrupted or destroyed (Figure 2.6).

2.6   Industrialization

2.6.1 Merchant capitalism

The early phases of capitalist industrial development in Europe, up to about 1770, are categorized in various ways. To some writers this is the period of **merchant capitalism**, to others this is the period of **proto-industrialization**. Both terms have generated a substantive academic debate as to precisely what processes of transformation were involved. The main characteristics of this period are shown in Box 2.6.

Textile manufacture is the most studied production process of this period and some of the transformations of this industry can be seen to begin in several parts of Europe as early

Figure 2.6(a) 'Unknown' Africa in 1847 as published in *Black's General Atlas* (Edinburgh).

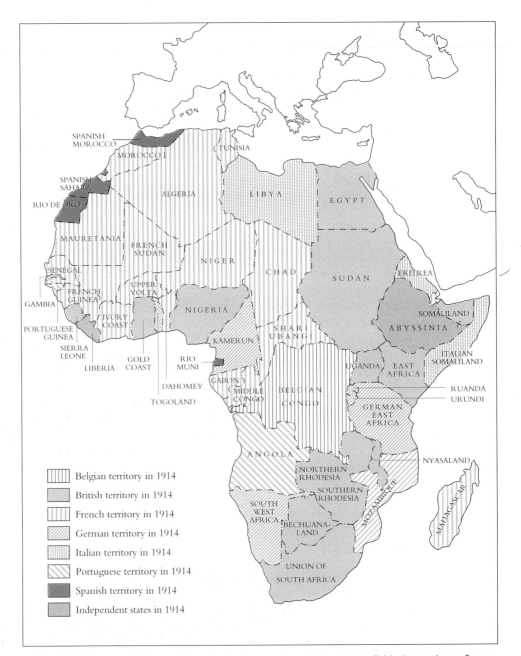

Figure 2.6(b) 'Unknown' Africa in 1914 after the European imperial contest to divide the continent. *Source:* Moore, 1981: 74

as the fifteenth century. By the early eighteenth century, in some branches of the industry, more and more of the production processes were being mechanized and industrialists were concentrating their machinery into single, multi-floor buildings. By the 1770s, the mechanization of spinning had been completed and weaving followed in the 1820s. The

<table>
<tr><td>

Box 2.6

</td><td>

Proto-industrialization

</td></tr>
</table>

- Many early industrial processes were located in the countryside.

- Most industries were characterized by some type of 'domestic' production system whereby the capitalist/merchant provided raw materials/machinery for producers who worked in their own homes.

- Producers generally worked in family units and were paid piece rates according to their production each week or month.

- Machinery was frequently water-powered.

logic of this transformation was completed by steam power and the factory was established as both a building (Figure 2.7) and a system of production (Box 2.7). Different sectors of the textile industry went through this transformation at different times, and some sectors declined absolutely as capital was diverted to the most prosperous sectors. The textile industry, more than any other, had a relentless tendency to geographical specialization and concentration with the result that many towns and regions were dangerously dependent on a single industry (Laxton 1986). By the 1770s in Britain, west Yorkshire was beginning to dominate light worsted woollen cloth; Lancashire and the north Midlands cotton cloth production; the Welsh borders flannel, and the east Midlands hosiery.

The complex interaction of resources, labour skills, technological innovation, and capital circulation in particular regional economies, which increased production and lowered prices, meant that new markets had to be developed and transport improved. Market opportunities were to be found both in the growing urban markets at home, and among the colonial populations overseas. Thus, cotton textiles were first imported to Britain from

Figure 2.7 Starkey's woollen textile mill, Huddersfield, c. 1850. Though this mill was steam-powered, by mid-century its powerlooms were still housed in multi-storey buildings rather than a single-storey shed, since the site was a restricted one. *Source*: Giles and Goodall, 1992: 102

Box 2.7

The factory system of production

Under the factory system the capitalist:

- Had complete control of the production process from receipt of raw materials to finished product.

- Had control of the labour force on whom a new, disciplined time geography could be imposed (early factories often worked day and night).

- Could apply capital to the development of new machinery to simplify processes and reduce labour costs through mechanization; and,

- By dividing the tasks of the production process, could reduce labour costs further by employing women and children.

India; the techniques of production were learnt; manufacture commenced; factory production lowered costs; cotton textiles were exported to India; there they undercut and destroyed the Indian industry.

A key transformation of the transport infrastructure, so essential to industrialization, was the improvement of links with growing towns, and with the ports. Canals and turnpike (toll) roads, which were the principal innovations before the 1820s, also lowered the costs of raw materials, especially of coal. This is graphically illustrated by the way in which urban populations celebrated the opening of new canals, not for their industrial potential, but because they dramatically reduced the cost of domestic heating. In Britain, the construction of better roads, and of the canal network, mostly between 1760 and 1815 (the growth phase of the first Kondratieff [Table 2.2]), was financed by private capital. The state was involved only in providing the necessary legislation to enable construction to take place and fees or tolls to be levied. Thus Josiah Wedgwood (china and pottery), Abraham Darby (iron manufacturer) (Figure 2.8), and Matthew Boulton (jewellery and silver plate), as well as the Duke of Bridgewater (landowner and coal owner), all invested some of their capital in canal companies to improve the distribution of their products (Freeman 1986). A similar pattern is observed in New England. In France and Germany, by contrast, new roads were seen to have a military function in the centralization of state control and were therefore largely financed by the state through taxation.

2.6.2 Agricultural change

It was not just industries such as textiles, metal manufacturing and pottery that were transformed by early capitalist production; food production needed to keep pace with a rising population that was increasingly employed full-time in manufacturing. Consequently, the eighteenth century saw the transformation of agriculture into increasingly capitalistic modes. In parts of Europe, especially in Britain, self-sufficient peasant farming began to come under pressure from landowners who used state legislation to remove common rights to land in favour of private ownership through enclosure. Ownership was increasingly concentrated into fewer hands, and farms were consolidated from scattered strips to single block holdings, again, especially in Britain and, rather later, in Scandinavia and

Figure 2.8 The world's first iron bridge, constructed across the River Severn, east Shropshire, in 1788 by Abraham Darby III's Coalbrookdale Company, was part of the transport infrastructure improvements financed by local industrialists. *Source*: The Ironbridge Gorge Museum

northern Germany. Elsewhere in Europe this process was delayed until the twentieth century. In northern and western Europe landowners invested in transforming production for the new urban markets, using technology and new techniques and crops first developed in the Low Countries in the seventeenth century. In eastern Europe, the semi-periphery, this period saw a reversion to serfdom and near-feudal relations of production. Capital was also deployed to increase the area of land under intensive, rather than extensive, production. Fenland and coastal marshes were drained; heathland soils improved for cultivation, and the moorland edge of improved pastures pushed higher up hill and mountain slopes.

In the nineteenth century, European agriculture continued to increase production but was unable to keep pace with the demand from the growing urban industrial population. In 1840, Britain repealed the Corn Laws, which had protected local grain producers, and opened its markets to colonial and American producers. New methods of extensive grain farming were used in mid-western North America, and in the new British colonies in Australasia and South Africa. The invention of the steam ship, allowing more rapid transport, and of refrigeration and meat canning from the 1870s, led to similarly extensive ranch grazing of sheep and cattle in these countries. Together with the Pampas grasslands of Argentina, all became an important part of the semi-periphery of Britain's global economy. This new export-dominated agriculture was reliant on overseas investment by British capital in new railway networks to transport these products to slaughterhouses and industrial packing plants at the port cities whilst, in the USA, Chicago was growing rapidly on the same economic foundations (Cronon 1991, Miller 1997). It was accompanied, too,

Table 2.3 European emigration 1881–1910

Source countries	Destination countries	Numbers of migrants
Great Britain	N. America, S. Africa, Australasia	7,144,000
Italy	USA and N. Africa	6,187,000
Germany	North and South America	2,143,000
Austria/Hungary	North and South America	1,799,000
Russia	USA	1,680,000
Scandinavia	North America	1,535,000
Spain	Central and S. America; N. Africa	1,472,000
Ireland	USA	1,414,000
Portugal	South America	775,000
S.E. Europe	USA	c. 465,000
France	Central and S. America	223,000
Low Countries	USA	171,000

Source: Moore (1981: 57).

by new waves of migration to the colonies and the mid-west of the USA (Table 2.3), especially from Ireland and Scotland, where the Great Famine and the Highland Clearances, respectively, were other manifestations of capitalist agricultural change in the British Isles. The United States absorbed enormous numbers of European migrants throughout the nineteenth century. Irish migrants predominated in the 1840s and 1850s, followed by Scots and English, Scandinavians and Germans through the second half of the century, with Russians, Italians and south-central Europeans from the 1880s onwards (Table 2.3) (Ward 1987).

This later colonization, and the internal colonization of the United States mid-west in particular, were grandly characterized as a moving frontier of settlement by the American historian, Turner, writing in 1894, who suggested that it also transformed the 'character' of frontier settlers, making them self-reliant, opportunistic, individualistic and democratic. Recent commentators have noted that Turner's hypothesis says little about the continued sweeping aside of the rights of the native populations of these lands (the land was regarded as 'open' or 'free' for white settlement); nor does it give any credit for the settlement process to women, who provided both the domestic labour and, often, especially in the initial phase of settlement, much of the farm labour too. The advancing frontier of capitalist agriculture was as much small-scale, incremental and domestic as it was wide-sweeping and large-scale.

2.6.3 Factories and industrial production

Some commentators have seen the 'Industrial Revolution' in Britain as a short period, between 1770 and 1830, of rapid transformation whereby the country's economy moved from an agricultural to an industrial manufacturing basis. More recent writers see the transformation as much more drawn-out, extending from the seventeenth century and into the early twentieth century. They also see it as much more regionally diversified and geographically uneven, with particular regions specializing in particular products which transformed their production systems at different times. Most attention has been given to the textile and iron and steel industries, both of which are characterized by large factories requiring massive capital investment in buildings and machinery, driven by steam power, with manufacturing processes increasingly vertically integrated, and employing a large, and increasingly disciplined labour force (Figure 2.7). The same forces were at work in New England. By 1855, there were 52 cotton mills in Lowell, Massachusetts, employing more than 13,000 people, two-thirds of them women (Groves 1987), and in Europe, where Lille (France), Ghent (Belgium) and the Wupper valley of Germany developed as centres of cotton textile manufacture (Pounds 1990: 402).

Many other industries were developed in the same way as new markets were developed on a national and international basis and production expanded. Food and drink, for example, is rarely thought of as a factory-based industry, but brewing, once the prerogative of almost every village inn, became increasingly an urban, large-scale industry with regional markets for its beers. In England, Burton-upon-Trent developed rapidly as the 'brewing capital' of Britain, with huge factory-scale breweries, on the basis of its colonial contracts to supply bottled 'India Pale Ale' to troops and civilians stationed in the Indian sub-continent as a substitute for local water. For other industries, however, including many manufacturing industries, 'factories' remained small-scale workshops employing fewer than 50 people (the Sheffield cutlery industry in northern England, and wire-drawing in the eastern Ruhr in Germany are good examples). Indeed, many industries remained almost domestic in scale well into the twentieth century (for example, the Birmingham jewellery and Coventry watch industries, both in central England).

The nineteenth century is characterized not only by this enormous variety in the scale of production, but also by an equally enormous variety of manufactured products. The tag 'The Workshop of the World' is as true of nineteenth-century Britain in general as it is of several of the country's manufacturing cities. London's 1851 Great Exhibition (Figure 2.9) was seen as a showcase of the huge variety of products from Britain's factories and workshops and those of its empire. In other British, European and American regions there was an increasing specialization so that local economies were dependent on maintaining market advantage on a single industry: the Polish textile city of Łódź (Box 2.8 see pages 65–6) and Lancashire's cotton mill towns were of this kind. Such places also produced distinctive working-class cultures which, in the case of Łódź, was also predominantly Jewish (Koter 1990). Again, in those same mill towns much factory employment was female (see the example of Lowell above), whereas in heavy industry centres, such as ship-building in Newcastle and Sunderland, in the north-east of England, waged employment was male-dominated and women's roles were primarily domestic. E.P. Thompson, in his classic Marxist interpretation of the *Making of the English Working Class* (1963), is especially sensitive to the variety of experience of working people in this period of aggressive industrial capitalism. By the second half of the century British labour had won the right to organize itself into trade unions to improve work and social conditions for members.

There are, however, some common features of nineteenth-century industrial development in Britain and, subsequently, in Europe and North America. First was the fact that the power supply for the majority of industries came to be dominated by steam. There was therefore a move by many industries to the coalfield regions to reduce the cost of fuel (Figure 2.12 see page 67). The development of the Ruhr region of Germany (Pounds 1990: 412–29), and of western Pennsylvania in the USA (Meyer 1990: 256–8), is a good example of this process. The reason for this change in fuel supply was the invention and commercial development of the rotative steam engine by James Watt and Matthew Boulton in 1769–75.

Figure 2.9   The Great Exhibition, held in Hyde Park, London, in 1851, in Joseph Paxton's Crystal Palace, was intended to display the manufactured products of industrial Britain and its empire to the rest of the world. (Popperfoto)

Secondly this, in its turn, was fundamental to the invention and development of railways that transformed the transport costs of both raw materials and finished products almost as dramatically as it reduced the friction of distance. The Western and, increasingly, the colonial world saw a new time geography from the 1840s onwards, whereby journeys requiring several days to accomplish previously could be undertaken in a few hours. This was especially important in the industrialization of North America with its vast transcontinental distances (the first transcontinental line was completed in 1869). It was also significant in Britain's imperial control of India, and of Russia's colonization of the interior of Siberia and central Asia. In all three cases the railways enabled industrial-scale exploitation of natural resources (from timber to metalliferous ores) to take place, the profits from which could, in turn, be invested in further industrial, urban or railway development. The rapid expansion of San Francisco following the California gold and Nevada silver discoveries in 1849 and 1859 respectively is a good example of this (Walker 1996). Railways were also major industries in their own right, of course; the Indian railways employed three quarters of a million people by the 1920s, for example.

Third was the fact that, as capitalists sought economies of scale, the transformations of the nineteenth century were largely urban. By 1851, Britain had become the first country

Box 2.8   The textile city of Łódź, Poland

- Łódź developed thanks to the political circumstances of the remnant Polish state in the period 1815–31. Rapid industrialization was required to increase export earnings. A new town was founded beside a small market settlement in 1820–1 for 200 migrant drapers (4) with separate gardens (6).

- In 1824, the long straight road to the south, Piotrkowska, was laid out with 272 plots for weavers and 167 plots for cotton and flax spinners. The long gardens, 1.68ha in size, encouraged the spinners to grow flax (1 and 2). In 1828 a further area was developed to the south-east for Silesian linen weavers (3) (Figure 2.10).

Figure 2.10   Plan of Łódź. *Source*: Koter, 1990: 111   (Box continued)

Box 2.8   (Box continued)

- One-storey houses were provided by the state, but the tenants could purchase them subsequently. There were water-powered fulling, bleach, dye and print works beside the river (5). The first textile factories were built in 1827 and 1838. By 1860 there were 35,000 inhabitants in the town.

- In the second half of the nineteenth century the population of Łódź increased to 100,000 by 1878; to 314,000 in 1897; and to 506,000 in 1913. This explosive growth was almost entirely supported by the new factory-based textile industry.

- Jewish capitalist entrepreneurs such as Karl Scheibler and Israel Poznański built huge factory complexes. They were steam-powered, with belt-driven machinery, and the buildings were fireproofed. Scheibler and Poznański built grandiose palaces for their own residence beside their factories since they maintained day-to-day management.

- They also built model dwellings for some of their workers. The first were built in 1865. The dwellings were three-storey, one-room apartment blocks with kitchens and toilets in separate blocks in the yards.

- A self-contained workers' settlement was built by Scheibler in front of his Księży Młyn factory in 1875. 144 flats in 18 two-storey brick-built blocks were provided along a private gated street. A school, canteen and hospital were added later (Figure 2.11).

- In the city centre, along Piotrkowska, the long plots of the weavers were developed with tall insanitary tenement blocks on either side of a dark courtyard. Access was through the often grand Neo-Classical street front building by means of a central passageway.

Figure 2.11   Łódź, Scheibler's factory and workers' housing. *Source*: Poplawska and Muthesius, 1986

Figure 2.12 The coalfields of north-west Europe saw a concentration of new industrial enterprises after the 1780s as steam-powered machinery became normal in textile manufacturing, iron and steel making, and metal goods manufacturing. *Source:* Pounds, 1990: 44

anywhere in which a majority of its inhabitants lived and worked in towns. By 1900 there were sixteen cities in the world with a population which exceeded one million, as against one (Peking) in 1800, and another 27 places had half a million or more (Lawton 1989).

2.7   Urbanization

The urbanization of capital has been theorized from a Marxist perspective by Harvey (1985b). He notes that, as capitalists over-accumulate, the surplus flows into secondary circuits of capital, of which the principal is the built environment (in its widest sense). As in the primary industrial circuit, these flows are cyclical according to the perceived profitability of such investment. Because of the longevity of the built environment these 'building cycles' (sometimes called Kuznets cycles) are much longer than the business cycles, somewhere between 15 and 25 years. Crises may lead to the loss of profitability of an investment, but they do not often lead to the loss of building fabric. Western cities are therefore made in the image of past capitalist decisions and subsequent adaptations to fit new circumstances. Harvey goes on to suggest ways in which class struggle is written into the landscape of the Western city (1985b: 27–31), which he explores in more detail in an analysis of Paris in the third quarter of the nineteenth century. More recently, Dodgshon (1998: 148–61) has built on these arguments to show how the built environment of towns and cities is a major source of inertia in capitalist economies and societies.

There are perhaps two major phases of development in Western industrial cities before 1900, and the beginnings of a third. The first phase, which began in Britain in the 1770s (Table 2.2), was characterized by the need to provide homes for workers once the production process was separated from home. Initially, such housing was provided by a process of 'densification' whereby more and more living spaces were crammed into the existing built-up area. Gardens and yards were used for increasingly high-density basic accommodation, much of which was multi-occupied. By the 1770s in larger British cities this process could go no further and housing began to be provided on the city fringe where it proved a profitable 'crop' for small landowners. Such housing was brick-built, basic, unplanned and unregulated, and regionally distinctive in its plan forms. The most notorious of these forms was the back-to-back dwelling originating in the Yorkshire textile towns (Beresford 1988). The lack of regulation meant that these housing areas lacked effective water supplies, sewage and waste disposal systems, social and educational facilities, and connected road systems. The consequences, when combined with the poverty of low wages, were ill health, disease and very high urban death rates, especially among children. European and North American cities were also characterized by unregulated slum housing areas as urban populations rose from the 1820s onwards (Box 2.9 see pages 70–1) (Pounds 1990: 368–91, Homburger 1994: 110–11).

The second feature of this first phase of industrial cities was the separation of the residences of the bourgeoisie from those of the workers. Industrial cities became class-divided cities, though the working classes, especially young females, were recruited to provide domestic services by the elite (Dennis 1984). In the southern states of the USA that servant class was African American and still slaves until the 1860s. Thirdly, the capitalist land market began to value more highly the accessibility of the city centre and so the central business district began to emerge, as did districts in which industry and warehousing were the predominant land use, often close to the port facilities of canal, river frontage or harbour. One of the classic portrayals of this phase of urban development in Britain is Engels's (1845) *The Condition of the Working Class in England*.

The second phase of development, from 1840 to the 1890s (Table 2.2), was one of increasing regulation in towns and cities. New forms of local government and the collection of statistical information led to building regulations; the provision of sewerage and better water supplies; the provision of new cemeteries and parks; and better urban transport, especially tramways. All this led in turn to the increasing suburbanization of the better-paid elements of the urban workforce in the last quarter of the nineteenth century. These transformations occurred in most European core economies, and in North American cities, but with temporal variations. Thus, in England, the characteristic late-nineteenth-century dwelling was a suburban, brick-built, tunnel-back, terraced house. In North America, excepting only the largest cities, the vast majority of urban housing was timber-built; in German and French cities (as well as in Scotland), the apartment or tenement block was the norm (Box 2.9). The North American city, from the 1840s, was also characterized by the ghettoization of immigrant populations: first Irish and Germans, then freed slaves from the southern states after the Civil War, then Italians, and finally, from the 1890s, Jewish migrants from the Russian empire of eastern Europe (Ward 1987).

Central business districts were modified by two other major transformations in the capitalist system in this second phase. First, banking services were increasingly required to enable surplus capital to be safely stored, efficiently invested and recycled towards new opportunities; such banks were but one aspect of developing office quarters, including, for example, insurance offices, land agents and other property services, and legal services. This tertiary employment sector grew rapidly in the second half of the nineteenth century and 'the office' developed as both a specialized building type and a means of production (Daniels 1975). It was increasingly characterized by female employees using typewriters (the key technological invention) supervised by male managers. In North America the building type showed an increasing propensity to both large floor areas and height at the end of the century (the skyscraper was born in Chicago) and relied on the development of steel-frame construction methods and the invention of the elevator (Gad and Holdsworth 1987). Second was the growth of consumption. This led to the development of a variety of new retail spaces in city centres, including department stores and arcades, as well as an enormous growth in the variety of specialized shops.

The incipient third phase of urban development reflects a minority interest in improving the living conditions of workers by industrialists. Model settlements have a history that dates to the beginning of the factory system. Modern commentators note both the generally higher standard of accommodation in these places, but also the capitalist control of the home life as well as the working life of their workers. At the century's end the experimental garden suburb settlements at Bournville and Port Sunlight in England were widely admired and imitated in Germany and the USA. At the same time, the first experiments in the large-scale provision of social housing were underway in London using apartment block housing. These two themes, the garden suburb and social housing, were to combine powerfully in the development of modern planning to shape the urban environment in the first half of the twentieth century in Europe and its empires, but not in North America.

Box 2.9

Tenement housing for industrial workers

- The 1865 New York *Report on the Sanitary Condition of the City* illustrated these tenements (A) on what had originally been four lots facing First Avenue. The apartments in the five-storey front block were of two or three rooms, only one of which was heated.

SKETCH PLAN A

EAST 37th STREET

Yd
P
Yd
P
Yd
Yd
P
Yd
P

FIRST AVENUE

▷ Door   Yd Yard
🏛 Stairs   P   Privy
◼ Fireplace

SKETCH PLAN B

Yd

MUSKAUER STRASSE

Figure 2.13 Plans of tenements in New York (A) and Berlin (B). *Sources*: A. Ward, 1989: 34; B. Borgelt *et al.*, 1987: 11

(Box continued)

Box 2.9	(Box continued)

- In the southernmost lots dark tunnels provided access to the yards where smaller, cheaper two-room apartments occupied four-storey buildings to the rear of the lots.

- On the northern lots, East 37th Street allowed for crossways sub-division of the property and a block of small two-room apartments had been built.

- Privy blocks occupy the yards which would also normally have been strung with drying washing.

- Berlin was expanding rapidly at the end of the nineteenth century as the capital of newly unified Germany. Workers lived in 'Mietskasernen' (single-room barrack blocks) earlier in the century.

- In the 1890s 'Mietshäuser' (tenement blocks) were the normal housing of workers and artisans.

- Some of these were 'L' shaped on the lot, some 'C' shaped, but many were built around a central courtyard, as here on Muskauer Strasse (B).

- Shops often occupied the ground floor on the street frontages of these blocks; a central passageway gave access to the yard where four stairwells accessed the apartments. Most were five or six storeys, so that the yards were dark and gloomy.

- Apartments were of between one and three rooms, but by the 1890s usually had integral sanitation.

2.8 Conclusion

Within three centuries, Western capitalism had utilized its ill-gotten gains from its first colonial adventures to develop a series of specialized industrial regions supplying worldwide markets. Food production, industrial manufacture, service provision, consumption and transportation were all radically transformed by the capitalist enterprise. Cultures, societies and governments were necessarily impelled to change, too. States became more centralized and powerful. The difference between rich and poor at the level of both individuals and between countries became more marked. For some individuals and groups capitalism brought prosperity, improved living conditions and greater freedom; for others it brought destruction of local cultures, impoverishment, degradation and slavery. At the end of the nineteenth century the South African Boer War gave the first glimpse of the industrialization of warfare which was to scar the twentieth century so deeply.

Learning outcomes

Having read this chapter, you should know that:

- Scholars have interpreted and theorized the development of capitalism in different ways.

- The European colonialist enterprise was critical in the evolution of capitalism.

- Industrial capitalism transformed all sectors of the economy, including the built environment and social relations between individuals and classes.

- Industrial capitalism was essentially urban.

Further reading

Blaut, J.M. (1993) *The Colonizer's Model of the World: Geographical Diffusionism and Euro-centric History*, Guilford Press, New York and London. This is a very readable and thought-provoking polemic of post-colonial writing providing an alternative explanation of the success of European capitalism.

Dodgshon, R.A. (1998) *Society in Time and Space: A Geographical Perspective on Change*, Cambridge University Press, Cambridge. A more advanced text which provides much more detailed arguments and evidence for the themes of this chapter.

Knox, P. and **Agnew, J.** (1994) *The Geography of the World Economy*, 2nd edition, Edward Arnold, London. A well-written and popular textbook which will give readers another perspective on this period and much else besides.

Langton, J. and **Morris, R.J.** (eds) (1986) *Atlas of Industrializing Britain, 1780–1914*, Methuen, London and New York. Historical atlases usually treat this period very well with innovative cartography and thought-provoking texts. This one deals with Britain in considerable detail.

Michell, R.D. and **Groves, A.** (eds) (1987) *North America, the Historical Geography of a Changing Continent*, Hutchinson, London. There are a number of good historical geographies of North America. This one is well written, thoughtful and copiously illustrated.

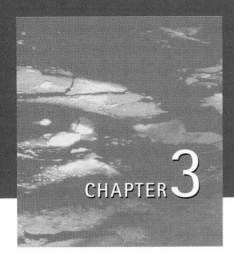

CHAPTER 3

DENIS SHAW

The making of the twentieth-century world

The twentieth century could be said to have been the period when capitalism finally triumphed over most of the globe. But it has been neither a straightforward triumph nor an unchallenged one, and capitalism itself has been changed in the process. This chapter is concerned with the various spaces created by and in response to twentieth-century capitalism – spaces of resistance and reinterpretation as well as spaces of adaptation and acceptance. The patchy and unequal world in which we now live reflects the erratic and conflict-laden nature of the processes which have produced it.

3.1   The changing capitalism of the early twentieth century

In the autumn of the year 1933 the writer and journalist J.B. Priestley set out by bus on a trip which was to take him the length and breadth of England and which he later described in his *English Journey* (Priestley 1937). As he left London by the Great West Road, Priestley noted how the road 'looked odd. Being new, it did not look English. We might have suddenly rolled into California.' What struck Priestley as particularly odd was 'the line of new factories on each side' of the road. 'Years of the West Riding', he explained (he was born and raised in Bradford), 'have fixed forever my idea of what a proper factory looks like: a grim, blackened rectangle with a tall chimney at one corner. These decorative little buildings, all glass and concrete and chromium plate, seem to my barbaric mind to be merely playing at being factories.' Armed with a copy of the now celebrated textbook, *The British Isles: Geographic and Economic Survey* by Stamp and Beaver (1933), which was later to be used by

Plate 3.1   The Hoover factory, Perivale, west London. This splendid example of Art Deco architecture, designed by Gil Wallace in 1932, reflects the new consumer industries which were being established in the years after the First World War. (Corbis UK/Angelo Hornak)

generations of geography undergraduates, Priestley went on to make some astute geographical and social points about these factories: 'Actually, I know, they are tangible evidence, most cunningly arranged to take the eye, to prove that the new industries have moved south. They also prove that there are new enterprises to move south. You notice them decorating all the western borders of London. At night they look as exciting as Blackpool. But while these new industries look so much prettier than the old, which I remember only too well, they also look far less substantial. Potato crisps, scent, tooth pastes, bathing costumes, fire extinguishers; those are the concerns behind these pleasing facades –' (Priestley 1937: 3–5).

In these few words, Priestley summarized some of the major ways in which the industrial world of the twentieth century was to differ from its nineteenth-century predecessor. The fact that he was making his journey by road was itself significant; thirty years before he would have had to go by rail. The new factories he observed were the products of the technological changes that had been transforming industrial capitalism since the late nineteenth century, and many had clearly developed to serve an expanding consumer market. And the location of the factories by the new arterial highway leading westwards out of London were the result not only of a revolution in transport and communications but also of the locational freedom deriving from the availability of electricity and other fuels. The textile industries which Priestley remembered from his childhood in Bradford were tied to the coalfields; the newer industries which were appearing in London by the 1930s no longer needed coal, and were much cleaner and brighter in consequence.

Of course, what Priestley saw along the Great West Road was by no means representative of all the technological changes that had been affecting the industrial economies of Britain and other countries for the previous few decades. What had been happening in these countries was that a whole series of new industrial branches had been developing to supplement, and eventually to eclipse, the traditional activities based on coal, iron and textiles. Not all of these were as pleasing to the eye as those observed by Priestley. In the second half of the nineteenth century, for example, the metallurgical industries had been transformed as a result of a series of inventions allowing the production of cheap steel. Next came the rapid development and proliferation of different branches of the chemical industry (alkalis, dyestuffs, explosives, lacquers, photographic plates and film, celluloid, artificial fibres, plastics). The electricity industry, which began to flourish by the end of the nineteenth century, was dependent on earlier inventions, like the steam turbine. About the same time came the rise of the motor industry, which was in turn associated with other industries like oil and rubber. When Priestley set out on his journey, society was already beginning to adjust to the impact of the many new activities catering to the consumer (most notably, domestic appliances) and to new means of transportation (car, bus, aircraft). Of course the full impact of such developments was to come later, after the Second World War, while some technologies, like regular TV broadcasting, the jet engine, nuclear power and the microchip, still lay in the future.

So profound were the technological and accompanying social changes which affected industrial capitalism from the late nineteenth century that some historians have described them as a 'second industrial revolution' (Landes 1969: 4). But it is important to remember that the older industries – coal, textiles, railways, some forms of engineering – did not die immediately or indeed quickly. One of the features of the changing industrial geography of the late nineteenth and early twentieth centuries was that countries like the USA and Germany, whose industrialization came later than Britain's, now began to forge ahead on the basis of the newer industries described above. Britain remained overdependent on the older and less dynamic branches (Figure 3.1).

Low furniture does not hinder the Hoover. The handle can be swung down to the ground if necessary.

Plate 3.2   Using a vacuum cleaner or Hoover, one of the new consumer appliances which were becoming widely available by the 1930s. Photos and advertisements of the period almost always depicted women performing such domestic tasks. (Mary Evans Picture Library)

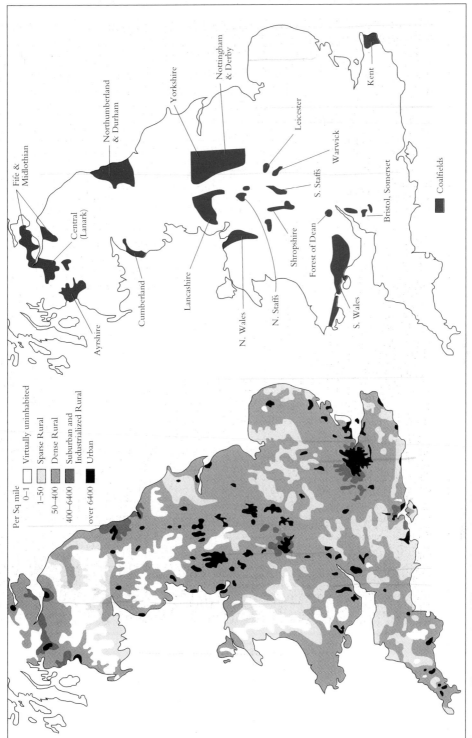

Figure 3.1 The geography of UK population and coalfields compared. With the significant exception of London, there is a strong correlation between the geography of population and that of coalfields, reflecting both the early stage at which Britain industrialized and urbanized, and the country's long dependence on nineteenth century industries. *Sources:* population, based on Mitchell, 1962; coalfields, Stamp and Beaver, 1963: 286

For the first half of the twentieth century, the industrial changes described above only directly affected certain parts of the world, notably Western Europe, North America, Japan, and by the 1930s the Soviet Union and some other areas. Much of what was later to become known as the Third World, or the developing world, was still agricultural. Yet, in continuation of earlier processes (see Chapter 2) many colonies and other regions were now being organized commercially to supply the industrial countries with raw materials and tropical products – for example, bananas and sugar from Central America and the Caribbean, Brazilian coffee, Indian tea, Malaysian rubber. They were thus being tied in to the capitalist world economy. Gradually certain of these countries began to adopt the technologies of the industrialized world – the rise of the Indian textile industry is one example – but only later in the twentieth century did industrialization become more widespread.

Thus the foundations of what is now known as a global economy were already being laid in Priestley's day, or even earlier. By the beginning of the twentieth century capitalism had become a world phenomenon, tying far-flung countries together by means of international trade and fostering international capital flows through major financial centres like London. The first multi-national corporations were already appearing. All this was aided and abetted by the new systems of communication and transport – telegraph, telephone, radio (from the 1900s), fast steamships, aircraft – which were beginning to span the globe. Of course, none of this bears comparison with the forces of globalization which were to become so significant later in the century. Yet the world was already becoming a smaller place. J.B. Priestley himself suggested this when he compared the Great West Road to 'California'. What might this mean to the average English reader in the 1930s? The answer is, a great deal. The Hollywood film industry was in its heyday and the people, homes and landscapes it portrayed were being viewed, and copied, the world over.

3.2 Organized capitalism

It would be a mistake to suppose that the advance of capitalism in the twentieth century was a story of unmitigated triumph. On the contrary, its fluctuations and misfortunes were such that one historian felt constrained to call the period the 'Age of Extremes' (Hobsbawm 1995). The first half of the century – Hobsbawm's 'Age of Catastrophe'- was particularly disturbed, with two world wars (1914–18, and 1939–45) and a deep world economic depression (1929–33). By contrast, the years between 1945 and 1973 were ones of growing prosperity across much of the globe (Hobsbawm's 'Golden Age') to be followed once more by a disturbed period in the wake of the shock oil price rises of the 1970s.

It was Karl Marx who originally emphasized the unplanned, competitive and even chaotic nature of capitalism's development. But one of the features of the twentieth century has been the attempt, by both government and private agency, to regulate and even to control it. The reasons for this phenomenon are many, but they are no doubt linked both to the precipitate nature of technological change during this period and to the severe fluctuations mentioned above. Attempts to 'organize' capitalism have taken a number of different forms. For example, already in the late nineteenth century, and especially in the USA and Germany, there were moves towards the formation of inter-firm agreements, cartels and larger companies and corporations. Large corporations could more easily marshal the huge capital resources which modern industry requires, and also influence their markets more effectively. As noted already, the first multinational corporations appeared at this time, those based on the USA being most notable. Nineteenth-century examples include the German electrical firm, Siemens, and the US Singer sewing machine company. Twentieth-century examples include US-based Hoover, Ford, Coca-Cola, Pepsi Cola, Nabisco (Shredded Wheat), and Kelloggs, all of which invested in Britain in the 1930s. Most of the major international oil firms also date from this period.

A further way in which capitalism became more 'organized' was the phenomenon of mass production, linked especially to the growing consumer market. Here two Americans are regarded as particularly significant. F.W. Taylor (1856–1915) is especially associated with time-and-motion studies, whereby complex tasks on the factory floor could be completed more efficiently and productivity increased. The other is Henry Ford (1863–1947) who organized car production in his Dearborn, Michigan, plant using modern methods like the assembly line and interchangeable parts. The result of his centralized approach was a significant reduction in the cost of producing cars, which could now be manufactured on a mass basis. The ensuing era of mass production methods and associated patterns of mass consumption is frequently referred to as **Fordism**.

Governments were also affected by the desire to 'organize' capitalism and to tackle the many problems to which it seemed to give rise. At the international level, the USA, which emerged after the Second World War as the undisputed leader of world capitalism,

Plate 3.3 The Ford assembly line, first organized by Henry Ford in his Dearborn, Michigan, car plant in the years before the First World War. Ford did not invent such mass production methods, but certainly helped to popularize them. (Mary Evans Picture Library)

took the lead in establishing a series of institutions like the World Bank, the IMF and the General Agreement on Tariffs and Trade (GATT) to ease international monetary payments, promote trade and encourage economic development. At the national level, numerous countries, especially some of the west European ones, pursued democratic agenda of various kinds such as attempts to construct 'welfare states' to tackle such social problems as unemployment, ill health, old age and social inequity. Arguably the experience of government planning and controls in wartime helped pave the way for the optimistic belief in planning and large-scale social engineering which characterized the post-war period. In western Europe especially this was the era of bold experiments in new town and city development, slum clearance and ambitious social housing schemes, regional planning, and extensive controls over land use (Hall 1996). No doubt the success of these schemes was dependent on the spreading affluence which accompanied Hobsbawm's 'Golden Age'. Across western Europe and North America a tide of suburbanization signalled not only a growing ability to own one's own home in a desirable location but also the availability of the social and physical infrastructure, the private cars and the many new consumer products which now made such a goal possible for many.

The writers Scott Lash and John Urry have described the era ushered in by the methods of Ford and Taylor but reaching its apogee during the period 1945–73 as 'organized capitalism' (Lash and Urry 1987). Some of its principal features (which were especially characteristic of the developed world) are described in accordance with their views in Box 3.1.

Plate 3.4 A suburban house built in the Modern style in the 1930s, Essex, England. The suburbanization of the inter-war period was a product of the affluence and increased mobility of some middle-class groups during this period. (Corbis UK/Gillian Darley/Edifice)

Box 3.1

Organized capitalism (after Lash and Urry 1987)

- Extractive and manufacturing industries are the dominant economic sectors.

- There is an accent on economies of scale, leading to the importance of large industrial plants. Such plants may structure entire regional economies around themselves. Examples might include (in the UK): the West Midlands, based on cars and engineering; Lancashire, based on textiles; the North-East, based on mining, shipbuilding and heavy engineering; and (in the USA): Detroit, based on cars and engineering; and Philadelphia, based on textiles and port-related activities.

- Manufacturing plants are controlled centrally by big industrial corporations – there is an accent on mass production and standardization.

- There is state regulation of the economy to overcome problems generated by the market, for example regional unemployment problems.

- Big industrial cities are the spatial expression of large-scale manufacture.

- There is state-controlled welfare provision to even out social inequalities, address unemployment problems, and raise health standards.

- In culture and social provision, there is an accent on mass provision, for example in housing, consumer goods, TV programming, and newspapers. The emphasis on mass coverage and standardization leaves relatively little choice, reflecting a modernist perspective.

Needless to say, such generalizations would be more or less true, depending on the time and location being considered. Capitalism had different histories in different places, and the exact form it took had much to do with the long-term evolution of each society affected by it.

3.3 Challenges to liberal capitalism

Come, bombs, and blow to smithereens,
Those air-conditioned, bright canteens,
Tinned fruit, tinned meat, tinned milk, tinned beans,
Tinned minds, tinned breath.

Mess up the mess they call a town –
A house for ninety-seven down
And once a week a half-a-crown
For twenty years

From 'Slough' by John Betjeman (1937)

John Betjeman's famous fulminations against the town of Slough, situated just west of London and experiencing developments similar to those observed by J.B. Priestley a few years earlier, are in fact a hymn against modernity. Betjeman was railing against many of the social repercussions of the profound twentieth-century changes noted earlier in this chapter. In this he was by no means alone. Indeed, according to some commentators (e.g.

Wiener 1985), Britain has been characterized by an anti-urban and anti-industrial culture which has made it very difficult for the country to adapt to twentieth-century change. However, rather similar problems have also occurred in other countries where modernity (perhaps best defined as the spectrum of economic, social, political and cultural changes associated with twentieth-century capitalism) has brought problems of adaptation.

It has already been noted that the twentieth-century cannot be described as a century of uninterrupted progress for capitalism, particularly in its Western 'liberal' form. Capitalism has been subjected to a series of challenges and political struggles that have greatly affected the course of twentieth-century history, and in various parts of the world there have been attempts to create spaces in which alternatives to liberal capitalism can flourish. Interestingly enough, in terms of Wallerstein's world systems theory discussed in Chapter 2, most of these attempts have been associated with countries outside the core, or with those like Germany after its defeat in the First World War struggling to rejoin the core states. Particularly important for the political geography of the twentieth century were the attempts by Marxists and others to reject the capitalist development model entirely and to reconstruct society on a new basis. This issue will be discussed in the next section. This section will focus on two movements that have also challenged Western-style capitalism in profound ways, but without discarding it entirely: Nazism (National Socialism) with its close relative, Fascism; and the Islamic revival.

Although they had nineteenth-century antecedents, both Fascism and Nazism were essentially products of the inter-war years. Fascism, under its leader Benito Mussolini, ruled in Italy from 1922 until its final defeat in the Second World War in 1945. Nazism under Adolf Hitler ruled Germany from 1933 until it too was defeated in 1945. Various Fascist or neo-Fascist groups ruled or were active elsewhere in Europe, and in some other regions, during this period, and to a lesser extent since.

There is no doubt that Nazism was by far the most influential Fascist movement after 1933, and so the following remarks will be devoted to it (Kershaw 1993). One of the problems of discussing Nazism or other forms of Fascism is their lack of a consistent ideology or philosophy. However, certain general points can be made. Like certain Western intellectuals and others, the Nazis were moved by a dislike of facets of capitalist modernity, such as commercialism, materialism, individualism, threats to the traditional family, and similar tendencies that they associated with the 'decadent' Western democracies. They also despised Western-style parliamentary democracy, with its plurality of political parties and class divisions. In its place, they advocated the concept of a single national community, a *Volksgemeinschaft*, headed by a single Leader or Fuehrer (Hitler), who was regarded as representative of, and chosen by, the people (this was the Nazi concept of 'democracy'). The Leader's power was absolute. Such a creed, however, seemed exceedingly unlikely to come to power in Germany in the early 1930s had it not been for the extreme circumstances reigning there. One was a general sense of resentment at Germany's defeat in the First World War and subsequent national humiliations. As extreme nationalists, the Nazis promised to avenge this defeat. Another was the dire straits to which many of the middle class had been reduced by post-First World War inflation and the Depression which began in 1929. There was also the fear of the many strikes and disorders perpetrated by communists and other left-wing groups (as well as by the Nazis) which were, of course, encouraged by the selfsame economic difficulties.

One of the oddities of Nazism was that it was, at one and the same time, both reactionary and modernizing (Herf 1984). On the one hand, the Nazis looked backwards to

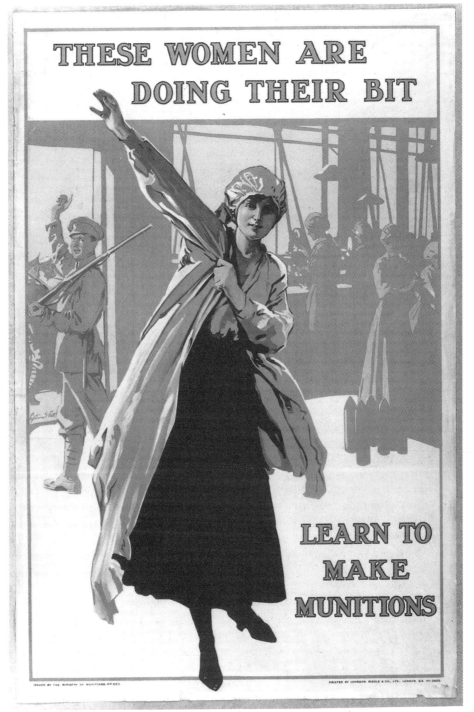

Plate 3.5   Women workers in armaments plants during the First World War. Although the Nazis and some other political groups were opposed to female employment, the labour shortage during the First World War first gave women job opportunities which were increasingly taken as the twentieth century advanced. (Imperial War Museum)

Plate 3.5 (continued)

an imagined heroic and rustic Germany of the past, to Nordic myths, Germanic land-scapes (see Rollins 1995) and happy peasants tilling the fields in traditional costume. They tried to bolster the 'traditional' family: women were to remain at home and raise children for the fatherland. There were even attempts to build villages and garden settlements to reflect such ideals. On the other hand, the Nazis were also modernizers who built the autobahns, fostered industry and spent vast sums on the military. Their aim was to turn Germany into a Superpower, able to dominate the European continent and regions beyond (see Chapter 15). To this end, the economy was expected to play its full part and, though the Nazis never rejected private capitalism, their economic controls became ever more pervasive, especially with the onset of war in 1939 (many Marxists have argued that Nazism was capitalism's most degenerate form). Similarly, while the Nazis went out of their way to provide for the leisure time and health needs of the German masses, the ulti-mate goal was to transform them into a race of heroes, ready to fight and die for their Fuehrer.

One of the things that distinguished Nazism and Fascism from other right-wing ten-dencies was that they were mass movements. The masses were to be organized and captivated by ceremonial and display, by mass rallies and triumphal processions (rather than, for example, by rational persuasion) (Mosse 1975; see also Atkinson and Cosgrove 1998). The landscape of Germany was dotted with places where such ceremonials could be performed, the Nuremberg Party Rally grounds being perhaps the most notable. Had Nazism survived, the whole of central Berlin would have been reconstructed on a grandiose scale to reflect the supposed glories of the Third Reich (Speer 1995: 195–223).

Hobsbawm and others have argued that Nazism represented a rejection of the values of the eighteenth-century Enlightenment and the French Revolution, the values upon which modern Western notions of science, rationality and democracy are based. The Nazis wished to replace these values by irrational and Romantic ones, stressing the importance of the will, of faith, even of occult powers. Nazism thus claimed to be a religion in its own right, aiming to compete with and ultimately to replace rival faiths.

Nazism's central and most notorious feature (and one which largely distinguished it from Italian Fascism, at least until 1938) was its racism. The German people were imagined to be descendants of an Aryan race, equipped by nature to be rulers of the earth. This belief was related to racist ideas which were all too common in late nineteenth- and early twentieth-century Europe (and which were, no doubt, related in turn to Darwinism). The Nazis took them to their illogical extreme. All non-Aryan peoples were regarded as inferior, particularly the Jews who, because of their culture, religion and cosmopolitan ways, seemed to represent all that the Nazis feared and hated. As time went on, it became clear that the Nazis meant to exterminate the Jews (they killed six million of them) as well as others (gypsies, homosexuals, the insane, certain religious groups) who could have no place in the world they intended to reconstruct. Had Nazi Germany triumphed in the Second World War, it would no doubt have entirely rearranged the ethnic map of Europe – exterminating some peoples, reducing others to slavery, and resettling yet others. The Nazi contribution to the twentieth century must be regarded as wholly negative and morally repugnant.

Dislike of some of the features associated with Western-style capitalism has also characterized the other movement to be considered here: the Islamic revival. Islam, of course, is quite different from twentieth-century secular (and essentially 'Western') movements like Fascism. Islam is one of the world's ancient religions, with many millions of adherents in a swathe of territory stretching from North Africa, across southern Asia and down to Indonesia, with many others living beyond (Figure 3.2) (Esposito 1995, Park 1994). It has a rich cultural heritage, and indeed for many centuries in the medieval period the Islamic (or Muslim) world easily outstripped Europe in terms of intellectual, scientific and cultural achievement. The Islamic revival is essentially a product of the twentieth century, particularly of its second half, and, in the opinion of many, represents a reaction against the inroads of secularism, Western colonialism, and some forms of modernization which have affected the Islamic world during this period. Parts of the Muslim world, for example, were colonized by one or more of the European powers several centuries ago, whilst others experienced Western colonialism only during the twentieth century. All such colonies are now politically independent, but the colonial experience naturally had a cultural impact. Furthermore, beginning with Kemal Ataturk's founding of the modern, secular state of Turkey in the early 1920s, many Muslim political leaders have embraced Western-style political, social and economic ideas in the hope of modernizing their countries.

In the opinion of some, the Muslim faith recognizes no divisions between the realms of the sacred and the secular and exhorts its followers to obey its injunctions in every aspect of their lives. The whole of life and society, in other words, is to be regulated in accordance with Islamic teaching (though there is room for debate and disagreement over how far modern social changes can be reconciled with those teachings). Some devout Muslims were particularly offended by the programmes of Westernization and technological development adopted by several Muslim states in the 1960s and 1970s (Scott and

Figure 3.2 The geography of Islam. Based on *Times Atlas of World History*, 1989: 105.

Simpson 1991, Park 1994). Secular education, 'immodest' dress and liberal sexual attitudes, the use of alcohol, Western-style entertainment, growing social inequalities and
profligate spending by new social elites (offending against Muslim teachings on charity and
social justice) have all been subject to criticism. In the minds of many Muslims, such tendencies are associated with the West and its 'degenerate' society. This is hardly surprising
in view of the colonial legacy, and the non-core nature of most of the Muslim countries.

Across the Muslim world, there have thus been movements for Islamic revival and
renewal, often taking a political form and demanding a return to the original purity of
Islam. Sometimes these movements take a particularly militant and anti-Western stance,
though this is by no means always the case (Saudi Arabia is an interesting case of a pro-
Western state whose society is nevertheless built on strict Islamic principles). Such movements may sometimes threaten existing regimes that are supported by the West. However,
the Islamic world is so variable that the political consequences of the revival are very difficult to predict. Perhaps the best known case of a country that experienced an Islamic
revolution is Iran, whose previous pro-Western ruler, the Shah, was overthrown in 1979.
Here the Islamic clergy took direct control of the government in order to build an Islamic
republic. This entailed, among other things, the implementation of strict Islamic (Sharia)
law that was intended to nullify some of the less acceptable Western cultural influences
(Stempel 1981).

One feature of the Islamic revival that has attracted much attention in the West is the
veiling of women. This has often been regarded in the West, and in the Islamic world
also, as a return to the past when women were exploited and confined to a life of domes-
ticity. An interesting point, however, is that in parts of the Muslim world some educated

Plate 3.6 Demonstrators carry posters of Ayatollah Khomeini, leader of the Iranian Islamic revolution. (Popperfoto)

women have taken the veil voluntarily, seeing it as a protest against the sexual and commercial exploitation of women which is sometimes regarded as a product of Western-style capitalism. Another interesting but less publicized feature of the revival has been the attempt to move to an Islamic system of (zero-interest) banking. This is regarded as not only in accordance with Islamic ethical norms but also much fairer than Western-style banking in spreading risk. It suggests an attempt to construct an economic system based on Islamic rather than Western-style commercial principles.

The Islamic revival can be said to challenge Western-style liberal capitalism both ideologically and geopolitically. Ideologically, it challenges some of the ethical and social values of the West. Geopolitically, it may threaten Western interests in the Middle East and elsewhere in the Islamic world, especially where spaces are created, as in Iran, where Western-style liberal capitalism is unwelcome. Whether a genuine alternative to the latter can be sustained over the long term remains to be seen.

3.4 Communism and the command economy

Perhaps the most significant challenge to capitalism in the twentieth century has come from Marxist-style communism (Calvocoressi 1991). By the 1960s and 70s up to one third of humanity was living under communist governments which explicitly rejected capitalism as an acceptable way of organizing society. The reasons for that rejection and why it largely failed must now be considered.

An outline of some of the principal features of Marxism is given in Box 3.2. An

> ## Box 3.2   Marxism
>
> - Marxism, which derives from the teachings of Karl Marx (1818–83), is related to other forms of socialism which seek to moderate or reform the injustices of capitalism.
>
> - Unlike many forms of socialism, Marxism regards capitalism as an innately unjust and exploitative system.
>
> - According to Marxism, capitalism divides society into two antagonistic classes: those who own the main sources of wealth (the capitalists) and those who must live by selling their labour to the capitalists (the proletariat).
>
> - Marxism teaches that the capitalists are forced by the very nature of capitalism to maximize the profits they obtain at the expense of the workers.
>
> - Marx thought that eventually capitalism would become so exploitative that its downfall was inevitable.
>
> - It would then be replaced by a much more just, classless society (socialism, gradually maturing into communism).
>
> - In the meantime Marx exhorted the world's workers (especially the industrial workers) to organize politically to hasten capitalism's downfall.

important point is that Marx's teachings failed to change those societies at which they were initially aimed – the industrial societies of Western Europe and North America. The point is that Marx himself had expected that communism would find support among the growing industrial working classes of countries like Germany and Britain where factories were bringing such groups together in increasing numbers. However, it was by no means obvious, as the years passed, that the workers of those countries were necessarily being increasingly exploited, as Marx seems to have expected (later, these countries were accused by Marxists of exporting exploitation to colonies and other less developed regions). Instead, Marxism triumphed in Russia (in 1917), in what was in fact the least industrialized of Europe's great powers. Thus, whereas in terms of Wallerstein's world systems theory Marxism was expected to find favour in the core countries, in fact it initially triumphed in a semi-peripheral one. That it did so changed the character of Marxism, which was now faced with the challenge of building socialism in a peasant society, and in virtual isolation from the rest of the world.

What happened after 1917 in Russia (or the Soviet Union as it was now to be called) was of profound importance for the other countries which later adopted communist systems, if only because Russia was the pioneer. What happened there began to assume something of the character of orthodoxy (Hosking 1992). In view of the difficulties they faced, and in all likelihood because of their own inclinations, the Bolsheviks (as the Russian or Soviet communists were called) adopted a highly centralized political system which brooked no opposition. Eventually, from the late 1920s, they implemented a fully centrally planned economic system. This involved the abolition of private enterprise and virtually all forms of market relations, and the collectivization of agriculture. The purpose of this extraordinary economic system was both to speed the process of economic development and to build up the country's military resources. As the 1930s advanced, it became clear that what had earlier seemed a rather vague threat from the outside capitalist world was beginning to take a concrete and menacing form in the guise of Nazi Germany. The centrally planned or command economy (which involved much suffering on the part of the Soviet people) proved equal to this challenge. In the ensuing war with Germany (1941–5), the Soviet Union emerged victorious, but only after sustaining enormous losses.

Soviet victory in the Second World War greatly enhanced that country's prestige. Moreover, because the Soviet armies were now in occupation of much of central and eastern Europe, they were able to ensure that regimes friendly to the Soviet Union (that is, communist regimes) would assume power in those regions. Communism soon spread into other countries, notably China (in 1949), south-east Asia and beyond: in other words, into Wallerstein's 'periphery'. All these countries initially followed the Soviet development model, but soon found that it was necessary to adapt it to their own needs. In the meantime the spread of communism, and Soviet ambitions, excited the suspicions of the capitalist West. From the 1950s, therefore, the world was split into two armed camps, both equipped with nuclear weapons. The ensuing confrontation, known as the Cold War, profoundly influenced both sides and encouraged their militarization. However, many countries, especially in the developing world, tried to avoid taking sides, while China, though communist, began to pursue its own version of communism outside the Soviet sphere.

While the actual form that communism took in the Soviet Union and other countries may have been a modification of Marx's own ideas, it did represent a radical departure from the capitalist development model. Not only was the command economy a very different, state-centred approach to economic development (and one which was copied in many parts of the developing world after the Second World War, with varying degrees of success), it regarded itself, and was regarded, as a threat to the whole idea of capitalism (Nove 1987). Internally, it led to attempts to reconstruct society on a different basis, particularly in terms of society's spatial structure, but it did in fact give rise to new forms of spatial inequality (Bater 1986). Interestingly enough it also had certain spatial features in common with Nazism and Fascism, such as the emphasis on creating urban spaces specifically for the purpose of mass ceremonial and display (Plate 3.7). The creation of 'spaces of terror', such as concentration camps for the incarceration or elimination of those deemed unacceptable to the regime, was also a feature of the two systems.

In the end, communism failed to prove itself a successful viable challenger to capitalism. Especially from the 1970s, the Soviet Union and its eastern European allies fell behind their capitalist rivals in terms of productivity, flexibility and innovation. Whether this was because of problems inherent to command economies as such, or whether it has more to do with mistakes made by the various political leaderships, is hard to say. Whatever the reasons, by the end of the 1980s practically the whole of the communist world was in a state of economic and political crisis. The subsequent fall of communism in eastern Europe, and the splitting of the Soviet Union into fifteen separate states, signalled the end of the Cold War and opened up that part of the globe to capitalist penetration.

So what was the true character of this most radical challenge to Western-style capitalism? Was it, as many commentators have claimed, merely another form of capitalism which was destined to fail, perhaps because Lenin and the other Soviet leaders should never have tried to build socialism in what at first was still in many ways a peasant society?

Plate 3.7 Central Moscow in the communist period, showing the open spaces created or enhanced by the Soviet dictator Joseph Stalin for official communist demonstrations and ceremonial display.

– or perhaps because socialism is an impossible dream anyway? Or was it a genuine attempt to follow Marx's teachings in difficult circumstances, which was eventually crushed by the forces of international capitalism? Opinions are bound to differ, perhaps above all because the answers we give to such questions will inevitably reflect what we think about the societies in which we ourselves live.

3.5   The end of imperialism?

Father, Mother, and Me
Sister and Auntie say
All the people like us are We,
And everyone else is They.
And They live over the sea,
While We live over the way,
But – would you believe it? – They look upon We
As only a sort of They! –

From 'We and They' by Rudyard Kipling (1912: 763–4)

At the beginning of the twentieth century Britain and several other European powers sat proudly at the centre of a series of empires which spanned the globe (Figure 3.3 see page 92).

As noted in Chapter 2, these empires were the products of a long period of European explo-ration, settlement, economic exploitation and imperial rivalry. Something of the compla-cency and condescension with which Europeans commonly regarded their empires at this period is nicely captured by Rudyard Kipling in his comic poem 'We and They', quoted above.

The early years of the twentieth century were not in fact the high points of empire. After the First World War (1914–18), Britain, France and other victorious powers helped themselves to the former German colonies, while Italy invaded the independent African state of Abyssinia (Ethiopia) as late as 1935–6. However, the future of European imperialism was already being questioned even before the war. The English liberal J.A. Hobson, and later the Russian revolutionary V.I. Lenin, popularized the idea that Europe's overseas colonies were being economically organized and exploited mainly for the benefit of the European 'mother countries', forming an undeveloped periphery to the European core. Lenin taught that imperialism was an inevitable consequence of cap-italism – its 'highest stage'. In the meantime various rumblings of discontent were being felt in various parts of the European empires. Britain, for example, felt constrained to grant home rule to a number of its colonies where white settlement had been signifi-cant – Australia, New Zealand, Canada, South Africa. It was also considering doing something similar in Ireland (Box 3.3 see page 94), Egypt and India by 1914. Russia's defeat by Japan in the war of 1904–5, the first time that a 'non-white' people had hum-bled a major European power, suggested that European imperialism might be challenged successfully.

It was, however, the three great episodes of the first half of the twentieth century – the First World War, the Great Depression (1929–33) and the Second World War – that gave conclusive evidence of European weakness and fatally undermined European imperialism. Nationalism, which had had such an impact on the political geography of Europe (see Chapters 15 and 16), had also influenced the colonial world where 'national liberation movements' began to demand independence for their countries. Starting with the inde-pendence of India in 1947, the next quarter of a century witnessed the break-up of all the European empires (though the Soviet Union, successor to the old Russian empire, finally disappeared only in 1991). Many new, independent states appeared on the map of Africa, Asia and other regions though not, unfortunately, without considerable turmoil in some cases. The world's political geography was transformed (Figure 3.4 see page 93).

In giving (or being forced to give) independence to their colonies, the former imperial powers hoped that they would adopt European-type political systems and capitalist econ-omic systems, partly because these appeared the best basis for future development, and partly because they seemed a reasonable way of upholding European influence. The elites who were now to hold power in the new states were often sympathetic to these aims since they had frequently been educated by Europeans and wished to see their countries mod-ernized along European or Western lines. However, the adoption of Western models was not always a suitable response to their problems. For example, many of the new states attempted to copy the European idea of the nation-state, hoping to unite their peoples around a common sense of national identity. But the old colonial boundaries had gener-ally been drawn up to suit imperial convenience rather than that of local communities. Thus these boundaries, now the boundaries of independent states, frequently grouped peoples into one state who had no common culture or history while dividing others who did so. This contributed little to the political stability or unity of the new states. Another

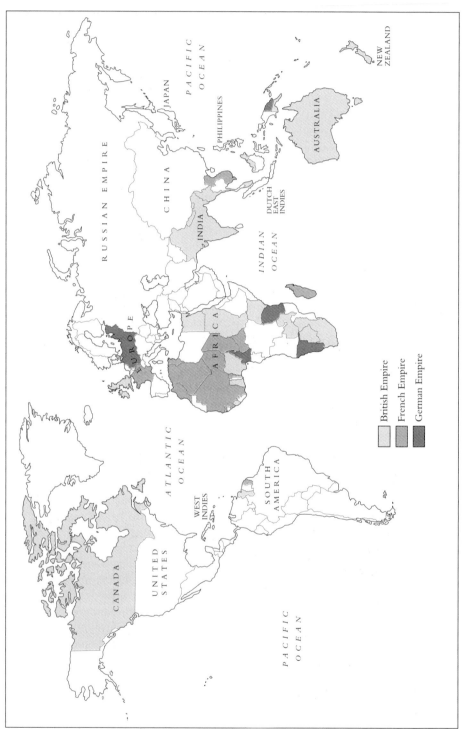

Figure 3.3   The world in 1914 showing the British, French and German empires.

Figure 3.4 The world in the late twentieth century.

Box 3.3   **Ireland in the twentieth century (see Foster 1989)**

Ireland's history in the twentieth century illustrates many of the problems bequeathed by imperialism to newly independent countries.

Although the English had begun to penetrate Ireland as early as the twelfth century, they were never able to assimilate the territory into the English state. Differences between the two countries were exacerbated at the time of the Reformation when England and Scotland became Protestant but most Irish remained Catholic. Sporadic resistance to English/British rule was met by repression. In the nineteenth century the rise of Irish nationalism stimulated several unsuccessful attempts to secure autonomy (home rule). Only after the violence of the Easter Rising in Dublin in 1916 and revolution in 1919–22 did most of Ireland achieve dominion status under the British Commonwealth, becoming the Irish Free State in 1922.

Irish autonomy, however, did not appeal to the Protestants who formed a majority population in the north-eastern part of the island. Most of these people were descendants of settlers who had been deliberately introduced to the region by the English government in the seventeenth century in an attempt to subdue the Irish. Protestants, who had traditionally had a largely privileged status in Ireland, feared incorporation into a Catholic state and demanded continued union (unionism) with the British crown. As a result of various manoeuvres the northern Protestants were enabled to establish their own statelet of Northern Ireland as a continuing part of the United Kingdom in 1921. Unfortunately the six counties of Northern Ireland contained a substantial minority of Catholics, and the partition of Ireland was widely resented by nationalists on both sides of the new border.

After independence governments of the Irish Free State became concerned to secure a greater measure of freedom from Britain and also to assert a post-colonial Irish national identity. Policies such as economic protectionism, attempts to revive the Irish language and Gaelic culture, and imposition of a Catholic moral code and censorship, were thus designed to build a new Ireland, free from the influence of the former imperial master. These policies culminated in Irish neutrality during the Second World War, and declaration of an Irish Republic and secession from the Commonwealth in 1949. However, this also had the effect of strengthening partition, as northern Protestants clung ever more tenaciously to the British connection whilst Northern Irish Catholics often suffered discrimination.

The violence and terrorism which partition has abetted must cast doubt on the extent to which it is a suitable answer to Irish problems. Fortunately recent developments on both sides of the border provide hope of better times in the future. Thus the Irish Republic now seems to be moving away from the narrow nationalism of the past and is evidently seeking a broader and more inclusive identity within the ambit of the European Union and the wider world. Meanwhile Protestant unionists and Catholic nationalists north of the border, divided and segregated for so long, seem at last to be striving for a long-term agreement.

---

problem was the feeling among many citizens of the new states that the capitalist economy was responsible for the underdevelopment of their countries. This set off a search for socialist or communist alternatives, much to the annoyance of the West, and encouraged debates about the meaning of 'development' (see Chapter 10). Finally, fundamental questions were frequently posed about how far modernization, as generally understood, was compatible with the traditions of the former colonial peoples. The Islamic revival, discussed above, can be seen as one response to this dilemma.

Just as imperialism had a profound impact on the political and social geographies of enormous areas in Africa, Asia and the Americas, it equally affected the imperial countries themselves. One of the most important manifestations of this in the twentieth century was the flow of migrants from colonies and former colonies, especially in the tropics and sub-tropics, to take up jobs in the former imperial states. This naturally had a far-reaching cultural impact in cities and regions in western Europe. In fact the whole experience of imperialism led to the mixing of peoples and cultures on a grand scale. It also led to widespread questioning of long-held assumptions about European (and often male) cultural dominance (see Chapter 6).

Many scholars have argued that the end of European colonialism did not mean the end of exploitative relationships between the core countries of the world economy (including the former imperial powers) and what was now increasingly referred to as the 'Third World'. According to such thinkers, the formal imperialism of the colonial era had merely been replaced by a more 'informal' alternative, but the basic situation of the core exercising hegemony over the periphery had not really changed (Frank 1969; Wallerstein 1980). From the 1970s, however, some fundamental changes seemed to affect the world economy which, in the opinion of certain scholars, demanded that international relationships be viewed in a new way. Lash and Urry have described these changes as 'the end of organized capitalism' (Lash and Urry 1987).

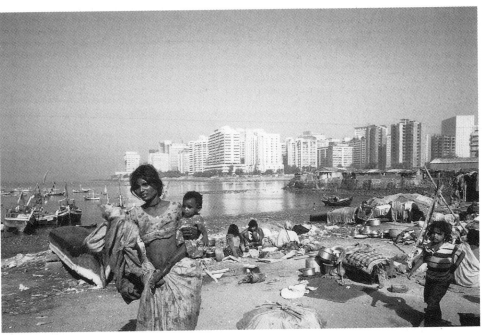

Plate 3.8 A Third World shanty town contrasts with the high-rise blocks of the modern city in the background. Scenes such as these have led many to question the meaning of development in peripheral and postcolonial countries. (Topham Picturepoint/UNEP)

3.6   The disorganization of capitalism

In 1960 the industrialized areas of Western Europe and North America produced almost 80 per cent of the world's industrial output. Even Japan accounted for only around 4 per cent. Much of the Third World remained agricultural. Only after this time did industrialization spread beyond its traditional centres (which since the 1930s had included the Soviet Union). Meanwhile many of the older industrial countries began to lose industries, even some of those which had arrived with the twentieth century.

Of course, some areas of the 'Third World' benefited far more from this industrial spread than did others. Most spectacularly, the Newly Industrializing Countries of east Asia, Brazil, Mexico and certain others soon seemed set to join the industrial core. Yet others, like certain Middle Eastern states, earned huge revenues from their energy exports. But there remained many areas which missed out on the new developments (although the latter part of the twentieth century was a period of unprecedented population growth and urbanization across much of the developing world – see Chapters 4, 7). The term 'Third World', used to group together countries with such disparate economic characters, seemed increasingly redundant, and the world as a whole seemed as unequal as ever.

How is one to explain the changes affecting the world's economic geography in the last third of the twentieth century? Geographers and others argued that such changes are part of the process of globalization. Speedier communications meant that the world was becoming a much smaller place as the twentieth century drew towards its close (Box 3.4). Capitalism itself was now a truly global phenomenon as markets were internationalized and finance became fully mobile. Before 1960, despite the importance of international trade, the world economy was structured around individual states. After 1960, the world economy became in effect transnational as the boundaries of individual states became ever less important to its functioning. Thus this period witnessed the rise to global importance of the transnational corporations, commercial conglomerates which became major players on a world scale. Because of the wealth and political influence they wielded, such huge companies became increasingly free to switch their operations from country to country as economic circumstances dictated. States, which had previously seemed unchallenged within their own frontiers, found it ever more difficult to control their own economies and began to bid against one another to attract footloose investment and the favours of the transnational corporations. Many industries now began to locate in parts of the Third World, where costs were cheaper, whilst core industrial countries began to experience deindustrialization and a switch into services and 'control' functions (the headquarters of the transnational corporations still tended to be located in the traditional core countries). A further important result of the development of information technology was that production became much more flexible than before and more geared up to highly specialized markets, changing fashion and the whim of the individual consumer. This new, more flexible approach to production is sometimes known as **post-Fordism**.

Later chapters of this book will explore some of the economic and social implications of these changes. However, it is worth stressing here how unsettling such developments were for the late twentieth-century world, and especially for the core countries. The years after 1945 had been ones of unprecedented affluence for these countries as commercial activity boomed, welfare states provided unheard-of security, and leisure opportunities seemed endless. These were the years of rampant suburbanization, ever-rising car

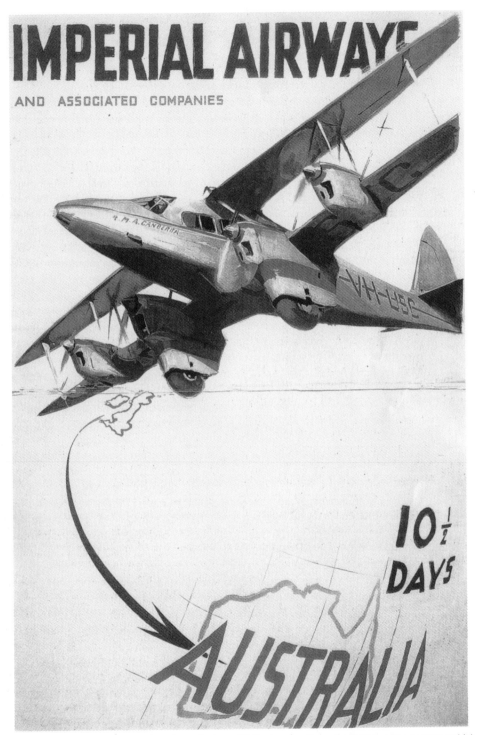

Plate 3.9 Imperial Airways flies to Australia. Imperial Airways was the major British airline company which operated at ever more ambitious international and intercontinental scales during the 1930s. (Vintage Ad Gallery)

Box 3.4   Air travel in the twentieth century

One of the most important agents of globalization has been air travel. In many ways air travel can be said to symbolize the technological achievements of the twentieth century.

While flight by balloon and airship predates the twentieth century, the first successful powered flight is usually taken to be that by the American brothers Wilbur and Orville Wright at Kitty Hawk, North Carolina, in December 1903. Within a few years, in July 1909, Louis Blériot had made the first successful cross-Channel flight between France and England, taking just over 35 minutes to fly the 37 kilometres. The first non-stop transatlantic flight was that by J.W. Alcock and A.W. Brown in June 1919. They took just over 16 hours to fly from St John's, Newfoundland, to Clifden in Ireland. This trip would have taken several days by sea.

In the early days powered flight had a serious competitor in the airship. Zeppelin dirigibles began the first passenger services between German cities in 1910. By the 1930s airships regularly crossed the Atlantic, the ill-fated *Hindenburg* taking some 43 hours at least to complete the trip between Frankfurt and New Jersey. Unfortunately, the hydrogen-filled airships were also very risky and the *Hindenburg*'s spectacular destruction at Lakenhurst, New Jersey, in May 1937 with the loss of 36 passengers out of 92 largely put an end to this form of passenger travel.

Commercial passenger services by aircraft developed very quickly after the First World War, the regular service between London and Paris inaugurated in 1919 being one of the first. Although flight speeds were slow by today's standards and aircraft range was limited, it soon became possible to cross the world in several days instead of the weeks and months it had taken previously. Thus by the 1930s the Dutch firm KLM was running fortnightly trips between Amsterdam and Jakarta. It took 12 days to complete this journey of 9,500 miles, with 18 stops (including overnight ones). Night flying was dangerous until the development of radar in the Second World War, but overland night flights became possible (as across the USA) with guidance from powerful, land-based beacons. Where water was available, the relatively slow but comfortable flying boats became popular by the late 1930s because of a lack of runways. On the whole, however, air travel remained a luxury enjoyed by the few.

The technological advances of the Second World War were a great boon to air travel. Speeds were greatly increased with the introduction of the jet engine. The first commercial passenger jet was the De Havilland Comet, which began operations in 1952, followed by the Boeing 727 in 1958. Other advances included fully pressurized cabins and advanced instrumentation, allowing aeroplanes to fly higher and faster, the introduction of wide-bodied aircraft from 1970 (the Boeing 747), which carried more passengers, and longer ranges. Air travel thus became possible for many more people and by the 1970s it had replaced sea travel for all but the shortest distances.

International air travel was not just the result of technological advance but also of advances in international co-operation. The International Civil Aviation Authority, established in 1947, sets international standards for safety, reliability and navigation, and facilitates the setting up of routes.

By the 1990s it had become possible to fly by scheduled airliner between Europe and Australia in less than 24 hours with one refuelling stop. The supersonic Concorde can fly passengers between London and New York in less than three hours. Quite an advance from Kitty Hawk in under a century!

Plate 3.10 Deindustrialization in the 1980s afflicted many parts of the world in the wake of the 1970s oil price shocks and of major changes in the world economy. (Format Photographers/Raissa Page)

ownership levels, international travel and television, as the benefits of mass production truly took effect. They were replaced, from the mid-1970s, by a period of greater uncertainty as deindustrialization set in, unemployment rose, insecurity became more widespread, and social inequality became more apparent. The world was no longer the self-confident, cosy, Eurocentric place it had been at the beginning of the century. Certain writers, notably Francis Fukuyama, celebrated liberal capitalism's victories over several of its twentieth-century rivals, particularly Fascism and communism (Fukuyama 1992). But this was a mood shared by few others. Capitalism had become 'disorganized' and its ability to solve the fundamental problems of our day could still be questioned and challenged as it had been throughout the previous hundred years.

Just as Lash and Urry have used the term 'organized capitalism' to describe the years down to about 1973 when Fordism reached its apogee, so they have described the last years of the century as those of 'disorganized capitalism'. Some of its more prominent features are listed in Box 3.5, once again paying particular attention to how the changes have affected the core countries of the world economy.

3.7 Conclusion

The twentieth century was a period of rapid economic and social change over most parts of the world. It was also a period when the world seemed to become smaller and most regions were gradually drawn into an ever more embracing global system. Yet it would

Box 3.5   Disorganized capitalism (after Lash and Urry 1987)

- The onset of 'disorganized capitalism' is marked by a decline in the relative importance of extractive and manufacturing industries.

- There is a relative increase in the importance of service and consumer industries, especially in employment.

- The use of flexible technologies encourages a reduction in the average size of manufacturing plants with more accent on labour-saving investments and more flexible employment processes, all induced by competition.

- Because of the need for flexibility and cost-cutting, industrial firms tend to 'hive off' many of the services and supporting activities they need to other firms and organizations. There are thus more opportunities for small firms, changing the traditionally specialized nature of the regional economy.

- Regional economies are also affected by the greater emphasis on non-standardized production – traditional regional specializations become less marked.

- The global economy reduces the effectiveness of state attempts at economic regulation – from the state's point of view, the economy becomes less predictable.

- Rising costs, demands for reduced taxation, and growing social inequality challenge the idea of a centralized welfare state.

- Smaller, more footloose industries, the rise of services, better communications and other factors reduce the traditional importance of big, industrial cities by comparison with small towns and rural areas.

- There is a rise in importance of the educated social strata needed to work in the new administrative, control, service and related activities – the so-called 'service class' – with more sophisticated and individualized tastes in consumption and other areas. The age of mass cultural provision is replaced by greater cultural fragmentation and pluralism (sometimes referred to as post-modernism). There is a commensurate decline in faith in large-scale planning and similar activities associated with modernism.

be a mistake to imagine that this was an uncontested process, or one which threatens to bring about a global uniformity. The legacy of the twentieth century is a world that is both dynamic and uneven, and therefore uncertain. The rest of this book tries to grapple with this uncertainty.

Learning outcomes

- Capitalism is inherently dynamic and unstable. It has been so throughout the twentieth century, and is likely to continue to be so in the future.

- The concepts of 'organized' and 'disorganized' capitalism are ways of trying to make sense of the changes which affected capitalist societies during the twentieth century.

Plate 3.11   The fall of the Berlin wall in 1989 signalled the end of communism in the Soviet bloc and the removal of a major twentieth century challenge to world capitalism. (Popperfoto/Reuters)

Learning outcomes (continued)

- 'Fordism' and 'post-Fordism' (concerning production and consumption) and 'modernism' and 'postmodernism' (concerning culture) similarly try to make sense of twentieth-century change. No concepts, however, can do justice to the complexity of change during this period.

- Western-style liberal capitalism is only one variant of capitalism. It has been challenged in various ways in the twentieth century. Future challenges might prove more successful than past ones.

- The twentieth century has been an era of nation-states and of nationalism. Towards the end of the twentieth century the role of the nation-state seemed increasingly challenged by globalization. Nationalism, however, may yet flourish as a response to globalization.

- Modernity may not lead to the disappearance of traditional cultural practices. The Islamic revival, and similar religious revivals across the world, are a case in point.

- Despite the view of some that globalization is leading to the emergence of a global culture, it might actually increase the differences between places and hence the importance of geography.

Further reading

Fieldhouse, D.K. (1999)   *The West and the Third World*, Blackwell, Oxford. The first attempt to provide a comprehensive survey of the relationship between the West and the Third World, and the debate over its effects, during the twentieth century. Combines theoretical discussion with empirical evidence.

Godlewska, A. and **Smith, N.** (eds) (1994)   *Geography and Empire*, Blackwell, Oxford. Discussions of the interplay between the rise of geography as an intellectual discipline and the development of the European empires culminating in the decolonization of the post-war period and the postcolonial experience.

Hall, P. (1996)   *Cities of Tomorrow*, updated edition, Blackwell, Oxford. A splendid account of urban development in the late nineteenth and twentieth centuries in different parts of the world. The accent is on urban planning and design, but there are many social insights.

Hobsbawm, E. (1995)   *Age of Extremes: The Short Twentieth Century 1914–1991*, Abacus, London. What is rapidly becoming a classic account of twentieth-century history, written by a doyen of British Marxist historians. Any chapter is worth reading, but geography students will find those dealing with economic, social and cultural change especially revealing.

Kennedy, P. (1988)   *The Rise and Fall of the Great Powers*, Random House, New York. Ironically published just before the collapse of the Soviet Union (which the book failed to predict), this is a survey of economic change and military conflict among the world's great powers over the last 500 years, with the accent on the twentieth century. A useful background for students of geopolitics and political geography.

Knox, P. and **Agnew, J.** (1998)   *The Geography of the World Economy*, 3rd edition, Edward Arnold, London. A survey of geographical change in the world economy which includes an account of twentieth-century developments.

SECTION 2

EDITED BY DENIS SHAW

Society, settlement and culture

Geographers are interested in the ways places differ from one another and also in the similarities and interconnections between places. But the complexities of the changes which human societies have experienced since the late twentieth century make the geographer's task of making sense of the world increasingly difficult. Old dichotomies, like that between 'urban' and 'rural', and the lines which geographers have customarily drawn on their maps to divide country from country, region from region, and culture from culture, seem to make less and less sense in a globalizing environment. Even so the results of such processes seem not to be a world which is more uniform. While distant places interconnect and interact with increasing intimacy, new spaces appear where people are excluded from such developments, and almost everywhere some individuals who share spaces with others are denied the rights and advantages that the latter enjoy.

It is such complexities, making themselves felt in urban and rural social spaces and in the sphere of culture, which form the subject of this section. The emphasis is on the multifaceted effects of globalization on human societies and the way people think. Thus the world has become more urban – most of the world's population now lives in cities and only three major regions of the world (sub-Saharan Africa, south and much of south-east Asia, and China) remain predominantly rural (though this still represents a substantial proportion of the world's population). But the experience of urbanization differs radically between cities, spaces within cities, and individuals. The corollary is that the world has become less rural – rural spaces may be urbanized, linked into global networks, or lose population to the cities – but again those spaces, and the individuals within them, are affected in very different ways by such processes. And the world has become more 'cultural' in the sense that present-day changes are as significant for their cultural effects as for any others. Issues like identity and the meanings and ways of life which are shared by, or which divide, individuals and groups become matters of concern, contestation and even conflict.

Chapter 4 on cities considers how the term 'urban' is to be defined and the implications of urbanism for the ways in which people live and interact. The significant issue of the role cities now play in global and regional economic and power networks is addressed before the chapter turns to examine the issue of urban governance and how urbanization processes are and can be regulated. Chapter 5, entitled 'Rural Alternatives', is similarly concerned with definitions of the 'rural', and analyses the alternative consequences for both rural spaces and individuals of the processes of urbanization and globalization. For rural spaces such consequences are the product of their dynamic and varying relationships with broader social processes. For individuals and social groups they are partly the result of constraints imposed by socio-economic structures (for example, income levels) and partly of the types of rural spaces in which they are located. Chapter 6 on culture defines the latter as a process and as a system of shared meanings. The argument is that globalization is not producing a single global culture although the idea of a global culture is becoming important. Global cultural processes interact with processes occurring at national and regional levels to produce a complex world in which it is necessary to foster approaches which will welcome and celebrate cultural differences rather than shun and fear them.

This section makes clear the fact that the geography of a globalizing world is as much a geography of exclusion as it is one of interconnection and interlinkage. It shows that it is a dynamic world where social and cultural change is very much the order of the day, and yet it is a world punctuated by continuities. Finally it suggests that it is a world in which people can make a difference, where they can be more than mere ciphers dominated by impersonal forces like global financial and power networks, and where ideas, attitudes and relationships matter.

CHAPTER 4

ALLAN COCHRANE

Cities

Most of the world's population now lives in cities.

It is no longer possible to divide the world into a developed world, with big cities, and an 'underdeveloped' or 'Third' world, without cities. A neat line can no longer be drawn on the globe that divides a world of urbanites from a world of peasants. Some of the biggest cities in the world are in countries that might previously have been described as 'underdeveloped'. São Paulo, Mexico City, Shanghai, Mumbai (Bombay), Buenos Aires and Beijing are all among the ten cities with the biggest populations in the world, while Rio de Janeiro and Calcutta are not far behind (see Table 4.1).

Table 4.1   Cities with the largest populations in the world (millions of people)

1 Tokyo/Yokohama	26.84
2 São Paulo	16.40
3 New York	16.33
4 Mexico City	15.64
5 Mumbai (Bombay)	15.09
6 Shanghai	15.08
7 Los Angeles	12.41
8 Beijing	12.36
9 Calcutta	11.67
10 Seoul	11.64

Source: Philip's Atlas of the World. Comprehensive Edition, 8th edn, London: George Philip, in association with the RGS with the IBG, p. 26.

Box 4.1   The growth of cities

Patterns of global population growth and migration are discussed in Chapter 7 of this book, and it is important to remember that: 'The steepest growth rate has been in cities. In 1950 29 per cent of the world's population was urban. In 1965 it was 35 per cent, in 1990 50 per cent, and by 2025 it could be at least 60 per cent. The world annual growth rate of urban population between 1965 and 1980 was 2.6 per cent; but between 1980 and 1990 it was 4.5 per cent. Nearly all the current increase is in poor countries' (Tickell, in Rogers 1997: vii).

As Massey (1999: 114) notes: 'about 3,000 million people are living in urban areas, each of which is home to many millions of people. In the late 1990s between 20 and 30 *million* people were leaving the countryside every year and moving into towns and cities. We have never faced a situation like this before. We are becoming, increasingly, an urbanized world.' Between 1900 and 1990 the number of cities with populations of over half a million grew from 20 to nearly 600 (Short 1996, 37).

The dramatic global growth of cities makes it very important for us to understand them and the ways in which they work. It represents a dramatic shift in the spatial organization of society – in the geographies of everyday life. As Harvey (1997: 23) emphasizes: 'The urban and city are not simply constituted by social processes, they are constituitive of them.' In other words, the changing spatial organization of our lives represented by global urbanization also fundamentally changes the ways in which we live our lives and the social relations of which we are a part. But, in what ways does it do so?

This chapter starts out by looking for ways of defining cities. It explores the features of urban life that distinguish it from other forms of human settlement, and other forms of socio-spatial organization. It considers the implications of some of these features for the ways in which people live their lives in cities, before going on to discuss the key role of cities within wider global and regional networks of social and economic power. Finally the chapter turns briefly to the challenge of urban governance, asking how the complex and fluid social relationships of urban life are governed and regulated in practice.

4.1   Trying to define cities

Recognizing the importance of urban development in these general terms is, however, only a starting point. Most of us probably feel that we have a reasonable grasp of what makes some places cities, while others are towns, villages or rural areas. But it is important to explore rather more systematically what it means to call some places cities. This should help us to understand what is shared between cities (and the social relationships associated with them) as well as being able to assess how they differ, for example, playing different roles in global systems of power and offering different opportunities to and placing different constraints upon their residents. So, for example, a large population may make a place a city (even a 'megacity' in Castells's terms, Castells 1996: 126–7), but only some cities (such as Tokyo, New York and London) are key nodes of financial and economic power within the global economic system, to the extent that they are often referred to as global cities (or world cities) (see, for example, Sassen 1991). As a result, one might also expect a higher concentration of wealth and of members of global elites in those cities, so that their social structures are likely to be quite different from those of other large cities, such as Mexico City or Mumbai.

So, what *are* the key – defining – features of cities?

In the 1930s, Lewis Wirth – a sociologist based at the University of Chicago – set out to specify them systematically, in ways that incorporate what are probably still among the dominant common-sense understandings of urban life. Wirth's definition was timeless and could, apparently, be applied to all societies. For him the city was: 'a relatively large, dense and permanent settlement of socially heterogeneous individuals' (Wirth 1938: 50). In other words cities could be defined through four main characteristics. They are:

- large;
- densely populated;
- permanent;
- socially heterogeneous.

But what are the implications of drawing such conclusions? Wirth himself pointed to a series of more or less logical consequences that he saw as flowing inexorably from these characteristics.

Size. The large population size of cities was held to mean that contact between urban residents would generally be impersonal, superficial and transitory, since no one person could know all the others in the city. Wirth's arguments build on a longer tradition which sought to analyse the rise of industrial society. Toennies (1887), for example, distinguished between what he called traditional and industrial society by associating **Gemeinschaft** (community) with traditional society and **Gesellschaft** (association or society) with industrial society. Closeness and stability, in which there was a high degree of informal personal contact within primary groups, characterized the former and everyone understood their responsibilities within a stable society. By contrast, the latter was characterized by transitory and superficial relationships, in which individuals only come together in groups to achieve particular ends (or defend particular interests). Pre-urban (rural or peasant) society has often tended to be associated with *Gemeinschaft*, while urban society has tended to be seen as the prime example of *Gesellschaft*. According to Wirth, the size of cities meant that contact between individuals was usually restricted to those occasions in which there was some mutual advantage to be gained. As a result, he argued, human relationships were segmentalized according to the services that each individual might perform for another.

Density. Similarly, the density of population was said to produce increased differentiation between individuals, an increasingly complex social structure, social segregation and a dissociation of workplace from residence. It fostered a spirit of competition, aggrandizement and mutual exploitation, requiring complex rules to manage the resulting friction.

Permanence. It is the permanence of cities which creates the spaces within which the social processes identified by Wirth work themselves out. In a sense it is also that permanence which confirms the importance of 'urbanism as a way of life' (the title of Wirth's seminal article). Urbanism gains its status by being the next stage of social organization, the dominant form of the industrial (or modern) era.

Heterogeneity. In some ways this is the most important (and distinctive) feature of cities for Wirth. He argues that: 'The city has ... historically been the melting pot of races, peoples and cultures. It has not only tolerated but rewarded individual differences. It has brought together people from the ends of the earth *because* they are different and thus useful to one another, rather than because they are homogeneous and like-minded' (Wirth 1938: 50). The heterogeneity of the individuals making up the city's population and the social interaction between different personality types, said Wirth, resulted in the creation of a complicated caste and class structure reinforcing the effects of high population density. No single group had the full loyalty of individuals, since different groups were important for different segments of each individual's personality.

Wirth's arguments have been criticized for failing to capture some key features of urban life and exaggerating others. So, for example, Gans (1968) has argued that it drew too much on the experience of Chicago in the 1920s and 1930s and, in particular, that of the city's inner city, which was indeed characterized by having a highly transient population. By contrast, Gans points to the significance of the suburbs. These, he argues, meet many of Wirth's broader criteria, without exhibiting quite the same symptoms in their social relations. He claims (of US suburbs, at least) that 'there is little anonymity, impersonality or privacy' of the sort identified by Wirth (Gans 1968: 40). Gans also identifies what he calls urban villages (Gans 1962) and Young and Wilmott (1957) make similar points about working-class community in Bethnal Green in London. In the latter case, the focus was not on any distinction between 'urban' and 'non-urban', because 'traditional' community was associated with the (inner city) urban of Bethnal Green, which was itself deemed to be under pressure from social change (symbolized to some extent by the threat of more individualized suburbanization).

Garreau (1991) (again in the US context) has also argued that what he has identified as the new models of urban living in the form of 'edge cities' represents quite a different 'way of life' (see also Soja 1996 Ch. 8 on 'exopolis'). Garreau starts with straightforward

Plates 4.1, 4.2, 4.3, 4.4 Images of 1930s Chicago: the slums (Corbis UK/Bettmann); the stockyards (Corbis UK/Hulton-Deutsch Collection); Gangsters (Corbis UK/Underwood & Underwood); the Wrigley Building (Corbis UK/Hulton-Deutsch Collection)

– even prosaic – criteria, although it is less clear that he sticks to them very consistently as his arguments develop. For him, an edge city is any place that: (a) has more than five million square feet of leasable office space, (b) has six hundred thousand square feet of leasable retail space, (c) has more jobs than bedrooms, (d) 'is perceived by the population as one place' (i.e. because of its mix of jobs, shops and entertainment), (e) was nothing like a 'city' as recently as thirty years ago (Garreau 1991: 6–7). Garreau calls these places edge cities because, he says, 'they contain all the functions a city ever has, albeit in a spread out form that few have come to recognize for what it is. Edge, because they are a

Plates 4.5, 4.6, 4.7   Living in the suburbs: Ruislip, UK (Format Photographers/Maggie Murray); Fairfax, USA (Corbis UK/Bettmann); Leavittown, USA (Corbis UK/Bettmann)

vigorous world of pioneers and immigrants, rising far from the old downtowns, where little save villages or farmland lay only thirty years before' (Garreau 1991: 4). He places great stress on the role of the market, the developer, and the individual choice of consumers, whether in their liking for malls or their preference for detached single-family housing.

By contrast, Pile is less sympathetic to the suburbs. He is rather more sympathetic to Wirth's analysis (1999b: 42–8), seeing them as attempts to escape from the realities of urban life. 'Suburbs', he says, 'are an attempt

- to eliminate or marginalize the tensions of urban life, and
- to separate out different geographies and histories
- and also to *de-intensify* urban social interactions' (Pile 1999a: 30).

In other words, for him, suburbs cannot be used directly as a means of defining cities, since they explicitly define themselves by not being cities. Pile (1999a) and Massey (1999) argue that while some of the features of urban life can be found outside cities it is the *intensity* with which they are experienced in cities that is significant.

If one set of criticisms of Wirth's approach relates to weaknesses it may have in dealing with the complexities of urban life in the United States (and – by implication at least – Western Europe where similar developments can be identified), then another might relate to its lack of engagement with the massive new cities of the Third World. In some respects, of course, his criteria can be seen to fit particularly well with these cities – their populations are even larger than those of most US cities, their populations are dense and heterogeneous, too. The notion of permanence is, perhaps, rather more ambiguous, because of the extensive temporary and informal housing and a fast-changing population, both in the core of cities and in extensive shanty towns on the outskirts. However, Castells's description of an emergent megacity around the Pearl River Delta in China – what he calls the Hong Kong-Shenzen-Canton-Pearl River Delta-Macau-Zhuhai metropolitan regional system – highlights some of the difficulties in uncritically pursuing Wirth's approach (Castells 1996: 407).

Even if matters are not quite as straightforward as he suggests, nevertheless Wirth successfully identifies some of the important features which are characteristic of urban life. His assumption that there are certain more or less automatic linkages between urban form and the lived experience of cities is difficult to sustain. Similar urban forms may be associated with rather different experiences of urban life, while apparently similar (urban) experiences may be associated with significantly different urban forms. Nevertheless, it is possible to build on his arguments in ways that highlight how cities are actively defined and redefined by those who live in them and those who move to them and between them. This is probably particularly clear in the case of heterogeneity.

4.2 Some implications of heterogeneity

Cities bring together (juxtapose) people from different classes and different cultural backgrounds as well as combining a wide range of activity spaces. According to writers such as Jacobs, it is precisely this that gives them their strength as sources of innovation and dynamism. In her classic book, *The Death and Life of Great American Cities* (published in 1961), she argues that:

Plates 4.8, 4.9   Making a megacity: Hong Kong and the Pearl River Delta. (Format Photographers/Maggie Murray, and Geophotos/Dr Tony Waltham)

| Box 4.2 | The Pearl River Delta, China |

'In 1995, this spatial system, still without a name, extended itself over 50000km², with a total population of between 40 and 50 million, depending on where boundaries are defined. Its units, scattered in a predominantly rural landscape, were functionally connected on a daily basis, and communicated through a multimodal transportation system that included railways, freeways, country roads, hovercrafts, boats, and planes ... [it] is not made up of the physical conurbation of successive urban/suburban units with relative functional autonomy in each one of them. It is rapidly becoming an interdependent unit, economically, functionally, and socially.... But there is considerable spatial continuity within the area, with rural settlements, agricultural land, and undeveloped areas separating urban centers, and industrial factories being scattered all over the region' (Castells 1996: 407–9).

The architect Rem Koolhas (1997) argues that 'In the Pearl River Delta, we are confronted with a new urban system. It will never become a city in the recognisable sense of the world. Each part is both competitive with and has a relationship to each other part. Now these parts are being stitched together by infrastructures so that every part is connected, but not into a whole. We call this new model the City of Exacerbated Difference (COED). A city that does not imply the stability of a definitive configuration because each part is fixed, unstable and in a state of perpetual mutual adjustment defining themselves in relation to other parts' (quoted in Hanrou and Obrist 1999: 10–11).

To understand cities, we have to deal outright with combinations or mixtures of uses, not separate uses, as the essential phenomena....

Although it is hard to believe, while looking at dull grey areas or at housing projects or at civic centres, the fact is that big cities *are* natural generators of diversity and prolific incubators of new enterprises and ideas of all kinds....

The diversity, of whatever kind, that is generated by cities rests on the fact that in cities so many people are so close together, and among them contain so many different tastes, skills, needs, supplies, and bees in their bonnets. (Jacobs 1961: 155–9)

Unlike those whom she accuses of seeking to impose a deadly uniformity on the city, Jacobs emphasizes the extent to which complexity and heterogeneity may create order out of apparent chaos. Like Wirth, she believes that urban form significantly influences the ways in which people live their lives. She positively celebrates those areas of cities that are characterized by mixed uses – commercial, retail and residential – mixed ages of buildings, and low-rise housing on streets of relatively small blocks. Because of the intricacy of interactions created by this, she suggests that it should be possible to ensure that order is maintained under the watchful gaze of a series of residents and workers at different times, in a sidewalk symphony or 'an intricate ballet in which the individual dancers and ensembles all have distinctive parts which miraculously reinforce each other and compose an orderly whole' (Jacobs 1961). She fundamentally criticizes those planners (and architectural theorists) who seek to create more unified areas, which separate traffic and employment from residence and encourage standardization of dwelling units.

Jacobs's vision of the city is an inspiring one – and has indeed inspired recent writing on the city (see, e.g., Donald 1997, Sennett 1990). But it can also be criticized for itself being rather romantic, failing to recognize the ways in which the social life of cities may generate inequality and division, power and powerlessness as well as diversity. Berman (1983: 324) notes the extent to which Jacobs's 'inventory of the people in the neighbourhood has the aura of . . . a Hollywood version of a World War Two bomber crew: every race, creed and colour working together to keep America free for you and me'. But then he goes on to highlight the absence of black and Hispanic people from her imagery, from her diverse neighbourhoods.

This is highly significant both in the context of developments which took place in the 1960s, but – more important – also because cities are characterized by substantial change and shifting population, particularly because of the ways in which they attract new populations through migration. As Berman notes:

> In the course of the 1960s . . . millions of black and Hispanic people would converge on America's cities – at precisely the moment when the jobs they sought . . . were departing or disappearing. . . . Many of them found themselves desperately poor, chronically unemployed, at once racial and economic outcasts, an enormous *lumpenproletariat* without prospects or hopes. In these conditions, it is no wonder that rage, despair and violence spread like plagues – and that hundreds of formerly stable urban neighbourhoods all over America disintegrated completely. . . . [I]t was clear by the late 1960s that, amid the class disparities and racial polarities that skewered American city life, no urban neighbourhood anywhere, not even the liveliest and healthiest, could be free from crime, random violence, pervasive rage and fear. (Berman 1982: 324–5)

Social heterogeneity and the juxtaposition (or physical proximity) of different groups

Plate 4.10   Making protected space in South London. (Format Photographers/Maggie Murray)

also ensure that cities become arenas within which different groups and classes struggle for dominance, or search for forms of security in the face of challenge. Cities both bring different people together, requiring them to live together, and they also provide ways of reinforcing and developing new forms of segregation and division. Urban spaces become sites across which conflicts take place and accommodations are negotiated, with particular places taking on symbolic meanings, which crystallize (for a time at least) the outcomes of those negotiations, sometimes as borders, sometimes as protected or open spaces. In their extreme forms, cities can become defined by what Davis (1992) has called an 'ecology of fear', in which a range of protected enclaves is constructed by the middle classes to exclude those who may challenge their security. In other words, explicit attempts are made to create order by the spatial policing of heterogeneity (see also McLaughlin and Muncie 1999: 117–25).

Some of the consequences of the apparently insatiable search for 'defensible space' (Newman 1972) seem rather paradoxical, however. Instead of creating greater feelings of security in suburban areas, or in areas with neighbourhood watch schemes, or even in shopping malls overseen by closed-circuit television, the creation of more protection

---

Box 4.3   The creation of defended enclaves

Davis (1990) vividly describes the creation of protected enclaves in Los Angeles: 'The security-driven logic of urban enclavization finds its most popular expression in the frenetic efforts of Los Angeles' affluent neighbourhoods to insulate home values and lifestyles ... new luxury developments outside the city limits have often become fortress cities, complete with encompassing walls, restricted entry points with guard posts, overlapping private and public police services, and even privatised roadways. It is simply impossible for ordinary citizens to "invade" the cities of Hidden Hills, Bradbury, Rancho Mirage or Rolling Hills without an invitation from a resident.... In the once wide-open tractlands of the San Fernando Valley, where there were virtually no walled-off communities a decade ago, the 'trend' has assumed the frenzied dimensions of a residential arms race as ordinary suburbanites demand the kind of social insulation once enjoyed only by the rich' (Davis 1990, 244–6).

Even in its most extreme form – that is the literal creation of gated areas, accompanied by private policing – Los Angeles is not unique. Similar arrangements are emerging in most of the world's major cities, in luxury developments in London, as well as Detroit and Johannesburg (Robinson 1999). In some cities in less developed countries the contrast is particularly sharp, since the physical distance created by suburbanization is often absent, so that the enclaves may be right up against the shacks or slums of the dispossessed. In São Paulo, for example, Caldeira (1996) describes one development (Morumbi) which was constructed in the 1980s. The new housing offers 'the amenities of a club, [they] are always walled, have as one of their basic features the use of the most sophisticated security technology, with the continual presence of private guards.... [The] luxury contrasts with the views from the apartments' windows: the thousands of shacks of the *favelas* on the other side of the high walls which supply the domestic servants for the condominiums nearby ... for residents of the new enclosures, the inconvenience seems to be more than outweighed by the feeling of security they gain behind the walls being exclusively among their equals and far from what they consider to be the city's dangers' (quoted in Amin and Graham 1999: 39).

seems to have been accompanied by increased concerns about crime and increased insecurity, potentially reinforcing 'a decline in communal trust and mutual support'
(McLaughlin and Muncie 1999: 135). It may, therefore, be more appropriate to look for
ways in which other sorts of spaces – which thrive on heterogeneity and the juxtaposition of a range of different activities – may be constructed or defended. In this context,
the example of Bombay's bazaar may be instructive:

> In the 1960s, the bazaar swept across Bombay, sprawling along transport lines, on slopes,
> underutilized land, undefined or unpoliced pavements, and on any other interstitial space
> it could find and occupy. In the process, it blurred beyond recognition the physical
> segregation inherent in the colonial city structure that had survived until then. The
> bazaar completely altered the exclusivity of the two domains, the Indian town (in the
> north) and the Western city (to the south). (Mehrotra 1997: 40, quoted in Massey 1999:
> 128)

In his powerful history of post-war Detroit, Sugrue (1996) explores some of the ways
in which a city is constructed out of a complex series of interactions – battles and accommodations. Sugrue is critical of those who simply see matters as predetermined, instead
emphasizing the active role of the city in constructing the lived experience of its residents:

> In the post-war city, blackness and whiteness assumed a spatial definition ... the
> completeness of racial segregation made ghettoization seem an inevitable, natural
> consequence of profound racial differences. The barriers that kept blacks confined to
> racially isolated, deteriorating inner-city neighbourhoods were largely invisible to white
> Detroiters. To the majority of untutored white observers, visible poverty, overcrowding,

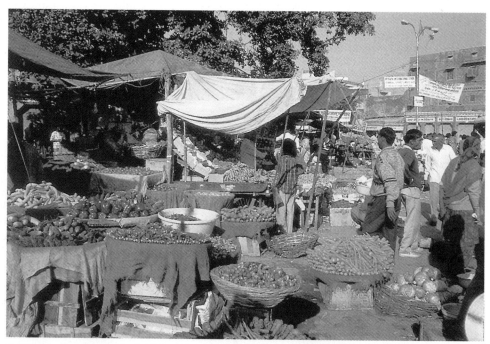

Plate 4.11   The bazaar, Mumbai (Bombay). (Images of India Picture Agency/Richard Galloway)

and deteriorating houses were signs of individual moral deficiencies, not manifestations of structural inequalities. (Sugrue 1998: 9)

The social divisions that characterize cities are sometimes approached through notions like social polarization, particularly in discussions of global cities such as New York, London and Tokyo, which operate as key nodes in a global system of finance and economic management – of what Sassen (1991) calls 'global control' (see Chapter 12 of this book on the emergent global production system, in which a second tier of global cities, the 'metaregional' capitals of Paris, Frankfurt, Sydney, Hong Kong, São Paulo and Mexico City, is also identified). Within these cities it is argued that there is an expanding gap between rich and poor, powerful and marginalized, as there is a growth in employment in financial and other specialist services, alongside a decline in manufacturing and traditional industry. At the same time as there is a growth in highly paid employment, there is a parallel growth in low-paid, low-skilled and casualized employment at the bottom end – servicing the high end service sector and those who work in it.

There is certainly evidence that points in these directions – with some sections of the urban population linked into key networks of global economic power, while others are explicitly excluded, to be managed by or to provide services to the powerful. But there is also a danger of taking a mechanical view of these relationships, in which urban social relations may be read off from the position of cities within global systems of one sort or another. While it may be possible to highlight tendencies, or point to the contexts within which social interaction takes place, it is also important to recognize that the ever-changing resolutions of tensions and conflicts may point in rather different directions. So, for example, Hamnett (1995: 125) argues that something rather different has happened in London, where, he says, there has been a trend 'towards professionalization ... with an overall increase in the number of professional and managerial jobs and a fall in the number of jobs at the bottom end of the occupational ladder'. Similarly, while acknowledging the significance of international migration for both New York and London, Cross and Waldinger (1992) highlight differences in the make-up and significance of that migration for the two cities – with the scale of recent migration being far smaller in the case of London. As indicated above, Sugrue (1998) argues that it is important to explore the precise forms of negotiation and battle between groups if the processes by which cities define themselves are to be understood (see also Davis 1990).

4.3   Cities of connection and disconnection

The discussion of cities often focuses on internal processes – the divisions and connections within them. But it is also important to recognize that cities are the products of flows and interconnections across space which link them into wider networks of economic and political relationships. They are, as Massey (1999) argues, 'the intersections of multiple narratives' or multiple space-times. Allen (1999) explores the ways in which cities of power are defined through their positioning in global networks. So, for example, he argues that the importance of cities such as New York and Tokyo stems 'from the fact that the world of banking and finance seems to have little choice other than to go through the financial districts of those cities. They stand at the intersections ... of all that matters in global economic terms. In that sense, both cities are locations for something far greater, far more significant in scale, than their actual size would have led us to believe.' They are 'the points at which the lines of authority and influence meet', drawing on a concentration of human resources and the local mix of skills and infrastructure (Allen 1999: 187). Summarizing the arguments of Castells, he notes that: 'global cities – or powerful cities – would not exist but for the connections which constitute them' (Allen 1999: 204). It is important to recognize that just as their positioning within these networks serves to define their power (through connection), so it may also confirm its absence (through their disconnection). So, to use an example from the UK, Liverpool was an important node in transatlantic trade during Britain's colonial era, but is now disconnected from major networks of trade, and this is reflected in its sharp decline even within the UK's urban hierarchy.

Castells has identified the growth of 'megacities' as a key feature of the network society that he sees as defining the contemporary world. Although including the cities usually labelled 'global cities', this is a wider notion which also incorporates a wide range of cities of significant world regional importance. They have very large populations, but their key defining characteristic for Castells is that 'They are the nodes of the global economy, concentrating the directional, productive, and managerial upper functions all over the planet' (Castells 1996: 403). They include Shanghai, Mexico City, Buenos Aires, Seoul, Calcutta, Osaka, Jakarta and Moscow among many others. As well as being linked into global networks, they are at the heart of massive urban regions which pull in huge numbers of people – hence the term 'megacity'. If the megacities and some of their populations are connected into these networks, then – according to Castells – a further defining characteristic is that large sections of their population, and significant spaces within them, are disconnected and explicitly excluded. '[W]hat is most significant about megacities', says Castells (1996: 404), 'is that they are connected externally to global networks and to segments of their own countries, while internally disconnecting local populations that are either functionally unnecessary or socially disruptive.'

In that sense, the world identified by Castells is not only one characterized by what he calls 'a space of flows', but also one in which access to that space defines what is possible – those who are disconnected are effectively excluded from participation in the new global economy. Castells (1998: 70–165), himself, uses the term the 'Fourth World' to highlight the significance of this. This, he says, is 'made up of multiple black holes of social exclusion throughout the planet'. Some of these 'black holes' are familiar enough as examples of what might otherwise have been called the Third World or 'less developed countries' (e.g. much of sub-Saharan Africa, the poor rural areas of South America and Asia), but Castells is making a stronger point. The Fourth World, he argues, 'is also present in literally every country, and every city, in this new geography of social exclusion. It is formed of American inner-city ghettoes, Spanish enclaves of mass youth unemployment, French banlieues warehousing North Africans, Japanese Yoseba quarters and Asian mega-cities' shanty towns. And it is populated by millions of homeless, incarcerated, prostituted, criminalized, brutalized, stigmatized, sick, and illiterate persons' (Castells 1998: 164–5). In other words it is a fundamental aspect of contemporary city life.

The implications of this are explored by Amin and Graham (1999) who stress that 'physical proximity' increasingly co-exists with 'relational difference'. In other words, the connected live alongside the disconnected – the favelas and shantytowns and informal living survive and develop alongside the gated homes of the rich. As we have seen, they use the example of São Paulo in Brazil, where they depict 'luxury enclaves rising above

Plate 4.12   Poverty in the midst of plenty in a German city. (Format Photographers/Silke Gester)

the roofs of the surrounding shanty town' (Amin and Graham 1999: 19). Similar examples could be drawn from cities across the globe, particularly – but not only – in the Third World.

Even cities that seem culturally and socially homogeneous, argue Amin and Graham, are made up of increasingly diverse and overlapping social worlds. Cities, they say, are 'places which bring together and superimpose diverse connections and disconnections' (Amin and Graham 1999: 9). In the case of Bangalore in India, for example, they note (drawing on the work of Madon 1997) that there has been a dramatic growth of software production linked into global networks and markets, which has generated a local concentration of skilled and relatively well paid professionals in the city, as well as creating strong development pressure (see also Chapter 11). But this growth is juxtaposed with in-migration from rural areas which has led to the creation of shantytowns. These migrants 'struggle to find poor quality work, whilst facing the constant threat of harassment from city agencies and property developers who are keen to "modernize" Bangalore's fabric by developing shopping malls and corporate work spaces in the place of informal settlements' (Amin and Graham 1999: 29).

Amin and Graham use the apparently trivial example of high-class restaurants as a very effective way of illustrating the nature of interactions between groups within cities. Zukin (1995) presents a highly complex picture of the globalized social relations that are reflected in the local experience of New York restaurants. 'In New York restaurant cuisine', she

Plate 4.13   A shanty town in São Paulo. (Panos Pictures/Julio Etchart)

notes, 'the local reterritorializes the global' through the identification of neighbourhoods with particular national cuisines – e.g. in the forms of Little Italy and Chinatown – as restaurants produce an increasingly global product tailored to local tastes (Zukin 1995: 156). More important, perhaps, the restaurants 'bring together global and local markets of both employees and clientele.... While it seems to be among the most "local" of social institutions, a restaurant is also a remarkable focus of transnational economic and cultural flows. As an employer, a restaurant owner negotiates new functional interdependencies that span local, regional, and global scales' (Zukin 1995: 158–9).

The heavy dependence on low-wage immigrant labour in restaurants is a reflection of the role cities play within global flows of migration, while the market for luxury restaurants is itself a reflection of cities' importance for global elites within worldwide networks of power. In these restaurants, the global elites (who, Castells argues, are linked to networks, rather than place) find themselves being served – in 'fleeting ties of local interdependence' – by those who are in rather a different position in the global pecking order – often migrant labour in casualized employment whose relationship to global networks is rather different but no less significant (Amin and Graham 1999: 14).

The emphasis in much discussion of cities and their role in a changing world often focuses on their economic role and on the role of elites. It is all too easy to forget the importance of the global networks that link the poor as well as the rich, the migrants as well as the global elites. Gilroy's discussion of the 'black Atlantic' is one useful corrective to this emphasis (Gilroy 1993). He carefully charts the emergence of a 'black Atlantic' political culture through music (soul, funk, reggae, hip hop and rap) which actively links black people in the cities of Kingston, New York and London. This is a global movement with its own networks and its own nodes. So, for example, Gilroy argues that hip hop emerged from the interaction of a DJ style drawn from Kingston in Jamaica, with the emergent dance music styles of the Bronx in New York (alongside some borrowings from Hispanic styles, too) (Gilroy 1993: 103). In other words, one of the most successful forms of global contemporary popular music came together from the active interaction of people and music through urban networks created by migration (and diaspora), even if it ended up as part of more familiar networks of urban power through major media corporations, from Sony to MTV.

4.4 Governing the ungovernable

One crucial aspect of the changes that have been taking place within and between cities is the more (explicitly) globalized context that is emerging. Some (such as Ohmae 1990) have argued that the process of globalization has fundamentally changed the role of urban governance, to the extent that we may be moving towards a 'borderless world' in which regionally based city-states (rather than nation-states) are likely to be the dominant form of political organization across the world (see Chapter 15). The economic challenge to nation-states may be exaggerated in this argument (as Hirst and Thompson (1996) forcibly suggest), but there is little doubt that the emphasis of urban governance has shifted towards forms of 'growth politics' based around negotiated alliances between major interest groups. Urban politics has increasingly become a politics of growth (Cochrane 1999b). This has meant that cities have had to become more entrepreneurial and more attuned to global market forces. They are increasingly required to define themselves in

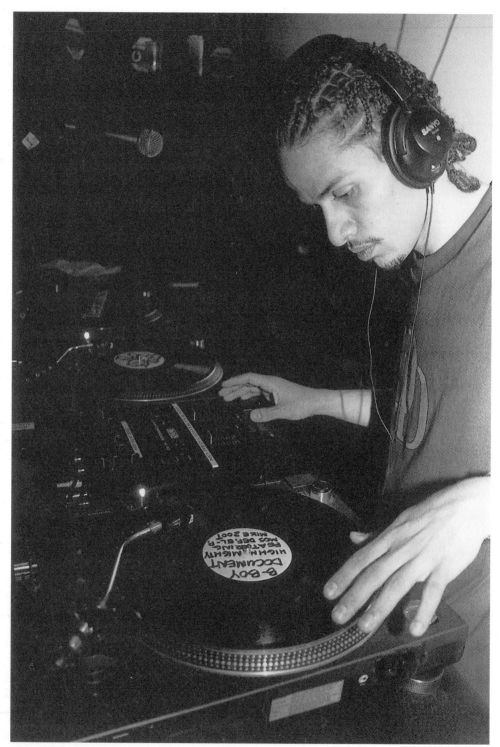

Plate 4.14   Mixing in New York. (Camera Press/Andrew Cantillon)

ways that are distinctive. They are engaged in a process of place marketing, and cannot rely on simply spreading outwards. The rise of urban entrepreneurialism can be seen in the intense inter-urban competition for high-profile events and feature developments (see, e.g., Cochrane *et al.* 1996). More generally, it involves the marketing of cities on a global scale.

The emphasis of place marketing is increasingly on redefining – or re-imagining – each individual city in ways which fit with dominant perceptions of urban success (see, e.g., Kearns and Philo 1993). The fashions of recent years have been clear: even the most land-locked cities have done their best to find some sort of waterfront and we have seen the emergence of some rather surprising places as cities of culture. It sometimes appears as if the power of imagination should be enough to transform what used to be accepted as the stubborn realities of urban-industrial decline. The construction of a Multimedia Super Corridor in Malaysia, stretching out from the massive Petronas Towers in Kuala Lumpur (explicitly constructed to be the tallest in the region), as well as the development of a strategy to construct and promote Singapore as 'an intelligent island', suggests that this is a global urban agenda, rather than one restricted to the cities of the West.

Box 4.4   **Redefining Berlin for a new century**

At the end of the twentieth century, Berlin sought to re-enter 'normal' urban history, having been excluded from it as a divided city, at least since 1945. West Berlin was an island of capitalism in the sea of East German communism, while East Berlin was the capital of the German Democratic Republic cut off from Western Europe by the Cold War. It is now in the process of re-imagining and repositioning itself within a changing Europe, as a world city, national capital or regional centre (see Cochrane and Jonas 1999). A key element in this process of redefinition has been the use of symbolic imagery of city centre redevelopment (particularly in the Potsdamer Platz), with its office blocks, cultural quarters and shopping experiences. In the 1990s, there was even a popular glorification of major building sites with a series of pamphlets guiding tourists around them (e.g. Schneider and Schubert 1997). In some respects the symbolism of redevelopment is clear enough, bringing in global imagery to highlight Berlin's role as a world city, but it coexists alongside attempts to impose local styles on the global developers.

Although there were building sites all over the new core of the city in the 1990s, the Potsdamer Platz was certainly the biggest. It is now back at the centre of Berlin, having been divided by the Wall as an area of dereliction until 1990. It was one of the historical centres, its busy street-life a feature of inter-war postcards in much the same way as that of Piccadilly Circus or Trafalgar Square was in London. The redevelopment was led by Sony and Daimler Benz (who between them hold over 80 per cent of the available land, with Daimler Benz holding around 75 per cent of this) but ABB (Asea Brown Boveri), Bewag, and Deutsche Bahn were all involved in the development. International architects (such as I.M.Pei, Richard Rogers and Helmut Jahn) were commissioned to design a series of prestige office buildings, constructing a vision which set out to define Berlin as a world city, with buildings like those of other world cities. The buildings incorporate large numbers of shops, hotels and offices, as well as residential space and what looks like a remarkable number of cinemas (including two big-screen, IMAX, cinemas). According to Dr Manfred Gentz, a member of the Daimler Benz

(Box continued)

Box 4.4   (Box continued)

Plate 4.15   Potsdamer Platz, Berlin. (Popperfoto/Reuters)

AG Board, the company 'has harnessed its entrepreneurial energy and vision of the future to the task of replacing this jewel in the old and new capital – with its old lustre but a new refinement'. It is 'a symbol of starting out into a common European future'. Whatever the accuracy of this statement, it nicely reflects the new urban boosterism.

The process of development itself became a means of defining Berlin as a place of the future. The popular pamphlets on Berlin's building sites proudly described the Potsdamer Platz as the biggest building site in Europe. A bright red exhibition centre (the Info Box) placed on stilts at the centre of the building sites on the Potsdamer Platz became a major tourist site, attracting 2 million visitors between its opening in October 1995 and January 1997. The visitors to the Info Box were explicitly expected to accept the imagination of architects, planners and developers for the reality of the city in the process of becoming something greater (Info Box 1996). The Potsdamer Platz was finally opened in late 1998 with the help of a firework display, even while the cranes continued to work on some of the buildings.

Attempts have been made to capture the new urban politics through terms such as 'urban growth machine' (Logan and Molotch 1987), 'growth coalitions' (Cox and Mair 1991) and 'urban regimes' (Stoker 1995). All of these notions are helpful in highlighting the significance of forms of urban governance that incorporate a range of major economic interests around a shared agenda of economic growth. But it is also important to remember that locally based political regimes have to be understood in the wider global contexts of urban development. Urban politics represents an attempt to bring together a range of different place-based (city-based) interests, which are themselves linked into wider networks of one sort or another. As we have seen, the urban elites may be defined by their connections to global networks, but other 'local' actors may also be members of their own networks. The language of 'growth' politics can make it difficult for the voices of those who are not members of the elite to be heard. Thus, there is a danger of losing the significance of the urban sphere as a site for pursuing a politics of welfare (in housing, education and health) or of culture (in terms of public space or identity). An exception is where this feeds into the language of place marketing (as each city develops its cultural strategy or celebrates diversity as a means of claiming metropolitan sophistication) (see, e.g., Bianchini and Parkinson 1993, Watson 1999).

Urban politics is not and cannot be monolithic – directed from the top down. Any attempt completely to control the apparent disorder and chaos of urban life through planning and careful 'administration' is ultimately doomed to failure (see Cochrane 1999a). The search for the 'good' city has been a recurrent theme of debates about cities. As a result, cities have been the focus of a great deal of dystopian and utopian thinking, which has incorporated different understandings of the ways in which cities are constituted. Thinkers such as Howard (1902/1965) and Le Corbusier (1929/1987) have tended to emphasize the extent to which cities are unruly places that need to be managed; thinkers such as Jacobs (1961) and Sennett (1990) have stressed the importance of fluidity, uncertainty and mixing as defining the urban condition. They not only see the attempt to capture social relations in plans and urban design as hopeless but also argue that it serves to damage the vitality of those social relations. The challenge of urban governance is to capture both aspects of the urban experience, in ways that permit the expression of vitality, while allowing for forms of popular control of urban development.

4.5 Conclusion

With the help of Wirth, this chapter set out by trying to identify some key defining characteristics of cities. It has stressed the importance of recognizing the complex, contested and overlapping, internal and external, networks of relationships that help to define cities. It is finally important to remember that cities cannot simply be explained in terms of those networks (or in terms of the flows that characterize them). They are more than simply nodes in a system. On the contrary, it seems increasingly clear that the density of activities that is concentrated in cities remains of vital importance, not only in sustaining the networks but also for the people who live in them. They continue to be defined by the intensity of the social relationships within them, not only the excitement generated by the juxtaposition of people and groups with different cultures and understandings, but also the possibility of interaction within relatively large concentrations of people who share particular (sometimes specialist) skills, cultures and understandings. The lives of the world's major cities are simultaneously characterized by the experience of closeness alongside extreme processes of social differentiation.

Learning outcomes

Having read this chapter you should be able to:

- Develop a geographical approach to cities which emphasizes that they are socio-spatial entities. Cities are the products of living social relationships, which have clear expressions both internally (as people come together within cities) and in terms of their changing position within wider economic and social networks of connection. Cities as socio-spatial forms are not just outcomes of the ways in which social processes work themselves out, but themselves play a major part in shaping social processes.

- Understand that the world is becoming increasingly urbanized – most people now live in cities. But, since cities are not all the same, recognize that it is important to understand both what links them together (what makes it possible to talk of 'cities' as recognizable phenomena) and what helps to differentiate them.

- Identify and critically explore key defining characteristics of cities, building on features drawn from Wirth (for whom cities are defined as being large, densely populated, permanent and socially heterogeneous).

- Recognize the significance and implications of social heterogeneity and diversity for the ways in which people live their lives in cities, including the consequences of the existence of increasingly diverse and overlapping social worlds.

- Understand some of the complexities of urban governance in a global context, highlighting some of the implications of bringing together the local politics of cities with their positioning within wider global systems.

Further reading

Allen, J., Massey, D. and **Pryke, M.** (eds) (1999)   *Unsettling Cities: Movement/Settlement*, Routledge, London. The chapters of this book focus on the open geography of cities, that is the interconnections between what happens within them and the cross-cutting networks which link them into wider social and economic relationships. Global economic change is linked to the experience of social groups and individuals living their lives within cities.

Hall, P. (1998)   *Cities in Civilization: Culture, Innovation, and Urban Order*, Weidenfeld & Nicolson, London. This book takes a historical view of cities. With the help of a series of fascinating case studies, from classical Athens to contemporary London, it presents an overwhelmingly positive vision of cities as the fundamental basis of civilization.

LeGates, R.T. and **Stout, F.** (eds) (1996)   *The City Reader*, Routledge, London. A well-chosen collection of contemporary and classic readings on cities that provides a useful introduction to important debates about cities and city life.

Massey, D., Allen, J. and **Pile, S.** (eds) (1999)   *City Worlds*, Routledge, London. Provides a relatively brief (185 page) introduction to and overview of geographical ways of thinking about cities. It highlights some of the tensions and ambiguities of urban life, exploring both the problems and potential of cities.

Pile, S., Brook, C. and **Mooney, G.** (1999) *Unruly Cities? Order/Disorder*, Routledge, London. Explores and questions the dominant image of cities as necessarily unruly and disordered. Sets out to identify both those aspects of urban life which are inherently fluid and diverse as well as those features which serve to regulate it – to make it manageable for both governed and governors. The book presents dystopian as well as utopian visions of cities, while also charting the practices of urban life.

Short, J.R. (1996) *The Urban Order: An Introduction to Cities, Culture and Power*, Blackwell, Oxford. A valuable textbook which provides a comprehensive and up-to-date introduction to the field of urban geography. It is well informed and uses case studies very effectively to illustrate the ways in which cities develop, as well as exploring some of the tensions within them.

Understanding Cities, DD304, Open University. A series of books prepared as part of an undergraduate course. Each of these edited texts incorporates chapters covering and developing key themes relevant to the topic of this chapter. Each is also accompanied by readings from selected texts, which contribute both theoretical and empirical support to the main arguments. These are helpful, challenging and sometimes even exciting books.

And finally, just because everybody should read it at least once:

Davis, M. (1990) *City of Quartz: Excavating the Future in Los Angeles*, Verso, London. Despite its critics (see, e.g., Soja 1997: 27), this is an inspirational book, bringing together politics, society and economics in a very powerful analysis of contemporary Los Angeles and its development. Most people who read it wish they could do what Davis does for Los Angeles for their own favourite city – unfortunately not all of us have quite the same style or detailed (and sympathetic) understanding to succeed.

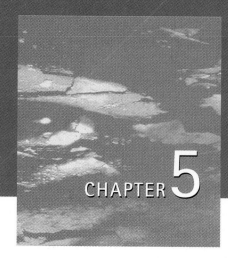

CHAPTER 5

IAN BOWLER

Rural alternatives

Topics covered

- Definitions of 'the rural'

- A typology of the 'rural' as socially constructed spaces

- Implications of encroachments of urban on rural spaces

- Rural depopulation

- The concept of counterurbanization

- Service provision and rural development

- Planned rural development

A variety of socio-economic trends can be observed in rural spaces around the world and two main processes are implicated, namely urbanization and globalization. On the one hand, rural spaces are being drawn into increasingly complex social and economic relations with national and regional urban systems through the interchange of capital, land, labour, raw materials and services. On the other hand, rural spaces are being drawn into competitive global relations through the increased international trading of their products and services and the restructuring of capital (McMichael 1996). This chapter is concerned mainly with the impact of urbanization on rural spaces in developed and developing countries alike.

While urbanization is experienced throughout the world, its impact varies significantly *between* rural spaces; consequently a number of 'rural alternatives' can be identified in terms of the uneven development trajectories of rural spaces. In addition, social groups and individuals *within* rural spaces also have varying experiences of urbanization and 'rural alternatives' can be considered in terms of varied 'life-styles' within the rural population (i.e. housing, employment, income, services and transport – Cloke 1996). This chapter explores the two meanings of 'rural alternatives' to reveal the varied human experience between and within rural spaces.

5.1   From rural functions to rural networks

5.1.1 Defining 'the rural'

A useful starting point for this chapter is a consideration of three constructs of 'rurality' as applied to rural spaces. At one level, the rural comprises those spaces left over once the urban has been accounted for; this *functionalist approach* assumes that rural space forms an analytical category containing non-urban, socio-economic structures and processes having causal or explanatory powers. For example, compared with the urban, rural structures include more extensive land uses (e.g. agriculture, forestry), lower population densities and employment based on natural resource exploitation (e.g. landscape, minerals, water). Processes include overcoming the longer time and cost distances between locations of economic activity and service provision. A number of descriptive typologies of rural spaces have been based on these assumptions, including urban–rural dichotomy, rural–urban continuum (urban fringe, agricultural heartland and rural periphery) and multi-dimensional models (Cloke and Edwards 1986). Box 5.1 illustrates a recent attempt for the United Kingdom (UK) to identify 'indicators' which delineate the countryside in functionalist terms.

A second construct of the rural is based on the interpretation that, with the possible exception of greater rural distance costs, all socio-economic structures and processes are common to both rural and urban spaces (Hoggart 1990). This *critical political economy approach* links rural areas to the dynamics of the national and international political economy (e.g. through urbanization and globalization), with the causes of 'rural' structures and processes lying outside 'rural' areas. In other words, while societies are defined by the dominant mode of production, and space comprises a means of production, *rural* space is denied any causal or explanatory power of its own. Rather, the rural comprises a distinctive spatial category within which the varying impacts of universal structures and processes can be examined. But the rural is also differentiated so that spatially contingent effects can

> ### Box 5.1
> ## Indicators for the countryside
>
> Countryside indicators measure socio-economic impacts on rural spaces and identify targets for reducing those impacts. However, they tend to reflect the often unstated interests of those who create them, are dependent on measurable characteristics and are selectively inclusive/exclusive of what is measured. The countryside (i.e. rural) indicators in the study include:
>
> - Level of tranquillity (distance from sites of noise)
> - Presence of landscape features
> - Darkness of the night skies
> - Sense of wilderness and isolation
> - Amount of derelict and vacant land and buildings
> - Sense of community
> - Diversity of local dialects
> - Level of rural services
> - Diversity of rural skills
> - Beauty of landscape
> - Frequency and availability of rural public transport services
>
> This list of indicators, whilst providing a sense of rural space, can also be interpreted in terms of a range of socially constructed representations of 'the rural' (see text).
>
> *Source*: Council for the Protection of Rural England 1995

result from the specificities of localities, for example their distance from urban markets or the quality of particular natural resources. This rural locality interpretation (Halfacree 1993) is considered to be most appropriate for the present discussion of 'rural alternatives'.

A third construct of the rural is based on a *social representation approach* (Mormont 1990). In this interpretation, the rural is constituted by lay, rather than academic, discourses of various social groups such as administrators, planners, service providers (e.g. tourism, health, electricity), farmers, recreationists and consumers. Here rurality becomes a social construct and, in Cloke's (1996: 435) words, 'reproduces the rural as an important intellectual category because human behaviours and decisions are influenced by the social constructs that indicate a place as rural.' This approach, therefore, focuses attention on the often divergent representations of rural space between the various lay discourses and the 'actors' who represent the rural. In addition it confirms the rural as comprising a multiplicity of social spaces which overlap the same geographic area. Insights from this approach inform the following discussion.

5.1.2 Constructing 'the rural' as a system

With rural space treated as a context for analysis, a conventional approach has been to examine its separate functions – such as food production, housing, transport, industry and service provision. Many researchers adopt this 'systematic' approach and commonly employ administrative boundaries to examine variations in human experience between rural spaces. More recently, however, given contemporary concerns over 'sustainability' and 'rural development', the various functions of a rural space have been conceptualized as comprising an open 'rural system'. Three components can be identified: space users, used space and space use. Figure 5.1, for example, shows Veldman's (1984) attempt to represent the interconnections between components of the physical environment, economic activities and the resident population within two types of rural space (peri-urban and peripheral). This approach recognizes the artificiality of examining, for example, economic activities separately from housing provision in rural spaces.

5.1.3 Constructing 'the rural' as a network

One problem of the rural systems approach, however, is conceptualizing space as a number of 'surfaces' with boundaries, each surface being constructed at a particular global, national or local scale. Just how these bounded surfaces at different scales are connected together tends not to be addressed and this is significant for the present interpretation based on the interaction between local rural spaces and a macro socio-economic process, such as urbanization. However, if the rural is constructed as an actor-network (Murdoch 1995), the local can be connected to the national, and even the global, through the connectivity of human *actors* (individual people, firms, agencies, institutions, corporations or government departments) and non-human *intermediaries* (scientific technologies, organizational forms, technical artefacts, money or knowledge in its material forms, for example texts and skills). With actors and intermediaries having equal status in defining each other, they comprise formal and informal networks that connect through space to link the local with the global. More powerful actors enrol other actors and intermediaries into a network so as to 'act at a distance' in influencing local events. In relation to agriculture in the countryside, for example, the regulations (documents as intermediaries) made by politicians and civil servants (actors) under WTO and the CAP impact from a distance on how farmland is used in a particular locality.

5.1.4 A typology of 'the rural' as socially constructed spaces

From the discussion so far we find that the rural cannot be equated merely with the agricultural, since rural spaces fulfil many other functions. Rural spaces and their populations are no longer autonomous and divorced from urban-oriented actor-networks; and the traditional economistic centre/semi-periphery/periphery typology of space fails to account for the multiplicity of social spaces which comprise the rural. What is needed is a typology of rural spaces based on social structures and processes in which actors feature prominently. Marsden *et al.* (1993: 187–9) offer one such typology which can be adapted for global application:

- Preserved rural spaces
  Locations where development pressures are regulated by the state through planning legislation and control agencies, including the designation of spaces to be preserved, such as green belts, national parks, game parks and sites of special scientific interest. Social groups on the basis of their economic or environmental interests commonly contest the mechanisms of control.

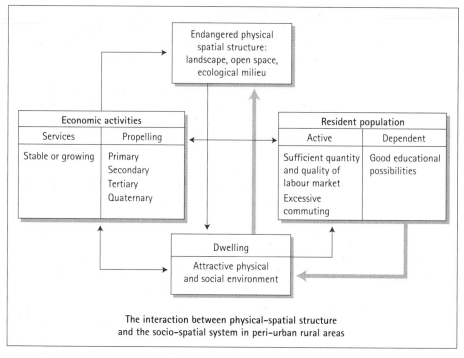

The interaction between physical–spatial structure
and the socio–spatial system in peri–urban rural areas

The interaction between physical–spatial structure
and the socio–spatial system in peripheral rural areas

Figure 5.1   The 'rural' system conceptualized. *Source*: After Veldman, 1984

- Contested rural spaces
  Locations where local economic development priorities are represented and contested between socio-economic groups but in the context of production and consumption dominated by markets rather than regulations.
- Paternalist rural spaces
  Locations where the state or large, privately-owned estates dominate the land-tenure pattern and rural development is shaped by the interests of such landowners. In these localities contestation between groups is based predominantly on social class structure.
- Clientilist rural spaces
  Locations dependent on financial provision by the state for their socio-economic well-being. These are commonly peripheral rural areas of nation states with processes of development dominated by corporatist relations between state agencies and local capital.

This typology of 'ideal types' of rural space implies no locational regularity: rural spaces are shaped differently within and between countries and over time by the forces of urbanization and globalization. Thus each type of space is interpreted as a situated combination of economic (structure of the local economy), social (population characteristics), political (representation and participation) and cultural (property rights and sense of community) processes. Nevertheless, three locational contexts are commonly recognized (Organization for Economic Co-operation and Development 1993), namely: (1) rural spaces within commuting distance of urban centres, (2) rural spaces at the peripheries of national economies, and (3) rural spaces intermediate between the previous two.

5.2 Encroachment of urban on rural spaces

The previous chapter examined the process of urbanization and here a rural perspective is taken on the consequences of the two main forms of urban expansion: development at the edge of existing urban areas and dispersed development better described as 'extended suburbanization'. The second of these forms is considered in the following section; here attention is turned to the transformation of the 'rurban fringe', that is the rural space at the edge of a city (Bryant and Johnson 1992). In this often *preserved* type of rural space, a continuum of state controls on land-use competition can be observed, from direct state planning in former socialist societies, through the development control process in designated green belts around west European cities and the land zoning ordinances around many North American cities (Bunce 1998), to the relatively permissive policies around cities of less developed countries. Whatever the cultural context, rurban fringes comprise a distinct 'rural alternative' both as a development trajectory and for the 'life-style' experiences of their inhabitants. Nevertheless, rurban fringes are heterogeneous: not only can 'inner' and 'outer' fringes be detected, but the dynamism of fringes varies significantly between cities and at different locations around individual cities. Studies by researchers such as Healey *et al.* (1988: 99–127) provide detailed empirical evidence on these generalizations.

5.2.1 Direct impacts of urban encroachment

Urban encroachment has direct and indirect consequences for 'rural alternatives' in rurban fringes. Looking first at direct impacts, three main features can be identified: farmland loss, the land market and rural land use.

Farmland loss

Much of the land in rurban fringes can be classified as 'prime farmland' and its loss to urban development represents an irreversible reduction in the physical resource base for food production. In developed economies there is a long-standing debate over the scale and speed at which such 'foodland' is transferred to non-agricultural uses, leading in some cases to national enquiries, as in the US in the 1970s (National Agricultural Lands Study 1980) and in other cases to national debates, as in the UK in the 1990s (Council for the Protection of Rural England 1993). Land budget estimates of the short-term requirement for 'foodland' in developed economies tend to produce findings of a land 'surplus', although estimates rely heavily on continued gains in agricultural productivity through technological progress. To understand the continuity of state controls for farmland preservation, and the farming life-styles they maintain in the rurban fringe, Bunce (1998) directs our attention to four ideologically-based, lay discourses: (1) farmland as a finite productive resource (productivism), (2) food production in the national interest (progressive agrarianism), (3) farmland preservation as environmental protection (ecological environmentalism), and (4) farming as a connection between nature and society (romantic agrarianism). He concludes that, in the case of the latter two lay discourses, 'the language of farmland preservation articulates ideals for which farmland itself acts as a physical symbol and thus elevates the meaning and significance of agricultural life and landscape above that of a basic productive resource'.

In developing countries, by comparison, the urban fringe is far more dynamic in terms of the transfer of farmland into urban-industrial developments. Around the city of Aligarh in Uttar Pradesh, India, for example, Khan (1997) has reported an increase in the urban area of 149 per cent between 1951 and 1991, leading to the loss of 30 sq. km. of agriculturally productive farmland. This scale of farmland loss has prompted a renewed interest in agricultural production within the rurban fringes of cities in developing countries as regards the maintenance of both farm families and local urban food systems (Gbadegesin 1991).

The land market

With the exception of compulsory purchase or appropriation by the national or local state, the transfer of land from rural to urban uses takes place mainly through the land market. Commonly there is a dual market between agricultural and non-agricultural (development) land in both developed and developing countries. The inflated sale value of land for urban development allows farmland owners a range of reinvestment opportunities either within agriculture at other locations or outside agriculture for alternative types of business investment, or retirement from economic activity. Clearly those with more land to sell are advantaged in relation to those with only small areas of land, or who are tenant farmers or farm workers with no land assets, with consequences for their subsequent life-styles.

Box 5.2   The development of New Bombay, Maharashtra, India

During the early 1970s, the City and Industrial Development Corporation (CIDCO), on behalf of the Government of Maharashtra, purchased land for the development of New Bombay. The land and houses of 33 villages, and the land but not the houses of a further 62 villages (a total of 117,000 people), were acquired at agricultural land prices by voluntary sale or compulsory purchase. At the time, households in the villages were employed by combining agriculture, saltpan work and fishing; only 22 per cent of households had access to industrial employment. Over 40 per cent of landholdings were less than one acre in size, so that the amount of monetary compensation was relatively small compared with the important role of subsistence food production in supporting the households.

By the early 1990s, household surveys showed that the money from compensation and the new employment opportunities provided by CIDCO (e.g. rickshaw-driving, quarries, small shops/stalls, industry) had increased the social and economic differentiation among the villagers to the advantage of village leaders (*sarpanch*) and larger landowners (*patil*). Employment in agriculture had been almost completely eliminated, together with fishing and salt-making; only 33 per cent of households had been able to obtain urban-related employment; and unemployment had increased from 48 to above 60 per cent. As a result, between 1983 and 1992 the proportion of households with a monthly income of less than 80 rupees increased from 9 to 29 per cent, while comparable figures for those with more than 280 rupees a month increased from 5 to 27 per cent. The impact of the loss of fishing was particularly damaging for women in the villages: previously they had been able to obtain an income by retailing dried fish and they had become marginalized in the local labour market. In

Plate 5.1   An urban fringe landscape, India. (Geophotos/Dr Tony Waltham)

(Box continued)

> **Box 5.2** (Box continued)
>
> addition, a majority of the villages had failed to receive promised amenities from CIDCO, such as sec-
> ondary schools, water supply, toilets and approach roads. But the average level of material posses-
> sions of the villagers had increased (e.g. consumer durables such as a TV) and some households had
> used their compensation money to build homes of bricks and cement. As Shaw (1994) concludes:
> 'the gradual outward spread of the city is deadly in its impact on pre-existing agricultural systems,
> social structures and life styles.'
>
> *Source:* Parasuraman 1995

Rural land use

Urbanization has direct consequences for related changes in land use in the rurban fringe, including fragmentation of farmland by transport routes and the provision of sites for waste disposal (e.g. landfill, incineration and the spreading of sewage sludge) and mineral extraction (e.g. construction materials such as sand, gravel and brick clay). The creation and operation of such sites have consequences for local land, air and water pollution and the sites are often contested by social groups in relation to their vested interests in maintaining livelihoods and residential quality of the rural locale. Research has shown how rural communities in countries as disparate as the US (Furuseth 1992) and India (Parasuraman 1995) mobilize to resist land use changes using a range of actions, from participation in the planning process, thorough lobbying decision makers, to direct action which, at the extreme, has resulted in physical violence and the deaths of protesters. Similar processes of resistance have been observed at sites for transport, residential and industrial development where often the class interests of private capital are supported by national development priorities (Healey *et al.* 1988: 103).

5.2.2 Indirect impacts of urban encroachment

Turning now to the indirect rural life-style impacts of urban encroachment, two main elements can be identified: agricultural market opportunities and rural labour markets.

Agricultural market opportunities

There is a well-researched gradient of increasing agricultural intensification with decreasing distance to urban markets, a phenomenon observable in developed and developing countries alike. Contributory factors include reductions in transport costs in purchasing farm inputs and marketing farm products, and access to urban markets for perishable crops and livestock products such as fruit and vegetables and liquid milk. In addition products and services that exploit proximity to urban consumers characterize the rurban fringe; these include farm stalls/shops, direct marketing and farm diversification. The last form of development allows farm families to combine employment in agriculture with non-farm work *on* the farm, including agricultural contracting and the provision of recreational activities, such as horse riding, and farm accommodation (Ilbery *et al.* 1997).

The rural labour market

Urban encroachment also changes the structure of rural labour markets in rurban fringes, especially by raising the opportunity cost of working in agriculture. The highest rates of transfer of farm workers to full-time urban-industrial occupations occur in the rurban fringe; in addition income from agriculture can be combined with employment *off* the farm (termed 'other gainful activity' or OGA), such as work in factories, offices, schools, hospitals and transport. This phenomenon, also termed 'part-time' farming, is observable in the rurban fringes of developed, developing and former socialist countries alike (Cavazzani and Fuller 1982). The categories of farm diversification and OGA, when taken together comprise 'pluriactivity' and this modification of the rural labour market is particularly prevalent in the rurban fringe. However, entry into a pluriactive life-style in developed and developing countries alike, and its various forms, is selective within the rural population to the advantage of individuals who are younger, have more advanced levels of educational attainment and occupy larger farms. In addition, women have a varied experience with pluriactivity: for example, those with prior, non-farm work experience often develop and manage the diversified enterprise on farms thereby enhancing their status in the farm family. Also women with higher educational attainment more frequently follow urban-based managerial and professional employment off the farm, while those with a lower level of educational attainment tend to take up poorly paid manual OGA within their locales.

5.3 Depopulation and counterurbanization

One common historical experience of most rural spaces has been the loss of population through rural-urban migration. The age (younger), gender (higher proportion of males) and educational (more advanced) selectivities of rural depopulation are well known, as are the causal 'push' and 'pull' processes. The former processes include relative or absolute poverty, the restructuring of the farm sector, the decline of resource-based industries (e.g. quarrying and textiles), the paucity of non-farm employment opportunities, falling levels of service provision, civil/military unrest and natural catastrophes (e.g. drought and earthquake). Amongst the 'pull' processes are the growing employment opportunities of urban-industrial locations, the enhanced social opportunities of city life and the rising levels of service provision, including access to food supply. That depopulation acts as a 'vicious circle' in rural spaces is well documented (Cloke 1985). The consequences include higher levels of unemployment relieved by out-migration, an ageing population, rising levels of poverty, erosion of the extended family and multiple deprivation as regards access to services such as health and education (Figure 5.2).

The greatest rates of rural depopulation occur during the transition from an agricultural to an urban-industrial economy. This was the experience of countries such as the United Kingdom and Germany in the eighteenth and nineteenth centuries (see Chapter 2), whereas for many former socialist countries and most developing countries it remains a contemporary process. In the People's Republic of China, for example, rural-urban migration is a matter of state policy as a vehicle for modernizing the economy and society (Shen 1995). In the 'reform' era since 1984, it has been possible for individuals from 'rural' areas to obtain permanent registration in towns and townships provided they engage in industrial and commercial activities. But depopulation can also be observed as

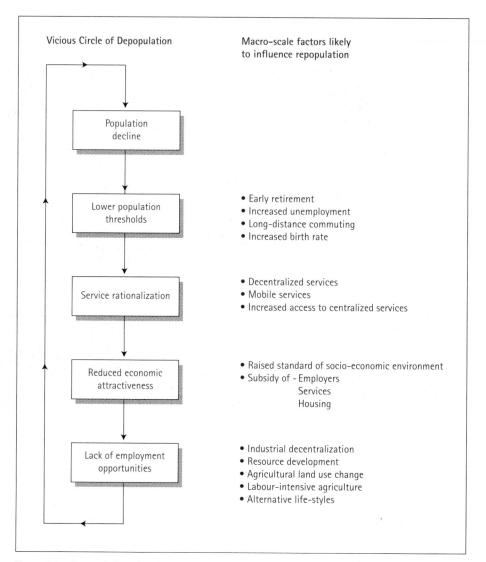

Figure 5.2   Factors influencing depopulation or repopulation. *Source*: After Cloke, 1985

a contemporary process in some rural spaces within most developed countries: for example, in Australia between 1976 and 1981, 46 per cent of small inland settlements in New South Wales lost more than 20 per cent of their population, while 12 per cent of such settlements experienced similar levels of population increase (Sant and Simons 1993). Thus contemporary depopulation is not characteristic of particular types of rural space, although it is most prevalent within developing countries and in *paternalist* and *clientilist* spaces at the peripheries of developed countries.

Plate 5.2   Poverty and depopulation: rural Romania. (Magnum Photos/Jean Gaumy)

5.3.1 Evidence of counterurbanization

Beginning with the USA in the 1970s, evidence has been emerging of population increases in rural spaces in most wealthy, highly urbanized countries – a process now termed 'counterurbanization' (Vining and Kontuly 1978). Resurgence of the rural population has been based on net in-migration rather than natural rates of increase (i.e. the excess of births over deaths) and Figure 5.3 shows this distinction for départements in France. Between 1975 and 1982, Paris and industrialized areas of north-east France lost population through net out-migration, whereas peri-urban areas around Paris and much of rural France, especially in the south, experienced population gains through net in-migration. Some researchers have written of a 'clean break' in rural population trends: Berry (1976), in the context of the US, concluded that 'counterurbanization has replaced urbanization as the dominant force shaping the nation's settlement patterns'. However, the strength and timing in the onset of counterurbanization has varied between countries while, in the 1980s, varying national rates in a widespread downturn of rural population increase were encountered. In the 1990s some countries, such as the US, showed evidence of a return to the historical experience of higher net in-migration rates for metropolitan areas, but in others, such as the UK, counterurbanization continued (Murdoch 1998).

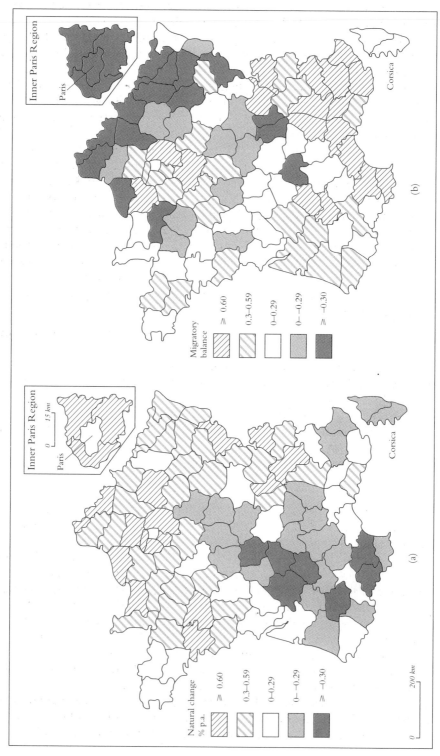

Figure 5.3 France: components of population change, 1975–82, by département: (a) natural change; (b) net migration (per cent per annum). *Source:* After Ogden, 1985

A precise definition of counterurbanization is problematic, not least because of varying national experiences with the phenomenon. Firstly, most researchers exclude from the term those population increases in rural spaces brought about by suburban expansion in the rurban fringe. For developing countries in particular, such expansion is associated with the movement of the rural population to the margins of urban centres. In population census data, the 'statistical underbounding' of cities is reflected in rapid rates of population increase in rurban fringes. Secondly, some but not all researchers exclude 'extended sub-urbanization' (urban decentralization or urban spillover – Robert and Randolph 1983) from a definition of counterurbanization. This term is used to describe the further out-migration of commuters into residential developments within *contested* rural spaces in the peri-urban hinterlands of large metropolises and provincial cities. Much of the population increase in the rural spaces of densely settled countries, such as the UK (Spencer 1995) and the Netherlands (Serow 1991), has been produced by this process: in effect long-distance commuting has extended the daily urban system. Thirdly, a potentially confusing range of terms has been applied to population increases in rural spaces beyond the commuting hinterlands of large urban centres. In addition to counterurbanization, these terms include: rural population turnaround, rural regeneration, population dispersal, core-periphery migration, population reversal, non-metropolitan population growth and rural repopulation. Nevertheless, a consensus can be identified in the published research on the main characteristics of counterurbanization:

- higher population growth rates at progressively lower levels of the settlement hierarchy
- buoyant rates of population growth in peripheral rural areas
- population shifting from urban-industrial areas to locations more favoured in environmental terms
- a variety of causal processes
- the varied experience of rural spaces in the timing and magnitude of population increases.

5.3.2 The 'period' explanation of counterurbanization

A number of competing explanations of counterurbanization have been developed. In a 'period' explanation, counterurbanization is interpreted as an historically contingent, short-term aberration in the long-term trend of urbanization. Summarizing, the 1960s and 1970s comprised an era of rising real incomes, increased personal mobility through car ownership, higher retirement rates, improved long-distance communications (e.g. rail, telephone, telematics) and rural infrastructure (e.g. roads, shops, housing). Taken together, these elements facilitated a relocation of the population into rural spaces during a 'mobility transition'. This explanation, however, is unable to account for the observable downturn in counterurbanization in the 1980s in some but not all countries, nor does it offer an insight into the motives of the population relocating from urban to rural spaces. Even so, Geyer (1996) has generalized this historical experience into a 'differentiated urbanization' model in which counterurbanization forms a late stage in the formation of a mature urban system. But with counterurbanization increasingly interpreted as a cyclical process, Champion (1988) argues for a changing balance through time between processes favouring the concentration or deconcentration of the population.

5.3.3 The economic explanation of counterurbanization

An alternative explanation is offered by writers who interpret counterurbanization as one facet of the periodic restructuring of capital and labour (Sant and Simons 1993). Again summarizing, in the 1960s and 1970s regional sectoral specialization was replaced by a new functional spatial division of labour, including the ruralization of manufacturing and service-based industries (e.g. local government, hospitals, schools, tourism) in many developed countries. In Italy, for example, between 1971 and 1981, 63 per cent of non-agricultural jobs created by private firms were situated in rural spaces (Lowe *et al.* 1998). In other words, as metropolitan areas lost employment in manufacturing and, to a lesser extent, service industries, so medium and small-sized settlements in rural spaces experienced gains in those sectors of employment (see Box 3.5). Population increased not only in the larger rural settlements, where employment growth was concentrated, but also in the surrounding villages as a result of increased commuting within local labour market areas.

For researchers such as Fothergill and Gudgin (1982), an understanding of new start-up or relocating firms in rural areas is to be found in the economic constraints of metropolitan locations. Such constraints are expressed mainly through the increasing space needs of firms, including the physical congestion of existing metropolitan sites, the high cost of land for expansion, and the quality of the work space in terms of access for loading/unloading and staff parking, and the contemporary requirement to locate production within a single-storey building. Fielding (1989), however, prefers a capital restructuring explanation based on the new hierarchical organization of production, where lower-cost, less unionized and militant labour in rural areas carries out routine labour processes (often in branch offices and factories), leaving long-established industrial areas with skill-based labour processes and metropolitan areas with specialized management functions and allied activities such as financial services. But not all rural spaces have been affected equally or at the same time, nor has all new employment been in routine labour processes or financed by external capitals. New communications corridors, as provided by motorways, science parks associated with universities and biotechnology/information technology-based industries and services, and interacting networks of small, locally-financed firms in 'industrial districts' have all contributed to varying employment growth in different *contested* rural spaces.

5.3.4 The socio-environmental explanation of counterurbanization

While new employment has been a major component of counterurbanization, two socio-environmental motivations for urban–rural migration – Geyer (1996) summarizes them as 'environmentalism' – have also been recognized. The first of these is concerned with *retirement*, including the prior establishment of second homes for some migrants and return migration for others who, earlier in their life-cycle, left rural areas for urban employment. For example, Perry *et al.* (1986) found that, in the mid-1980s, 36 per cent and 25 per cent of the inflow of population into Finistère (France) and Cornwall (UK) respectively was return migration. Coastal, mountain and other 'high quality' rural environments are favoured by retirees, for example the 'Gold Coast' of New South Wales in Australia (Walmsley *et al.* 1995). But whereas young adults are most often implicated in rural–urban migration, and families with young children in 'extended suburbanization', retirement migration tends to occur late in the life-cycle after children have left the parental home.

Plate 5.3 Polders repopulation, Netherlands. (Geophotos/Dr Tony Waltham)

The second socio-environmental motivation in counterurbanization is access to a *rural life-style*, including the desire of members of the 'service class' (professional and managerial occupations) to acquire the material or 'positional' good of a rural residence. The search for a 'rural idyll' is also expressed in social terms: being part of the social life of a rural community. These are significant motivations in counterurbanization, as found amongst households in 'extended suburbanization' (Spencer 1995), the founders of new firms in rural areas (Keeble 1993), and the relatively small proportion of households which abandon urban-industrial employment and life-styles for high environmental amenity living, often in *clientelist* rural spaces. This last phenomenon, for example, has been reported by Jones *et al.* (1986) in relation to long-distance migrants from south-east England to northern Scotland.

5.3.5 Localized experiences of counterurbanization

Counterurbanization has contributed to the uneven development trajectories of rural spaces by providing a variety of localized experiences for population net in-migration. There is insufficient space to investigate the bases of these experiences in detail and a listing of the factors identified by Cloke (1985) must suffice:

- land market (price and ownership)
- physical environment (landscape amenity)
- settlement quality (heritage, conservation)
- housing market (stock and price)
- employment market (opportunities)
- social and community structure (services and prestige)
- accessibility (from urban centres).

As shown by Perry *et al.* (1986) for Finistère, for example, the operation of these factors produces a range of localized experiences of counterurbanization: varied combinations of employment (tourism, food processing, ferry services), retiree and alternative life-style migrants interact with situated characteristics such as the land and housing markets. To this list should be added the planning system which, as experienced in Finistère in Quimper, Brest and Morlaix, and shown by Spencer (1995) for lowland England, can distribute new employment and housing selectively between settlements planned for growth.

Individuals also experience a range of alternative life-style changes under counter-urbanization. On the one hand, the operation of the housing market is socially selective to the advantage of those with high disposable incomes. Consequently 'service class' in-migrants gain from access to low density housing in smaller settlements set within rural environments. Moreover the 'newcomers' defend their new life-styles against further economic and residential development (termed 'nimbyism' – not in my backyard) and obtain further prestige and status by taking over the organization of local social institutions such as youth groups, interest societies and local government. On the other hand, 'newcomers' commonly exclude other social groups (defined by their lower incomes, ethnicity or family structure – Lowe *et al.* 1995) from local housing, as well as from professional and managerial posts in newly created employment and from organizational positions in social institutions. Counterurbanization, therefore, redefines the meaning of 'rural community' and brings about new forms of social exclusion for individuals in rural society, including many of the pre-existing inhabitants of rural spaces.

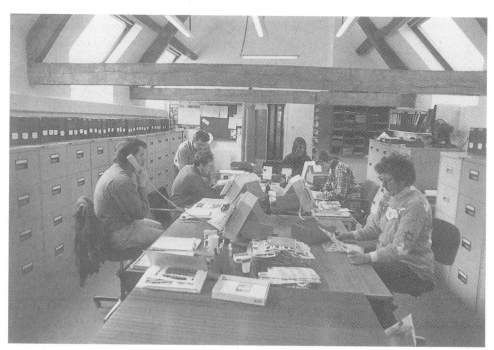

Plate 5.4 New small rural business: an example of the conversion of farm barns into business premises. (Popperfoto)

5.4 Service provision and rural development

In both developed and developing countries, access to services and employment form important but interrelated components of life-style but rural spaces and social groups have a varied experience of such access. In services, the main difficulties lie in the incidence of low incomes as a constraint on access and the distances separating locations of provision from a dispersed and historically diminishing population. In employment, most traditional rural industries, such as agriculture and natural resource extraction, yield relatively low incomes and have been in decline through national and global competition, without significant replacement until recent years. Consequently the state tends to intervene through the planned provision of services and programmes of rural development. In developing countries attention tends to be focused on services such as drinking water, health care (including family planning), education and employment; in developed countries, by comparison, more concern is expressed about transport, retailing, housing and care for the elderly. Nevertheless, individuals and social groups, particularly at the peripheries of national economies, experience life-styles that vary as regards access to services and employment.

5.4.1 The meaning of 'services' in a rural context

The term 'services' encompasses the social goods that comprise the fabric of non-production life, ranging from educational and health, through transport and utilities (e.g. water, electricity), to retail, religious, legal and financial (e.g. banks, post offices) services. Depending on the national or regional context, some services are provided entirely by public agencies, some entirely by the private sector in competitive markets, some through self-help organizations, and others through varied combinations of the three previous modes of delivery. While the private sector is driven by commercial imperatives, the public provision of services is underpinned by considerations of need, social justice, allocation and distribution. At issue are social and political judgments on (Humphreys 1988):

- the 'need' for various services, such as residential care for the elderly;
- what is 'fair' and 'just' in society, for example as regards access to public transport;
- the amount and quality of social (merit) goods to be provided, for example drinking water;
- who will get how much of scarce resources and at what cost, for instance health care.

5.4.2 Locational access to services

Three main concepts underpin 'locational accessibility' to services, namely threshold, service area and order (Honey 1983). Threshold relates to the population size required to maintain the economic viability or economic efficiency of a service. Characteristically, the per capita cost of providing a service declines with the increasing number of people served (i.e. economies of scale – Powe and Whitby 1994), but begins to rise beyond an optimum size. Service area is configured as the division of space between service centres so as to equalize accessibility by the population to a particular type or quality of service. Order concerns the cost/value or quality of a service, with the implication that lower order services are used more frequently than higher order services. For example, the (low order) surgery of a general practitioner can be compared successively with the (higher order) facilities of a 'cottage' hospital and metropolitan hospital. The three concepts interact in the provision of rural services: for instance, service areas vary in size according to population density and the order of the service, with provision commonly structured within a hierarchy of 'central places'.

Under rural depopulation, increasingly large service areas are required to reach threshold levels that maintain economies of scale, with a progressive reduction in the number

---

Box 5.3   Health care provision in Bangladesh

Health care is provided in a hierarchical structure of centres distributed throughout the country (Table 5.1). By the mid-1990s 57 District Hospitals had been built, together with 351 rural Upazila Health Complexes (UHC), 4,500 Union Health and Family Welfare Centres (which contain a surgery), with support from visiting Health Assistants (UHC staff) and Family Welfare Assistants in village-based Health Posts.

Even so, with continuing high birth and infant mortality rates, and with life expectancy hovering around 52 years, problems remain. These include the 0.6 per cent of GDP allocated to health care by governments, the low health status ascribed to women, a shortage of qualified staff, the allocation of funds to buildings and salaries rather than medicines and equipment, corruption at the local level in the purchase and supply of medicines, and the preoccupation of salaried doctors with private patients at the expense of the public sector.

Table 5.1   The health care delivery system in Bangladesh

Level	Facilities	Services
Tertiary	Medical College Hospital/ District Hospital	(i) Referral hospital for the secondary level for specialized medical care (ii) Education, research and professional development in medical science
Secondary	Upazila Health Complex (UHC)	(i) Referral hospital for primary level (ii) Support service to primary health care
Primary	Union Health & Family Welfare Centre	(i) Treatment of common and minor diseases (ii) Prevention of contagious diseases (iii) Maternal and child health care
Domiciliary	Visiting staff attached to the UHC	(i) Basic health education (ii) Immunization (iii) Family planning (iv) Oral rehydration (v) Control of contagious diseases (vi) Maternal and child health care
Community	Health Post	(i) Motivational work (ii) Preventive and curative services

Source: Siddiquee (1996).

of service outlets through either formal planning by public agencies or competition in the market. Most rural settlements show evidence of this rationalization process, perhaps as a redundant church, shuttered bank or disused shelter at a stopping place for a now withdrawn bus service. Allocation-location models can be used to calculate the most economically efficient spatial distribution of fixed resources for access by a changing population (Askew 1983), including the required adjustments to the existing locational pattern of service provision. Oppong and Hodgson (1994) demonstrate such a model for health care facilities in the Suhum District of Ghana, taking into account resources available from the state, private medicine, charities and foreign aid. The optimum spatial pattern for access involves closing and relocating 10 of the 30 funded health centres.

It might be expected that counterurbanization would create a 'virtuous circle' of population growth and so reverse the decline of rural services (Figure 5.2). However, the threshold sizes and service areas of most services have tended to increase through time under the economic imperative of reducing per capita delivery costs (e.g. through the privatization or reduced funding of public services), while the new mobile population involved in 'extended suburbanization' has tended to obtain services in the urban centres of their daily commuting with damaging consequences for services such as the village shop. Indeed a general feature of recent years has been the increased centralization of services into urban centres to the disadvantage of communities in peripheral rural spaces.

5.4.3 Effective access to services

The normative assumptions on equality of access used in allocation–location models have to be suspended when the actual consumption of services is investigated. Income acts as the main constraint on access to most services and a succession of surveys has revealed a higher level of relative and absolute poverty in rural as compared with urban communities (e.g. Tickamyer and Duncan 1990). Within England, for example, Cloke (1996) estimated the incidence of relative rural poverty in a range from 47 per cent of households in Devon to 35 per cent in Northamptonshire. Within rural communities, the lowest incomes can be traced to four main life-style contexts, namely low-paid occupations (e.g. in agriculture, tourism, catering and self-employed small businesses), unemployment, old age and poor health/physical or mental disability. In each context, individuals and families experience *'multiple deprivation'* in terms of inequitable access to a range of services, such as housing, health care and education but also, critically, transport (McLaughlin 1986). Disadvantage in paying for the cost of public (bus, train) and private (car) transport limits personal spatial mobility and reinforces the difficulty of access to both services and employment. But such disadvantage can also be experienced at the community level. Porter (1997), for example, reveals how isolation and inequality have increased in 'off-road' communities on the Jos Plateau of Nigeria – that is, villages without connection to the upgraded road system. With 'deprivation' now a contested term within the discourse on service provision, increasing attention is being paid to *'social exclusion'* (Shucksmith *et al.* 1994). This perspective emphasizes how the exercise of systematic economic and political power in society excludes some individuals from life-styles open to the majority. Such a perspective also brings 'neglected rural others' into focus (Philo 1992), such as single parents, ethnic minorities, New Age travellers, rural women, unqualified youth, the elderly and the disabled.

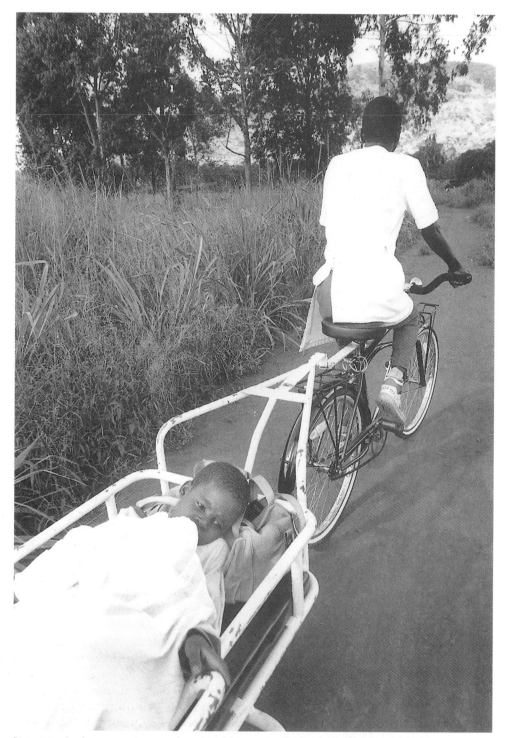

Plate 5.5   Getting access to services in developing countries – transfer to a health clinic. (Panos Pictures/ Penny Tweedie)

Table 5.2   Percentage rates of general practitioner (GP) consultations and outpatient attendances (Norwich, England)

Group	GP consultations		Outpatients	
	Accessible rural	Remote rural	Accessible rural	Remote rural
Age: 18–44	22*	9*	16*	5*
45–64	23	11	23	11
65+	21	27	17	13
Sex: male	18	13	18*	4*
female	25*	12*	18	14
Car: owning	22*	18	18*	8*
not owning	26	20	18	13
Social: manual	22	12	19*	6*
not manual	21	12	18	12

* difference significant at the 0.01 level

Source: Joseph and Phillips (1984).

Social exclusion is aspatial and not confined to any particular rural context. Even so, Table 5.2 shows how, using health care provision as an example, the longer travel distances in 'remote' rural areas limit effective access to services compared with the shorter travel distances in more 'accessible' areas. Here the significance of low income for access to personal transport is a defining issue, especially where public transport provision is itself problematic. Space-time budgets reveal how individuals, such as rural women with families, attempt to compensate for reduced personal mobility by combining access to a range of services during a single journey away from the home. Typically a journey taking children to/from a centralized, urban school, by either public or private transport, is combined within a short time period in a sequence of visits to services such as a shop, doctor and bank which can by-pass local places of service provision. Despite such coping strategies, and given the social exclusion experienced by those without personal transport, the state tends to intervene by either (1) increasing personal mobility by subsidizing public transport or self-help organizations (Banister and Norton 1988), (2) subsidizing services at locations from which they would otherwise be withdrawn on economic grounds, (3) concentrating the population into a limited number of larger settlements (e.g. key villages or growth centres) where services can be maintained at economic cost, or (4) facilitating the mobility of the services themselves (e.g. mobile shop, mobile health service) (Organization for Economic Co-operation and Development 1991).

5.4.4 Planned rural development

Huillet (1991) has observed that 'the appearance or disappearance … [of services] … is the best indicator of a region's economic health'. Unsurprisingly, therefore, state intervention on service provision often occurs within the context of programmes of rural development that include the diversification of employment opportunities. Such programmes are decreasingly sectoral (e.g. just rural electrification) but increasingly based on integrated development (e.g. jobs, housing, social facilities and transport taken together) with objectives beyond economic growth. While development programmes are commonly constructed for defined (particularly *clientelist*) rural spaces – for instance, the Objective 1 (lagging behind) and 5b (problem) rural regions of the EU – Shucksmith *et al.* (1994: 213) have observed incisively that 'people not areas suffer rural disadvantage: … area-based approaches are more convenient because … they do not challenge the fundamental processes which structure disadvantage.'

Two modes of rural development are commonly identified, namely 'top-down' and 'bottom-up'. The former tends to be driven by statutory agencies placed by the state within a hierarchy of responsibility from the national, through the regional to the local. Development tends to be based on exogenous capital and decision making, for example state investment in modernizing rural infrastructure, such as electricity supply, and foreign inward investment by urban-industrial corporations seeking low-cost labour for the establishment of agricultural production for export or branch factories. This form of development has been experienced in developing and developed countries alike and is prone to instability through changes in state funding or global demand for a factory product during times of economic recession.

The failure of 'top down' rural development programmes (Schramm 1993), particularly as official aid programmes to developing countries, has been influential in shaping alternative approaches. 'Bottom-up' development, for example, is community-driven and based on endogenous development processes in which local people participate (Lowe *et al.* 1998). Such development can be interpreted, on the one hand, as the resistance of marginalized communities against the pressures for assimilation into wider social and economic structures or, on the other hand, as the rural expression of 'local economic development' that is usually associated with cities. Common characteristics of this form of development are the maximization of locally available resources, including human capital through training programmes, the retention of added value through production and processing in locales, and local networking to share experiences and collaboration through joint projects which can gain economies of scale and access to wider non-local markets. The development of local entrepreneurship, the production of high-value products or services for niche markets, and the regional association of a product (such as furniture) or a service (such as rural tourism) further characterize this form of rural development.

However, this top-down/bottom-up dichotomy is largely illusory for in practice the two modes of development tend to be combined: local and non-local actors and

Box 5.4   Local rural development in Kenya

In Kenya since 1983, Sub-Locational Development Committees (SLDCs) have operated within the government's District Focus of Rural Development Strategy. Administrative Districts have autonomy in setting their development priorities, while various development committees are responsible for the initial identification, prioritization and design of projects at District, Divisional, Locational and SLDC levels. Central government appraises and finances projects through a combination of funding from governmental and donor organizational sources. Membership of each SLDC comprises the assistant chief, the local head teacher, local civil servants and representatives of village residents chosen at public meetings. In Siaya District, for example, 16 SLDCs operate with 17 to 72 members, each holding 3 to 12 meetings a year with attendance rates varying from 50 to 90 per cent.

Source: Schilder and Beove 1996

institutions are linked together in complex networks of collaboration. Garmise *et al.* (1996) go so far as to claim that, taken together, networked institutions comprise a new form of economic governance. The 1991 'Liaisons Entre Actions de Développement de l'Economie Rurale' (LEADER) programme of the European Union provides a good example of network-based rural development (Barke and Newton 1997). Over 800 selected Local Action Groups (LAGs) in their 'territories' bring together local representatives of the community, entrepreneurs, service providers and government officers to work with officials from the Directorate General VI (Agriculture and Rural Development) of the European Commission. Together they produce and implement, with varying degrees of success, innovative, low-cost local development plans, partly financed from the Structural Funds of the EU. The LAGs have a fragmented distribution within the EU but are sensitive to the needs of local communities. Programmes such as LEADER also provide a structure for experimenting with new forms of rural development, for example through the opportunities provided to small businesses by telematics for overcoming some of the locational disadvantages of rural spaces (Ilbery *et al.* 1995). Nevertheless, network-based development still creates uneven benefits through the conflicting interests and power of actors seeking to (1) protect the rural environment, (2) pursue development of rural resources, and (3) support the rural community. Social exclusion persists and requires 'local capacity building' to strengthen knowledge, skills and attitudes of all members of the community for their participation in the development process.

5.5 Conclusion

This chapter has considered the theme of how rural spaces experience differing development trajectories as a result of their dynamic relationship with the processes of urbanization and economic development. Trajectories vary both between and within countries: experience of employment, access to services and quality of life vary between, for example, the *protected* rural spaces of rurban fringes, the *contested* spaces of extended suburbanization, and *clientelist* spaces characterized by programmes of integrated rural development. The varied life-style experiences of social groups and individuals within rural spaces can be traced partly to constraints imposed by socio-economic structures, for instance as expressed through income levels, and partly to the characteristics of the types of rural space in which they are located. In this conclusion two brief observations are made on the challenge of the new millennium as regards these themes.

The first observation concerns the changing functions of rural space. While many rural spaces will continue to serve a *production* function in economy and society, for example in developing countries in terms of producing food, minerals, timber and water, it is evident that other rural spaces, for instance in developed countries, are taking on enhanced *consumption* functions as regards residence, recreation, tourism and environmental conservation (Lowe *et al.* 1993). Estimates on the future use of rural land in developed countries predict a 'surplus' of farmland, with a reduced role for food production and its associated employment (Netherlands Scientific Council for Government Policy 1992). Conversely enhanced roles for urban land, environmental conservation and recreation are anticipated, together with their associated employment. The previously anticipated expansion of forest and woodland on to 'surplus' rural land now seems unlikely, except in marginalized, peripheral rural spaces. Periodic surges in urban expansion and counterurbanization can also be anticipated, including in developing and former socialist countries, thus ensuring a continuing variety of development trajectories in which centrality and marginality in relation to urban centres are likely to remain important. Variations in such factors as land ownership and the skills base of the population, social trends as regards long-distance commuting from urban centres, and the investment behaviour of different capitals will add localized experiences to the broad trend towards more differentiated rural spaces. A challenge of understanding for the new millennium concerns the localized experiences of Philo's (1992) 'neglected rural others'.

The second observation concerns the implications for rural spaces of the increasing emphasis being placed on *sustainable* development. Gaining political prominence through the 1992 Rio de Janeiro 'Earth Summit' and subsequent Agenda 21 national development programmes, sustainable development has environmental, social and economic dimensions, albeit to a large degree chaotic and contested. Sustainable development offers both constraints on and opportunities for life-style experiences of social groups in rural spaces. For example, sustainable development can validate intervention by the state in maintaining sustainable rural communities, especially in marginalized spaces (Bryden 1994); it has the potential to shape urban development processes at the rurban fringe to the disadvantage of extended suburbanization. Sustainable development contextualizes the emerging struggle between central and local political and economic power in rural governance as evidenced in the tension between local, endogenous, community-driven rural development and its assimilation into the wider framework of capitalism.

Learning outcomes

Having read this chapter, you should understand:

- The problems of identifying and classifying rural spaces throughout the world.
- The varying development trajectories of different types of rural space.
- The varying life-style experiences of social groups and individuals within rural spaces.
- The impact of urbanization on all types of rural spaces through processes such as counterurbanization and rural development.
- The significance of the changing functions of rural spaces and sustainable development for human geography in rural spaces in the new millennium.

Further reading

Bernstein, H., Crow, B. and **Johnson, H.** (1992) *Rural Livelihoods: Crises and Responses*, Oxford University Press, Oxford. An Open University text that concentrates on the human experience of poverty, making a living, health and development in rural spaces within developing countries.

Ilbery, B. (ed.) (1998) *The Geography of Rural Change*, Longman, Harlow. A collection of essays by different authors on a wide range of contemporary changes in rural areas (land use, economy and society) but limited to the context of developed countries.

Marsden, T., Lowe, P. and **Whatmore, S.** (eds) (1990)   *Rural Restructuring: Global Processes and Local Responses*, Wiley, London. A collection of research-oriented essays by a variety of authors drawn from different disciplines which examine the theoretical basis of global restructuring processes but also the varying experiences of local spaces. Most of the discussion is for developed countries.

Simpson, E.S. (1994)   *The Developing World: An Introduction*, 2nd edition, Longman, Harlow. A book divided into two contrasting parts: the first part considers the theoretical basis of the causes and consequences of the socio-economic problems of developing countries; the second part takes a country-by-country approach in examining the varied experience of both the problems and attempts to resolve them.

CHAPTER **6**

CHERYL MCEWAN

Geography, culture and global change

Culture is a word on everybody's lips these days. Hardly a moment seems to pass when we do not hear on the radio or television, see in newspapers and magazines, or read in academic texts some account of the world of the cultural. Governments at all level announce cultural policies and provide funding for cultural activities; intellectuals announce the need for cultural initiatives or bemoan the loss of traditional cultural values; famous cultural icons – musicians, artists, novelists – themselves culture makers, are increasingly sought out for their opinions on the state of the world.

(Kahn 1995: ix)

Topics covered

■ A definition of culture

■ The 'cultural' and 'spatial turns' in the social sciences

■ An evaluation of the extent of cultural globalization

■ The impacts of globalization on local cultures

■ An exploration of the concepts of multiculturalism and hybrid cultures

■ Progressive ways for geographers to think about culture

6.1 What is culture?

This chapter explores some of the challenges posed by and for culture in the new millennium. However, before considering these challenges, it is first necessary to define what is meant by culture. This itself poses a challenge, since this is an extremely complex and difficult task. By the 1950s, for example, there were over 150 different definitions of culture in circulation among academics. As Mike Crang (1998: 1) argues, it sometimes sounds the most airy of concepts but, however defined, 'can only be approached as embedded in real-life situations, in temporally and spatially specific ways'. Students of geography often imagine studying cultures to be either about looking at exotic peoples in other places, or about high culture, such the arts. However, the work of cultural geographers in recent years has demonstrated that culture is broader-ranging than these narrow definitions. Cultures are part of everyday life, and are **systems of shared meanings** that can exist on a number of different spatial scales (local, regional, national, global, among communities, groups, or nations). They are embodied in the material and social world, and are dynamic rather than static. It is important to remember that cultures are socially determined and defined and, therefore, not divorced from power relations. Dominant groups in society attempt to impose their definitions of culture and these are challenged by other groups, or **sub-cultures**. The latter might include various types of youth culture (e.g. rave culture and the dance club scene), gang cultures, and different ethnicities or sexualities, where identities are organized around different sets of practices and operate in different spaces from dominant cultures (Crang 1998; Skelton and Valentine 1998). Culture makes the world meaningful and significant. As Philip Crang (1997: 5) argues, we should think of culture as a *process* we are all involved in rather than as a *thing* we all possess (see Box 6.1).

6.1.1 The 'cultural turn'

As Kahn suggests, the subject of culture has generated a great deal of interest in recent years, for academics, policy-makers, and at the popular level. This can be explained by the

---

Box 6.1   Cultures: a summary

- The systems of shared meanings which people who belong to the same community, group or nation use to help them interpret and make sense of the world, and to reproduce themselves
- These systems of meanings include language, religion, custom and tradition, ideas about 'place'
- Cultures cannot be fixed, but shift and change historically
- Culture is a process, rather than a thing, but is embodied in the material and social world
- Cultures include those social practices which produce meaning, as well as those practices which are shaped by those shared meanings
- Cultures give us a sense of 'who we are', 'where we belong' – a sense of our own identity and identity with others
- Cultures are, therefore, one of the principal means by which identities are constructed
- Cultures are not divorced from power relations

Source: adapted from Hall 1995

phrase 'the cultural turn' – put very simply, the world has become more 'cultural'. The social sciences have witnessed a 'cultural turn' since the late 1980s. Geographers have turned their attention towards cultural explanations of global, national and local phenomena. They have explored issues such as the cultural embeddedness of economic processes (e.g. McDowell 1997), the relationship between identity and consumption (e.g. Jackson 1993), and more sophisticated understandings of the construction of social relations of gender, ethnicity and class (e.g. Radcliffe and Westwood 1996). However, the current popularity of culture is not simply a trend in academe, but is reflective of a broader cultural turn in (Western) society as a whole.

The world has changed fundamentally in the past two decades; the modern student of geography lives in a world very different from that of their parents at the same age, and these changes are deeply cultural in character. For example, huge economic changes have occurred in Britain since the early 1980s (the decline in manufacturing, the growth of services, the feminization of the workforce, increased flexibility – all characteristic of 'post-Fordism' as discussed in Chapters 3 and 12). However, as Hall (1996: 233) argues, if 'post-Fordism' exists, then it is as much a description of cultural as of economic change. For example, modern consumption depends overwhelmingly on image (for example, the marketing of food and drink products and fashion clothing). The intensifying movement around the world of images and symbols and modes of thought and communications are unparalleled at the end of the millennium. As the age of computer technology, video imagery and electronic music demonstrates, the material world of commodities and technologies is profoundly cultural. In addition, culture has become increasingly commodified; in other words culture is being translated into material goods that can be marketed and sold. Zukin (1995: 263) refers to the example of the annual Gay Games in New York to highlight what culture has come to mean. The games are now more than an athletic competition and a political event. In the publicity surrounding the games, homosexuality is presented as if it were an ethnicity with its own traditions and roots, and as a life-style, with its own forms of entertainment and consumption patterns. The Gay Games have drawn the support of large corporate sponsors eager to capture the pink dollars of a supposedly affluent gay population. The mainstream press covers the games as a cultural, rather than either a sporting or a political event. As Zukin argues, culture now has great appeal – *culture sells*.

6.1.2 The 'spatial turn'

If both the world we live in, and the way that geographers (and other academics) interpret it, have become more cultural, it is also important to note that there has also been a 'spatial turn' in explanations in cultural and social theory. It is commonly accepted that the world is changing fast, and the rate of this change is probably greater than ever before. New technologies such as the Internet and satellite communications mean that the world is becoming more global and more interconnected. The increased speed of transport and communications, the increasing intersections between economies and cultures, the growth of international migration and the power of global financial markets are among the factors that have changed everyday lives in recent decades. There is no historical equivalent of the global reach and volume of 'cultural traffic' (Held *et al.* 1999) through contemporary telecommunications, broadcasting and transport technologies. The challenge for geographers and other social scientists is to find ways of understanding and interpreting these changes.

Plate 6.1   Image marketing is central to popular consumption: the world of commodities is profoundly cultural (DKNY billboard – Trip/C. Trotman; Ron Jon Surf Shop – Trip/J. Livermore; Pizza Hut, Benetton – Trip/H. Rogers)

Culture can be said to operate at three spatial scales: local, national and global. Two main interpretations have dominated discussion. The first highlights the *global* aspects of change. At its simplest, this approach suggests that it is possible to identify processes of cultural homogenization – the idea that everywhere is becoming the same – dominated by the USA and most easily recognized in terms such as 'Coca-Colaization', 'McDonaldization' and 'Hollywoodization' (Cochrane 1995: 250) (see also Chapter 14). This *cultural globalization* involves the movement of people, objects and images around the world, through telecommunications, language, the media industries, radio and music, cinema, television and tourism. The second interpretation places emphasis on the *local* and the localization of people's everyday lives and experiences. Instead of homogenization, stress is placed on the diversity of culture, on the ways in which global icons such as Coca-Cola or McDonalds are reinterpreted locally so that they take on different meanings in different places (this idea is discussed in greater detail below). Social scientists are responding to these changes by exploring the *interconnectedness* between global and local processes. For example, although the same event can be witnessed simultaneously around the world (e.g. an incident broadcast in a CNN news report, or an international sporting event), this event will be interpreted differently in different places. Furthermore, locality does not necessarily refer to the opposite of globality. For example, some environmentalists imagine the world as a locality, a 'global village'. As McDowell (1994b: 147) argues, cultural theorists have a growing interest in how increasing globalization, especially of cultural production and consumption, affects people's sense of identity and place at both local and national levels. Thus a geographic or spatial perspective has become central to studies of culture more widely. Some of the challenges faced by those studying contemporary cultures include:

- finding ways of understanding the cultural impacts of globalization
- rethinking how both the global and the local affect cultures
- finding productive, progressive and non-exclusionary ways of thinking about culture in the contemporary world
- re-conceptualizing cultural identities in increasingly global times.

These are some of the concerns that form the focus of this chapter. Subsequent sections explore in more detail ideas about a global culture, examine ways of rethinking local cultures, and explore progressive way of thinking about cultures in the contemporary context.

In summary, culture can be defined as a process and a system of shared meanings. The world has become more 'cultural' in recent years, and there has been a 'spatial turn' in explanations of culture, making geography increasingly significant. The global and the local, and the interconnectedness of the two, are important in studying cultures.

6.2   Towards a global culture?

6.2.1 Imagining a global culture

Processes of cultural globalization have a very long history and are not peculiar to the beginning of the new millennium. Through global patterns of trade and migrations, and through the spread of world religions and empires, people, objects and ideas have been circulating for centuries (see Chapters 1–3). However, contemporary globalization is distinctive in extent, form, rapidity of change, intensity and impact. Today, the *idea* of a global culture is in the process of becoming as meaningful as the idea of national or local cultures. This is a clear break with the past, and is clearly influenced by debates about globalization, especially the cultural impacts of economic globalization. Proponents of the idea of an emerging global culture suggest that different places and cultural practices around the world are converging and becoming ever similar. As Shurmer-Smith and Hannam (1994: 76) argue, a global culture might be the product of two very different processes:

- The export of supposedly 'superior' cultural traits and products from advanced countries, and their worldwide adoption ('Westernization', 'Americanization', 'modernization');
- The mixing, or hybridization of cultures through greater interconnections and time-space compression (the shrinking of the world through transport links and technological innovation), leading to a new universal cultural practice.

The first is a process where Western time-frames (for example, Greenwich mean time, the twenty-four hour day and the Gregorian calendar) become global, and where space and distance are bridged by connections (travel, trade, colonization, markets, capital, the flow of goods, labour and profits). It is a view of cultural globalization that sees Western culture being globalized alongside capitalism. As economic activity has globalized, so this has been accompanied by a wave of cultural transformation. The manufacture of universal products (e.g. the motor car or television) and the development of global media empires (e.g. satellite TV) are believed to create global cultural convergence – people around the world are becoming increasingly similar in terms of consumption, life-style, behaviour and aspirations (see Box 6.2). This process is inseparable from historical processes of exploration, imperialism, colonialism, the creation of a 'world market', the subsequent generation of forced and free migrations, and the creation of powerful spheres of cultural influence by Western countries. It can be perceived positively (as 'modernization' or 'development') or negatively (as 'cultural imperialism', where 'we' assume that others in the world should aspire to be like 'us').

Box 6.2

The globalization of culture: some examples

1 Music

Music lends itself to globalization because it is one of the few popular modes of cultural expression that is not dependent on written or spoken language for its primary impact (Held *et al.* 1999: 352). The relative ease of cultural diffusion has led to the spread of many genres and major artists all over the world. Madonna, Michael Jackson and the Spice Girls are associated with global products, images and messages. The production, distribution and consumption of these products have a particular geography. The global music industry is dominated by transnational corporations, with the US and UK dominating domestically generated popular music. 'World music' is now a significant part of the marketing strategies of these corporations, and exposes global audiences to local musical traditions from around the world. Migrations of people have also had cultural impacts on music, with increasingly 'hybridized' forms (see Box 6.5).

2 Television

Until recently, television programmes tended to be produced primarily for domestic audiences within national boundaries, and could be subjected to rigorous governmental control. However, with the advent of cable, satellite and digital technologies, in addition to political and legal deregulation in many Western and developing states, several television channels are now globally disseminated, and can to some extent circumvent national restrictions. The USA, France, Germany and the UK are the major exporters of television programmes, but Brazil, Mexico, Egypt, Hong Kong, Spain and Australia are increasing their output (Held *et al.* 1999: 360).

3 Sport

Sports are forms of cultural expression that are becoming increasingly globalized, as well as increasingly commodified. Football/soccer is the most obvious example. The sport has become increasingly globalized in the past few years. The World Cup was held in the US in 1994 in an effort to expand

(Box continued)

Box 6.2   (Box continued)

its global popularity; similar motives underlie the decision to hold the 2002 World Cup in Japan/South Korea. The English Premier League is increasingly internationalized, with some clubs (like Chelsea) fielding teams made up almost exclusively of overseas players. Manchester United (plc) is a globally recognized icon, is now a global company, and opts to play in world club competitions at the expense of prestigious domestic cup competitions. Similar trends can be observed in US Major League Baseball. The New York Yankees are a global icon; many major league players hail from countries such as Cuba, the Dominican Republic, Puerto Rico and Costa Rica; the sport is becoming increasingly globalized through television coverage and its inclusion as an Olympic sport. Although football/soccer is increasingly globalized, it has different meanings around the world. In the UK, football was, until recently, a working-class sport. It remains an overwhelmingly male sport, despite the increasing numbers of women playing the game. The same can be said for Latin America, where very few women play the game because of social constructions of gender, which determine that it is 'unfeminine' for women to play football. In the USA, soccer is a sport played predominantly by white, middle-class people. The recent success of the US women's team in the Women's World Cup is likely to have implications there for the popularity of women's soccer. In South Africa, football is an overwhelmingly black sport, unlike cricket and rugby, which remain overwhelmingly white, despite the collapse of apartheid. The World Cup-winning French side of 1998 – made up primarily of second-generation immigrants from the African continent – was momentarily a symbol of national unity in a former imperial power struggling to come to terms with its multiculturalism. Therefore, football/soccer is globalized, but it is also fundamentally localized.

4 Tourism

Tourism is one of the most obvious forms of globalization. Once again, the geography of tourism is skewed, since it is dominated by people of all classes from developed countries (North America, Western Europe and, increasingly, Japan and Australia). It can also be exploitative, particularly through the growth of international sex tourism and the dependency of some developing economies on the exploitation of women. However, it is a form of international cultural exchange that allows vast numbers of people to experience other cultures and places. It also locks specific places (tourist destinations) into wider international cultural patterns (Urry 1990).

The second process assumes that globalization and the mobility of peoples around the world is producing a mixing of cultures, as opposed to unidirectional 'Westernization', to create a global commonality. This process allows us to consider how Western cultures have and are being greatly influenced by this mixing of cultures. For example, former colonized cultures have influenced the cultures of the former imperial powers, as well as vice versa. The diversity of restaurants in any European city, or the 'ethnic' influences on European street fashions are testimony to this. Ideas of a global culture centred on the developed world are distorted when we begin to consider the great waves of migration and the impact of these on global culture. Flows of music, food, ideas, beliefs and literature continue to percolate from around the world into the cultures of the West (the current popularity of Feng Shui is but one recognizable example).

In reality, both these processes are flawed explanations for what is happening today. If a global culture exists, it is far from a product of unidirectional 'Westernization'.

Plate 6.2   Tourists in Amsterdam: European cities are increasingly sites of cultural diversity. (Topham Picturepoint)

However, alternative ideas about cultures mixing to produce a universal global culture are also problematic. Cultures are mixing, but this does not necessarily mean we are all becoming the same.

6.2.2 Debunking global culture?

A different departure point for discussing global culture is that there is no such thing. Ideas about a global economy, a global politics, and a global culture imply some sort of world-wide commonality which simply does not exist. While there clearly are global processes at work, the idea of a global culture is more complex and problematical. Firstly, as discussed above, ideas about 'Westernization' or 'Americanization' are flawed. The image of rampant cultural imperialism by the West, and especially the US, is problematic since apparent cultural sameness is rather limited in scope, located only in the consumption of products and media images. The possibilities of this eroding centuries of local histories, languages, traditions and religions are perhaps rather far-fetched. Besides, we have a very limited knowledge of how people in different parts of the world respond to these images and products, and they are always translated locally. Many millions of people around the world do not have access even to a television. (In 1992, there were only 38 TV sets per 1,000 people in Africa, and 73 per 1,000 people in Asia, compared with 381 per 1,000 in Europe, and 800 per 1,000 in the US (UNESCO *Statistical Yearbook* 1994).) There is no single global culture because of the unevenness of globalization.

Secondly, some theorists would argue that national cultures remain stronger than global cultures. This is borne out when we consider the many conflicts occurring throughout the world along the geopolitical fault-lines of national cultures (the recent conflicts between Pakistan and India are but one example of this). The late eighteenth century saw the emergence of nation-states, national cultures and national cultural institutions, which displaced older forms of cultural globalization (such as mercantilism and trade). For the past 200 years, nation-states and national cultures have monopolized cultural power (state television is but one example of a national institution influencing national cultures within national territorial boundaries). At the end of the twentieth century, this balance has begun to change, with international telecommunications and media corporations challenging the centrality and importance of national cultures. However, it could be argued that despite these changes a great deal of cultural life is still organized along national and territorial lines.

Thirdly, if culture is a system of shared meanings, then looking at the world today we must acknowledge that there are many systems of shared meanings, and many different cultures. People in different places use different techniques and technologies to reproduce culture, such as oral histories, literature or television and cinema. These techniques have different patterns of dispersion, penetration and scale. Therefore, some cultures are more able to become globalized than others – those reproduced through television and cinema have a greater range and speed of dispersion than those reproduced through oral histories. However, this does not mean that globalized cultures completely erode localized cultures; the ways in which these different cultures intersect is important. Global 'cultural signifiers' – McDonalds, the Hollywood movie, Coca-Cola, Nike, MTV, Marlboro, Ford – might suggest that globalization equals Westernization. However, this vision falls down when Sony, Nissan, Toyota, Mitsubishi, Honda, Suzuki are added to this list. Japanese consumer goods do not sell on the back of exporting Japanese culture. If the term 'Japanization' means anything it is in terms of a market strategy based around the concept of *dochaku* ('glocalism'). This involves a global strategy not of imposing standardized products but of tailoring Japanese consumer products to specific local markets. These goods are, therefore, both globalized and localized. This is now a popular strategy for multinationals in other parts of the world. Therefore, how intersections between cultures are played out at a local level are of significance (this is discussed in more detail below), and this suggests that imagining a universal global culture is quite problematic.

6.2.3 Rethinking global culture

What does this mean for thinking about global culture? On the one hand, it makes speaking about a single global culture a relative impossibility; on the other, it opens up possibilities of speaking about culture in a fashion that recognizes its interrelationships with broader political, social and economic factors, and acknowledges that there is no single global culture. Those 'things' (products, symbols, corporate entities) that have become global signifiers are clearly globalized (they are recognized the world over), but the ways in which people around the world make cultural responses to them are complex and multiple. It might be more useful to think about a global culture in the restricted sense of what Featherstone (1995: 114) calls 'third cultures' (Box 6.3). Instead of imagining a

---

Box 6.3   **Cyberfeminism: an example of 'third culture'**

Transnational, or 'third' cultural networks, are not only generated by multinational corporations and business interests. Third cultures also include international flows of knowledge. The women's movement, the peace movement and the environmental movement are all global movements, and are cultural and intellectual networks of communication, discussion and political activity. They operate within many nation-states, but are transnational in terms of their organization, reach and activities. Because they are networks that operate beyond the nation-state, global movements can be both powerful and effective in bringing about change. Increasingly, these movements make use of new technologies, such as electronic communications, to sustain and expand their networks. Cyberspace is becoming increasingly important in building new institutions and campaigning for democracy and human rights.

The international feminist movement is one example of a potentially powerful third culture, which has taken up new information and communications tools to support global networking. The fourth UN World Conference on Women held in Beijing in September 1995 was a catalyst for women's electronic activism, particularly in setting an agenda for a Global Communications Network for women (Harcourt 1999). There are obvious problems that women in poorer countries might be further marginalized by these new technologies, since they do not share the same access. However, by linking women's practical experience with an increasing need to influence national and global policy, the new communications provide a means through which women across the world can improve and enhance their attempts to bring about a more gender-equitable global culture.

One example of this exists in South Africa. It has been said that sexism was one aspect of apartheid that was not racialized – women of all ethnicities were discriminated against in various ways, although black women were doubly oppressed. Today, even though apartheid has ended and the constitution guarantees that all women should be free of discrimination, (black) women still constitute the majority of the poor and violence against women is widespread (it is estimated that one woman is raped in South Africa every 20 seconds). Here, the Internet (through sites such as womensnet.org.za) is being used to encourage women to participate in constructing a stable democracy and in designing and implementing models of economic development. It allows ordinary women in civil society to organize and keep pressure on the government to uphold the guarantees of equality pledged in the constitution (McEwan, forthcoming). Cyberspace is an important site for the development of communications, a sense of trust and solidarity among women in South Africa who have different life experiences according to their ethnicity, social class, culture and geographical location. It is also an important site in linking women in South Africa to other women's groups in Africa, and to broader international feminisms. It empowers women in civil society, allowing them to claim a voice in both national and international policy frameworks. The Internet, therefore, can be used to empower women to engage in global and local debates, and to inform and change global cultures (Harcourt 1999). [Examples of other web-sites include FEMNET (Kenya), Famafrique (for francophone African women), Isis (Asia-Pacific), ALAI (Ecuador), Fempress (Chile), AWID (North America), DAW and Womenwatch (UN) (see Walker 1999).]

global culture that is erasing local and national cultures, we can think of local, national and global as three important spatial levels at which culture operates. Those aspects of culture that operate at the global level are 'third cultures'. As we have seen, the beginning of the new millennium is marked by the fact that national institutions are no longer in complete control of cultural globalization. 'Third cultures' are sets of practices, bodies of knowledge, conventions and life-styles that have developed in ways which have become increasingly independent of nation-states, and which are global. Phenomena such as patterns of consumption, technological diffusion and media empires are part of these third cultures, and transnational and multinational corporations are the institutions that make them global. In this sense, global cultures exist but only as third cultures, outside national and local cultures, yet intersecting on both these scales in different ways around the world. Rather than imagining a single global culture, we should instead think about the different spatial scales at which culture operates and the intersections between these various scales (e.g. between the global and the local). Acknowledging that 'our' global view might be very different from that of people elsewhere, living in very different contexts, is also important. (Indeed, the very notion of the millennium is culturally specific, and meaningless to millions of people around the world whose cultures are not centred on Christianity and Western time-scales.) It is clear that multiple global cultural networks exist, such as those connecting the overseas Chinese with their homeland, or those linking Islamic groups around the world. These networks disrupt any notion of a singular global culture.

Power is a very significant issue in discussing global cultures. As Massey (1993) argues, a **power geometry** exists, which gives people with different access to power different notions of what global means. Culture is not simply a spontaneous event, but depends upon such factors as tradition, heritage and history. These are often produced through institutions, such as state governments. As discussed above, new institutions for the production, transmission and reception of cultural products have emerged recently, including global media and communications industries. These institutions are creating infrastructures supporting cultural globalization, including electronic infrastructure (radio, television, music, telecommunications), linguistic infrastructure (the spread of bi- and multi-lingualism, particularly the dominance of English), and corporate infrastructure (producers and distribution networks). These new institutions often operate at scales beyond the nation-state, and they are sites of power in the production of culture. The ownership, control and use of these institutions remain uneven across and within countries (Held *et al.* 1999: 370), thus creating a 'power geometry' that is centred very much on the West. People have very different experiences of culture because of their different locations in the world and their relationship to these sites of power. The challenge for geographers is to map this power geometry, to identify sites of power, and to reveal the marginalization of some peoples around the world by cultural globalization. We might also suggest ways in which marginalized peoples might be empowered by engaging with, and perhaps transforming, the new institutions driving cultural globalization (see Box 6.3 for an example of this).

Plate 6.3   BSkyB: part of a global media empire and agent of cultural globalization. (Sky B)

In summary, the *idea* of a global culture is becoming increasingly important. Global processes are occurring, but they do not produce a universal global culture, nor are they distributed evenly around the world. Global cultural processes are not simply a result of a unidirectional 'Westernization', since culture flows transnationally. There is no singular global culture, but a number of different global cultures. Global cultures exist as 'third cultures' – in patterns of consumption, flows of knowledge, diffusion of technologies and media empires that operate beyond, but connecting with, the local and national scales.

6.3 Reinventing local cultures?

6.3.1 Locality and culture

In the past it has often been assumed that there is a simple relationship between local place and local culture. Places were thought of as having a distinct physical, economic and cultural character. Places were, therefore, unique, with their own traditions, local cultures, and so on that made them different from other places. It is clear, however, that processes of globalization are also posing serious challenges to the meaning of place. Places and cultures are being restructured. According to Massey and Jess (1995: 1), 'on the one hand, previous coherences are being disrupted, old notions of the local place are being interrupted by new connections with a world beyond'. The appearance of 7-Eleven stores in the rainforests of northern Thailand is one example of how even the remotest of places are becoming increasingly internationalized, in this case through tourism. 'On the other hand', Massey and Jess continue, 'new claims to the – usually exclusive – character of places, and who belongs there, are being made.' A dramatic example of this is the revival of place-based ethnic identities in the former Yugoslavia, and the conflicts that have resulted. Therefore, modern life is characterized both by decentralization and globalization of culture and by the resurgence of place-bound traditions. Following this, the impact of the new global context on local cultures has two, possibly contradictory, outcomes.

6.3.2 Negative sense of culture

Where global processes are perceived to pose a threat to local culture there might be an attempt to return to some notion of the exclusivity of culture. At the extreme, this might take the form of exclusivist nationalism or even 'ethnic cleansing'. Reactions to the perceived threat to local cultures include nationalistic, ethnic and fundamentalist responses, which also entail a strong assertion of local cultures, such as reviving or inventing local traditions and ceremonies. These can create a level of local fragmentation, with a parochial, nostalgic, inward-looking sense of local attachment and cultural identity. In this sense, cultures are thought of as **bounded**, with very clear definitions of 'insiders' and 'outsiders' in the creation of a sense of belonging. As Sibley (1995) argues, this mode of thinking about cultures produces **geographies of exclusion**. An example can sometimes be found in the English countryside. Rural areas are often conceptualized as the preserve of English culture and identity. This idea is mobilized through the myth of the 'rural idyll', where the English countryside is thought of as timeless, unchanging and unaffected by global processes. This cosy vision of a peaceful countryside excludes many people who live in rural areas but do not fit this stereotype. Gypsies, travellers, environmental pro-testers, hunt saboteurs, people with alternative life-styles and people from minority ethnic communities are deemed to be 'outsiders' and a threat to local cultures (Cloke and Little 1997). As Halfacree (1996) argues, the 'rural idyll' is a selective representation. It is exclusive in its class, race and status connotations, is profoundly conservative and demands conformity. It is based on a very inward-looking sense of place and culture.

Conservative reactions to change can be thought of as a kind of cultural fundamental-ism through which the process of cultural change is often bitterly contested. Gender plays an important role in this. As Yuval-Davis (1997) points out, women are often considered as guardians of the borders of culture. Women not only bear children for the collectivity, but also reproduce it culturally. In closed cultures, the control of women's sexuality is seen as imperative to the maintenance of the purity of the cultural unit. It is important that women do not marry outside their cultural and ethnic group. Ethnicity and culture, therefore, are seen to be one and the same. In addition, symbols of gender play an import-ant role in articulating difference between cultural groups. Women's distinctive ways of dressing and behaving very often come to symbolize the group's cultural identity and its boundaries. Women are often also the ones chosen to be the intergenerational transmit-ters of cultural traditions, customs, songs, cuisine and the 'mother' tongue, primarily through their role as mothers. This is especially true in minority situations where the school and the public sphere present different and dominant cultural models to that of the home.

Societies do not evolve smoothly from these closed, bounded perceptions of culture to more open, dynamic notions. The question of cultural power, identity and resistance also needs to be considered. For many groups cultural survival has always been deemed to depend on a closed idea of culture, with strongly marked boundaries separating it from 'others'. An example of this can be seen in the continuing sectarianism in Northern Ireland, where two different cultures, religions and national identities collide with each other. Another example of this is found in contemporary South Africa. Here, the tran-sition from an apartheid state where racialized cultures were very rigorously bounded and demarcated, to a post-apartheid state where an acceptance of cultural pluralism (the so-called 'Rainbow Nation') is key to the survival of the nation-state, has not run smoothly. On the one hand, the new government has attempted to create a non-exclusive, pluri-cultural society, with eleven official languages and constitutional guarantees for the pro-tection of cultural difference. Television programmes are made and broadcast in all major languages, and the new national anthem combines verses in English, Afrikaans and Xhosa. On the other hand, some groups (e.g. some Afrikaaners, some Zulus) still claim a separ-ate, culturally and ethnically defined homeland, and see this as the only way that their cul-tures can survive. Cultural fundamentalism is still a powerful force in post-apartheid South Africa. Elsewhere the mixing of cultures under the impact of globalization is often seen as threatening and as weakening the sense of cultural identity. This has created a revival of ethnicity that cuts across the political spectrum. Examples include the 'little England' reaction to European monetary union (characterized by a strong attachment to the sym-bolic importance of the pound), the rise of neo-fascism in Europe (characterized by anti-immigration and racism), and the strengthening of Islamic and Christian fundamentalism around the world. These phenomena are not all the same, but they do share a response to globalization that involves a closed, fixed, bounded and place-specific definition of cul-ture, and a strong resistance to changes heralded by cultural globalization.

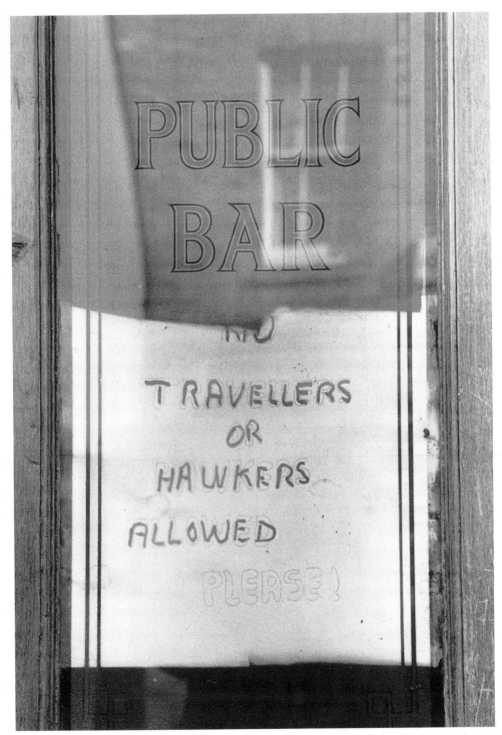

Plate 6.4 Exclusion in the countryside: travellers and other minority groups are often excluded from shops and public houses, marking their position as 'outsiders'. (Format Photographers/Joanne O'Brien)

6.3.3 Positive sense of culture

A more positive response to global processes would be to imagine cultures as fluid, ever-changing, unbounded, overlapping and outward-looking – akin to Massey's (1994: 151) 'progressive sense of place'. This involves people being more cosmopolitan (free of prejudice and tolerant of difference). Increasing *interconnectedness* means the boundaries of local cultures are seen to be more permeable, susceptible to change, and difficult to maintain than in the past. Rather than everywhere becoming the same, nation-states seem to be reconstituting their collective identities along pluralistic and multicultural lines, which take into account regional and ethnic differences and diversity. In Europe, this involves re-creation and invention of local, regional, and sub-state or new 'national' cultures (for example, the cultural renaissance of the Scots, Basques and Catalans, or the cultural assertiveness of minority ethnic communities in cities such as London, Paris and Berlin). Thus, what can be perceived as destruction of local cultures by globalization might in fact be the means of the creating new senses of locality. This still involves notions of local identity, but recognizes both the differences between cultures and their interconnectedness, taking account of the positive aspects of cultural mixing and increased cosmopolitanism. This new sense of local identity is based on notions of inclusion rather than exclusion.

This is not to say, of course, that all people within the same place will share the same culture and the same sense of locality. Within these more culturally pluralistic and cosmopolitan locales, different class fractions, ages, genders, ethnicities and regional groupings mingle together in the same sites, consuming the same television programmes and products, but in highly uneven ways. These groups often possess different senses of affiliation to places and localities, possess different cultural identities and belong to different cultural groupings (Featherstone 1995: 97). A progressive sense of culture does not foresee the locale as a 'melting-pot', where everything becomes the same, but rather recognizes the different experiences of people, and that increasing interconnections might create new, dynamic and exciting cultural forms (see below). An understanding of this is crucial to the creation of a progressive notion of place and culture, which demands the recognition of cultures as fluid and open and interconnected, and an acceptance that those older local cultures might decline as new local cultures emerge.

In summary, localities are important in maintaining cultural difference, but can also be sites of cultural mixing and transformation. Ideas about culture can be negative (bounded, fixed, inward-looking) or positive (progressive, dynamic, outward-looking). Bounded, fixed notions of culture can lead to localized resistance, racism, nationalism, and even 'ethnic cleansing'. Progressive ideas about culture involve the recognition of differences between cultures, the interconnectedness of cultures and their constant evolution.

6.4   Multi and hybrid cultures?

6.4.1 Hybridity

One of the major contemporary challenges concerns what we do with the concept of culture in the changing global scene, where nation-states in the West have been forced to tolerate greater diversity within their boundaries. Some people want to see national identity as homogeneous and assimilatory – in other words different cultures are subsumed into

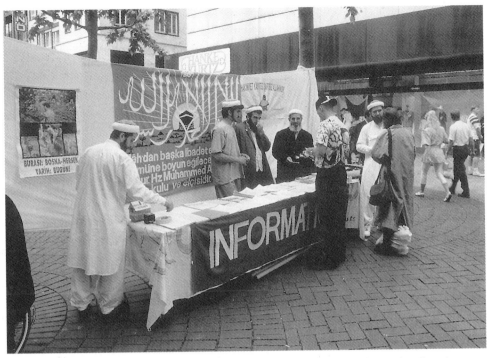

Plate 6.5 Turkish people in Germany have increased their cultural assertiveness through a variety of means, including a radio station that broadcasts to Turkish communities around Europe. (Trip/B. Turner)

the dominant culture (the 'melting-pot' idea). Others call for the acceptance of ethnic pluralism and the preservation of minority ethnic cultures as a legitimate part of the national project. This is the politics of **multiculturalism**, which instead of thinking of different cultures as being absorbed by dominant cultures, relies more on a notion of a cultural mosaic, or a 'patchwork quilt' of cultures. Each culture is recognized as different and distinct, but these differences are understood and valued. Multiculturalism might seem progressive, but it can actually 'essentialize' and reinforce difference. In other words, culture is seen as essentially connected to 'race', where racial difference is rooted in biological difference. (These ideas are no longer considered acceptable; anti-racists have demonstrated that 'race' is socially constructed and has little basis in biology (see Kobayashi and Peake 1994). We could just as easily have 'races' of blue-eyed and brown-eyed people.) Multiculturalism, therefore, still relies on a negative notion of bounded cultures. It might suggest tolerance, but often results in segregation and ghettoization. The same argument is relevant when discussing attitudes towards homosexuality in the West. The appearance of 'gay spaces' in cities such as San Francisco, London or Manchester might appear to signal greater tolerance and cultural diversity, but it points to segregation rather than integration. Indeed, in the US politically-correct accommodation agencies now ask applicants to tick a box for 'gay' or 'straight' so that they can be housed suitably. As Reeves (1999) argues, 'Homophobia, overt and covert, is herding lesbians and gay men into ghettos – which are then celebrated as signs of a tolerant society.' It is OK to be gay in these segregated spaces; it is still not OK to be openly gay in other spaces of the city.

One way in which a more progressive idea of culture might be developed is through the concept of **hybridity** (mixing). Hybridity is a powerful new way of thinking about the manifestations of culture such as ethnicity, gender and sexuality. Hybridity breaks down barriers. It adheres to neither the 'melting-pot' nor the 'mosaic' idea of cultural mixing, but rather sees different cultures coming together and informing each other in different ways to produce something *entirely new*. This process is not peculiar to the end of the twentieth century, but has a long historical trajectory. Indeed, some would argue that cultures have always been hybrid forms – they have never existed in isolation from other cultures, and thus have always been subject to change and influences from elsewhere (Werbner 1997, 15).

One of the most obvious places we can observe hybridity is in popular mass culture, an immediate example being popular music. Innovations in music have always involved the fusion of different styles to create new sounds and rhythms. Rock-and-roll, rhythm-and-blues and Latin jazz are obvious examples; we might also think of contemporary forms of music that fuse rap and hip hop with other styles (e.g. UK bands Massive Attack and Portishead, or US band The Fugees). Authors such as Barthes (1972), Bourdieu (1984) and Bakhtin (1984) see popular hybridity as an exciting challenge to, or subversion of, dominant cultures and the exclusive life-styles of dominant elites. Such popular mixings and inversions, like the subversive elements of youth cultures (see Hebdige 1979; Skelton and Valentine 1998), are hybrid in the sense that they bring together and mix languages and practices from different and normally separated domains. They have the potential to disrupt dominant cultures by their 'out-of-placeness'.

In many ways hybridity is related to the notion of **transculturation**. Transculturation describes one of the key cultural processes that operate between hitherto sharply differentiated cultures and peoples who are forced (usually by the processes of imperialism or globalization, and primarily through migration) to interact. This interaction often takes place in profoundly asymmetrical ways in terms of relative power between different groups. Despite the illusion of boundedness, cultures evolve historically through borrowings, appropriations, exchanges and inventions. Cross-fertilization of cultures is endemic to all movements of people throughout history (see Box 6.4). However, for those who aspire to bounded notions of culture and refuse this idea of perpetual hybridity, cultural mixing is felt to be threatening and a deliberate challenge to social order. In reality there are no fixed cultures in modern nation-states, but some people cling to ideas of pure or impure cultures. For others, however, hybridity remains the site of revitalization, progression and fun (see Box 6.5).

6.4.2 Diaspora

Related to the idea of hybridity is the notion of **diaspora**. This term was originally used to refer to the dispersal of Jewish peoples. However, it is now used in reference to the long-term settlement of peoples in 'foreign' places that follows their scattering or dispersal from their original homeland. It refers to a modern condition where a sense of belonging is not derived from attachment to territory, and where different peoples mix together through the processes of migration (forced or free). European imperialism and associated processes of globalization have set many of these migrations in motion. Diasporas are classic **contact zones** (spaces in which two cultures come together and influence each other) where transculturation or hybridization takes place. Diasporic

Box 6.4	Englishness: bounded or hybrid culture?

Nationalists around the world cling to a notion of bounded cultures that make them distinct from others. One effect of globalization has been resistance in the form of increased nationalism to what is perceived to be the erosion of national cultures. Increased mixing of cultures is seen to pose a threat to the survival of national cultures. Nationalists seek to preserve the symbols of nationhood, such as language, life-styles and cultural forms in the face of what are perceived to be sweeping changes. However, cultures are not unchanging, and a fundamental flaw in nationalist ideology is the adherence to a notion of static culture that has been free from change, and its reliance on a mythical history of the origin of the nation. We can use the example of English nationalism to demonstrate this point. English nationalists define Englishness as distinct, which is used to justify anti-immigrationist ideas, anti-Europeanism and, in some cases, racism. But who are 'the English'? After the last ice-age many communities settled in the British Isles from all over Europe. They lived and fought with each other and in a short space of time produced a mixed group of people who eventually chose to call themselves English. The British Isles have been subject to waves of invasion and settlement (e.g. Celts, Vikings, Romans, Anglo-Saxons, Normans). The point is that England and the British Isles have always been hybrid; peoples and cultures have mixed and evolved together. Some English nationalists avoid thinking about this point by arguing that the final invasion (by William the Conqueror in 1066) marks the origin of England and Englishness (Anderson 1982). This myth is also flawed. William the Conqueror spoke no English. Whom did he conquer? He conquered 'the English'. For many nationalists the founding father of England is French! It is also no small irony that one of the symbols of Englishness, the monarchy, changed its official name to Windsor in 1917 from Saxe-Coburg-Gotha, thus hiding its German origins. Even protectionist policies towards language are flawed; as with culture more generally, language is always hybridized and evolving. Most sentences in 'English' contain words that derive from German, French, Latin, Nordic and Celtic languages; increasingly US words have also entered the vocabulary (hi, OK). Similar myths of origin and notions of bounded cultures exist elsewhere in the world. In some places they emanate from deeply contradictory ideas (anti-immigrationist views in the USA and Australia, for example); in other places they have led to conflict (recent events in Bosnia and Kosovo, for example). The idea of nationhood, based on fixed, bounded and unchanging cultures is an ideological creation that masks profound cultural divisions of gender, race, class and region within a nation-state, and ignores the fact that, in reality, all cultures are hybrid and dynamic.

identities are important because they are at once local and global, and are based on transnational identifications encompassing both 'imagined' and 'encountered' communities (e.g. Irish-Americans belong to an imagined international community of people who have 'Irishness' in common, but whose identities are also informed by the communities in which they live in the US). In other words, diasporas are a direct challenge to the idea that there is a simple relationship between place and culture. They transgress the boundaries of the nation-state and provide alternative resources for constructing identity and fashioning culture. Therefore, the concept of diaspora can be useful in rethinking culture.

Brah (1996: 208) refers to the many processes of cultural fissure and fusion that underwrite contemporary forms of transcultural identity. They are never fixed and

Box 6.5	British Bhangra: an example of hybridity

British Bhangra developed out of the more traditional forms of folk dances and music of the people of the Punjab. Traditional Bhangra arrived in Britain in the late 1960s with the migration of people from South Asia and East Africa. British Bhangra emerged in the mid-1980s among second-generation South Asian-British musicians, primarily in Birmingham and west London. This latter style combined the robust and energetic punctuated rhythms of double-sided drums (*dhol*), modern technology and urban sounds with Punjabi poetry, traditional Punjabi lyrics, acrobatic dance skills and celebrations of the Punjabi harvest festival (*bhaisakhi*). It played an important role in the proclaiming of a cultural identity for certain groups of youths defining themselves as South Asian-British. Although it appeals to the Punjab as a specific region of origin, Bhangra, with its fusion of instruments and musical styles from South Asia and urban Britain, has a large following among diverse groups of South Asian-British youths. Throughout the 1990s, Bhangra has been characterized by ongoing processes of fusion with broader ranges of musical styles (ragga, reggae, soul, jazzfunk, rock, hip-hop, pop), both to reflect the lived experiences of South Asian-Britons in the 1990s, and as an attempt to 'cross-over' into mainstream charts. Despite huge sales of relatively cheaply priced music, however, Bhangra remains marginalized within the broader music industry, being virtually ignored (or stereotyped) by the mainstream music press and handicapped by poor retail and distribution.

Bhangra can be understood in many different ways in many different contexts. Some commentators differentiate between the politicized lyrics of late 1980s Bhangra, which highlighted the racial and social inequalities of British society, and the more celebratory styles of the 1990s. Although the Bhangra scene is organized primarily by young men (through car entertainment systems and urban cruising, daytime discos, live band gigs and wedding celebrations), there are increasing numbers of popular women Bhangra artists. Bhangra also has a broad appeal to both men and women on dance floors at wedding parties and other social gatherings. Bhangra exists in many different contexts (daytime discos, live gigs, familial settings of weddings and parties) and is, therefore, consumed in many different ways. However, in each of these contexts is the underlying element of a transformative South Asian music taking shape in the light of a British experience. Bhangra, then, is quintessentially a British phenomenon, as British as the Royal Family. It is also the antecedent to newer and more fashionable forms of post-Bhangra and 'Asian Kool' which have been less marginalized by mainstream culture, and are part of an emergent and vibrant British-Asian urban youth culture. Bhangra is an example of those popular, hybrid cultural formations seen as critical to the social, cultural and political realizations of immigrant communities struggling to locate and establish themselves as legitimate participants in Britain and elsewhere.

Source: adapted from Dudrah 1999

homogeneous, but are always in a state of flux. The concept of **diaspora** space allows us to think of 'culture as a site of travel' (Clifford 1992), which seriously problematizes the idea of a person being a 'native' or an 'insider'. Diaspora space is the point at which boundaries of inclusion and exclusion, of belonging and otherness, of 'us' and 'them', are contested. As Brah (1996: 209) argues, diaspora space as a conceptual category is 'inhabited', not only by those who have migrated and their descendants, but equally by those who are constructed and represented as indigenous or 'native'. As Brah argues:

The similarities and differences across the different axes of differentiation – class, racism, gender, sexuality, and so on – articulate and disarticulate in the diaspora space, marking as well as being marked by the complex web of power. In the diaspora space called 'England', for example, African-Caribbean, Irish, Asian, Jewish and other diasporas intersect among themselves as well as with the entity constructed as 'Englishness', thoroughly reinscribing it in the process.

Like notions of hybridity, the concept of diaspora is important since it allows for the recognition of new political and cultural formations that continually challenge the marginalizing impulses of dominant cultures.

6.4.3 Selling hybridity and the commodification of culture

In today's world, culture sells. Hybrid cultures, in particular, sell. Cities are now constructing themselves as cosmopolitan, and hybridity has become a form of 'boosterism' – where city authorities create marketing images to attract investment in the form of business and tourism. As the example of Birmingham (UK) (Box 6.6) suggests, hybrid culture is perceived as creating economic advantage. With increasing deindustrialization in Europe and North America, cultural strategies have become key to the survival of cities. Examples include the international marketing of the Mardi Gras in Sydney, or the importance of 'Chinatowns' and other 'ethnic' districts to tourism in cities throughout the developed world. Ironically, hybridity is in danger of becoming just another marketable commodity. For example, treating the political work of some British-Asian bands as marketable hybridity trivializes black political activity and leaves problems of class exploitation and racial oppression unresolved. As Hutnyk (1997: 134) suggests, to focus on hybridity while ignoring (or as an excuse for ignoring) the conditions in which this phenomenon exists (the commodity system, global economic inequality, inequitable political relations), is problematic in that it maintains the status quo. Hybridity and difference sell, but in the meantime the market remains intact, power relations remain unequal, and marginalized peoples remain marginalized. Moreover, as culture is subsumed into capitalism, those marginalized peoples who might be capable of oppositional politics are also subsumed under the rubric of hybridity.

The notion of hybridity, therefore, can be problematic. In some Latin American countries, cultural elites and the nation-state have appropriated the hybrid *mestizo* (mixed) identity, making it dominant. This has been seen to be oppressive of 'Indian' populations, who have in turn been accused of ethnic essentialism (or emphasizing their racial difference) because of their desire to protect their cultures (Radcliffe and Westwood 1996). In the West, ideas of hybridity are currently popular with highly educated cultural elites, but ideas about culture, ethnicity and identity that develop in poverty-stricken underclass neighbourhoods are likely to be of a different nature (Friedman 1997: 83–4). Evidence of racial tensions in many North American and European cities points to the fact that class and local ghetto identities tend to prevail, with little room for the mixing pleaded for by cultural elites. The global, cultural hybrid, elite sphere is occupied by individuals who share a very different kind of experience of the world, connected to international politics, academia, the media and the arts. In the meantime, the world becomes more polarized in terms of wealth, and heads towards increasing balkanization where regional, national and ethnic identities are perceived as bounded, threatened, and in need of protection. As Bhabha (1994) reminds

Box 6.6

Diaspora space: the example of Birmingham

In a globalizing world, cities are in competition with each other to make their mark as the next global city. Birmingham, the UK's second city, has long harboured aspirations of becoming a global city and, in a bid to rid itself of the images associated with the industrial decline of the 1980s, embarked on the largest city centre redevelopment plan in Britain. This comprises the National Exhibition Centre, the International Convention Centre, the National Indoor Arena, Symphony Hall, and the corporate redevelopment of its city-centre canals. The flagship projects, whilst of great civic pride to many of Birmingham's residents, are also extremely divisive. They have created an international enclave that is increasingly divorced from its regional and local context. They say nothing about Birmingham's distinctiveness as a truly multicultural city (21 per cent of its population in 1991 comprised minority ethnic groups).

In an attempt to rectify this failure to recognize Birmingham as diaspora space, Birmingham City Council announced the Winterval'98 promotion of the city's lights, entertainments, events and exhibitions for the festive season and, of course, the array of high-quality shopping on parade in the city centre. In a deliberate policy to broaden the period of winter festivities, the Council sought to highlight that Christmas and the (Gregorian) New Year were merely examples of a number of celebrations over this period. (Other festivals include Hannukah (Judaism) and Ramadan (Islam), but the promotional period curiously managed to avoid both Diwali (Hindu) and the Chinese New Year.) This can be considered ruthless marketing, and even cultural appropriation, by the City Council of events that take place annually in Birmingham, irrespective of 'Winterval'. In response, the national headlines shouted 'Birmingham cancels Christmas'. However, in some respects the negative responses that this received were unfortunate given that the council's previous economic strategies have lacked any recognition of the cultural and ethnic diversity of its residential population.

'Winterval' was a flawed but valiant attempt to reconceptualize multiculturalism in a positive light, which might also bring economic benefits. The fact is that Birmingham does not have to strive to become global – it already is. It is famous for the Birmingham Balti, a hybrid dish of immense popularity. Many hybrid versions of Islam have been created in the city. Transnational capital flows into the city, symbolized by the construction of the Pagoda at the gateway to the Chinese Quarter (a gift to the city from a prominent Chinese businessman). Birmingham is less a 'melting pot' than a site of hybridity and diaspora. The city's flagship boosterism has failed, as yet, to construct Birmingham as a place of both international recognition *and* a city centre as a living, breathing heart of a vibrant multicultural city, but the signs are that a more progressive version of Birmingham as a place is starting to emerge.

Source: adapted from Henry 1999

us, hybridity seems an insufficient basis on which to consolidate new forms of collectivity that can overcome the embeddedness of prior antagonisms. Hybridity sounds nice in theory, but does not necessarily exist outside the realms of the privileged.

As Hall (S. Hall 1996: 233) argues, however, we should not view the current fashionability of hybridity in a wholly negative light. The celebration of multiculturalism and hybridity might indeed have a hidden agenda that is all about the pursuit of profit. However, as cultures are increasingly commodified, we should not forget the potential for the democratization of culture in this process, the increased recognition of difference and

Plate 6.6 Chinatown in San Francisco: 'ethnic' districts are promoted as major tourist attractions in many 'global' cities. (J. Allan Cash)

the diversification of the social worlds in which women and men now operate. This pluralization of social and cultural life expands the identities available to ordinary people (at least in the industrialized world) in their everyday working, social, familial and sexual lives. As Hall (1996: 234) argues, 'these opportunities need to be more, not less, widely available across the globe, and in ways not limited by private appropriation'. For Bhabha (1994: 9), it is the interconnections of different cultural spaces and the overlapping of different cultural forms that create vitality and hold out the possibility of a progressive notion of culture.

A challenge for geographers, then, is to think about the place and meaning of cultural hybridity in the context of growing global uncertainty, xenophobia (fear of foreigners) and racism. We need to think about why it is that cultural hybridity is still experienced as an empowering, dangerous or transformative force. Why is it that on the one hand difference is celebrated through a consumer market that offers a seemingly endless choice of identities, sub-cultures and styles, yet on the other hand hybridity continues to threaten and shock? Conversely, why do borders, boundaries and 'pure' identities remain important, producing defensive and exclusionary actions and attitudes, and why are the latter so difficult to transcend? It could be that the sheer pace of change in cultural globalization is producing these reactions.

To summarize, hybridity and diaspora are examples of more progressive ways of thinking about culture. It could be argued, in fact, that all cultures are always already hybrid; they are never pure, have always evolved and changed through time and through contact with other cultures, and they continue to evolve. In today's world, hybridity is being commodified, which might make it less radical. However, despite the problems with the notion of hybridity, it has the potential to democratize culture, and to allow us to rethink culture in ways that are more tolerant of difference. Finally, geographers need to think about the meaning of cultural hybridity in the context of growing global uncertainty, xenophobia and racism.

6.5 Conclusion

In this chapter it is suggested that there are two apparently contradictory tendencies in thinking about cultures – the attempt to secure the purity of a culture by conceptualizing it as strong, fixed, bounded, permanent and homogeneous, and the hybridity of most cultures. We should therefore think of culture as a *contested concept*. A progressive way of thinking about culture is to reject the idea of boundedness and internal cohesion. In the modern world especially, culture is a meeting point where different influences, traditions and forces intersect. There is, therefore, a continual process of change in cultural practices and meanings. Globalization is undermining closed, fixed ideas of culture and leading to new ways of conceptualizing cultures (transculturation, contact zones, hybridity and diaspora). However, the fact that cultures are not fixed or homogeneous does not mean that we will stop thinking of them in this way. As Hall (1995: 188–9) argues, this is because some people need 'belongingness' and the security that closed conceptions of culture provide.

Despite this, recent years have witnessed a de-centring of culture, with nation-states increasingly superseded by transnational institutions that are producing cultural globalization. This is producing greater cultural diversity. There has also been a shift in the awareness of the cultural capital of the West, and an understanding of the cultural dominance of developed countries. At the same time, there are now more voices talking back, reflecting the cultural assertiveness of marginalized groups and making us aware of new levels of diversity. Thus,

> if there is a global culture it would be better to conceive of it not as a common culture, but as a field in which differences, power struggles and cultural prestige contests are played out.... Hence globalization makes us aware of the sheer volume, diversity and many-sidedness of culture. (Featherstone 1995: 14)

This points to a 'more positive evaluation by the West of otherness and differences' (*ibid.*: 89). For Massey and Jess (1995: 134), globalization is not simply a threat to existing notions of culture, but a 'stimulus to a positive new response'. What does this mean for geography, and how should geographers rethink ideas about culture?

One possibility is to consider the impact on national and local cultures of cultural globalization. As Held *et al.* (1999: 373) argue, the most that can be claimed is that in the West there has been some degree of homogenization of mass consumption, particularly among the young, and that this is spreading among more affluent peoples in the developing world, especially in Latin America and east Asia. Thus, similar fashions, popular music, film and television are being consumed simultaneously in many different places. However, it is less clear that homogenization of other cultural practices, beliefs and identities is taking place. Products may be globalized, but it does not follow that they are consumed and translated in the same way the world over. For example, the broadcasters of the popular music channel, MTV Europe, were clearly aware that they could not simply replicate the tone and content of the US original; MTV Europe is marketed to a very different audience. National cultures in many places around the world do seem to have weakened. Moves towards devolution, regionalism and independence around the world are evidence of this (e.g. the Legga Lombardo in Italy, Basque, Catalan and Galician regionalism in Spain, the Bloc Québécois in Canada, and the many right-wing anti-statist groups in the US). However, it is not clear that cultural globalization is responsible for challenging and eroding national cultures. Furthermore, it seems that hybrid and trans-national cultures, which might be important in rethinking local identities, have yet to make any significant impact on mainstream national cultures and national identities.

Culture is a process that operates at a number of different levels − local, national and global. Therefore, rather than conceptualizing a singular cultural landscape we should study the ways in which different cultural forms and processes come together in particular locations, creating both fragmentation and fusion. These different cultural forms (which Appadurai (1990) terms 'ethno-scapes' (cultural maps of ethnic identity), 'media-scapes' (representations of cultures in various media), 'ideo-scapes' (the range of ideas that people have about the world), 'techno-scapes' (the impact that technologies have in changing relationships over time and space), and 'finance-scapes' (the flows of money and capital at global and local levels)) might be the subjects of geographical inquiry. As Mike Crang (1998: 174) suggests, Appadurai's ideas are useful in that they produce a complex array of different cultural maps that intersect in many different ways to produce unique cultural terrains. In this way, it is possible to analyse the complexities of cultural processes at different spatial scales. Even though most people remain physically, ideologically and spiritually attached to a local or a national culture and a local place, these various networks or cultural maps ensure that it is becoming increasingly impossible for people to live in places that are completely isolated and disconnected culturally from the wider world.

Finally, in reconceptualizing culture, we must also consider *cultures of knowledge*. People around the world have different cultures and systems of meaning. We cannot avoid reading the world from within our own cultures and interpreting it through our own systems of meaning. Understandings of global culture for the majority of the readers of this book are filtered through the logic of the West. Spivak (1985) argues that one of the key aspects of Western knowledge is 'worlding' – the power to describe the world. This has its origins in imperialism, with the increased abilities of Western imperial powers to write about and culturally represent what became known as the 'Third World'. This imperial 'worlding' means that areas beyond the West were, and still are, described and represented in European terms. This is how those areas that were culturally different from Europe came to be perceived by Europeans as 'barbarous' and 'uncivilized'. Western ideas and cultural forms are still considered superior. 'Development' means incorporating non-Western economies into global capitalism. Forms of government not based on Western forms of democracy are deemed invalid. Western cultures, therefore, have become hegemonic, or dominant. Similarly, one's own cultural positioning (on the basis of gender, ethnicity, class, location, sexuality, stage in life cycle, ability) also influences understandings of local cultures. The same processes operate at local levels; dominant cultures marginalize others on the basis of ethnicity, sexuality, gender, and so on. However, as we have seen, those dominant cultures also produce resistances that have the potential to create new ways of thinking about culture.

The real challenge is to confront the limits of 'our' knowledge, to recognize other worlds, to acknowledge the legitimacy of other cultures, other identities and other ways of life. There is a need to consider 'cultural translation' (Bhabha 1994), which involves taking seriously the questions about our relationship to others – other cultures, other histories, other states, other experiences, traditions, peoples and destinies. Geographers should perhaps conduct transnational studies of culture, which highlight the evolutionary and hybrid nature of culture and demonstrate the influence of marginal cultures on dominant cultures. These need to negotiate between the local, the national, the global, and the historical, as well the contemporary diasporic. In this way, dominant cultures can also be subjected to analysis, rather than taken as the norm. It should be recognized that people in marginal cultural systems at local, national and international levels are also active in creating their own systems of meaning; they do not simply absorb ideas from, or become

Learning outcomes

Having read this chapter, you should understand that:

- Culture is complex; it is a process, rather than a thing, and is therefore subject to change over time.

- Cultures operate at the local, the national and the global levels.

- Global processes are occurring with relation to culture, but these are always filtered through localities – there is no singular global culture.

- Local cultures can be either negative (closed, bounded), or positive (progressive, hybridized).

- A progressive way of thinking about culture might be through new concepts such as hybridity and diaspora. These terms can be problematic (hybridity exists more in theory than it does in reality) but they hold potential for transforming how we think about culture in changing times.

- We must consider our own cultures of knowledge, and the ways in which these condition our ideas about global, national and local cultures.

absorbed into, more dominant cultures. We might then develop a cosmopolitanism in the new millennium that is global, sensitive to cultural difference, and dynamic.

Further reading

Crang, M. (1998) *Cultural Geography*, Routledge, London. A useful introduction to the ways in which geographers think about culture, exploring the global and the local, cultures of production and consumption, and ideas about place, belonging and nationhood.

Massey, D. and **Jess, P.** (eds) (1995) *A Place in the World? Places, Cultures and Globalization*, Oxford University Press, Oxford. A comprehensive first-year text, which explores the conceptualization of place, and how this relates to identity, cultures and contestation.

Shurmer-Smith, P. and **Hannam, K.** (1994) *Worlds of Desire, Realms of Power: A Cultural Geography*, Arnold, London. An introduction to the 'new' cultural geography covering themes such as senses of place, and gender, ethnicity and class. Has a useful final chapter on potential undergraduate projects in cultural geography.

Skelton, T. and **Valentine, G.** (eds) (1998) *Cool Places: Geographies of Youth Cultures*, Routledge, London. Focuses specifically on youth cultures, exploring ideas about gender, ethnicity, class, sexuality, ability and so forth. A useful example of how geographers are rethinking ideas about culture at a slightly more advanced level.

SECTION **3**

EDITED BY MICHAEL
BRADSHAW

Population, resources and development

The issues of population growth, resource depletion and environmental degradation have come to dominate the policy agendas of international agencies. In the 1970s the political manipulation of energy supplies and subsequent price increases contributed to recession in Europe, North America and Japan and aggravated the economic problems of the so-called developing world. In the 1980s famine struck the Sahel region of Africa and concerns were raised about the ability of the world's poor nations to feed themselves. In the 1990s the mounting evidence that humankind was damaging the earth's ecosystem finally resulted in collective action to reduce greenhouse gases. Citizens in the developed world now protest against their national governments and organizations such as the International Monetary Fund and the World Trade Organization, urging them to forgive Third World debt and to ensure that free trade is fair trade. The concern is not just that the world will run out of resources; but that those resources are produced, exchanged and consumed in such a way that benefits the rich nations at the expense of the poor and, ultimately, the natural environment.

The four chapters in this section examine the interrelationships between population dynamics, resource production and consumption and economic development and inequality. They all adopt a global perspective. Global in the sense that the processes producing the geographies of population change, energy and food production and consumption and, ultimately, development are shown to operate and interact at a number of different scales. They all stress the importance of the 'history' of human development in shaping the current geographies of population distribution, resource production and consumption and economic development. Thus, the division of the world between 'developed' and 'developing' countries, or alternatively, the 'North' and the 'South' are understood as the product of historical processes (see Section 1) that have impacted upon different regions in different ways. Such a crude two-fold division of the world hides a set of complex and contested issues; the notion of the developing world is defined in terms of its relationship with the developed world. If we (the majority of people reading this book) live in the developed world, the rest of the world must, by definition, be underdeveloped or developing (these issues are pursued further in Chapter 10).

All of the chapters stress that the twentieth century witnessed a growing imbalance in life chances between the developed and developing worlds. Chapter 7 analyses the various factors influencing population dynamics. The analysis reveals that while in the developed world the population has stabilized, and in some countries even declined, population growth remains the norm in the developing world. Consequently, the balance between North and South is shifting. In 1998, the developed world accounted for 20 per cent of the world's population; by 2050 its share could have declined to 13 per cent. However, that 20 per cent of the world's population generates the majority of its economic output (80 per cent of GNP in 1998) and consumes most of its food and energy. Thus, there is a growing mismatch between the geographies of production, consumption and development. However, the processes of globalization (see Section 4) are generating a complex set of interdependencies between North and South, often orchestrated by transnational corporations (TNCs). In the case of energy (Chapter 8), the developed world is becoming increasingly dependent upon supplies from the developing world. Similarly, certain parts of the developing world are being drawn into a global food supply system that delivers exotic foods to supermarkets in the developed world (Chapter 9). However, the presence of the TNCs can mean that the local economies see limited benefit from their incorporation into the global economy. In fact, such production may deprive the local population of food and other resources and promote environmental degradation.

In assessing the prospects for the future, all of the chapters question the practice of seeing the experience of Europe as a universal model for the evolution of societies elsewhere. Concepts such as the demographic transition and modernization suggest a predetermined progression through a set of stages that describe the European experience. If the world is to manage the problems currently posed by the conflicting geographies of population, resources and development, it must accept that the European experience does not constitute a universal development process. Furthermore, current dynamics indicate that it will be very difficult to maintain the living standards of the developed world *and* meet the increasing demands of the developing world, unless there are radical changes in the nature of the relationships between population, resources and development.

CHAPTER 7

TONY CHAMPION

Demographic transformations

Topics covered

■ Scale and distribution of world population growth

■ Transitions taking place in life expectancy and fertility

■ Developments in international migration

■ Population ageing

Though it is theoretically possible for a population to retain a particular size and structure, in practice most populations have been changing substantially over recent decades and seem set to carry on this way well into the twenty-first century. This is especially the case at a global scale as the total population now surges past the 6-billion mark, but it is also true for all major world regions and indeed for most countries around the world. Nor is it just a matter of population size, for also evolving rapidly are age distribution, social structure, ethnic composition, the way people group into households and the life-stage experiences people go through. People born in the last few years are exposed to life-chances and behaviour norms that are very different from those of a generation ago, let alone those prevailing at the beginning of the twentieth century. Similarly, it is pretty clear that, demographically speaking, the world will be a very different place in another 25 years' time. It is therefore enormously challenging to speculate on the way things might be by the year 2050, let alone the end of the century. In this chapter, the focus is on the scale and distribution of world population growth, the transitions taking place in life expectancy and fertility, the developments in international migration, and the single most important element of demographic restructuring, namely population ageing.

7.1 World population growth

The challenge presented by these transformations can most readily be illustrated by reference to total population numbers. While it is not easy to provide accurate statistics on the

---

Box 7.1

Population 'doubling time'

This refers to the period of time that it would take a population to double in size, given a particular rate of annual growth. The important point to remember is that, as Malthus recognized in his classic (1798) treatise on population, change takes place in the form of a geometric function or – in everyday parlance – at a compound rate, just as money accrues in an investment account if the interest is not paid out. As a result, a growth rate of 1 per cent a year will lead to a doubling of population in roughly 70 years rather than 100. Doubling times can be approximated by dividing 70 by the growth rate, so that at 2 per cent it will take only about 35 years for the population to double, at 3 per cent about 23 years, and so on. The fact that world population doubled from 3 to 6 billion between 1960 and 1999 means that its annual rate of increase over this period averaged just under 2 per cent.

Normally, doubling times are calculated with respect to the rate of natural increase without considering the effect of migration. This is partly because the concept is often used in relation to global population growth and partly because in the past population projections for major regions and countries have been based very largely, if not entirely, on trends in births and deaths. This approach is probably still valid when talking about major regions and where natural increase rates remain very high; for instance, for Africa as a whole, where the late-1990s rate of natural increase is equivalent to a doubling time of barely 27 years. It makes less sense for an individual country like Oman, where the current doubling time is currently just 18 years but where actual population change may be significantly affected by migration. Similarly, it is less relevant in relation to likely population trends for countries with very low rates of natural change; for instance, for the UK where a doubling would take 433 years at the rate prevailing in the mid-1990s but where natural increase was then responsible for only half of overall growth.

Source: Daugherty and Kammeyer 1995, PRB 1998

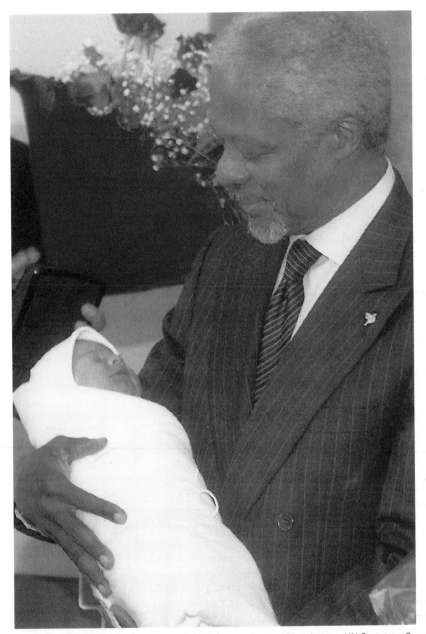

Plate 7.1   World population reaches six billion according to UN estimates. UN Secretary-General Kofi Annan holds the boy born in Sarajevo on 12 October 1999 who has symbolically been declared the 6 billionth addition to the planet's population. (Popperfoto/Reuters)

world's population for any given year (even with censuses taking place in most countries), we can say for certain that the number of people alive in the world today is greater than it has ever been. It is also undeniable that its growth over the past half-century has been

larger than at any previous time. The latter is the case whether considering absolute increments or percentage rates of increase.

The total population of the world is estimated to have passed 6 billion during 1999,
representing the latest stage in a relatively short period of generally accelerating absolute
growth. It was not until the beginning of the nineteenth century that world population
attained its first billion (around 1804). Then, while it took some 123 years to add a second
billion (1927), the third took only another 33 years, the fourth only 14 years, the fifth 13
years and the sixth 12 years. Looking at it another way, the first millennium saw virtually
no net increase from the 300 million people estimated to have been alive around the time
of Christ's birth, whereas the second millennium has witnessed a 20-fold growth of population. Perhaps most dramatically, while the first 1900 years from Christ's birth added 1.3
billion people, the last 100 years have added 4.4 billion.

Moreover, the pace of increase accelerated greatly as the twentieth century proceeded.
Whereas world population growth averaged barely 0.5 per cent a year during the first two
decades, the rate rose to around 1.0 per cent a year during the 1920s, 1930s and 1940s. It
then leapt to around 1.8 per cent in the 1950s and reached 2.0 per cent in the 1960s.
Given the 'compound' nature of this growth, the latter figure represents a population
'doubling time' of only 35 years (Box 7.1). In absolute terms, the annual increment was
running at barely 10 million at the start of the century but rose to 20 million in the 1920s,
averaged 50 million in the 1950s and passed the 80 million mark during the 1980s (Figure
7.1).

By the close of the century, however, there is plenty of evidence that the rate of world
population growth has peaked and is now slowing, though the absolute volume of
increase remains very high by long-term standards. The annual rate of increase fell back
somewhat during the 1970s and, despite then sticking at around 1.7 per cent for a decade
or more, it is reckoned to have fallen to around 1.3 per cent by the year 2000. This latest
reduction in rate has been large enough to offset the effect of the growing population

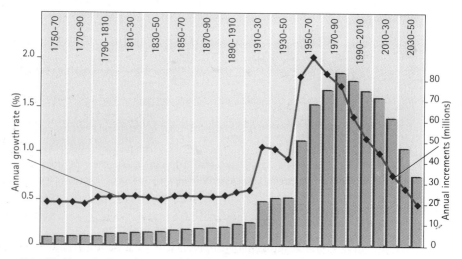

Figure 7.1   World population growth, 1750–2050. *Source*: UN (1999) and author's calculations. *Note*: Data
from 1990–2000 onwards are based on medium-variant projections

base, so that the size of the annual increments has fallen somewhat in recent years – down to an estimated 78 million a year in the second half of the 1990s (Figure 7.1).

These latest trends are expected to continue. According to the 1998-based projections prepared by the United Nations (UN 1999), the growth rate is likely to fall below 1.0 per cent during 2015–20 (almost exactly a hundred years since it first exceeded this rate). It is expected not to be much over 0.3 per cent by mid-century (a level not seen since around 1750). Because of the growing population base, the annual increments seem destined to stay above 70 million till around 2020, but are then expected to more than halve by 2045–50. On this basis, world population would reach 8.9 billion by 2050, almost 3 billion larger than now but with the majority of this addition occurring in the first quarter of the century and bringing up the 8-billion mark in 2028.

However, if there is one thing that demographers have learnt over the last few decades, it is that projections are never correct and indeed sometimes prove to be wildly inaccurate. Adopting a range of plausible alternative assumptions on fertility, therefore, the UN (1999) demonstrates that world population might be only 7.3 billion in 2050 and already contracting, but equally it could have risen to 10.7 billion, getting on for double the current size. It takes only relatively small changes in basic assumptions to produce very substantial differences over a period of decades.

This review of global trends suggests two major challenges for the new millennium. One is the very practical one of coping with the expected continuation of large-scale growth for the next few decades, further increasing the pressures and problems experienced since 1960 as the world added its latest three billion people. This takes on particular salience when it is recognized that, even more than in the past, the growth will be concentrated in those parts of the world that are economically the least equipped to cope. The other is the challenge posed to people involved in population studies as they try to understand better the factors affecting population trends and people's behaviour and thereby establish more firmly the likely future course of events.

7.2 Differential population change

The most important feature of the unevenness of population growth across the world is the disproportionate concentration of growth in Less Developed Regions (LDRs). Over the past half-century, these have accounted for fully 89 per cent of the world's 3.4 billion extra people, with their contribution rising steadily to reach 97 per cent by 1995–2000. The UN projections suggest that between now and 2050 the LDRs will account for the whole of world population growth because the More Developed Regions (MDRs) will experience overall decrease (Figure 7.2 and Table 7.1).

Within the LDRs, it is the 48 least developed countries that, in aggregate, are now witnessing the greatest upward surge in population (Figure 7.2). Though containing less than 8 per cent of world population in 1950, these have contributed one-eighth of world population growth in the past half-century. This proportion is expected to rise to well over a quarter (29 per cent) in the next half century, with their absolute volume of population increase almost doubling to 17 million a year (Table 7.1).

This unevenness in growth rates is producing a substantial shift in the distribution of population (Table 7.1). Whereas the MDRs accounted for almost one-third of all people in 1950 (32.2 per cent), the proportion was down to one-fifth in 1998 (20.0 per cent) and

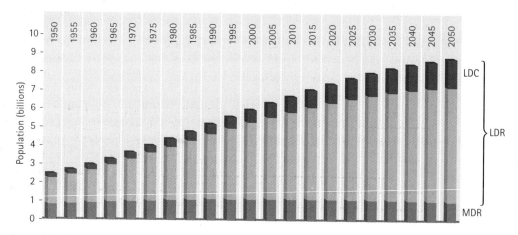

Figure 7.2 Population growth, 1950–2050, for development categories. *Source*: Data from UN (1999). *Note*: Rates of growth are annual average rates for 5-year periods, 1950–5 to 2045–50, estimated up to 1990–5 and medium-variant projections thereafter. The Least Developed Countries (LDCs) are a subset of Less Developed Regions (LDRs).

seems destined to fall to barely one-eighth (13.0 per cent) by 2050. Meanwhile, the share of the 48 least developed countries, while rising relatively slowly between 1950 and 1998, is projected to increase from 10.4 per cent in 1998 to 16.8 per cent in 2050, thereby contributing the vast majority of the LDRs' extra 7 percentage points.

This imbalance is also seen clearly in the changes in distribution between the major areas of the world (Table 7.1). Particularly symbolic is Africa's surpassing of Europe's total population size in the mid-1990s, in marked contrast to the situation in 1950 when the latter was more than twice the size of the former. By 2050 Europe's share is likely to have fallen to 7 per cent, while Africa's is expected to be nearly three times this. Over the period 1998–2050 Africa's population is projected to grow by just over one billion, representing fully one-third of global growth (33.8 per cent) compared to under one-sixth (15.6 per cent) for 1950–98. Meanwhile, Asia seems destined to remain the main contributor in absolute terms, adding a further 1.68 billion up to 2050, but this is a marked reduction in the pace of growth compared to its 2.18 billion gain in 1950–98.

Table 7.1 Population growth, 1950–2050, for development categories and major areas

Category and major area	Population (millions)			Share of world population (%)			Change (millions)		Share of world change (%)	
	1950	1998	2050	1950	1998	2050	1950–98	1998–2050	1950–98	1998–2050
World	2,521	5,901	8,909	100.0	100.0	100.0	3,380	3,008	100.0	100.0
MDR	813	1,182	1,155	32.2	20.0	13.0	369	−27	10.9	−0.9
LDR	1,709	4,719	7,754	67.8	80.0	87.0	3,010	3,035	89.1	100.9
LDC (within LDR)	197	615	1,495	7.8	10.4	16.8	418	880	12.4	29.3
Africa	221	749	1,766	8.8	12.7	19.8	528	1,017	15.6	33.8
Asia	1,402	3,585	5,268	55.6	60.8	59.1	2,183	1,683	64.6	56.0
Europe	547	729	628	21.7	12.4	7.0	182	−101	5.4	−3.6
L. America	167	504	809	6.6	8.5	9.1	337	305	10.0	10.1
N. America	172	305	392	6.8	5.2	4.4	133	87	3.9	2.9
Oceania	13	30	46	0.5	0.5	0.5	17	16	0.5	0.5

Note: data for 2050 are medium-variant projections. MDR More Developed Regions, LDR Less Developed Regions, LDC Least Developed Countries (subset of LDR), L. America Latin America and the Caribbean, N. America Northern America.

Source: UN (1999) and author's calculations.

At individual country level, China and India continue to dominate world population, together accounting for over one-third of world population. The number of large-population countries is, however, growing. In 1950 only nine countries (on the basis of 1998 boundaries) contained over 50 million people, roughly the size of the UK at that time. By 1998 this figure had grown to 23 and is projected to rise to 37 by 2050, with the UK dropping to thirtieth in the ranking. By that time, India is likely to have overtaken China, though both will be home to close on 1.5 billion people. Figure 7.3 shows what are

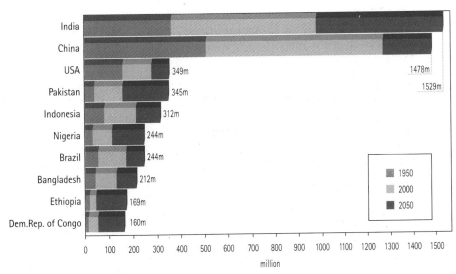

Figure 7.3 The ten countries projected to have the largest populations in 2050, and growth 1950–2050. *Source*: Data from UN (1999). *Note*: 2000 and 2050 data are medium-variant projections

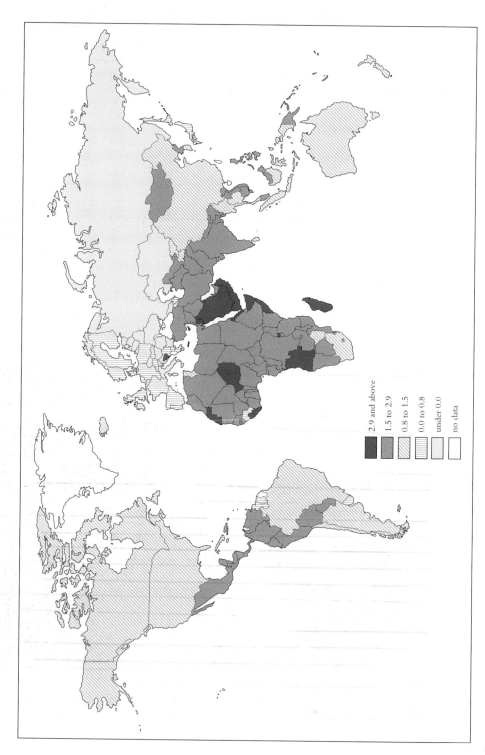

Figure 7.4 Annual average rate (%) of population change, 1995–2000. *Source*: Data from UN (1999)

expected to be the ten largest countries in 2050 and their scale of growth since 1950. One conspicuous feature of the recent past is the rapid growth of countries with a significant Muslim presence, with some of the highest growth rates occurring in the Arabian sub-continent and with the world's main zone of high population growth rates being dominated by the swathe of Islamic countries that stretches from Pakistan westwards to Africa's Atlantic coast (Figure 7.4). This is a trend that seems likely to become even more marked over the next few decades (see Box 7.2).

Box 7.2   Population growth in the Islamic world

According to Weeks (1988), 'Islamic nations are the world's fastest growing population group'. Only a small part of this growth is coming about through conversions from other faiths. The vast majority is due to the high levels of natural increase, resulting from high birth rate and youthful age structure. Data for 1988 put the fertility of Islamic nations at 6.0 children per woman, a third as much again as the 4.5 figure for other developing nations, while the proportion of people aged under 15 is put at 43 per cent.

Analysis of the latest UN population data indicates the scale of growth over the past half century and how things are expected to change over the next five decades. These data are for the 40 countries where Muslims made up at least half the population in 1988, thus including those of other

Plate 7.2   The Muslim population is now growing especially rapidly. Muslims are recording more rapid population growth than any other major group in the world, with their high average birth rates producing a big school-age population and the basis for strong future growth. (Panos Pictures/Neil Cooper)

(Box continued)

Box 7.2   (Box continued)

religions living there but excluding significant numbers of Muslims in other countries like Nigeria, Ethiopia, India, China and former USSR. The population of these 40 countries in 1950 was 302 million, but had more than trebled to 952 million in 1998 and is currently projected to reach 1,869 million by 2050. The latter increase represents almost one-third (30.5 per cent) of world population growth from now till 2050, compared with their contribution of less than one-fifth (19.3 per cent) in 1950–98. By 2050 these 40 countries would account for 21 per cent of world population, up from 12 per cent in 1950 and 16 per cent in 1998.

These projections, however, assume that the Islamic world moves to replacement-rate fertility (around 2.1 births per woman) before 2050. It is currently very difficult to know whether this is going to be possible. Given that many of these 40 countries are amongst the least developed in the world, much depends on whether they will come to enjoy the economic development normally associated with the desire for smaller families. Even if they do, there is the question of whether the cultural and political contexts will support such a desire. There is some evidence, notably from Iran and Afghanistan, that the recent rise of Islamic fundamentalism has served to stall, or at least slow, the fertility transition process.

Source: Aghajanian 1991, UN 1999, Weeks 1988

7.3   The progress of the demographic transition

For decades it has been customary to view trends in population growth rates in terms of the *demographic transition* model. The origins and nature of the demographic transition

---

Box 7.3

The demographic transition model

The 'demographic transition' concept has its origins in the work of the American demographer Warren Thompson, who in a paper in 1929 identified three broad groups of countries on the basis of birth and death rates. Countries of Western Europe and North America, which he termed Group A, were characterized by death rates that had fallen steadily over the previous few decades and by birth rates that were moving down rapidly towards the level of the death rates, indicating a future of very low population growth. Group B countries, found mostly in eastern and southern Europe at that time, had death rates that were falling quite quickly but birth rates that remained high or were declining only slightly, with the result that their populations were growing rapidly. Finally, Group C countries, accounting for most of Africa, Asia and Latin America, had high rates of both births and deaths and thus quite low overall growth.

Subsequently, mainly through the work of Notestein, these three groups of countries came to be seen as three different stages in a demographic transition. As evidence accumulated on the number of countries that appeared to be following this progression, the model was adopted by Western capitalist agencies as part of their Rostow-based 'stages of economic growth' strategy for Third World development. An early test of this approach was made by Chung (1970), who mapped the diffusion of the demographic transition across the globe from the early years of the century.

Chung's work was based on a very simple framework in which the two criteria for classifying

(Box continued)

Box 7.3   (Box continued)

countries were a death rate of 15 per 1000 and a birth rate of 30. His framework of three stages is shown in Figure 7.5, but modified by subdividing the third stage to reflect the substantial post-1960s reduction in birth rates in most MDRs. This forms the basis of an analysis of thsituation reached by the end of the twentieth century (see Figures 7.6 and 7.7, together with the text commentary).

While the model is still useful as a descriptive device, it has to be admitted that it lacks explanatory power, as it does not explicitly incorporate any causal factors nor even any direct linkage between trends in mortality and fertility. Moreover, its relevance as a descriptive tool is now much weaker. For one thing, a majority of the world's population is now living in countries that have passed through the transition or are at a late stage of it. The model says very little about natural increase in a post-transitional situation (but see Box 7.5 for a discussion of the 'second demographic transition').

Another criticism is that the model ignores the migration component of population change. According to Zelinsky (1971), it should from the outset have been called the 'vital transition' because it deals only with the vital events of birth and death. International migration has been an extremely important component of population change at particular times and is once again having substantial impacts on many parts of the world.

Source: Chung 1970, Daugherty and Kammeyer 1995, Jones 1990

Figure 7.5   The demographic transition model. *Source*: Modified from Chung (1970)

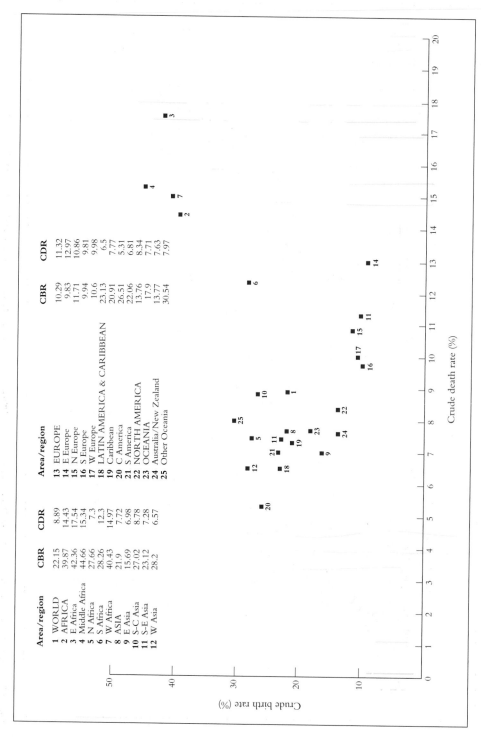

Figure 7.6   Crude death and birth rates, 1995–2000, for major regions and areas. *Source:* Data from UN (1999). *Note:* Shaded numbers refer to stages of the demographic transition model (see Box 4.3 and Figure 4.5)

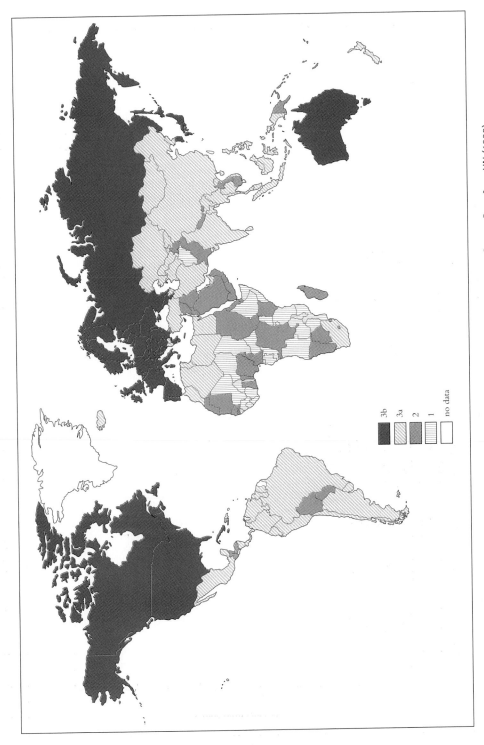

Figure 7.7   Countries classified on the basis of stage reached in the demographic transition by 1995–2000. *Source*: Data from UN (1999)

model are outlined in Box 7.3 and Figure 7.5. While it is an essentially descriptive rather than analytical tool, it provides a simple way of summarizing the state of demographic development reached across the globe. Here, this is done by adopting Chung's (1970) approach to studying the transition's 'diffusion' (see Box 7.3).

Figure 7.6 presents the 'vital trends' location of the world's major areas and regions in the form of a graph that also shows their relative position within the stages. At this level of geography, it can be seen that by the end of the twentieth century virtually the whole world had reached the third and final stage of the transition process. Only Eastern Africa remains clearly in the first stage with a crude death rate of over 15 per 1,000 and a birth rate well in excess of 30 per 1,000. Middle and Western Africa lie close to the border between the first and second stages, while Other Oceania (Melanesia, Micronesia and Polynesia combined) is on the verge of moving into the third stage.

At the same time, however, the regions now in the third stage exhibit a considerable degree of diversity (Figure 7.6). While their death rates are closely aligned around 6–9 per 1,000 (except for Europe), there is a much broader spread of crude birth rates. Some regions lie only a short way below the 30 per 1,000 threshold, notably Northern and Southern Africa, Western and South-central Asia and Central America. By contrast, Europe's four regions average barely 10 per 1,000, North America and Australia/New Zealand are both at 14 and East Asia is relatively low at just under 16. As mentioned in Box 7.3, Chung's criteria do not specifically indicate what constitutes a post-transitional situation, but on the basis of their distinctive birth rates these seven regions would seem to qualify for this description.

At the level of individual countries, a much more varied pattern is evident (Figure 7.7). In particular, at this scale there are found to be a substantial number of LDR countries placed in Stage 2 on the basis of the criteria used here. These include seven countries in Latin America and the Caribbean, three in Oceania and 14 in Asia. Africa, too, contains a fair number of the countries with death rates of below 15 per 1,000 but birth rates of 30 or more. Several of these Stage 2 countries can be found within wider regions that remain in Stage 1; for instance, Kenya and Madagascar in Eastern Africa and Cameroon and Democratic Republic of the Congo in Middle Africa. Even so, tropical Africa is dominated by countries that are still in the first stage, while at the other extreme sub-Saharan Africa contains only one country that has moved into the third stage, this being the Republic of South Africa. By contrast, Stage 3a dominates North Africa, Latin America and Asia. Clearly, outside tropical Africa and the Arabian sub-continent, the vast majority of countries have now reached the final stage of the transition (Stage 3a) or moved beyond it (Stage 3b).

7.4 Mortality and life expectancy

During the past half-century enormous strides have been made in reducing mortality and increasing life expectancy. This is the case even in those countries where death rates were already low in 1950. Yet there remains considerable scope for further improvement and new concerns have appeared in recent years. A key question is how quickly the parts of the world where death rates are still high, largely due to the persistence of infectious diseases, will be able to move through the epidemiological transition to a situation of low mortality rates linked mainly to the diseases of old age.

Across the world as whole, crude death rates (deaths per 1,000 people) have more than halved during the past 50 years – from 19.8 in 1950–5 to 9.3 in 1990–5. Even for the least developed countries, rates have almost halved over this period, dropping from 27.9 to 15.4, the latter being only marginally above Chung's criterion for moving out of the first stage of the demographic transition (see Box 7.3). The most impressive change has occurred in the more developed parts of the LDRs, because the LDRs in aggregate have seen their death rate plummet by two-thirds, ending up at only 9 per 1,000, lower than the average for the MDRs. These trends in crude death rate are, however, partly influenced by differences in age structure. The LDRs' current low rates are helped by their large numbers of younger people who are generally at least risk of dying while the MDRs' rates are now rising because of their increasing proportions of older people (see below).

In terms of life expectancy, a measure which is not affected by age structure and provides a direct indicator of life-chances, it is found that even the MDRs are continuing to experience gains, though the rate of improvement appears to be slowing (Figure 7.8). Between 1950–5 and 1970–5 life expectancy there rose from 66.5 to 71.2 years for both sexes combined – an increase of almost one year during every four years of elapsed time. Since then, however, the rate of progress has slowed, moving up by only around 3 years of extra life over the two decades. Part of the explanation for this slowdown lies in the situation in Eastern Europe, which not only emerged from the Communist era with below-average life-chances, but has been hit badly by the effects of the subsequent economic liberalization on employment opportunities and access to health care. Most alarmingly, the Russian Federation's life expectancy for men had slumped to 57 by the mid-1990s, markedly lower than the averages of 73 for Southern and Western Europe, 76 for Sweden and 80 for Japan. More generally, in many MDR countries, the social differential in life expectancy appears to be widening.

For the LDRs, the improvements in life expectancy have been considerably more marked, but here too the pace has been slackening (Figure 7.8). In the two decades from 1950–5, life expectancy rocketed from 41 to almost 55 years, but over the next 20 years

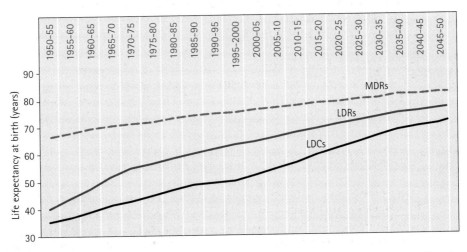

Figure 7.8   Life expectancy at birth (both sexes), 1950–2050, for development categories. *Source:* Data from UN (1999). *Note:* Data for 1995 are projections. Least Developed Countries are a subset of Less Developed Regions

Box 7.4

The impact of HIV/AIDS

The AIDS virus, unknown to the medical world a quarter of a century ago, has become a major cause of death and is one of the reasons why the UN has recently reduced its projections of future world population numbers. While its impact in the MDCs has been restricted largely to intravenous drug users, recipients of contaminated blood products and gay men, in parts of Africa and Asia its toll is much wider. In 1997 the UN estimated that 30 million people in the world were infected by HIV, with 26 million of these accounted for by 34 countries: 29 in sub-Sahara Africa plus 3 in Asia (Cambodia, India and Thailand) and 2 in Latin America (Brazil and Haiti). Up to 1998 some 14 million people are known to have died from AIDS.

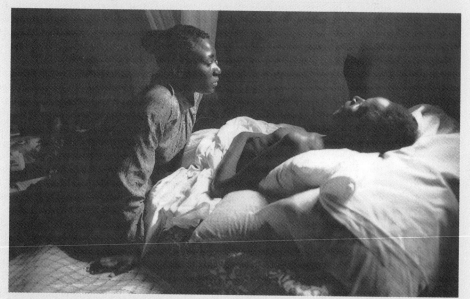

Plate 7.3   The AIDS epidemic continues to spread in Africa. While AIDS is now a worldwide disease, its greatest incidence is in Africa, where it is already a major cause of death and is bound to affect fertility behaviour in the longer term. (Panos Pictures/Crispin Hughes)

The impact of HIV/AIDS on mortality and population growth has been particularly severe in Africa and is expected to intensify in the future. Already it is estimated that life expectancy in the 29 hard-hit African countries is 7 years lower than it would have been without AIDS and that the gap is likely to widen to a deficit of 9 years in the first decade of the twenty-first century. Thus far the worst affected countries are Botswana, Kenya, Malawi, Mozambique, Namibia, Rwanda, South Africa, Zambia and Zimbabwe, in each of which the virus has infected at least 10 per cent of adults. For these 9 countries, average life expectancy in 2010–15 could be down to 47 years compared to 63 years in the absence of AIDS, representing 16 years of life per person lost to the epidemic.

The hardest-hit country so far is Botswana, where one in every four adults is now infected by HIV. It is estimated that life expectancy there is already 20 years below what it would have been without AIDS-related deaths, 47 years instead of 67. By 2000–5 life expectancy is expected to have fallen further to 41 years, which is 29 years less than the without-AIDS projection. Because of high fertility, its population will still rise, but considerably more slowly than previously expected; by 2015 it is likely to be 20 per cent smaller than it would have been in the absence of AIDS.

Source: Bongaarts 1996, Löytönen 1995, UN 1999

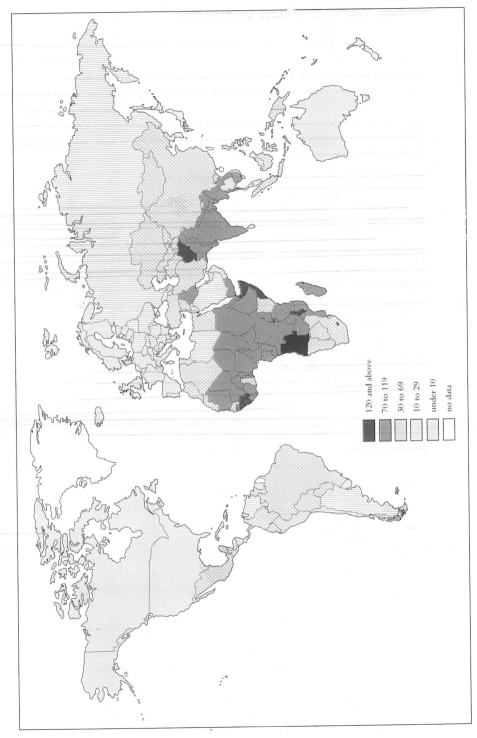

Figure 7.9 Infant mortality rate, 1995–2000. *Source:* Data from UN (1999)

average life expectancy advanced by a further 7 years only. The earlier progress can be attributed to the rapid reduction in malaria, principally due to the use of techniques developed to protect soldiers in the early 1940s fighting in south-east Asia, and to sustained programmes of curbing infectious and water-borne diseases like smallpox, tuberculosis and cholera. Within the space of 20 years, thanks to the importing of mortality control measures from the MDRs, many countries were able to pass through an epidemiological transition that had taken over a century for many European countries. Even so, life expectancy remains stubbornly low in the least developed countries in general and in sub-Saharan Africa in particular. For 1990–5, Rwanda's life expectancy was only 23.4 years, the lowest in the world, though admittedly aggravated by the civil war there and its attendant economic and social disruption. Next lowest were Sierra Leone (34.2) and Liberia (38.8), at that time relatively stable politically.

Two main challenges face those countries that currently have the lowest life expectancies. One is relatively new, but already appears to be having a significant effect on population growth rates in many countries, notably in Africa: the HIV/AIDS epidemic (see Box 7.4). The other is infant and childhood mortality. Whereas in the MDRs almost four-fifths of deaths in 1990–5 were accounted for by people aged 60 and over, the proportion was only two-fifths in the LDRs. The contrast is even greater for 0–5-year-olds, who make up under 2 per cent of all deaths in the MDRs, 28 per cent in the LDRs as a whole and 41 per cent in Africa. The main reason for this difference is in the numbers of babies that fail to reach their first birthday. Whereas the infant mortality rate (IMR) was under 10 in the mid-1990s in North America, most of Europe and Australia/New Zealand, it then averaged 36 in Latin America, 66 in Asia (excluding China) and 96 in sub-Saharan Africa. The IMR is particularly diagnostic of progress through the demographic transition, and of the pace of socio-demographic change more generally. Figure 7.9 thus acts as a fairly good delineator of the current extent of the so-called 'Developing World'. Given that the number of children likely to survive into adulthood is a prime consideration in deciding how many children a family should have, the IMR is also a key factor in fertility behaviour.

7.5   Fertility and family size

Despite some uncertainties about the future of mortality, it is primarily fertility behaviour that will determine future population numbers for the world as a whole and most of its major regions. Even quite small changes in fertility rate can produce major differences in population size in the longer term. In looking forward into the next century, the biggest issue concerns whether the least developed countries, especially in Africa, can continue the reductions in fertility that appear to have got underway over the last decade. In addition, however, it is also not clear whether the well-below-replacement fertility now being recorded by many European countries and some elsewhere will intensify or be subject to a rebound.

On the former issue, it is remarkable how much Total Fertility Rates (TFRs), which are broadly equivalent to the average number of births per woman, have fallen for most of the LDRs since mid-century. In 1950–5 their overall TFR was about 6.2 and indeed stayed around the 6 mark for the next 15 years, but since then it has virtually halved in three decades (Figure 7.10). Asia and Latin America have been the driving force behind this fall, for their TFRs have fallen from close to 6 to 2.6 and 2.7 respectively. Not all parts of these two regions have, however, been equally involved in this. For one thing, China has made a major contribution to Asia's progress, with its TFR estimated to be down to 1.8 by the late 1990s, while many countries of Western Asia still have rates above 4 – indeed two of around 7.0. Variation is somewhat less in Latin America, but even so TFR here ranges from 1.5 for some Caribbean states to 4.9 in Guatemala (Figure 7.11). Overall, however, this constitutes nothing less than a **reproduction revolution.**

The situation in Africa is currently very different, but the fertility transition appears to be underway (Figures 7.10 and 7.11). Most countries had TFRs of at least 5.0 in 1995–2000, with several of 6.4 or more. As a result, Africa's overall TFR was 5.1 at that time, with Middle Africa being the highest at 6.2, followed by Eastern Africa at 5.8 and Western Africa at 5.5. All three, however, are in the process of moving downwards: Middle Africa from a peak of 6.6 in the 1980s and Eastern and Western Africa from 7.0 in the 1970s (Figure 7.10).

The key questions for Africa, as well as for South-central and Western Asia, are how fast the transition will proceed in the future and what will constitute its eventual 'bottom line'. The UN's medium-variant projections assume that TFR will have fallen to replacement rate (2.1 births per woman) in all regions of Africa by 2040 (Figure 7.10). Some international precedents would seem to support this; for instance, the experience of Thailand where TFR was still as high as 6.1 in the late 1960s but below replacement rate by the early 1990s, only a quarter of a century later. Yet progress would seem to be much slower than this amongst the late comers to the transition. Pakistan, for example, has seen its TFR drop from a peak of 7.0 to 5.0 now, but this has already taken 20 years. Within Africa itself, Northern Africa has almost halved its TFR to 3.6, but this has taken 35 years (Figure 7.10).

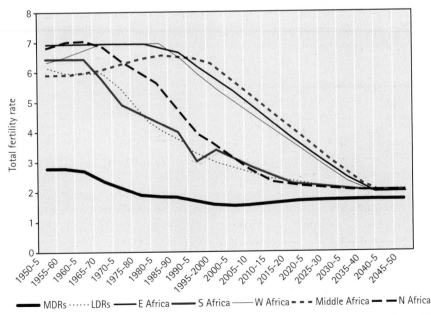

Figure 7.10 Total fertility rate, 1950–2050, for development categories and African regions. *Source*: Data from UN (1999). *Note*: Data from 1995 are projections.

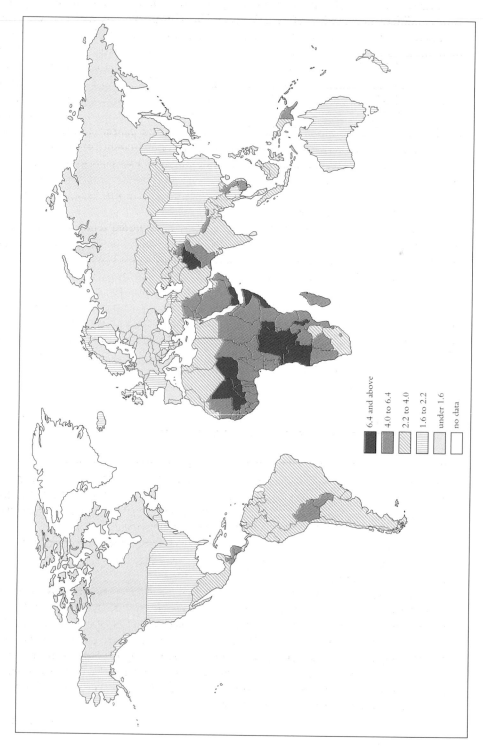

Figure 7.11   Total fertility rate, 1995–2000. *Source*: Data from UN (1999)

Progress towards replacement-rate fertility in these countries will no doubt require the same conditions to come into place as happened in other parts of the world. These include the provision of accessible family planning programmes, but before these can be effective, there needs to be the will to limit family size. This, in its turn, depends on a shift from a subsistence life-style to a consumer society, on the widening of education and career opportunities for women and on further progress in tackling disease and other causes of young-age deaths. In short, given their position at the end of the line of the development process and the strength of their Muslim or African traditions, nothing less than a combination of economic and cultural revolutions would appear sufficient to produce a repetition of the Eastern Asian and Latin American demographic 'success stories'.

At the other extreme, the situation in Europe might be expected to shed light on the eventual level of fertility, but in fact there now appears to be greater uncertainty about the significance of recent developments than there is in the least developed countries. Following the low fertility of the inter-war period and then the unexpected post-war 'baby boom', it came to be understood that fertility rates would tend to fluctuate in the post-transition regime. Yet, after the mid-1960s, most non-communist European countries saw their fertility plummet to unprecedentedly low levels and stay down there, though there has been some rebound, notably in Scandinavia. Falls were especially marked in Southern Europe, but have been paralleled by the experience of Eastern Europe since the late 1980s. Various epithets have been used to capture the essence of this process, including 'demographic revolution' and 'the twilight of motherhood', but perhaps the most challenging intellectually is the notion of a 'second demographic transition'. The latter suggests that there is taking place a shift to a new societal pattern that is just as fundamental as the original transition (see Box 7.5).

Box 7.5   **A second demographic transition?**

This thesis was put forward in 1986 by Dutch demographers Lesthaeghe and van de Kaa primarily to make sense of the dramatic fall in European fertility after the mid-1960s, but its rationale goes well beyond considerations of family size, arguing that a new era in demographic history has dawned. The new regime is seen to be associated with a complete shift in norms and attitudes that can be denoted as a switch from 'altruism' to 'individualism'. Whereas the first transition to low fertility was prompted by concerns for family and offspring, such as providing a high standard of living and ensuring good life chances for the next generation, the second is seen to emphasize the rights and self-fulfilment of individuals.

In particular, it is seen as giving women greater choice of roles. In the past high fertility restricted women's options, while the need for survival limited the variety of family and household forms. Improvements in life expectancy, and especially the reduction in infant and childhood mortality, paved the way to lower fertility, as too did the introduction of pensions to support parents in their old age. Combined with the increasing emancipation of women in higher education and the labour market, the development of easy-to-use and reliable contraceptives, and the rise of new attitudes about abortion, these have given women both the incentive and the opportunity for limiting fertility and achieving greater financial independence.

(Box continued)

Box 7.5   (Box continued)

Plate 7.4   The number of lone parents has increased greatly in the last 25 years. (Format Photographers/Brenda Price)

Associated with these changes is the rise of non-traditional life-styles, involving a huge diversity of living arrangements and a greater instability in household patterns. This is reflected in the growing number of people living on their own or in same-sex relationships or with other unrelated people, as well as in the growth of non-marital cohabitation, lone-parenthood, divorce and separation, and step-relationships. Most important for long-term demographics, however, are the rise of childlessness and the delay of childbearing into women's later reproductive years.

The key issue for the MDRs is the extent to which these tendencies are influenced by fashion or other factors that may be reversible. For the LDRs the main questions are whether the second demographic transition provides a valid explanation in countries where fertility is already below replacement rate and whether it is transferable to those where it is not yet as low as this.

Source: Cliquet 1991, Coleman 1996, Hall 1995, McLoughlin 1991, van de Kaa 1987

Whether or not this is the case, it cannot be denied that there are now many countries with fertility levels well below replacement rate (Figure 7.11). Indeed, TFR in a number of countries (Bulgaria, Czech Republic, Estonia, Germany, Greece, Italy, Latvia, Romania, Slovenia, Spain) is now 1.3 children per woman or lower – a level 'which a generation ago was considered highly improbable' (UN 1999). In 20 of the more developed countries TFR has been below replacement rate for at least two decades. Now that a number of LDR countries (notably in Eastern Asia and the Caribbean) also have TFR of 2.1 or below and there are a great many more where the rates are below 3.0 and still falling rapidly, it is little wonder that great attention is currently being given to whether recent European experience can provide an indication of what is in store for the rest of the world.

7.6 International migration

Migration is an increasingly important component of population change. This is not to say that it has not been significant in the past. In the century leading up to the First World War, upwards of 55 million people moved from Europe to the 'New World', most notably the USA which gained 30 million between 1860 and 1920 alone. Moreover, over that period migration played a substantial role in relieving population pressures during the long-drawn-out transition process in Europe; for instance, with emigration reducing Norway's population growth by over a quarter between 1860 and 1914. Nevertheless, migration has received little attention in demographic growth theory.

There are two reasons for the growing prominence of migration in countries' population trends. One concerns the reduction in natural increase rates in the MDRs. Here migration exchanges are now constituting a larger proportion of their rates of overall population growth than used to be the case and are increasingly determining whether their populations grow or decline. The other is that the volume and nature of migration is changing. According to Castles and Miller (1993), this is 'the age of migration'. While not denying the significance of migration in the past, they identify four ways in which migration is now developing and seems destined to play a major role over the next couple of decades (see Box 7.6).

The impacts of net migration on population trends have now become quite significant, at least as far as the MDRs are concerned (Table 7.2). At the broadest level there is expected to have been a net flow of almost 10 million people from the LDRs to the MDRs between 1995 and 2000. For the latter, because of their low rate of natural increase, this accounted for fully three-fifths of their overall population growth. For North America and Australia the proportion was around two-fifths, but in the case of Europe immigration was entirely responsible for the growth recorded, more than offsetting the effect of natural decrease. The relative impact on the source regions, by contrast, has generally been much smaller. The only LDRs where migration has removed more than one-twentieth of recent natural increase are the Caribbean, Central America and Middle Africa (Table 7.2). For the LDRs as a whole, the net emigration is equivalent to just 2.5 per cent of the population increase. The proportion is only half this for the 48 least developed countries despite the much faster build-up of demographic pressure on resources there.

At the same time, the scale of population movement within these broad regions is much greater and has been growing, despite the fact that losing and gaining areas may share similar demographic contexts. In 1995–2000 this is seen in the form of net inflows to Western Asia alongside net outflows from South-central Asia, as well as in net inflows to East Africa and outflows from Middle Africa (Table 7.2). To some extent, these reflect the geography of economic opportunities, but there have also been substantial refugee movements. Similarly, within Europe, the past decade has witnessed large-scale migration

---

Box 7.6

The age of migration

While 'migration' can be used to describe any form of residential mobility and the majority of such movement is fairly localized within countries, the epithet 'The age of migration' is used by Castles and Miller (1993) principally with reference to international migration, emphasizing how much this has grown and evolved in recent years.

They point to four general tendencies:

- Migration is becoming more 'global', in the sense that more and more countries are being affected at the same time and that the diversity of areas of origin is also increasing.

- Migration is accelerating, with the number of movements growing in volume in all major regions at the present time.

- Migration is becoming more differentiated, with no one type of movement dominating a country's flows but instead with combinations of permanent settlers, refugees, skilled labour, economic migrants, students, retirees, arranged brides, and so on.

- Migration is being feminized, with women not only moving to join earlier male migrants but now playing a much fuller part in their own right, notably being labour migrants themselves as well as often being dominant in refugee flows.

These trends greatly alter the policy implications of international migration. The growing scale of the process and the increasing involvement of women and families raise new challenges for providing for immigrants and, more generally, serve to increase native hostility in receiving countries. At the same time, increasing globalization and the growing diversity of migrants both make it harder for governments to implement effective policies for curbing migration or influencing its nature.

Source: Castles and Miller 1993, Champion 1994, Zlotnik 1998

---

from East to West, facilitated by the removal of the 'Iron Curtain' at the end of the 1980s but reinforced by economic turmoil and ethnic conflict in the formerly communist states, not least in the Balkans.

Within these changes, probably the single most important feature of the last 15 years is the increase in numbers of asylum-seekers and refugees. The average level was running higher in the 1990s than at any time since the first few years after the Second World War. Estimates for the early 1990s put the global figure at around 20 million, up from 6 million in 1978. Some 3 million asylum applications were lodged in Europe, North America and Australia during 1983–92, with the annual figure rising from just 100,000 to over 700,000 over the period. Europe was most affected by this increase, with the numbers applying for refugee status up from 20,000 in 1980 to 560,000 in 1992. In the following years the numbers of new arrivals in European countries fell in total, but the salience of the refugee problem has been intensified by the fall-out from the ethnic conflicts in former Yugoslavia. Some 340,000 Bosnians fled to Germany in the mid-1990s, while possibly over half a million ethnic Albanians left Kosovo in 1998–9, though in both cases the majority have subsequently been repatriated. Beyond this, in Europe (including 'hot spots' in the former USSR like Nagorno Karabakh in Azerbaijan and Chechnya in the Russian Federation) there are

Table 7.2 Net migration and population growth, 1995–2000, for development categories, major areas and regions

Development category, major area and region	Net migration 1995–2000 (thousands)	Annual net migration (per 1000)	Annual population growth (per 1000)	Growth due to migration (%)
More Developed Regions	9,853	1.7	2.8	60.7
Less Developed Regions	−9,853	−0.4	15.9	−2.5
Least Developed Countries	−930	−0.3	23.7	−1.3
Africa	1,434	−0.4	23.7	−1.7
Eastern	1,027	0.9	25.7	3.5
Middle	−1,237	−2.7	26.6	−10.2
Northern	−625	−0.7	19.6	−3.6
Southern	2	0.0	16.0	0.0
Western	−601	−0.5	24.9	−2.0
Asia	−6,035	−0.3	13.8	−2.2
Eastern	−326	0.0	8.7	0.0
South-central	−4,621	−0.6	17.6	−3.4
South-eastern	−1,484	−0.6	15.3	−3.9
Western	395	0.5	22.1	2.3
Europe	4,751	1.3	0.3	433.3
Eastern	1,778	1.2	−2.0	n/m
Northern	299	0.6	1.5	40.0
Southern	736	1.0	1.2	83.3
Western	1,938	2.1	2.8	75.0
Latin America & the Caribbean	−2,353	−0.9	15.7	−5.7
Caribbean	−436	−2.3	10.8	−21.3
Central	−1,745	−2.7	18.5	−14.6
South	−172	−0.1	15.2	−0.7
North America	4,652	3.1	8.5	36.5
Oceania	405	2.8	13.0	21.5
Australia & New Zealand	450	4.1	10.2	40.2
Other Oceania	−45	−1.1	21.4	−5.1

Source: UN (1999) and author's calculations from rounded figures. Note: n/m no meaningful figure can be calculated.

large numbers – estimated at 5 million in the mid-1990s – of 'displaced persons', people forced out of their homes and neighbourhoods but sheltering within their states of origin.

These figures, however, are small by comparison with the upheavals currently being contained within the LDRs. The UN High Commissioner for Refugees estimates that these form the reception areas of almost four-fifths of all international refugees. Through most of the 1980s, Asia hosted almost half of the world's refugees, with Palestinians constituting a long-term problem and with the other major sources at that time being Afghanistan, Vietnam and Cambodia. Another major refugee exodus in Asia was that of Iraqis during the aftermath of the Gulf War in 1991. By the mid-1990s, however, it was Africa where almost half the world's refugees could be found, reflecting the political and ethnic instability in that continent ranging from Sudan to Angola and from Rwanda to Sierra Leone. Not included in these figures are even larger numbers of displaced people. Estimates for the mid-1990s put the number of Africans who are long-term refugees within their own countries at around 16 million, much larger than the 7 million for Asia

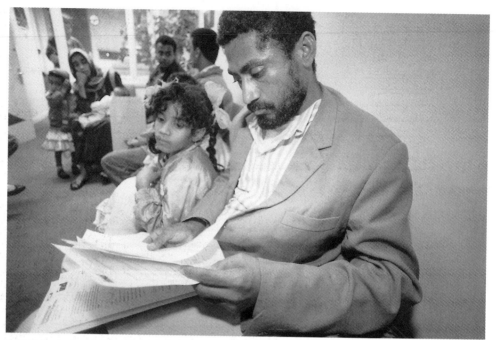

Plate 7.5 Asylum-seekers now form a major component of immigration for many countries. This family waiting with their forms near Heathrow were among 70,000 asylum-seekers filing applications in the UK in 1999, over ten times the average for the late 1980s. (Format Photographers/Karen Robinson)

and 3 million for Latin America. Partly overlapping with these figures no doubt are estimates that number the world's 'environmental refugees' at 25 million for 1998. This is believed to be the first year that this number exceeded those displaced by war and is seen as a worrying portent for the future, whether or not global warming leads to rising sea level and altered patterns of rainfall.

Nevertheless, despite their political salience, asylum-seekers and refugees make up only around one-fifth of all people living outside their country of citizenship and even fewer of those residing in a country that is not their birthplace. A second rapidly growing group of international migrants, but one that by contrast is often referred to as 'invisible', comprises professional, managerial and technical personnel. Most notably, these include staff transferred between postings by their transnational corporations (especially in manufacturing and financial services) and those recruited to work in MDRs or newly industrializing countries with shortages of medical, construction and other skills. A third group comprises the relatively unskilled labour that is prepared to work in low-wage sectors like agriculture, manufacturing and hotels and catering, often on a transient or clandestine basis. This process is advantageous not only to employers but also to national economies where demographic slowdown or strong unionization would otherwise have led to higher labour costs and weaker competitiveness. In addition to these three main types of international movement, there are a host of others including students, retirees and brides (see Box 7.6).

Plate 7.6   Civil war and ethnic cleansing are leading to a rising tide of refugees and displaced people. An ethnic Albanian refugee woman sitting next to a makeshift stove in the Stenovac refugee camp near Skopje, Macedonia, is one of thousands displaced by the conflict in Kosovo in 1999. (Popperfoto/Reuters)

In some cases, the impact of these non-refugee migrants on population distribution is relatively small. This may be because many of these people are relatively temporary visitors rather than permanent settlers or because there are flows in both directions between countries. Nevertheless, even in the latter case, this migration can lead to substantial differences in population composition. In the UK, for instance, the continuing pattern is of a net loss of British-born citizens and a substantial net gain of people born in the LDRs. For those countries where either the volume of inflows is considerable (like the USA) or where the rate of natural increase is very low (as across much of Europe), the impact of international migration can over time become very marked. The most conspicuous and politically contentious aspects relate to their economic and cultural effects. While employers tend to welcome the arrival of cheaper, malleable labour, the less skilled elements of the indigenous population and the previous cohorts of immigrants may feel their job prospects threatened and the governments are often only too conscious of the extra public spending needed for welfare and education programmes. Even more fundamentally, the arrival of people of different language, religion and especially skin colour may be seen by some as threatening not only the cohesion of local communities but the very integrity of the existing nation-states.

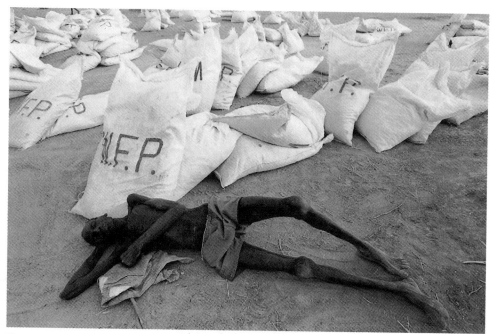

Plate 7.7   International food programmes are proving inadequate at coping with the effects of drought, conflict and famine. Tragic images such as this are commonplace in the western media. (Popperfoto/Reuters)

7.7   Population ageing

Current demographic trends are having a profound effect on population composition. As just seen, migration is increasing the ethnic and racial diversity of populations, while the second demographic transition is associated with marked changes in family and household arrangements. Perhaps the most fundamental transformation, however, is in age structure. Across most of the world, population ageing is now the dominant trend, as people are living longer and – normally the more important factor – falling fertility is reducing the proportion of younger people. Already, many MDR countries have moved a long way from the traditional 'pyramid' age structure and indeed are beginning to look distinctly 'top heavy', but the biggest growth in numbers of older people is now taking place in the newly industrializing countries. For the LDRs as a whole, coping with a 'greying population' constitutes one of the most serious challenges of the new millennium.

The UN's latest figures make extremely impressive reading, as do the simple age-structure diagrams in Figure 7.12. Over the past half century, the median age (in years) of the world's population has risen slightly from 23.5 to 26.6, but by 2050 it is expected to have shot up to 37.8. Much attention has been given to the ageing of the MDRs since 1950, and 1998 marked something of a milestone there as the number of older persons (60 and over) exceeded that of children (aged under 15) for the first time. Nevertheless, the rise in median age in the MDRs over the next half-century will be as great as for the last 50 years, rising from 37.5 in 2000 to 45.6 in 2050. Europe is, and will remain, the most affected region, with the projections indicating that older people will constitute one

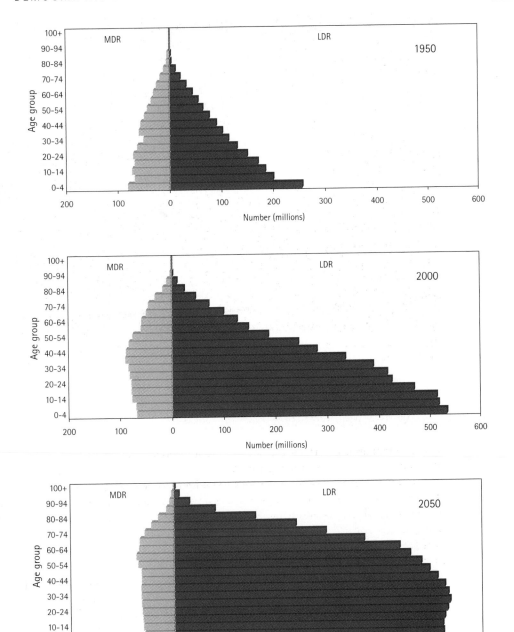

Figure 7.12 Age structure of More and Less Developed Regions, 1950, 2000 and 2050. *Source*: Data from UN (1999). *Note*: The 2000 and 2050 data are based on medium-variant projection

in three of the population by 2050, outnumbering children by two-and-a-half times. By that time, on current trends, Spain will be the oldest country in the world, with 3.6 persons aged 60 and over for each person below 15 years of age.

Ageing has been much slower until now in the LDRs in aggregate, with median age rising only from 21.3 to 24.4 between 1950 and 2000, but its impact over the next half century will be far greater than for the MDRs. By 2050 their median age is projected to rise to 36.7, an increase of 12.3 years compared to one of 8.1 years for the MDRs. Over the past half century the LDRs' number of people aged 60 and over has grown by some 264 million, a small fraction of their 3.2 billion total increase. But over the next 50 years the number of older people will rise by 1,221 million, accounting for over two-fifths of their total population increase over the period. By contrast, the number of children in 2050 is expected to be about the same as currently and, as a result, the LDRs will have reached the same situation as the MDRs in 1998, for by then the number of older people will have just outnumbered children. Of the three major LDR areas, Asia is the furthest down the ageing road, while Africa is the youngest, with its current median age of 18.4 years being lower than it was in 1950.

These trends constitute a big enough challenge for MDR countries, where governments have for some time been concerned about their ability to fund pensions and care provision through taxes imposed on a shrinking proportion of working-age people. Their recent and future problems, however, pale into insignificance in comparison with the sheer numerical growth of older people expected in the LDRs over the next half century

Plate 7.8 The elderly are starting to outnumber children in many parts of the Developed World. This familiar off-peak holiday scene is likely to become commonplace all year round and in all places, as the 'baby boomers' of the 1950s and early 1960s reach retirement age. (Camera Press/Chris Nevaro)

and take on even more frightening proportions when the fragile economic base of many of these areas is considered. Up till now, the most rapid ageing has taken place in the newly industrialized countries of East Asia, where very rapid economic growth (at least until 1998) helped to ease the burden. If similar growth fails to take root in the rest of the LDRs, and especially in the currently very poor regions of South-central Asia and above all sub-Saharan Africa, the direct consequences are likely to be tragic and the knock-on effects in relation to other aspects of mortality, fertility and migration are virtually impossible to gauge.

7.8 Conclusion

The world's demographic complexion has clearly undergone some major transformations in the past half-century. There are now nearly three times as many people living in LDRs than in 1950. The 'death control' element of the demographic transition, scarcely evident outside the MDRs at mid-century, has now affected all major regions and is largely complete in most of them. The past three decades have seen the fertility transition spread quite rapidly across many LDRs hand in hand with economic growth and family planning programmes. More recently still, international migration has moved to a new plane in terms of numbers of people involved, the range of countries affected and the variety of forms taken. All the time, populations around the world are becoming progressively older, with the traditional 'age pyramid' becoming increasingly obsolete as a description of age structure.

Looking to the future, the most likely scenario is that these trends, linked as they are to the process of demographic transition, will continue and eventually play themselves out. This is certainly what the latest UN projections suggest. This chapter has suggested a number of ways in which the assumption of continuity might come unstuck. First and foremost, there remain the Malthusian spectres of 'war, pestilence and famine', as exemplified by the many ethnic conflicts, the spread of HIV/AIDS and the latest food emergencies. Secondly, there would seem to be no guarantee that conditions conducive to small family size will appear in the remaining high-fertility countries, given their recent combination of rapid population growth and weak economic progress. Thirdly, while the factors that have led to the new 'age of migration' appear to show little sign of abating, international migration is notoriously volatile and unpredictable and, in addition, does not seem to be readily amenable to management by governments. To cap it all, talk of a 'second demographic transition' in Europe raises new uncertainties about trends there and about the eventual destination of the LDRs.

Nevertheless, some things seem very certain and raise extremely important issues. Barring a catastrophe, world population is very likely to rise by a further 3 billion in the next half-century. All this growth is destined to take place in the LDRs, leading to further shrinkage in the proportion accounted for by the MDRs. The number of old people is going to continue to increase rapidly, imposing ever greater strains on families and societies in both the MDRs and LDRs. The ethnic balance in the world will increasingly sideline the white, mainly Christian population, as Asian and African populations burgeon and as Islamic and other faiths expand. This is likely to increase tensions both between and within countries, particularly in areas coming up against the limits of their land and water resources. Nor will the MDR countries be immune, as their existing ethnic minority populations grow and as their case for strict immigration controls is subjected to further tests by the international community. Clearly, demographics now constitute a major force for global change and their combination of transformation and uncertainty represents a powerful challenge for the new millennium.

Learning outcomes

Having read this chapter, you should be able to:

- Recognize the dynamic nature of populations, not only in terms of size but also with respect to composition, notably age structure and ethnic make-up.

- Question the inevitability of the demographic transition process for those countries that have entered the mortality control stage fairly recently.

- Appreciate that the completion of the demographic transition no longer appears to herald the end of transformations in the population.

- Acknowledge the increasing importance of migration as a component of national population change alongside fertility and mortality.

- Assess the significance of the challenges which current demographic trends pose for individuals, families, communities and governments.

Further reading

Castles, S. and Miller, M.J. (1993) *The Age of Migration: International Population Movements in the Modern World*, Guilford Press, New York. This provides a comprehensive survey of past trends and current developments in international migration. Look especially at Chapter 6 on the globalization of international migration and Chapter 10 on migration in the New World Disorder.

Gelbard, A., Haub, C. and Kent, M.M. (1999) World population beyond six billion, *Population Bulletin*, 54, 1–44. This booklet, published by the Population Reference Bureau in Washington DC, reviews the history of world population, the causes and effects of population change, and the prospects for the next 50 years. See also the PRB's annual *World Population Data Sheet*, containing key statistics on population size, structure and change for all geographical entities with populations of 150,000 or more and all members of the UN.

Jones, H. (1990) *Population Geography,* 2nd edition, Paul Chapman, London. This is a clear introduction to geographical perspectives on population. Look especially at the chapters on population growth and regulation, international variations in mortality, fertility in developed countries, fertility in less developed countries, and international migration.

UN (1999) *World Population Prospects: The 1998 Revision*, United Nations Population Division, New York. This is the latest of the UN's biennial updates on world population, with estimates of population for 228 countries and areas of the world from 1950 to 1998 and projections through to the year 2050. This edition includes special reports on the demographic impact of HIV/AIDS, below-replacement fertility and ageing. The UN's *Demographic Yearbook* provides even more detail on the current situation and recent trends.

Useful web-sites

www.popin.org/pop1998   Provides access to extracts from the United Nations' *World Population Prospects: The 1998 Revision.*

www.census.gov/ipc/www/idbnew.html   US Census Bureau (on line) *International Data Base*, United States Bureau of the Census, Washington DC. This is a computerized source of demographic and socio-economic statistics for 227 countries and areas of the world from 1950 to the present, and projected demographic data to 2050, freely available on the World Wide Web.

www.populationaction.org   Home of Population Action International, an organization dedicated to advancing policies and programmes that slow population growth in order to improve quality of life for all people.

www.popindex.princeton.edu   The search interface for the entire Population Index database for 1986–99, maintained by the Office of Population Research at Princeton University. It contains over 40,000 citations of the papers and reports as covered in Volumes 52–65 of *Population Index*.

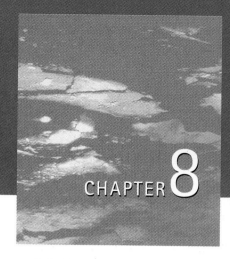

CHAPTER 8

MICHAEL BRADSHAW

Resources and development

Resources are defined by *society*, not by nature.

(Adapted from Rees 1985:11)

Resources *are* not, they become; they are not static but expand and contract in response to human wants and actions.

Zimmerman (Peach and Constantin 1972:16)

Topics covered

- The nature of natural resources

- Ways of defining and classifying resources

- The factors that determine resource availability

- The changing geography of energy production and consumption

- The relationship between energy consumption and economic growth

- Global energy dilemmas

The large-scale exploitation of the planet's resource base may well go down in history as a defining characteristic of the twentieth century. Today the scale of resource consumption threatens the entire global ecosystem. Throughout the century access to, and control over, natural resources have been a source of conflict. Equally, patterns of resource consumption are a major marker between the so-called 'developed' and 'developing' world (see Chapter 10 for further discussion). According to the World Bank (1998, 1999), in 1996 the so-called 'high-income countries' accounted for 16 per cent of the world's population, but produced 80 per cent of global Gross National Product (GNP) and accounted for 51.8 per cent of global commercial energy consumption. As we enter the twenty-first century, society is increasingly aware of the finite nature of the planet's resource base. Yet, a seeming relative abundance of supply has made us complacent, as has a blind belief that technological progress will solve the inevitable shortage of non-renewable resources. This chapter explores the nature of natural resources and considers the relationship between resource production and consumption, and economic development. The chapter is divided into three sections, each with a distinct task. The first section examines the meaning of the term 'natural resource', it evaluates the various ways of classifying resources and analyses the diverse factors that influence their availability. The second section analyses the specific case of energy resources and the changing geographies of production and consumption. The third, and final, section considers the interrelationship between energy consumption and economic development. The chapter concludes by discussing the resource dilemmas that currently face the different regions of the world.

8.1 Natural resources

At any moment in time, the planet Earth holds a finite stock of resources, the resource base. However, what human societies have considered a resource has varied through time and across space. Thus, the notion of what constitutes the Earth's resource base also changes. As both Zimmerman and Rees acknowledge above, something is a resource because human society attaches value to it. Many of the things we value today as 'resources' held no value in the past and may hold no value in the future. For example, petroleum was not an important source of energy until relatively recently in human history. Followers of science fiction programmes, such as *Star Trek*, see a future based on resources and technologies that have yet to be created.

8.1.1 Defining resources

In his seminal work *World Resources and Industries*, Zimmerman (Peach and Constantin 1972: 9) states: 'The word "resource" does not refer to a substance, but a function that a thing or a substance may perform, or to an operation in which it might take place.' He goes on to note that: 'resources are . . . as dynamic as civilisation itself.' In fact, one can suggest that each major human civilization was sustained by a particular set of resources and technologies for their exploitation (Simmons 1996). Thus, the archaeological record talks of the *Stone* Age, the *Bronze* Age and the *Iron* Age. In reference to the Industrial Revolution of the nineteenth century, Simmons (1996: 208) states that: 'industrialization based on fossil fuel energy represents a turning point in the history of human-nature relations.' If the nineteenth century was based on the exploitation of coal, then the

Plate 8.1   The use of firewood for cooking in a 'low-energy' society. (Panos Pictures/Sean Sprague)

twentieth century will been seen as the era of oil and gas. Thus, the changing notion of what constitutes a 'resource' is an important factor in shaping the relationship between human societies and the natural environment. Undoubtedly, the hunter-gatherer communities living in the Palaeolithic period had a very different relationship with nature from the industrialists of the nineteenth century (see Section 1). As Blunden (1995: 164) observes, 'Because definition as a resource depends on usefulness to human society, natural materials may be required as resources by societies in some times and places but not in others.' At the time that the industrial revolution was taking place in Europe there were societies located elsewhere on the planet that were still at the hunter-gatherer stage in terms of their concept of what constitutes a resource and in terms of their relationship with nature. This example adds a further dimension to the complexity of the term 'resource'. Its definition can vary across space at the same time. While it is still the case today that people in many parts of the world live in what Mather and Chapman (1995: 139) call 'low-energy societies' (dependent upon plants, animals and human labour), one of the defining characteristics of the twentieth century, and a consequence of the processes of globalization, is that the majority of societies now have at least some shared notion of what sort of things constitute resources. The Inuit living in northern Canada are as dependent on gasoline for their skidoos as the commuters in their cars in Los Angeles or London are. Furthermore, echoing Lenin's famous edict that 'Communism is electrification plus Soviet labour power', the provision of electricity is seen as a central component of the modernization process in all types of societies.

If the notion of resource is dynamic and intimately linked to the evolution of human society, it follows that so-called 'technological progress' both creates and destroys resources. As new technologies emerge, dependent on particular resources, so old technologies and their associated resources become redundant. For example, today we attach little value to flint (it has some value as a building material), yet in the Stone Age it was an essential resource for making tools. Because these resources had 'use value' they were also the subject of trade. On the other hand, many of the resources that were valued in the Bronze Age, such as copper and zinc, are still valued today but they are put to different uses. Thus, some resources retain their value as new ways of using them are discovered. The use value of resources also means that control over their supply is an important part of political and economic power. In theory at least, resource-rich regions are able to exploit this natural advantage in their dealings with resource-poor regions. Consequently, gaining control over particular resources has been at the heart of many wars and much conflict. The European colonization of what we now call the 'developing world' was in part motivated by a desire to discover and control new sources of resources. It is also no surprise that many of the world's largest corporations, until recently at least, were involved in resource development. Our discussion so far has implied that resources are a homogeneous thing. However, we know that there are many different types of natural resource and the various ways of classifying them is the subject of the next section.

8.1.2 Classifying resources

Natural resources are commonly divided into two types: **non-renewable** or **stock resources** and **renewable** or **flow resources**. Figure 8.1 presents a classification of resources on this basis. Stock resources are those, mainly mineral, that have taken millions of years to form and so their availability is finite. Hence, we also refer to them as non-renewable as there is no possibility of their being replenished on a time scale of relevance

Non-renewable/stock			Renewable/flow	
Consumed by use	Theoretically recoverable	Recyclable	Critical zone	Non-critical zone
Oil	All elemental minerals	Metallic minerals	Fish	Solar energy
Gas			Forests	Tides
Coal			Animals	Wind
			Soil	Waves
				Water
			Water in aquifers	Air

Flow resources used to extinction

Critical zone resources become stock once regenerative capacity is exceeded

Figure 8.1　A classification of resource types. *Source*: Rees, 1985: 15

to human society. For example, on the basis of the geological time scale, it is possible for new deposits of coal, oil and gas to be created; however, the time this will take means that we only have available to use the stock of already created fuel minerals. Within the category of stock resources, it is also useful to distinguish between those that are consumed by use, such as fuel minerals, those that are theoretically recoverable and those that are recyclable, such as aluminium. A further characteristic of stock resources is that they tend to be highly localized, that is they are found in relative abundance only in specific places, such as ore deposits and coalfields. Some stock resources are more abundant than others and their relative scarcity affects their value. For example, aggregate minerals, like sand and gravel, are relatively abundant, while precious metals, such as gold and silver, are relatively rare.

Renewable or flow resources are those that are naturally renewed within a sufficiently short time span to be of use to human society. Again we can distinguish between different types of stock resource. Figure 8.1 divides flow resources into 'critical zone' and 'non-critical zone' resources. The distinction here is between those flow resources whose continued availability is dependent upon management by society (critical zone) and those that will continue to be available independent of the actions of society (non-critical zone). As indicated by the arrows, it is possible for critical zone flow resources to become stock resources if they are mismanaged and their regenerative capacity is exceeded. Thus, the aim of resource management is to ensure that exploitation of a particular renewable resource does not damage its capacity to replace itself. In recognition of the increasing challenge, and the numerous failures, to manage renewable resources, Rees (1991: 8) has developed an alternative to the conventional two-part typology of natural resources: 'All resources are renewable on some timescale ... what matters for the sustainability of future supplies is the relative rates of replenishment and use ... it seems better ... to think in terms of a "resource continuum" than the conventional two-part typology' (Figure 8.2). The combined impact of population growth and industrialization is placing increasing stress on the planet's finite supply of fresh water and clean air. According to UNEP (1999: xxii), about 20 per cent of the world's population currently lacks safe drinking water, while 50 per cent lacks access to a safe sanitation system. Access to safe drinking water is likely to be one of the most important resource management problems in the twenty-first century, with implications not just for human health, but for the entire global ecosystem.

Figure 8.2   The resource continuum. *Source*: Rees, 1991: 9

The resource continuum recognizes that (clean) air and water supply now have scarcity value and should be considered as finite.

In our discussion so far we have highlighted the complexity of the notion of natural resource and considered the various ways of classifying resources, but what factors actually affect the availability of particular resources?

8.1.3 Resource availability

As was noted earlier, at any moment in time there is a finite stock of natural resources on the planet, the resource base. Each resource has in turn its own resource base, the total quantity of a substance or property on the planet, for example, the total amount of oil in existence today. However, that total resource base is *not* the amount available for human exploitation. Figure 8.3 illustrates the relationship between the resource base and the various sub-divisions of resource availability. The term **proven reserve** is applied to those deposits that have already been discovered and are known to be economically extractable under current demand, price and technological conditions. Thus, the extent of the proven reserves of a particular resource is highly dynamic and dependent upon a host of partly

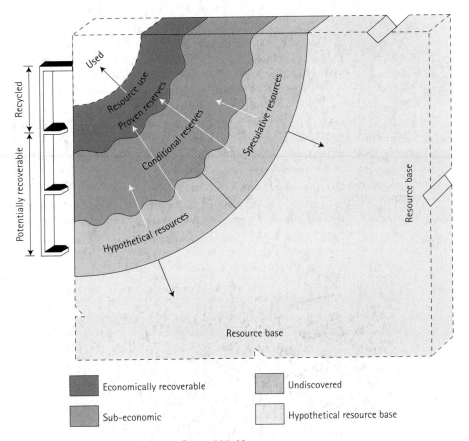

Figure 8.3 Resource availability. *Source*: Rees, 1985: 20

Plate 8.2   As offshore technology advances, so more remote resources can be exploited. (Popperfoto/Reuters)

interlinked factors. These include the availability of the technology and skills to exploit the resource, the level of demand, the cost of production and processing, the price it can command in the market place, the availability and price of substitutes, and the environmental and social costs of developing the resource. These factors determine whether or not a particular resource will be exploited in a particular place and the global level of supply of that resource. The extent to which each of these factors influences resource development also varies across space and time. Today we have the technological ability to recover resources in geological and environmental conditions that were previously uneconomic. For example, the exploitation of North Sea oil and gas in a physically challenging environment has been made possible by advances in offshore drilling and production technologies.

The category of **conditional reserve** refers to deposits that have already been discovered but that are not economic to work at present-day price levels with the currently available extraction and production technologies. The boundary between proven and conditional reserves is dynamic and bi-directional (that is resources that change from conditional to proven reserves can, if conditions change, revert to conditional status). The two remaining categories of resource are not readily available to society. **Hypothetical resources** are those that we may expect to find in the future. They are in areas that have only been partially surveyed and developed. They may be in regions, such as Antarctica, where the international community has placed a moratorium on resource development, for the time being at least. **Speculative resources** are those that might be found in unexplored areas which are thought to have favourable geological conditions. Finally, there

Table 8.1 The dimensions of resource scarcity

Type of scarcity	Concern
Physical scarcity	• Exhaustion of minerals and energy. • Human populations exceed the food production capacity of the land. • Depletion of renewable resources such as fish, soils or timber.
Geopolitical scarcity	• Use of minerals exports as a political weapon (e.g. sales embargoes). • Shift in the location of low-cost minerals sources to 'hostile' blocs of nations.
Economic scarcity	• Demand at current price levels exceeds the quantity supplied (therefore shortages). • Needs exceed the ability of individuals or countries to pay for resource supplies. • Rich economies can always outbid the poor for essential resources, creating unequal patterns of resource use. • Economic exhaustion of specific minerals or renewable resources causes economic and social disruption in producer regions or in nations dependent on them.
Renewable and environmental resource scarcity	• Distribution of essential biogeographical cycles (e.g. the carbon dioxide cycle and the greenhouse effect) threatening sustainability of life on earth. • Pollution loads exceeding the 'absorptive' capacity causing economic health and amenity problems. • Loss of plant and animal species and landscape values, with wide, but poorly understood, long-term consequences.

Source: Rees (1991: 6).

remains a large part of the earth about which we have no information on its potential resource base. The strength of this classification is that it stresses the highly dynamic nature of the concept of resource reserve. The danger is that it leads to the view that society will never run out of resources because there are always more to discover and technological progress will continue to make new resources available for exploitation. Even if the latter were the case, and clearly it is not, the planet now faces the additional problem that consuming resources, such as hydrocarbon fuels, is actually threatening the global ecosystem. Thus, there is a need to rethink the whole notion of resource availability to take into account the ecological cost of our current 'fossil fuels society'.

Given the highly dynamic nature of resources, it is very difficult to estimate at any moment in time the level of resource availability. It is even more difficult to speculate about future levels of production and consumption and, thus, the possibility of resource scarcity. This is not just because extrapolation on the basis of current trends is often misleading, but also because there are a whole variety of factors that can promote resource scarcity. Table 8.1 identifies a variety of different types of scarcity. The resource crisis of the 1970s was motivated by geopolitics, but by increasing the price of energy, it brought about other forms of scarcity. For example, the increased price of oil had a major negative impact upon those countries in the 'developing world' which had embarked upon industrialization and had become increasingly dependent upon imported oil. However, it also shocked the 'developed world' into the realization that energy resources were finite, energy conservation was worthwhile and that alternative sources of energy were required. The 'developed world', and particularly the United States, came to realize the strategic importance of securing access to energy supplies. Today, despite continued geopolitical

Plate 8.3   The energy crisis in 1973 brought lines at the petrol pumps and a 3-day week. (Associated Press)

tensions, we have a relative abundance of oil and prices are now lower in real terms that before the energy crisis of the 1970s. But, as noted above, society is increasingly concerned about the ecological consequences of increased energy consumption. The remainder of this chapter looks in detail at the critical question of energy.

8.2   Fuelling the planet

Of all the different types of resource that we have discussed so far, in recent history it is those that provide a source of energy that have been the most sought after. Figure 8.4 provides a simple classification of the different types of energy resource. For most of human history societies have utilized renewable sources of energy; flow resources such as wood that can be depleted, sustained or increased by human activity; and continuous resources, such as water to drive watermills or the wind to turn windmills, that are available irrespective of human activity. They have also used draught animals, horses, buffalo, etc., but these need to be fed and housed. The industrial revolution changed the way in which in certain parts of the world society powered their economies, while the rest of the world remained dependent upon renewable resources.

8.2.1 The dominance of fossil fuels

Since the invention of the steam engine about 200 years ago, much of human society has become ever more dependent upon the exploitation of non-renewable energy resources.

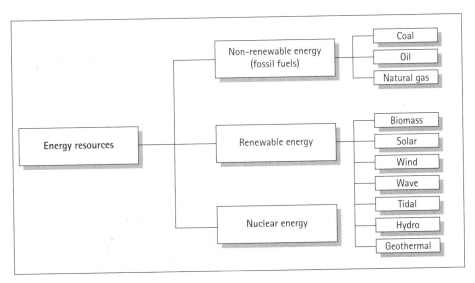

Figure 8.4   Categories of energy resources. *Source*: Jones and Hollier, 1997: 173

Today three fossil fuels account for over 85 per cent of the annual sale of the world's most important minerals. These resources were formed from the decomposition of organic materials millions of years ago and have been transformed by heat and pressure into coal, oil and natural gas. They represent the classic non-renewable, stock, resource, found in specific locations and in a variety of forms and conditions. For example, commercial coal deposits are of three types: lignite (brown coal), bituminous coal and anthracite. There are different grades of oil along a scale from light to heavy and individual refineries are designed to process particular grades of crude oil. Natural gas is often found in association with oil, but is also found on its own. Until recently, the natural gas associated with oil was simply flared off. Today it is often re-injected into the oilfield to enhance oil recovery or used as a hydrocarbon resource in its own right. The mineral content of natural gas also varies: for example, so-called 'sour gas' has a high sulphur content. This sulphur has to be extracted before the gas can be transported by pipeline. The advantage of fossil fuels as a source of energy is that they are readily accessible, easy to convert by using proven technologies, and are cost-efficient in production and use.

Of the three types of fossil fuel, oil is the most versatile. Today coal is used in many industrial processes and as a source of heat and remains the most important fuel for electricity generation. However, it is a relatively bulky resource and in recent years, particularly in Western Europe, it has been unable to compete with oil that is more versatile or cleaner natural gas. The great advantage of oil is its transportability. It can be transported over long distances in large volumes by pipelines, it can cross oceans in tankers, or it can be moved in smaller quantities by rail and road tankers. It has spawned a huge petrochemical industry that now produces a vast range of products. Many of these products, such as plastics and textiles, have substituted products based on renewable resources, such as wood and cotton, for products based on non-renewable resources. At the same time, oil has created many resources for which we have yet to develop commercially viable

Plate 8.4   Special LNG tankers enable natural gas to be delivered to markets where pipelines do not exist. (Quadrant Picture Library)

substitutes, such as jet engine fuel. This means that advanced industrial societies are totally dependent upon the continued supply of oil.

In recent years, natural gas has become an increasingly important source of energy. In many advanced industrialized economies it has even replaced coal and, to a lesser extent, oil as the favoured resource for electricity generation and domestic heating. However, natural gas does not have the same transport flexibility as oil. It is most economic when transported by pipeline, but the infrastructure investment needed requires secure sources of production and stable markets. Natural gas can also be processed to produce liquefied natural gas (LNG), but the liquefaction process requires substantial investment in plant and consumes large quantities of energy. Nevertheless, LNG is becoming an increasingly important fuel, particularly in the Asia-Pacific region.

8.2.2 Alternative sources of energy

Given the finite nature of hydrocarbon fuels, the problems of scarcity and the environmental consequences of burning fossil fuels, it is no surprise that alternative sources of energy have been sought. As Figure 8.4 shows, these alternatives take two forms: one, the development of new ways of harnessing renewable energy sources, and two, the development of nuclear power. At present all commercial nuclear power stations use so-called fission reactors in which atoms of uranium-235 and plutonium-239 are split by neutrons, thus releasing energy, mainly in the form of heat, which is then used to generate electricity. In the 1950s nuclear energy was heralded as a cheap and clean alternative to fossil fuels. However, the nuclear industry has failed to expand as expected (see Box 8.1).

Today, few countries are expanding their nuclear programmes and many now face the problem of decommissioning the older generation of reactors. Furthermore, the collapse of the Soviet Union has left the crisis-ridden post-Soviet republics with a nuclear energy system they cannot afford to maintain. In retrospect, the current nuclear energy industry

Box 8.1   The rise and fall of nuclear power

During the 1950s the peaceful use of nuclear power was heralded as the great hope for the world's energy needs. Harnessing the power of the atom promised a cheap and clean source of electricity. The energy crisis of the 1970s increased interest in a nuclear solution. By developing a nuclear power industry, the developed world hoped that it could reduce its reliance upon OPEC-supplied oil and at the same time reduce emissions of CO_2. The 1970s and 1980s saw a rapid expansion in the generating capacity of the world's nuclear power plants. Then, in the last decade, growth has stalled and there is the real possibility of a decline in total capacity. What went wrong?

In short, as the nuclear power industry developed, expanded and matured it became increasingly apparent that the supposed benefits were far outweighed by the environmental problems it posed. Reddish and Rand (1996: 79) have listed four criticisms levelled at nuclear power by environmental groups:

- That abnormal radiation levels from normal operations will cause cell damage, malignant cancers, genetic diseases, etc.

- That reactor operations and transport of irradiated fuels cannot be guaranteed safe against catastrophic accidents.

- That radioactive waste has to be disposed of safely and reactors have to be decommissioned, all of which is extremely costly.

- That the radioactive materials needed to fuel the reactors need to be kept safe and secure from theft or misappropriation, to make nuclear weapons.

Over the years public trust in nuclear power has been significantly eroded by a number of serious accidents such as those at Windscale, Cumbria, in the UK in 1957, Three Mile Island in the USA in 1979 and Chernobyl in the Ukraine in 1986. Furthermore, as the true cost of building, maintaining and decommissioning nuclear power plants has become apparent many have questioned the economics of nuclear power.

Despite these problems, in 1998 nuclear power accounted for 16 per cent of global electricity generation; but it has been very much an option for the developed world, partly due to cost and technology, but also due to concerns for the non-proliferation of nuclear weapons. In 1996, OECD member states accounted for 86.6 per cent of total nuclear power production. The top three countries were the United States 29.6 per cent, France 16.4 per cent and Japan 12.5 per cent. In the same year, nuclear power accounted for 78.2 per cent of total domestic electricity generation in France and 52.5 per cent in Sweden. But Sweden has decided to phase out its nuclear power stations and to switch to conventional thermal power stations. Prior to the Asian financial crisis, nuclear power was seen as a solution to Asia's shortage of indigenous energy resources. However, the high cost of constructing nuclear power stations, rising public opinion against them and the increasing availability of cheaper alternatives, such as LNG, means that substantial expansion is now unlikely. In

(Box continued)

Box 8.1   (Box continued)

North America, Western Europe and the former Soviet Union, decommissioning of old reactors is not being mirrored by construction of new capacity. Thus, while nuclear power will continue to provide an important source of electricity, it is no longer seen as the obvious solution to our energy problems.

Plate 8.5   The Chernobyl disaster spelled the end for nuclear power as a clean and safe alternative source of energy. (Science Photograph Library/Novosti)

may be seen to have created as many problems as it has solved. There is still the possibility that a new generation of fusion or breeder reactors can fulfil that early promise, but few forecasters see a significant increase in the role of nuclear power in the short term.

The other alternative to fossil fuels is the development of 'new' renewable sources of energy. Many of these renewable sources involve applying new technologies to historic sources of energy supply.

- *Biomass energy* involves the burning of plant and animal residue to produce heat. In much of the world this is still the traditional source of energy. According to the IEA's Biomass Project, the share of biomass energy in Africa, Asia and Latin America averages 30 per cent. However, in some of the poorest countries of Asia and Africa it can be as high as 80–90 per cent. It is estimated that for some 2 billion people biomass is the only available and affordable source of energy (see Box 8.2).

- *Solar energy* usually involves capturing the sun's rays in order to warm buildings and heat hot water. As the efficiency of solar panels increases and their cost declines, solar energy is becoming a viable small-scale local solution. However, it is subject to daily or seasonal variations and is weather-dependent, but it can provide an efficient supplemental source of energy for heating domestic hot water, etc.

- *Wind power* is of historical significance and has now made a return in the form of 'wind farms'. Improved design has increased the efficiency of modern windmills and they are becoming an increasingly common sight in exposed coastal and upland locations in the developed world. Wind farms can generate electricity for local use or for the power grid. However, some would argue that modern wind farms lack the aesthetic qualities of traditional windmills and therefore they are seen by many as a source of visual pollution.

Plate 8.6   Windfarms: a sustainable energy source for some, an eyesore for others. (Popperfoto)

Box 8.2   Biomass energy: problem or solution?

By its very nature, the contribution of biomass energy to global energy consumption is very difficult to measure. It largely takes place outside the market economy and is the result of unpaid (female and child) labour. Because it is not part of the commercial energy mix, it seldom figures in official statistics. Yet, the IEA has estimated that it currently represents 14 per cent of world final energy consumption, a higher share than that of coal (12 per cent) and comparable to those of gas (15 per cent) and electricity (14 per cent) (*www.iea.org/ead/biomass.htm*). Biomass energy in the form of fuelwood for cooking and heating is particularly important in countries in Asia and Africa, but there are considerable variations between countries in the two regions. Take Africa as an example. According to the UNEP (1999: 64) in Africa as a whole 'traditional' sources of energy account for 36.9 per cent of total consumption and 'commercial' energy 63.1 per cent. However, in Eastern Africa and the Indian Ocean islands traditional sources account for 83.9 per cent of energy use and in Western and Central Africa 74.2 per cent. By contrast, in Southern Africa commercial energy accounts for 79.2 per cent of energy use and in North Africa 96.8 per cent. Thus, problems related to fuelwood are concentrated in the Sahel regions and the tropical rainforest of equatorial Africa. In the 1970s and 1980s the international agencies were concerned with the so-called 'fuelwood crisis'. They saw the collection of fuelwood in the Sahel as a major cause of deforestation, desertification and hunger. Subsequent research has shown the situation to be far more complex. Undoubtedly, the collection of firewood is a contributory factor, but the clearing of land for agriculture and the encroachment of urban areas are just as, if not more, problematic and are increasing as the population continues to grow. Thus, biomass energy places added stress on an already threatened and diminishing forest ecosystem. The problem is that there are so many conflicting demands being placed on the forest. In Africa, commercial forestry is of limited significance at present, but in Latin America and Asia it is a major threat. Everywhere, the pressure to clear land for agriculture and settlement is ever increasing. Ultimately, removal of the forest means a reduction in the supply of fuelwood.

The UNEP (1999: 57), reviewing the environmental situation in Africa, has concluded: 'Africa's forests are threatened by a combination of factors including agricultural expansion, commercial harvesting, increased firewood collection, inappropriate land and tree tenure regimes, heavy livestock grazing, and accelerated urbanization and industrialization.' The same report notes that the consumption of fuelwood nearly doubled during the 1970–94 period and that consumption could rise by 5 per cent by 2010. This pressure on the forest resource is unsustainable, but what should be done? One solution would be to encourage the use of commercial sources of energy based on fossil fuels, but this would require people to earn money and would likely increase deforestation to expand agricultural production; it would also promote further urbanization. An alternative approach would be to manage the development of firewood plantations, on both a commercial and non-commercial basis. Middleton (1999: 256–7) advocates the development of dedicated energy plantations on land that has been deforested. This has the added benefit of arresting problems caused by the removal of the forest. He also notes that such a use of biomass fuels produces no net emissions of carbon dioxide since the amount released into the atmosphere by burning is then taken up by growing plants. Thus, in the future sustainable development (see Box 8.4) of biomass energy could offer a solution to the problem of deforestation, rather than contribute to the problem, as well as meet the energy needs of the poorest inhabitants of the developing world.

- *Hydropower generation* also pre-dates the industrial revolution, but the water-mills that powered the early machines have now been replaced by dams and hydro-electric schemes. In many regions of the world hydro-electricity has become an important source of energy, but, in all but a few cases, it cannot offer a large-scale solution to a country's energy needs. The building of dams to store water and build a 'head' to drive the turbines that generate the electricity floods large areas of land; it also starves river systems of flow and damages deltaic and riverine environments. The construction of Three Gorges dam on the Yangtze river in China involves planning the displacement of 1.2 million people and the creation of a reservoir occupying 111,000 hectares. Other forms of waterpower, such as wave and tidal power, are also being developed; again, they cannot provide a viable alternative to fossil fuel power generation.

- The final form of renewable energy is *geothermal energy*. As the name suggests, this form of energy uses naturally occurring heat in the earth's crust and is only really viable in areas of volcanic activity. There are two forms of geothermal energy. One is wet-rock geothermal energy where steam or hot water is trapped from boreholes or surface vents and used to heat buildings or generate electricity. The other is hot-dry geothermal energy, which is accessed by boreholes drilled into hot dry rocks. Water is then forced down and the steam used to generate electricity. Geothermal energy, by its very nature, is a highly localized energy source and requires careful management. For example, there is now evidence from New Zealand that excessive extraction of geothermal energy has reduced the level of geyser activity, creating a conflict between the desire to generate power and heat houses and the need to maintain the revenues from tourists attracted by the volcanic springs.

In nearly all of the above cases, renewable energy seems capable of providing localized, small-scale solutions to energy needs. At present, renewable energy sources do not provide a viable alternative to fossil fuels. Furthermore, it could be argued that most of these alternatives require considerable capital investment and access to advanced technology; therefore, they do not offer a solution to the fast-growing energy needs of the 'developing world'.

8.2.3 The changing energy mix

It follows from the discussion above that the relative importance of the various sources of energy has varied across time and space. The balance between the various sources of energy is known as the **energy mix**. During the last century, there was a substantial change in the global energy mix, but as Table 8.2 shows, there still remains substantial regional variation. The evolution of the energy mix tends to be linked to the development of the industrialized world, simply because it accounts for most of the world's energy consumption (Table 8.3). At the beginning of the twentieth century, coal was by far the most important energy resource in the industrialized world. The oil industry was still in its early stages; the first modern oil well was drilled in Pennsylvania in the United States in 1859. Natural gas was not recognized as a resource at all. It was not until after the Second World War that the dominance of coal was threatened. Advances in transportation technology, such as the invention of the jet engine, and the emergence of a petrochemicals industry led to a rapid increase in the demand for oil in the post-war boom

Table 8.2   Regional variations in energy balance, 1999 (per cent of total consumption)*

	Oil	Natural gas	Coal	Nuclear	HEP	Total (mtoe)
OECD	42.9	22.3	21.3	11.2	2.3	5,060.0
EU 15	44.6	23.1	14.5	15.7	2.0	1,416.8
North America	40.9	25.5	22.7	8.6	2.3	2,557.3
Europe	41.9	22.2	19.3	13.7	2.9	1,800.8
Former Soviet Union	20.0	53.1	18.9	5.8	2.1	908.1
S & C America	58.9	22.6	5.5	0.8	12.2	371.2
Africa	44.3	18.0	34.2	1.5	2.1	261.2
Middle East	56.5	41.5	1.8	0.0	0.2	380.3
Asia Pacific	41.2	10.7	40.5	5.6	2.0	2,254.7
World	40.6	24.2	25.0	7.6	2.7	8,533.6

* Includes commercially traded fuels only. Therefore excludes wood, peat and animal waste as fuels.

Source: BPAmoco, (2000: 38). (Also available at: http://www.bp.com/worldenergy/primary)

years. In 1950 coal accounted for 61 per cent of the world's commercial energy consumption and oil 27 per cent. By 1970 coal had fallen to 30 per cent, oil had increased to 44 per cent and natural gas to 20 per cent. The absolute level of coal production did not decline (in fact at a global scale coal production continues to increase). The rapid economic growth experienced by the developed world was instead increasingly underpinned by oil and, more recently, by natural gas. The recent 'dash-for-gas' in the developed world is being driven by a desire to reduce dependence on oil imported from the Middle East and by the fact that using natural gas to produce electricity produces less greenhouse gases than coal or oil. Thus, in the last hundred years, the energy mix in the industrialized world has gone through two major transitions, from coal to oil and from oil to gas.

These energy transitions have not been a universal experience. As Table 8.2 illustrates, the energy mix in the developing world is somewhat different. For the most part the developing world remains more reliant upon oil for its commercial energy and in some

Table 8.3   Global distribution of primary energy consumption, 1999 (per cent of total world consumption)*

	Oil	Natural gas	Coal	Nuclear	HEP	Total
OECD	62.7	54.7	50.7	86.7	51.4	59.3
EU 15	18.3	15.9	9.7	34.3	12.7	16.6
North America	30.2	31.6	27.3	33.7	25.7	30.0
Europe	21.8	19.4	16.3	37.8	22.7	21.1
Former Soviet Union	5.3	23.4	8.0	8.1	8.5	10.6
S & C America	6.3	4.1	1.0	0.4	20.0	4.3
Africa	3.3	2.3	4.2	0.6	2.5	3.1
Middle East	6.2	7.6	0.3	0.0	0.4	4.5
Asia Pacific	26.8	11.7	42.9	19.3	20.4	26.4

* Includes commercially traded fuels only. Therefore excludes wood, peat and animal waste as fuels.

Source: BPAmoco (2000: 38). (Also available at: http://www.bp.com/worldenergy/primary)

regions coal remains the dominant resource. For example, in China coal still accounts for 75 per cent of commercial energy production and China accounts for 30 per cent of global coal consumption. Elsewhere dependence upon oil has made the developing world particularly vulnerable to the sudden increases in oil price that have been experienced at times over the last three decades. But, unlike the developed world, these economies lack the capital and technology to develop their own energy potential or to diversify their energy mix. However, as the developed world has depleted its own sources of oil and gas it has become increasingly dependent upon supplies from the developing world. Thus there has been a marked change in the geography of energy production and, to a lesser extent, consumption. The transnational oil companies have been key actors in developing the energy resources of the developing world, but increasingly the governments of the oil-producing states in the developing world have seized control over their oil and gas industries. Now the leading oil majors, such as BP-Amoco, Exxon-Mobil and Shell, find themselves working in partnership with regimes whose actions in relation to human rights and environmental protection are condemned by governments and NGOs in the developed world.

8.2.4 The changing geography of energy production and consumption

Whereas in 1950 North America, Europe and the Soviet Union accounted for nearly 90 per cent of world energy demand, by 1990 their share had fallen to two-thirds (Jones and Hollier 1997: 181). In 1997 the member states of the Organization for Economic Cooperation and Development (OECD) accounted for 58.4 per cent of global energy consumption (Table 8.3). The redistribution of the global pattern of energy consumption has not been due to decline in the developed world, although during the 1990s this has been the case in the so-called 'transition economies', but to increased demand in the

Table 8.4   Energy production and efficiency

	Commercial energy use, average annual % growth 1980–95	Commercial energy use, per capita kg of oil equivalent 1995	GDP per unit of energy use (1987 $ per kg oil equivalent)	
			1980	1995
World	3.2	1,474	2.2	2.4
Low income	5.5	393	0.9	1.1
Excl. China & India	5.8	132	3.4	2.7
Middle income	5.8	1,488	1.2	1.1
Lower middle income	9.0	1,426	1.0	1.0
Upper middle income	1.9	1,633	1.7	1.5
Low & middle income	5.6	751	1.1	1.1
East Asia & Pacific	5.3	657	–	0.9
Europe & Central Asia	8.8	2,690	–	0.6
Latin America & Carib.	2.7	969	2.2	2.0
Middle East & N. Africa	5.2	1,178	3.3	1.8
South Asia	6.6	231	2.0	1.7
Sub-Saharan Africa	2.0	238	2.1	1.9
High income	1.7	5,123	2.9	3.4

Source: World Bank (1998: 144–8).

Table 8.5   Coal reserves at end of 1999 (million tonnes)

	Anthracite and bituminous	Sub-bituminous and lignite	Total	Share of total %	R/P ratio
North America	116,707	139,770	256,477	26.1	239
S. & Cent. America	7,839	13,735	21,574	2.2	474
Europe	41,664	80,368	122,032	12.4	161
Former Soviet Union	97,476	132,702	230,178	23.4	*
Africa & Middle East	61,355	250	61,605	6.2	268
Asia Pacific	184,450	107,895	292,345	29.7	164
Total World	509,491	474,720	984,211	100.0	230
Of which OECD	206,483	240,617	447,100	45.4	221

* More than 500 years

Source: BPAmoco (2000: 30). (Also available at: http://www.bp.com/energy/coal)

Table 8.6   Proved oil reserves at end of 1999

	Thousand million tonnes	Share of total %	R/P ratio
North America	8.4	6.2	13.8
S. & Cent. America	12.9	8.6	37.7
Europe	2.7	2.0	8.3
Former Soviet Union	9.0	6.3	24.2
Middle East	91.5	65.4	87.0
Africa	10.0	7.2	28.2
Asia Pacific	5.9	4.3	16.3
Total World	140.4	100.0	41.0
Of which OECD	11.3	8.3	11.8
OPEC	109.1	77.6	77.4
Non-OPEC*	22.3	16.0	13.6

* Excludes Former Soviet Union

Source: BPAmoco (2000: 4). (Also available at: http://www.bp.com/energy/oil)

developing world. Table 8.4 provides information on the average annual growth in commercial energy use between 1980 and 1995. During this period global energy consumption grew by 3.2 per cent, but growth in the high income economies (the developed world) was only 1.7 per cent a year, while growth in the low- and middle-income countries (the developing world) was 5.6 per cent a year. Thus, in relative terms, demand for energy in the developed world has been slowing, while it has been increasing in the developing world (compare this pattern with that of population growth discussed in the previous chapter). This differential growth rate, together with the decline in indigenous production in the developed world, has major implications for the geography of energy production and consumption and for global geopolitics.

Table 8.7   Proved natural gas reserves at end of 1999

	Trillion cubic metres	Share of total %	R/P ratio
North America	7.31	5.0	10.0
S. & Cent. America	6.31	4.3	66.2
Europe	5.15	3.5	17.6
Former Soviet Union	56.70	38.7	81.8
Middle East	49.52	33.8	*
Africa	11.16	7.7	98.2
Asia Pacific	10.28	7.0	40.4
Total World	146.43	100.0	61.9
Of which OECD	13.34	9.1	12.8
European Union 15	3.25	2.2	15.0

* Over 100 years

Source: BPAmoco (2000: 20). (Also available at: http://www.bp.com/energy/gas)

There are at least three elements to the geography of energy production: the global distribution of reserves, the distribution of energy production and consumption and the resultant pattern of trade between energy surplus and energy deficit regions. As we noted earlier, the potential to produce coal, oil and natural gas is geologically determined, but whether or not a given deposit is exploited is the consequence of a whole host of factors. The regional distribution of global reserves of coal, oil and natural gas are presented in

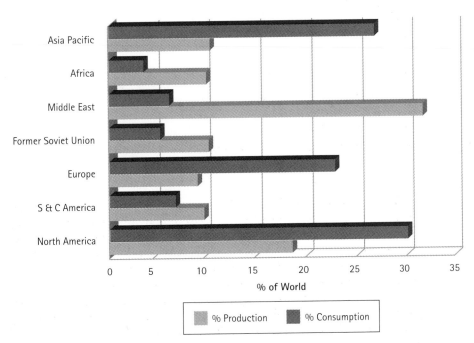

Figure 8.5   Distribution of oil production and consumption in 1998.

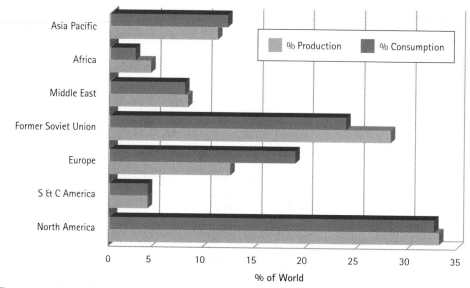

Figure 8.6   Distribution of natural gas production and consumption in 1998.

Table 8.8   Producers, exporters and importers of crude oil

Producers in 1997	Mt	% of world	Exporters in 1996	Mt	Importers in 1996	Mt
Saudi Arabia	439	12.8	Saudi Arabia	327	United States	432
United States	383	11.1	Norway	142	Japan	225
Russia	303	8.8	Russia	126	Germany	104
Iran	182	5.3	Iran	125	Korea	100
Venezuela	175	5.1	UAE	100	France	84
Mexico	168	4.9	Venezuela	96	Italy	82
China	160	4.7	Nigeria	85	Netherlands	62
Norway	156	4.5	United Kingdom	81	Singapore	55
United Kingdom	128	3.7	Mexico	79	Spain	55
UAE	120	3.5	Kuwait	64	United Kingdom	51
Rest of world	1,227	35.6	Rest of world	498	Rest of world	538
World	3,442	100	World	1,722	World	1,788

Source: www.iea.org/stats/files/keystats/stats_98.htm.

Tables 8.5, 8.6 and 8.7. Each table also includes the reserve to production ratio at the end of 1997. This is a measure of how many years the current proven reserves would last if current levels of consumption were maintained. From a comparison of the reserve to production ratios for the three fossil fuels, it is quite clear that coal is the most abundant fossil fuel, followed by natural gas and then oil. At the end of 1997 the world had sufficient reserves of coal to last 219 years at current rates of consumption, while current oil reserves would last 40 years. These tables also reveal some interesting regional variations in the level of reserves. While the OECD countries have more than sufficient coal to meet their

Figure 8.7 Movement of oil production in 1998 (million tonnes). *Source:* BPAmoco, 1999: 19

needs, it is these economies that have reduced their reliance upon coal the most. Thus, the developed world is dependent upon the very fuels for which it has the lowest reserves to production ratios. What this means is that the developed world is likely to become even more dependent upon imports of energy from the developing world. In the case of oil reserves, the Middle East still accounts for the vast majority of the world's oil. Natural gas reserves are provided by two dominant regions, the Middle East and the former Soviet Union (predominantly Russia). Thus, the majority of the world's oil and gas reserves are in regions that are presently politically and/or economically unstable.

The relationship between production and consumption is illustrated in Figures 8.5 and 8.6. These diagrams compare the share of world production and consumption by major region. The comparative distributions of oil production and consumption show that Asia–Pacific, Europe and North America are the major deficit regions with Africa, the Middle East, the former Soviet Union and South and Central America as the major surplus regions. The individual countries involved in the resultant trade are identified in Table 8.8 and the movements of oil are shown in Figure 8.7. Despite being the second largest producer in 1997, the United States still imports almost as much oil as it produces. Likewise, North Sea production falls well short of meeting Europe's demand for oil. The result is a movement of oil production dominated by Middle Eastern supply to Europe and North America. Figure 8.7 also reveals substantial movements to Singapore and Japan.

The major exporters and importers of natural gas are listed in Table 8.9, while Figure 8.8 shows the movements of natural gas. A comparison of the two figures reveals the less 'transportable' nature of natural gas; for the most part there is a balance within each region. The notable exceptions are Africa and the former Soviet Union (Russia and Turkmenistan) which have a surplus of natural gas and Europe that has a deficit. Natural gas supply into Europe is via pipeline from West Siberia (Russia) and North Africa or as LNG from North Africa. There is also a growing movement of natural gas in the form of LNG to Japan and South Korea. In 1995 LNG accounted for 14.3 per cent of Japan's mineral fuel imports (Japan Institute for Social and Economic Affairs 1996: 74–7). At present most of this LNG is supplied from within the Asia–Pacific region, but there is likely to be an increase in supply from the Middle East. For example, in 1995 the top

Table 8.9 Producers, exporters and importers of natural gas

Producers in 1997	Mm³	% of world	Exporters in 1996	Mm³	Importers in 1996	Mm³
Russia	560,895	24.4	Russia	196,474	Germany	87,669
United States	539,068	23.5	Canada	80,117	United States	83,178
Canada	167,828	7.3	Netherlands	48,823	Ukraine	67,600
United Kingdom	91,778	4.0	Algeria	43,459	Japan	64,973
Netherlands	84,595	3.7	Norway	37,922	Italy	37,086
Indonesia	74,233	3.2	Indonesia	35,724	France	35,218
Algeria	69,504	3.0	Turkmenistan	24,000	Belarus	14,345
Uzbekistan	47,252	2.1	Malaysia	19,137	Belgium	13,924
Norway	47,052	2.0	Australia	9,783	Korea	12,532
Saudi Arabia	46,558	2.0	Uzbekistan	9,076	Czech Republic	9,297
Rest of world	567,389	24.7	Rest of world	33,068	Rest of world	116,432
World	2,296,152	100.0	World	536,583	World	542,254

Sources: www.iea.org/stats/files/keystats/stats_98.htm.

Figure 8.8   Movement of natural gas in 1998 (billion cubic metres). *Source:* BPAmoco, 1999: 29

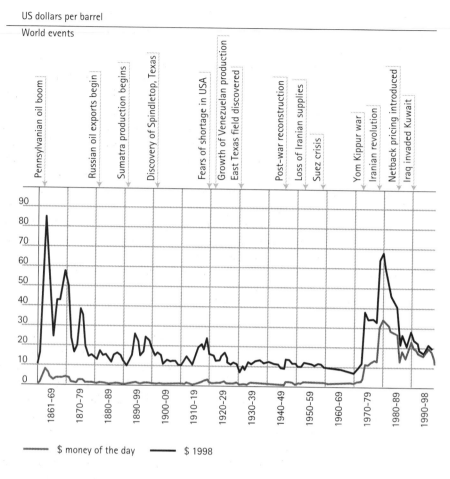

US dollars per barrel

World events

Labels (left to right):
Pennsylvanian oil boom
Russian oil exports begin
Sumatra production begins
Discovery of Spindletop, Texas
Fears of shortage in USA
Growth of Venezuelan production
East Texas field discovered
Post-war reconstruction
Loss of Iranian supplies
Suez crisis
Yom Kippur war
Iranian revolution
Netback pricing introduced
Iraq invaded Kuwait

——— $ money of the day   ——— $ 1998

1861–1944 US average
1945–1985 Arabian Light posted at Ras Tanura
1986–1998 Brent Spot

Figure 8.9 Historical trends in the world oil price. *Source*: BPAmoco. 1999: 14

three suppliers of LNG to Japan were Indonesia (41.2 per cent), Malaysia (18.4 per cent) and Australia (16.5 per cent). There is also the possibility of pipeline supply of natural gas from Russia to China, Japan and South Korea.

The geographies of production and consumption described above are the result of a complex interaction of economic and political factors. Not surprisingly, there are sub-stantial regional differences in the cost of oil production. The record-low oil prices of the late 1990s focused attention upon the actual cost of producing oil. A report by Cambridge Energy Research Associates puts the cost of production in the Middle East as low as US$ 2 a barrel (*The Economist*, 6 March 1999: 29). That same report estimates the cost of production in Indonesia at US$ 6, Venezuela and Nigeria at US$ 7, Mexico at US$ 10, the United States and the North Sea at US$ 11 and Russia at US$ 14. These production costs pose two questions: why develop high-cost fields when the Middle East has massive reserves and very low production costs, and what will happen to the high-cost producers if oil prices fall?

The answers lie in the realms of geopolitics as much as economics. Prior to the first oil shock of 1973–4 the industrialized world had become increasingly dependent upon supplies of cheap oil from the Middle East (Odell 1989). When the Middle Eastern oil producers formed OPEC (the Organization of Petroleum Exporting Countries, visit *www.OPEC.org*) and used oil as a political weapon to punish the industrialized West for supporting Israel in the Arab-Israeli war this brought about a rapid increase in the price of oil. The first oil shock was followed by a further shock in 1979–80 following the Iranian Revolution and a mini-shock in 1990–1 because of the Gulf War (see Figure 8.9). Following the second shock prices started to decline. There were at least three reasons: first, high energy costs promoted conservation; second, the high price of oil and the actions of OPEC promoted production by high-cost non-OPEC producers; and third, recession and economic restructuring reduced the growth of demand in the developed world. Most recently, the economic crisis in Asia has further dampened demand for energy. As a response to very low prices, OPEC members agreed to cut back production and this action, combined with global economic recovery, saw prices recover during 1999. Nonetheless, many analysts believe that OPEC's solidarity will not last and that there will soon be a return to low oil prices.

The consequences of cheap oil might be to drive high-cost producers out of the market, making the developed world even more dependent upon Middle Eastern oil. The geopolitical costs of such a development are too high and it is likely that high-cost production will be sustained. This is, firstly, because the producing countries are overly reliant upon oil production and unlikely to cut back on production and, secondly, because the consuming countries will wish to continue to diversify their sources of supply. That the developed world is prepared to pay more than it has to for oil to avoid dependence upon supplies from the Middle East is a clear illustration of how geopolitics influences the geography of energy production. But what does the current situation mean for the developing world? Much depends on whether or not the developing world follows the same pattern of economic development as the developed world; this is the subject of the final section.

8.3 Energy and development

What is the relationship between energy consumption and economic development? The experience of the developed world suggests that in the initial phases of industrialization there is a direct link between increased energy consumption and economic development. That is that as industrial activity grows it consumes more and more energy. However, in the last thirty years it has become increasingly apparent that in the developed world the relationship between economic development and energy consumption has changed. This was a direct consequence of the OPEC-inspired energy crisis of the 1970s and is largely due to the processes of de-industrialization and economic restructuring that have seen the less energy-intensive service sector replace heavy industry and, to a lesser degree, manufacturing as the major generators of wealth. At the same time, conservation and technological change has made industry more energy-efficient. Similarly, a change in

Plate 8.7   When 'muck was money'. (Format Photographers/Sheila Gray)

Plate 8.8   Not a smoke stack in sight on this post-industrial estate. (Popperfoto)

transportation technology has increased energy efficiency. Thus, today's post-industrial societies have decoupled the link between economic growth and increased energy consumption. Nevertheless, it is still the case that the developed world consumes the lion's share of the world's energy. Indeed, just one country, the United States, which accounts for only 5 per cent of the world's population, consumes almost 25 per cent of total world energy consumption (Mather and Chapman 1995: 140).

8.3.1 Energy consumption and economic development

The relationship between energy consumption and economic development is usually depicted as a scatter plot between the level of per capita energy consumption on the one hand, and GNP per capita on the other hand (Figure 8.10). The data do indeed suggest a clear relationship between energy consumption and GNP per capita; the higher the level of energy consumption the higher the level of GNP per capita. There are obvious outliers, but these are easily explained away. Some countries traditionally have higher levels of energy consumption than GNP. This was the case in the Soviet-type economies that had a bias towards heavy industry and were notoriously wasteful in their use of energy. Others, such as Japan, have low levels of energy consumption given their high GNP. This is because they have introduced energy-saving technologies and have exported energy-intensive industry offshore. Finally, there are some countries, such as Canada, where climatic extremes (both hot and cold) require a large amount of energy to be used for heating or for air conditioning. There are also other factors that affect the level of energy consumption: for example, the level of urbanization. But it has also been suggested that there is a similar relationship between economic development and urbanization as there is between the former and energy consumption. Finally, the size of a country may also affect the level of energy consumption, as larger amounts of energy are required to move between places. Such an argument could be marshalled to explain the very high levels of energy consumption in the United States, as could climatic factors.

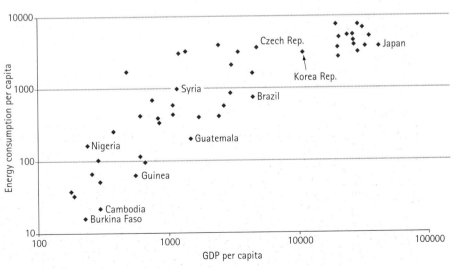

Figure 8.10   Relationship between energy consumption and economic growth.

Box 8.3	Global warming and the Kyoto Protocols

This discussion considers how international actions to reduce greenhouse gas emissions might impact on the global energy futures. While there is still a good deal of controversy surrounding the issue of global warming, there are a number of basic propositions that seem to be accepted. First, that there is a natural greenhouse effect that warms the planet's surface by some 33° C. Without this human habitation would not be possible. Second, since the onset of industrialization, some 150 years ago, there has been a substantial increase in the amount of CO_2 released into the atmosphere. Carbon dioxide is known to be a greenhouse gas, that is it traps long-wave radiation that is emitted by the warming of the earth's surface by the sun and, in doing so, heats up the atmosphere. In addition, industrialization has increased the release of other greenhouse gases, such as methane and nitrous oxide, and has introduced new compounds, chlorofluorocarbons (CFCs), which promote global warming and deplete the ozone layer. Third, because of deforestation, there has been a reduction in the ability of the Earth's ecosystem to absorb CO_2 through the process of photosynthesis. Fourth, these processes promoting an increase in the level of greenhouse gases in the atmosphere come at a time when the Earth is experiencing a warming trend. While it is impossible to prove that this global warming is due entirely to increased greenhouse gas emissions – there are natural processes that could also account for the trend – it is now accepted that increased CO_2 is a major cause of global warming. The single largest anthropogenic source of CO_2 is the burning of fossil fuels (for details on global warming and guidance on further reading, see O'Riordan 2000, Chapter 7). One solution to the problem of global warming is to reduce fossil fuel combustion. The question is how?

The 1992 Earth Summit in Rio (see Box 8.4) resulted in the signing of the United Nations Framework Convention on Climate Change (UNFCC, visit their web-site: *www.unfcc.de* for further details). The ultimate objective of the UNFCC is the 'stabilization of greenhouse gas concentrations in the atmosphere at a level that would prevent dangerous anthropogenic interference with the climate system. Such a level should be achieved within a time-frame sufficient to allow ecosystems to adapt naturally to climate change, to ensure that food production is not threatened and to enable economic development to proceed in a sustainable manner'. The so-called Conference of the Parties (COP comprise all the states that have ratified or acceded to the Convention – over 175 by May 1999) held their first meeting in Berlin in 1995. However, the third COP meeting held at Kyoto, Japan, in 1997 saw agreement on the so-called 'Kyoto Protocols', the first collective action to address carbon dioxide and other greenhouse emissions. An earlier Protocol on Substances that Deplete the Ozone Layer was signed in Montreal in 1987 and dealt with CFCs. The breakthrough at Kyoto was due to a flexible approach that allowed individual countries to set their own targets, while aiming at a 5 per cent global reduction in CO_2 over 1990 levels. The target date for individual countries was set as some time during the four-year period 2008–12. In addition, the developed countries agreed to reduce emissions in 2000 to their 1990 level, a target few will have met. The European Union has been set an emission reductions target of −8.0 per cent, the USA −7.0 per cent and Japan −6.0 per cent. The transition economies of central Europe and the former Soviet Union were granted concessions; for example, Russia has a zero change target. However, those central European countries seeking admission into the EU, such as Poland, accepted reductions. Not surprisingly, a complex set of rules and regulations surrounds the targets and it is even possible to trade 'emissions reductions' between states. Thus, for example, if Japan helped Russia to reduce its emissions below the 1990 level, that reduction would be set against Japan's target.

(Box continued)

Box 8.3	(Box continued)

At the same time as the developed world committed itself to emissions reductions, the developing world was persuaded to accept the concept of the *clean development mechanism* (CDM). The developing countries feel that emissions reductions in the developed world should be on such a scale as to compensate for increased future emissions due to economic growth in the developing world. The concept of CDM is aimed at enabling economic growth without a commensurate increase in greenhouse gas emissions. However, this will require transfer of a considerable amount of capital and technology. In other words, the developed world will have to help the developing world implement the CDM. Yet, it is unclear how this will happen. It is proposed that developed countries can earn credits towards their own reduction targets by assisting in the implementation of CDM in the developing world. O'Riordan (2000: 202) notes that: 'the CDM mechanism denies the right of Third World nations to select their own CO_2 future. ... This is a kind of "ecological colonialism" by another name. While the developed world has access to the capital and technology to achieve the Kyoto targets, it still requires the political will. In the developing world, political will amounts to little without the material requirements to balance the human need for economic growth with the needs of the Earth's ecosystem.'

On the face of it the relationship between energy consumption and economic development seems unproblematic. However, as Chapter 10 reveals, the whole question of what stands for 'development' is problematic and contested. The measurements used in Figure 8.10 systematically understate the relative position of the developing world. First, because the use of commercial energy as a measure of energy consumption ignores the role of non-commercial biomass energy sources, as a consequence it only measures the 'modern' sectors of the economy. Second, the monetary measure of GNP per capita is an inadequate measure of the 'human condition' in many of the world's poorest countries. That said, a re-evaluation of the relative position of the poorest countries would only move them slightly 'up the curve'. It would not alter the fact that there is a clear relationship between industrialization and energy consumption. Even so, industrialization does not necessarily equate to development.

The relationship between industrialization and energy consumption may seem to suggest that all the countries of the developing world will eventually follow the same energy and development trajectory as that experienced by the developed world (see Box 8.3). Thus, the patterns of energy and development can be equated to a stage model of the kind proposed by W. W. Rostow (this is discussed further in Chapter 10). Such a model is presented in Figure 8.11. This model examines the change in energy ratios over time as an economy develops. The energy ratio is the relationship between the rate of change in energy consumption and the rate of change in economic growth (Mather and Chapman 1995: 154). In the pre-industrial phase, the energy ratio is less than one as these are 'low-energy' societies, mainly dependent on subsistence agriculture. As discussed earlier (also see Chapter 2), the industrial revolution resulted in a change in the relationship between society and resource consumption, and the harnessing of new sources of energy was at the heart of this process. As economies industrialize, the energy ratio begins to exceed one. In the early phases of industrialization energy efficiency is low and economies are dominated by energy-intensive industries. As the industrial economy matures so the energy

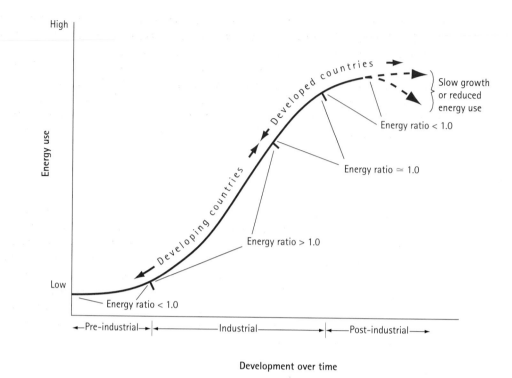

Figure 8.11   Energy ratios and economic development. *Source*: Mather and Chapman, 1995

ratio declines. This is due to increased efficiency, increased energy costs or to the actual decline in the level of economic activity. This stage is typified by the developed world during the 1970s. Eventually, the economy moves into a post-industrial phase and the ratio falls below one. If the model seems very familiar, it is much like the model of the demographic transition (see Chapter 7) because it describes the evolution of the energy economy as it actually occurred in Western Europe and the United States.

What evidence is there that the rest of the world will inevitably have to go through the same stages? It could be argued, with some justification, that the Newly Industrialized Countries (NICs) of Asia have followed this pattern of industrialization; however, the time taken to move through the phases here seems to have been compressed as the NICs have tried to diversify their economies and promote service sector growth. They have also tried to industrialize without the advantage of substantial indigenous energy resources, a fact that has made them vulnerable to the effects of price increases. Somewhat ironically, the financial crisis in Asia was largely responsible for the fall in energy prices in the late 1990. By contrast, the transition economies of the post-socialist world (East Central Europe and the former Soviet Union) pose a different challenge to the model. Prior to 1989–90 these economies were undoubtedly on the steepest part of the curve, consuming large amounts of energy and producing a relatively low level of economic growth. Isolated from the oil price shock by cheap supplies of Soviet oil, these economies had failed to increase their energy efficiency during the 1970s and 1980s. This failure was a major reason for the collapse of the Soviet-type economic system. Transitional recession has followed the collapse of the socialist economies and this has depressed energy consumption. The question now remains, what will the relationship be between energy consumption and economic recovery in the transition economies? Will they follow the path of post-industrialization (see Chapter 12) or will they re-industrialize? The likely answer is that individual economies will follow their own paths (OECD 1999). A much bigger question mark hangs over the developing world. Is it inevitable that it will have to industrialize to improve living standards? If it is, then the availability of energy resources will be very much centre stage for world economic development in the twenty-first century.

8.4 Conclusions: global resource dilemmas

This chapter has considered the nature of resources, through a detailed case study of the relationship between energy and economic development. The chapter concludes by considering the resource dilemmas that now confront the world. It is no longer the relatively simple matter of whether or not the world has sufficient resources to meet demand. In the short term, there is little danger of physical scarcity at a global scale. Rather the essential dilemma is whether the global ecosystem can absorb the consequences of continued increases in resource production and consumption (see Box 8.4). Furthermore, the exact nature of the resource dilemma varies across the globe. Therefore, there is no single resource problem, nor a universal solution.

Box 8.4   Sustainable development

The concept of sustainable development dates back to the early 1970s when the United Nations held the first global environmental summit in Stockholm in 1972. At the time there was increasing concern that humankind would soon run out of resources. Publications such as the Club of Rome's *Limits to Growth* predicted doomsday scenarios. We know those predictions to have been overly pessimistic; nevertheless they rang alarm bells and made us aware of the critical importance of the interrelationship between economy and environment. The energy crisis and the subsequent global recession that followed promoted increased concerns for efficiency and conservation. It became apparent, for the developed at least, that it was possible to maintain economic growth and reduce resource consumption per unit of wealth created. Nonetheless, the continued expansion of the world's population and the growing pressures from the desire for economic development meant that the environment was under ever-increasing stress. In 1980 the IUCN proposed a World Conservation Strategy to address global environmental problems. The strategy had three main objectives:

- To maintain essential ecological processes and life support systems;

- To preserve genetic diversity;

- To ensure the sustainable utilization of species and ecosystems.

The strategy introduced the notion of sustainability, but placed the needs of the environment ahead of those of human society. Consequently, it failed to get the political support necessary to realize its objectives.

The next milestone in the evolution of a global environmental policy was the report of the World

(Box continued)

Box 8.4 (Box continued)

Plate 8.9   The Rio Summit: global problems require global meetings. (Still Pictures/Mark Edwards)

(Box continued)

Box 8.4 (Box continued)

Commission on Environment and Development, known as the Brundtland Report and published in 1987. This report adopted the position that it was possible to pursue economic growth without compromising the environment. The report gave us the most widely accepted definition of sustainable development: 'development which meets the needs of the present without compromising the ability of future generations to meet their own needs.' Few would disagree with such an objective. The problem is that this definition is far too vague; for example, how do we determine present and future needs, and whose needs, anyway? Further attempts have been made to refine the concept and this has resulted in numerous, often contradictory, definitions. The result is that 'Sustainable Development' has become a bland soundbite with little real meaning, except to signal a recognition that the environment matters. Nevertheless, the Brundtland Report placed the relationship between economic growth, social conditions and environmental degradation firmly on the political agenda.

The Earth Summit in Rio in 1992 marks the next stage in the evolution of a global strategy for sustainable development. The summit brought together world leaders, international agencies and interest groups to discuss the Earth's environmental problems. Even before the summit started, the environmentalists had written it off as a publicity stunt. The leaders of the developing world were worried that environmental concerns were being used by the developed world to 'limit development [their development] for the sake of the Earth'. There is not the space here to delve into the achievements, or not, of the Earth Summit; for further reading and guidance on the literature see O'Riordan (2000). Whatever its failings, the point is that the Rio Summit did result in a number of agreements that now form the basis of a global strategy:

- The Framework Convention on Climate Change (see Box 8.3)
- The Convention on Biological Diversity
- Agenda 21
- The Rio Declaration
- Forest Principles
- The Convention to Combat Desertification.

Progress has been slow, but there have been agreements reached to address the problem of climate (see Box 8.3) and the concept of biodiversity has replaced sustainable development as the buzzword for the 1990s. However, few of the difficult political decisions have been made. It is clear that the fears of physical scarcity that prompted the resource crisis of the 1970s have now been replaced by environmental scarcity (see Table 8.1). O'Riordan (2000: 32) observes that the world is not running out of non-renewable resources; rather it is the renewable resource base that is under siege. This is largely due to our inability to balance the desire for economic growth with the needs of the environment. The question remains how to achieve that balance?

- In the advanced post-industrial societies of the developed world, the dilemmas relate to the economic and political costs of *geopolitical scarcity* (see Table 8.1). Increasingly, the economies of the developed world are finding themselves dependent upon what they perceive to be 'hostile' and 'unstable' sources of supply in the developing world. The response has been to develop high-cost resources close to home, to promote increased efficiency and conservation and to seek alternatives, such as renewable energy. This strategy imposes an economic cost as it forces the developed world to pay more for its energy than it needs to. However, decoupling economic growth from energy consumption, increasing energy efficiency and reducing carbon dioxide emissions compensate for the increased cost of energy. The developed world can afford to seek technological solutions to its resources problems, particularly in the area of energy supply.

- In the transition economies of Central Europe and the former Soviet Union resource dilemmas are somewhat different. The Soviet economic development strategy imposed a resource-intensive form of industrialization upon these countries that made them dependent upon cheap supplies from the Soviet Union (predominantly Russia); it also caused widespread environmental degradation. The economic transformation now taking place in the post-socialist world has closed many of the smoke-stack industries; however, the newly independent states in this region now face the problem of having to secure access to resources. The resource-rich states, such as Russia, hope that economic recovery will bring a resource bonanza as regional demand rebounds and prices increase. It remains to be seen if economic recovery in Central Europe and the former Soviet Union will create increased demand or whether these economies will follow the path of the developed world and achieve economic growth without an equivalent increase in resource demand. There is already evidence that the fast-reforming economies of Central Europe and the Baltic States are likely to follow the model of Western Europe; while the member states of the Commonwealth of Independent States will follow a different, and as yet undetermined, path.

- In the developing world there is considerable diversity in terms of resource dilemmas. The NICs of south-east Asia have succeeding in industrializing, but in doing so they have become increasingly dependent upon resource imports. Most recently, economic crisis in Asia has severely depressed resource demand. Elsewhere in the developing world a dualistic pattern of resource consumption seems to have developed. On the one hand, there remains a 'low-energy' society dependent upon biomass energy and subsistence agriculture, in which the majority of people exist. Here population pressure is leading to environmental problems as forests are stripped for fuel wood and land ploughed up or over-grazed. At the same time, there exists a growing 'high-energy' society linked to increasing commercial agricultural (see the next chapter for more on this) and industrial activity and the spread of urbanization. Much of this activity is aimed at supplying resources to the developed world. The two sectors of society combine to create an increasing demand for resources and pressure on the environment. The balance between the sectors also varies greatly among the countries of the developing world.

Given that the majority of the world's population lives in the developing world, this growth in demand will be the major factor contributing to increases in global resource demand. In its *Global Environment Outlook 2000*, the United Nations Environment Programme (UNEP 1999: 2) concluded that: 'A tenfold reduction in resource consumption in the industrialized countries is a necessary long-term target if adequate resources are to be released for the needs of developing countries.' It may be that in the future it is increasing demand in the developing world that raises the spectre of resource scarcity in any number of forms. However, the developed world already recognizes this threat and is using the issue of global warming to try and curb increased resource consumption in the developing world. For this to happen without curtailing the life chances of the people of the developing world, the developed world must help to shape an entirely new relationship between resources and development.

Learning outcomes

Having read this chapter you should understand that:

- The notion of what is a resource is socially constructed and varies through time and across space.
- There are a variety of different ways of classifying resources.
- There are a complex set of factors that influence the availability of resources.
- There are substantial regional variations in energy production and consumption.
- There is a complex and changing relationship between energy consumption and economic development.
- Different types of resource dilemmas confront the different regions of the world.

Further reading

Cloke, P., Crang, P. and Goodwin, M. (1999) *Introducing Human Geographies*, Arnold, London. See Section 3 'Environmental geographies' for more on sustainable development, environmental problems and management and environmental knowledges and environmentalism.

Jones, G. and Hollier, G. (1997) *Resources, Society and Environmental Management*, Paul Chapman, London. Provides more detailed discussion of the issues raised by this chapter and deals with some essential issues, such as water resources, which have not been discussed in any detail.

Mather, A.S. and Chapman, K. (1995) *Environmental Resources*, Longman, London. Provides a detailed, technical analysis of the resource issues raised by this chapter with numerous supporting examples.

Middleton, M. (1999) *The Global Casino: An Introduction to Environmental Issues*, 2nd edition, Arnold, London. Provides concise analyses of numerous environmental issues; each chapter contains a list of suggested readings and recommended web-sites.

O'Riordan, T. (ed.) (2000) *Environmental Science for Environmental Management*, 3rd edition, Prentice Hall, Harlow. An essential guide to the science behind the issues, plus up-to-date discussion of key challenges such as sustainable development and climate change.

Sarre, P. and Blunden, J. (eds) (1995) *An Overcrowded World? Population, Resources and the Environment*, Open University and Oxford University Press, Oxford. One of a series of books produced by the course team at the Open University in the UK. This volume deals with many of the issues raised in this section. Chapter 4 'Sustainable resources?' and Chapter 6 'Uneven development and sustainability' are of particular relevance to this chapter.

Useful web-sites

www.bpamoco/worldenergy　Home of the BP-Amoco *Statistical Review of World Energy*. This web-site, which is updated on an annual basis, has a wealth of statistical information on the energy sector, much of which is downloadable in Excel format or as power-point slides.

www.iea.org　The official site of the International Energy Agency. The site contains information on the Agency's operations and publications, as well as statistics on energy production and consumption. Also has links to other international organizations.

www.eia.doe.org　The official site of the US Government's Energy Information Administration. In addition to information on US energy matters, it contains reports on individual countries and a massive list of links to other US government agencies, international agencies, foreign governments and commercial company sites.

www.worldbank.org　The official site of the World Bank. The site contains information on World Bank operations and publications, as well downloadable statistics and briefing documents.

www.wri.org　The World Resources Institute is an independent centre for policy research and technical assistance on global environmental and development issues. Its web-site provides information on its own activities and publications, some of which are downloadable, as well as links to other organizations in the same area. The Institute, together with the UNEP, UNDP and World Bank, produces the biennial *Resource Report* that is an indispensable reference guide to the state of the earth's resources.

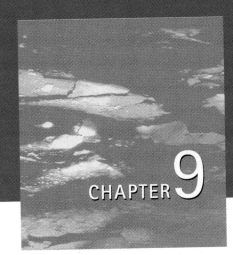

BRIAN ILBERY

CHAPTER 9

Changing geographies of global food production

Topics covered

- Population and food supply

- Globalization and food regimes

- Agricultural industrialization

- Regional speciality food products

- Future challenges for the supply of food

Food is a basic necessity for human survival and shapes people's lives in profound cultural, ideological and economic ways (Goodman and Watts 1997). Yet, rapid population growth, especially in the least developed countries (LDCs), creates a major challenge for the production of food on a global scale (see Chapter 7). Rising from just 1 billion people in 1800 to 4 billion in 1975 and a projected 7 billion by 2010, population growth is having a major impact on levels of food self-sufficiency around the world. Despite major technological advances in food production and the emergence of global agri-food systems, the reality is that the developed world has too much food whilst many parts of the developing world have too little.

9.1   Population growth and food supply

Although food supply has increased in most major world regions since 1950, too many of the world's population are still experiencing hunger. Defining hunger primarily in terms of undernutrition, the Food and Agriculture Organization (FAO) suggested that about 786 million people were undernourished in 1990. As Table 9.1 indicates, the world's worst-fed people are growing in numbers the fastest. The greatest number of undernourished people live in Asia, but the region with the highest proportion is sub-Saharan Africa (33 per cent). Whilst heavily concentrated in the LDCs, it is important to emphasize that certain groups within society in developed countries also go hungry. In California, for example, up to 5 million people are short of food whilst others are consuming 'designer' organic vegetables which have been shuttled around the world in sophisticated and refrigerated forms of transport (Buck *et al.* 1997, Goodman and Watts 1997).

Hunger and food insecurity are the result of population growth, conflict, poverty and environmental instability (Grigg 1985). The output of food per capita has been falling in the LDCs since the 1960s. Indeed, they now account for approximately 25 per cent of world food imports by value, a significant increase on the 1970 figure. Their falling levels of food self-sufficiency can, in part, be explained by three factors (Ilbery and Bowler 1996):

Table 9.1   Population growth and undernutrition, around 1990

World region*	Population growth rate 1985–95 (% per year)	Crude birth rate, 1985–95	Number under-nourished (millions) 1990	Percent population under-nourished
Sub-Sarahan Africa	3.0	45.6	168	33
Middle East	2.6	33.8	31	12
South Asia	2.2	32.7	–	–
Far East	1.4	21.6	528	18
Latin America	2.0	27.0	59	13
Europe/FSU	0.4	14.2	–	–
North America/Oceania	1.1	15.8	–	–
World	1.7	26.0	786	15

* Seven world regions based on standard United Nations regional classification of countries, 1992.

FSU: Former Soviet Union

Source: Based on various statistics in Dyson (1996).

- Population growth has outpaced the rate of increase in food production in many LDCs; often combined with environmental disturbances, this has meant that imports have been necessary just to maintain per capita food consumption levels.
- Many countries, for example Brazil and Kenya, have expanded their agricultural exports whilst neglecting the domestic food staples. This has been necessary to earn foreign currency to repay debts acquired to fund development programmes.
- Incomes have risen in some LDCs, for example in oil-exporting countries such as Saudi Arabia, Venezuela and Libya. Thus, the wealthier sections of the population have increased their consumption of imported wheat, rice and livestock products.

Consequently, many LDCs have become net food importers, increasing their dependence on developed world exports. In the 1950s and 1960s, considerable proportions of these exports were in the form of food aid (Tarrant 1980). Although falling significantly by the 1980s, food aid is still important to many low-income countries in sub-Saharan Africa and south Asia. For example, Egypt and Bangladesh are the largest recipients of cereal aid (mostly from the USA), accounting for 25 per cent of the world figure in 1991. However, such aid has been criticized on numerous grounds, not least because it increases dependency, distorts local markets and stifles domestic production (Maxwell and Singer 1979). By the 1990s, there had been a shift in emphasis, away from food aid and towards food trade. This has been encouraged by multinational organizations such as the World Trade Organization and the World Bank, and the increasing liberalization of world trade in agricultural products.

Considerable debate has developed since the 1960s over the extent of the 'world food problem' (Grigg 1985, Dyson 1996). Pessimists from the neo-Malthusian school of thought predicted a 7 per cent fall in per capita food production in the 1990s, leading to an absolute global shortage of food and widespread starvation and famine in developing countries (Myers 1991; Ehrlich *et al.* 1993). In support of their claims, they point to the fall in the global output of cereals per person since the mid-1980s (cereals are often used as a proxy for general food trends), the failure of the 1960s and 1970s 'green revolution' to continue to develop new and higher-yielding varieties of wheat and rice into the 1980s and 1990s, and the real likelihood of future global harvest failures due to drought, leading to a repeat of the 1972–4 'world food crisis' (Dyson 1996).

Although accepting that global food prospects are mixed and that there remain too many undernourished people in sub-Saharan Africa, the more optimistic anti-Malthusian school of thought believes that the world food problem is not deteriorating (Dyson 1996). The number of undernourished people actually fell by 108 million between 1980 and 1990, despite an increase of 841 million people during that period. Indeed, Dyson (1996: 75) suggests that

> the frequency and demographic impact of famines has declined greatly during recent decades. Modern food crises have had a diminishing and increasingly short-lived impact on population growth. The phenomenon of famine has become more and more restricted to Sub-Sarahan Africa.

Whilst accepting the general decline in per capita cereal production, optimists argue that population growth has not outstripped cereal production in all world regions. Indeed, cereal yields are not levelling off, neither are the effects of the **green revolution**, which in any case was not a revolution in the sense of creating a major upsurge in global cereal yields (except possibly in Pakistan). Any slowdown in cereal yields has occurred in developed rather than developing countries, in response to production control measures and prevailing international cereal prices. Finally, as will be demonstrated later in this chapter, food production and consumption patterns are becoming more diverse and so cereals may no longer be an adequate proxy for general food trends.

The aim of this chapter is not to explore the relationships between population and food (see Dyson (1996) for an excellent account), but to examine the continued globalization of the food supply system and its consequences for both developed and developing economies. More specifically, it describes the process of globalization and the 'industrialization' of the food supply system. In so doing, it emphasizes the dominant role played by transnational agribusinesses in developing globalized agro-food systems and in concentrating their activities in particular agricultural sectors and regions. The chapter then focuses on both the transfer of industrialized farming systems to particular regions in the LDCs and the relocalization of agriculture in the developed world. Case studies are included to illustrate the key points. Finally, the implications of these trends for global food production are considered.

9.2　Globalization and food regimes

Defined as the 'degree of purposive functional integration among geographically dispersed activities' (Robinson 1997: 44), globalization coincided with the emergence of global money and markets, the increase in transnational economic diplomacy, and global cultural flows in terms of tourism, guestworkers and consumerism (Amin and Thrift 1994). Global integration has been guided by powerful transnational corporations (TNCs), institutions and actors and has led to a new political economy of agriculture in which agro-food industries are epitomized by the mass production of manufactured food (for more on globalization and its many facets see Section 4).

Whilst there remain large inequalities in food supply between developed and developing countries, they are increasingly linked together in 'highly industrialized and increasingly globalized networks of institutions and products, constituting an agro-food system' (Whatmore 1995: 37). McDonalds is the classic agro-food expression of globalization (Ritzer 1996), and Friedmann (1982) used the term **agro-food complex** to describe the relations of production and consumption in particular food products, especially for highly industrialized agricultural commodities like salad crops (Friedland 1984; FitzSimmons 1986). Indeed, Friedmann and McMichael (1989) argued that relations between agriculture and industry have historically been more global than generally thought. They used the concept of **food regimes** to link international patterns of food production and consumption to the development of the capitalist system since the 1870s. It is possible to identify three food regimes (Box 9.1), with each representing for its time the modern food system. The first involves settler colonies supplying unprocessed and semi-processed foods and materials to the metropolitan core of Western Europe and North America. The second relates to the 'productivist' phase of agricultural change, focused on North America, and the development of agro-industrial complexes and grain-fed livestock. Finally, the third involves greater flexibility, including the production of fresh fruit and vegetables for the global market, the continued reconstitution of food through industrial and bio-industrial processes, and the supply of inputs for 'elite' consumption in the DCs. Unlike the second regime, TNCs have been dominant in the evolution of the third food regime.

The globalization of agriculture is thus presented by these authors as a logical progression, when in reality it is a contested, partial, uneven and unstable process (Whatmore and Thorne 1997). The global presence of McDonalds and Pepsi Cola, for example, is not typical of the complex and highly uneven process of globalization which has reshaped food production and consumption in the post-war period. Indeed, Le Heron (1993) highlights the importance of global–local interactions, leading to uneven development. Thus the role of locality is still important and 'globalization does not imply the sameness between places, but a continuation of the significance of territorial diversity and difference' (Amin and Thrift 1994: 6).

Globalization tendencies are, therefore, mediated by regional/local relationships and TNCs can play an increasing role in the social and political development of individual regions and nation-states (Marsden 1997). Indeed, giant TNC food companies and retailers are transforming the global agro-food economy; they have become dominant actors in the reconstruction of international and national food systems, based around the increasing demands of groups of consumers in the developed world for high quality, natural

---

Box 9.1   Three food regimes since the 1870s

First regime: pre-industrial (1870s–1920s)

This involved settler colonies supplying unprocessed and semi-processed foods and materials to the metropolitan core of North America and Western Europe. Characterized by *extensive* forms of capital accumulation, the main products involved were grains and meat. The regime gradually disintegrated as agricultural production in DCs competed with cheap imports and trade barriers were erected.

Second regime: industrial (1920s–1970s)

This regime relates to the productivist phase of agricultural change, focused on North America and the development of agro-industrial complexes based on grain-fed livestock production and fats/durable foods. Characterized by *intensive* forms of capital accumulation, the second regime incorporated developed and developing nations into commodity production systems. The resulting agricultural surpluses and environmental disbenefits began to undermine productivist agriculture in the 1970s.

Third regime: post-industrial (1980s onwards)

This developing regime relates to the post-productivist phase of agricultural change and involves the production of fresh fruit and vegetables for the global market, the continued reconstitution of food through industrial and biotechnological processes, and the supply of inputs for 'elite' consumption in DCs. Characterized by a *flexible* form of capital accumulation, the third regime is dominated by the restructuring activities of agribusiness TNCs and corporate retailers.

Sources: Based on Le Heron 1993, Robinson 1997 and Ilbery and Bowler 1998

foods. As FitzSimmons (1997) suggests, TNCs have become the primary agents of globalization in the agro-food sector; they sit at the centre of webs of relationships that link farming, processing and marketing. Indeed, corporate retailers play an essential role in developing regulatory systems that ensure their dominance over the supply of major food products.

9.3   Industrialization of agro-food systems in the developed world

Agricultural production in the developed world has become increasingly concentrated on a limited number of large-scale farms (Ilbery and Bowler 1998). At the same time, there has been an increase in capital expenditure on major agricultural inputs such as chemicals and machinery, and a growth in the processing and manufacturing of food. As a consequence, it becomes vital to view farming in the wider context of an **agro-food system**, in which the production sector itself is inextricably linked to various 'upstream' (input supplies) and 'downstream' (processing, distribution and marketing) industries. The food supply system is dominated by large **agribusinesses** which, according to Davis and Goldberg (1957) and Wallace (1985), are the sum of all operations involved in the manufacture and distribution of farm supplies, the production operation of the farm, storage, processing and distribution of farm commodities and items made from them. These agribusinesses often develop agricultural commodity chains beyond national boundaries.

As Le Heron (1993: 45) suggested, 'industrialisation is progressively obliterating the farming base'. However, it is especially the non-farm sectors of the agro-food system which have become highly industrialized and dominated by TNCs. This has occurred through two discontinuous processes (Goodman *et al.* 1987):

- **Appropriationism**, whereby certain agricultural *inputs* are appropriated and replaced by 'industrial' alternatives. Good examples would include synthetic chemicals replacing manure and tractors replacing animal power.
- **Substitutionism**, which focuses on *outputs* rather than inputs and is concerned with the increased utilization of non-agricultural raw materials and the creation of industrial substitutes for food and fibre. Examples would include the substitution of sweeteners for sugar and nylon for cotton.

The rate and direction of appropriationism and substitutionism are determined largely by prevailing levels of technological change (e.g. witness the current controversies over genetically modified crops), state support policies for agriculture, and consumer tastes and fashions. Thus the industrialization of agriculture proceeds in successive rounds of appropriation and substitution. This leads to increasing concentration in most segments of the agro-food system, from input suppliers to supermarkets. For example, in the UK over 70 per cent of the agro-machinery market is concentrated in just four companies, as is over 50 per cent of turnover in food retailing (Whatmore 1995).

However, agro-food systems are different from other production systems because agriculture is bound by biological processes and cycles. So, agribusiness TNCs attempt to reduce their dependence on 'nature' and increase their influence over farming indirectly through a process of *formal subsumption*, where arrangements or contracts are made with farmers to provide 'raw materials' for their value-adding food manufacturing activities. However, farmers are not 'passive in this logical progression of ... market forces'

(Whatmore 1995: 45) and the family farm has a capacity to survive through various forms of self-exploitation. This hinders the industrialization process and the farming sector remains quite diverse, predominantly family-based, and geographically uneven.

The restructuring of farming, therefore, is dominated by geographical differences in the integration of agriculture into a global and industrialized agro-food system. Indeed, differentiation and diversity are key features of capitalist agriculture. Thus agricultural industrialization becomes concentrated in particular sectors and regions characterized by large farm businesses which adopt intensive farming methods and integrate into global networks. In the European Union (EU), for example, 80 per cent of agricultural production is concentrated on less than 20 per cent of farms in particular 'hot spots', including East Anglia, the Paris Basin, southern Netherlands, and Emilia Romagna in north-east Italy (Whatmore 1995). The remaining 80 per cent of farms are marginal to the global agro-food system and are located in 'lagging' rural regions, where farming is less market-dependent.

Overall, agricultural industrialization is increasingly global in scale, where producers become 'distanced' from consumers. Such a system is unstable and uneven. It leads to increasing differentiation over space, overproduction, increased indebtedness as farmers invest in more technology, environmental disbenefits, and concerns over food security and safety. One reaction of TNCs and major retailers is to relocate the production function of the agro-food system to new agricultural spaces, often in LDCs. This is usually associated with the intensive production of high-quality horticultural products where little consideration is given to environmental degradation.

9.4 Reorientation of developing world agriculture

For many years, food production in the developing world was often of a subsistence nature. Early attempts in the late eighteenth and early nineteenth centuries to transform agriculture were through the creation of foreign-controlled and operated plantations, which provided raw and semi-processed materials for industries in the developed world. Thus many European settlers in such African colonial countries as Kenya, Mozambique, Angola and Southern Rhodesia began exporting crops such as tea, coffee and tobacco to the developed world, especially to European imperial states. Encouraged by colonial governments, African smallholder farmers also grew cash crops for sale to European traders at major ports, even though local people often had too little protein. A dependency culture thus developed on the basis that foreign earnings from export crops could be used by government agencies for development programmes. By the time these countries had become independent in the post-Second World War period, agricultural policy was focused on exports, with a neglect of food staples. Consequently, LDCs became net importers of staples and processed foodstuffs.

In more recent years, many former plantations, often run by agribusiness TNCs, have applied intensive and industrialized systems of farming, including the employment of wage labour and the application of capital and advanced technology to their production activities (Loewenson 1992). Many of these estates specialize in one crop, such as tea, coffee, sugar, bananas, green beans or flowers, destined for markets in the developed world. Estate managers often put in place infrastructure and marketing arrangements; these can attract government support because of perceived local development advantages. Indeed, estates, as earners of export income, importers of technology and employers of labour, are seen as one answer to the challenge of poverty and population growth in many LDCs.

As Goodman and Watts (1997) demonstrate, the classical export commodities (coffee, tea, sugar, cocoa) have now been complemented by high-value foods, including fruits, vegetables, poultry and shellfish. Throughout the 1980s, the value of world trade in high-value foods increased by 8 per cent per year. By 1990 they accounted for 5 per cent of global commodity trade, one-third of which came from the LDCs (far exceeding the value of LDC exports of coffee, tea, sugar, cotton, tobacco and cocoa) (Goodman and Watts 1997). By the early 1990s, 24 low–middle income countries annually exported over US$500 million of high-value foods; indeed, just five countries accounted for 40 per cent of such exports from the LDCs. Friedmann (1993) labelled these the newly agricultural-izing countries, namely Brazil, Mexico, China, Argentina and Kenya.

The major growth in high-value food exports in the 1980s and 1990s reflects many factors, including technical changes in the food industry and the liberalization of world agricultural trade. Significantly, however, dietary changes in the developed world have been instrumental and Goodman and Watts (1997) highlight how high-value foods are produced to satisfy the tastes and life-styles of particular groups of consumers in developed countries. Agribusiness TNCs are attracted to high-value food production in newly agri-culturalizing countries because of low-cost labour, government support (e.g. through Structural Adjustment Programmes), good international communication links, and the ability to produce high-quality, fresh products for niche markets in the developed world.

Marsden (1997) explains how recent agricultural development in the newly agricultur-alizing countries is based on the use of intensive production techniques, which are under-pinned by quality conditions and control (from planting through to consumption). Design and quality are vital ingredients in high-value food production. Responding to consumer demand and tastes for size, shape, colour and content of products, the concern for 'qual-ity' emanates from the strategies of corporate retailers. Storage and transhipment are other important functions, often involving exporters and importers, acting under instructions from the retailers. Produce is rejected if quality and appearance are not good. In turn, exporters have to satisfy quality control procedures specified by importers and retailers. As Marsden (1997: 177) states, 'those actors and agencies who are closest to the definition and implementation of quality conditions begin to accumulate power in the food net-work. This leads to a growing social and economic differentiation in the region, with smaller producers prone to exclusion in globalised food networks.'

So, corporate retailers in the developed world play a key role in translating quality defi-nitions of foods globally. Indeed, they develop their own regulatory systems that ensure a dominance over the supply of main food products; it is highly competitive and vulnerable to consumer reactions. Production for export markets leads to a decline in production for the local population that are reliant on increasing amounts of food imports. There is little doubt that large-scale agribusinesses and corporate retailers have successfully linked regional economies and crop sectors of LDCs to a global system of food production and consumption. However, their behaviour has attracted a number of criticisms as regards their economic, social and environmental impacts on host countries (Espiritu 1986, Ilbery and Bowler 1996):

- They contribute to sharp inequalities in income, productivity and technology compared to the sector producing domestic staples.

- Labour-saving innovations by agribusiness mean that numbers employed are reduced, with many becoming deskilled, part-time or seasonal. Specialist export experts are often brought in.
- The benefits of intensive agriculture accrue disproportionately to foreign investors and, in the case of nationally owned estates, to urban-based people and local elite groups.
- Large estates, especially those of agribusiness TNCs, often take the best land for export crops. This land absorbs the most inputs, investment and expenditure, even though it occupies only a small proportion of the total agricultural area. Unable to compete, localized agricultural production often begins to break down.
- Many of the agricultural practices of agribusiness TNCs are unsustainable. Environmental degradation is a real problem and there are reported instances of groundwater depletion and agribusinesses moving on to new areas when soils have become depleted.

9.4.1 A Kenyan case study

Since 1990, an increasing number of African countries have entered the export trade in high-value foods and cut flowers (Barrett and Browne 1996). The competitiveness of African suppliers in this sector reflects low costs of production, complementarity to European seasons, relatively short flight times and the ability to supply produce of the quality and quantity required in international markets all year round. Kenya is the only **newly agriculturalizing country** in Africa; it is also an established exporter of horticultural products to the UK market.

Between 1991 and 1996, Kenyan exports of horticultural produce increased 58 per cent to reach a total output of 84,824 tonnes; by the year 2000, this was expected to rise to over 100,000 tonnes (HCDA 1997). During this period of economic liberalization in Kenya, huge investments have been made in the industry, especially by the 10 largest producers in the cut flower and pre-packed vegetable sectors. This has mainly been to attract and keep lucrative contracts with large British supermarket chains. Kenya specializes in the export of high-quality and often pre-packed green beans, mange touts, Asian vegetables (mainly okra, chillies and aubergines), avocados and cut flowers. The export of cut flowers, especially roses and carnations, has been increasing at about 20 per cent per annum since the early 1990s. Europe is the main customer for Kenyan fresh horticultural produce; in 1994, 85 per cent went to just four EU countries: UK (30 per cent), The Netherlands (29 per cent), France (16 per cent) and Germany (10 per cent).

New marketing chains have evolved to facilitate the Kenyan export trade in high-value foods and cut flowers (Barrett *et al.* 1999). First, there is an older, more fragmented *wholesale chain*, linking small- and medium-scale growers into the export market through a series of agreements and contracts between growers, agents, exporters, freight agents, importers and retailers (Figure 9.1). Many of the links in this chain are based around complex family-based connections where, for example, Kenyan-Asian exporters deal with British-Asian importers and retailers. The produce for this market is generally imported and sold loose; significantly, it does not have to meet supermarket specifications. Consumers are interested in variety, quality, flavour and value for money rather than presentation and packaging. Although still accounting for approximately 30 per cent of

Plate 9.1   High value foods and cut flower exports from Kenya.

Kenya's horticultural exports to the UK, the wholesale trade is declining. This reflects the growing importance of a second chain – the *supermarket chain* – which accounts for at least 70 per cent of horticultural exports. High-value fruit, vegetables and cut flowers require speed in harvesting and packaging, as well as in transportation. Quality, and thus price, is dependent on the environment in which the produce is stored and shipped. Certain very delicate crops such as green beans, mange touts and cut flowers are better preserved if travelling along a 'cool chain', where most or all stages of transit between harvest and retail point are temperature-controlled. Not surprisingly, this chain mainly comprises large-scale commercial growers, geared to supplying the expanding supermarket trade in trimmed and packaged products.

The supermarket chain consists of *partially integrated* and *fully integrated* systems (Figure 9.1). Whilst the former have developed primarily for large-scale producers who are also exporters, the latter see production, exportation, freight handling and often importing being controlled by one company. The fully integrated system is heavily associated with the cut flower and supermarket trades; the large-scale farms involved use modern technologies, mostly imported, to achieve high yields of uniform-quality produce and high standards of presentation and packaging. Seed stock, irrigation, floodlighting and other technologies encourage intensive, 'productivist' farming systems, with which small and medium growers cannot compete. The partially-integrated chain supplies about 25 per

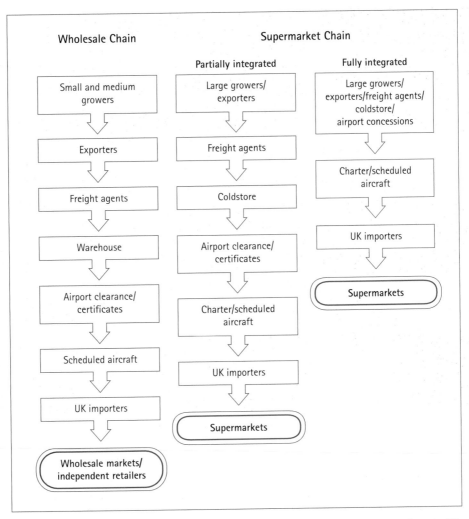

Figure 9.1 Marketing chains and the export of high-value foods and cut flowers from Kenya into the UK.
Source: Based on Barrett *et al.*, 1999: 165

cent of exports, but differs from the wholesale chain in that a small number of large-scale growers have seen the benefits of integrating production and exporting aspects of the business, thus bypassing agents in Kenya. This integration allows supermarket demands for quality and traceability to be satisfied. To remain competitive, the producers have started to trim and package vegetables and to diversify into baby vegetable production.

Developing since the early 1990s, the fully integrated chain is a response to the specific needs of the cut flower industry and supermarket demands. Quality, freshness and speed of delivery are paramount. In order to achieve this and maintain control of the 'cool chain', the three largest producer/exporters of horticultural produce in Kenya have set up their own export and freight companies; they proudly claim that within 48 hours of being harvested, produce can be on the supermarket shelves in Britain. They now account for as much as 45 per cent of Kenya's horticultural exports. Most of the produce is sold in pre-packs which are bar-coded and priced in Kenya, as directed by the supermarkets who provide the stickers. To maintain the loyalty of the supermarkets, and the quality requirements of customers, producers have invested heavily in EU-standard packing stations, refrigerated trucks and coldstores (including at Nairobi airport). The 'farms' have research and development laboratories where quality is monitored, packaging tested and new products screened. Once inspected by customs and excise officials at the airport, the produce is loaded directly onto specially chartered planes. The produce hardly ever leaves the 'cool chain'.

EU importers sourcing from Kenya are a buffer between exporter and supermarket. In the fully integrated system, they are often part of the same group that includes producers, exporters and freight agents. Importers are responsible for forwarding the produce to warehouses and undertaking necessary health, safety and quality checks as required by regulators. Most importers deal with the major supermarkets through informal rather than formal contracts, which are 'worked out' on an annual or six-month cycle. These are then confirmed on a monthly, weekly or even daily basis, with trade increasingly moving towards a more flexible 'just-in-time' principle. It is thus the importers who take the greatest risk in the marketing system. The supermarkets' demands for a high standard of packaging and presentation, including a significant movement towards pre-packs in the country of origin, favour the sourcing of produce from large-scale 'productivist' agri-business companies.

It is quite clear that the supermarket retailers are the most powerful 'actors' in this agro-food system. They have flexible arrangements with different importers, and exert their power through 'remote control' (Whatmore and Thorne 1997). Only when inspected and approved by the retailers' quality control systems will produce be accepted and paid for. Overall, therefore, horticultural exports from Kenya are increasingly concentrated in the hands of large, highly capitalized producers. The requirements for documentation about sourcing and traceability militate against small-scale producers who are likely to become increasingly marginalized. Yet, rather than produce much-needed staple products for the indigenous population, they attempt to participate in the export sector. The real paradox of this whole system is that it has embraced intensive 'productivist' methods to satisfy 'post-productivist' demands from 'new consumers' in regions of the developed world rather than internal demands for more staple foods.

9.5 Relocalization of developed world agriculture

The 'productivist' model of agriculture in developed market economies, based on the principles of efficiency, rationality and high levels of government support, has been under challenge since the early 1980s (Ilbery and Holloway 1997). This challenge has been stimulated by the environmental and socio-economic disbenefits of agricultural industrialization and, as a consequence, alternative systems associated with environmental protection, sustainability, food quality, reduced price support, and the integration of agriculture within broader rural economic, social and environmental objectives have arisen.

The movement away from an essentially productivist ethos in developed world agriculture has been conceptualized as the **post-productivist transition** (Shucksmith 1993, Lowe et al. 1993, Ilbery and Bowler 1998), the exact nature of which has yet to be defined and theorized. However, Bowler and Ilbery (1997) see this transition as a progressive

---

Box 9.2   The post-productivist transition in agriculture

1 Major characteristics
   - A reduction in food output and greater emphasis on food quality
   - Progressive withdrawal of state subsidies for agriculture
   - The production of food within an increasingly competitive and liberalized international market
   - The growing environmental regulation of agriculture
   - The creation of a more sustainable agricultural system

2 Pathways of farm business development
   - Extension of the industrial model of productivist agriculture, based on traditional farm products
   - Redeployment of farm resources into new agricultural products on the farm (agricultural diversification)
   - Redeployment of farm resources into new non-agricultural products on the farm (business structure diversification)
   - Redeployment of farm resources into employment off the farm (other gainful activity)
   - Maintaining the traditional model of conventional farm production (extensification and/or adoption of agri-environmental programmes)
   - Winding down to hobby or semi-retired farming
   - Retirement from farming

---

movement away from agricultural intensification, concentration and specialization and towards a more extensive, diversified and dispersed pattern of farming. Thus a number of known characteristics can be associated with the post-productivist transition (Box 9.2). It embraces the whole of the agro-food system, from production and processing to consumer choices and the marketing systems that link them. Government policy is central to the post-productivist transition, but this time responding to rather than stimulating change. So, for example, the gradual withdrawal of subsidies for farmers under the Common Agricultural Policy (CAP) of the EU is a response to processes such as the globalization of the agro-food system, trade liberalization and the growing international significance of the environmental movement. Further reforms are an almost inevitable consequence of the next round of the World Trade Organization negotiations beginning in 1999. Continued liberalization of world trade in agricultural products and moves towards market orientation will be key issues.

Similarly, and in response to the environmental disbenefits of 'productivist' agriculture, there has been an increasing environmental regulation of agriculture and a willingness to embrace the principles of sustainable agriculture. Again within the context of the EU, agri-environmental programmes were introduced as part of the CAP reforms in 1992 (Robinson and Ilbery 1993). These enabled member countries to introduce a series of voluntary schemes to reduce the application of chemicals and fertilizers in specified areas and to encourage the development of, for example, environmentally sensitive areas, extensification, organic farming, moorland restoration and habitat improvement. Although still relatively new, there is already debate as to whether or not such programmes provide real environmental benefits (Morris and Potter 1995, Potter 1998).

However, one of the key drivers of the post-productivist transition has been a shift in consumer demand and the increasing interest in, and knowledge of, food (Coakley 1998). In response to concerns over food safety, there has been a movement towards 'healthy eating' and fitness, and a growing demand for high-quality regional speciality food products with authenticity of geographical origin and *traceability* (Marsden and Arce 1995, Gilg and Battershill 1998, Ilbery and Kneafsey 1998). Bell and Valentine (1997) argue that food is an indicator of socio-cultural status and Chaney (1996) sees the consumption of food as part of a life-style image. Thus some consumers (the 'new consumer') are giving more emphasis to aspects of quality and convenience than to price and quantity (Barrett *et al.* 1999).

As suggested earlier in this chapter, the shift from productivist to post-productivist farming systems does not signal the end of agricultural industrialization. Indeed, both types of farming system are likely to co-exist and become more spatially differentiated. So, on the one hand, an intensive, high-input, high-output system emphasizing food *quantity* and continued integration into the agro-food system may continue in 'hot spot' regions in both developed and developing countries. On the other hand, a more extensive, low-input, low-output system which emphasizes *quality* and sustainability may develop in more marginal agricultural areas in developed countries (those left out of the globalized agro-food system). The latter may present opportunities for a relocalization of agricultural production, whereby farmers in such areas develop alternative strategies by, for example, concentrating on niche markets and responding to consumer demand for either high-quality regional speciality food products and/or ecologically-sound farming practices.

Undoubtedly, the post-productivist transition will lead to the emergence of new agricultural land uses. However, which ones emerge in which areas will depend on the different adjustment strategies adopted by farm households. Bowler (1992) and Bowler *et al.* (1996) referred to these as *pathways of farm business development* (Box 9.2). The different pathways fall neatly into three categories:

- Maintaining a full-time, profitable food production element to the farm business (pathways 1 and 2).
- Diversifying the income base of the farm business by redirecting resources into non-farm enterprises and occupations (pathways 3 and 4).
- Surviving as a marginalized farm business at a lower level of income, perhaps supported by investment income, pensions and agri-environmental payments (pathways 5 and 6), or retiring from farming altogether (pathway 7).

The pathways are not mutually exclusive and farm households may combine different elements, through for example organic farming and 'adding value' by processing and marketing. However, if households begin to make the transition from pathway 1 towards pathway 7, it is highly likely that they will become marginalized for ever. Much will depend on the nature of the farm household and whether its members can be described as *accumulators* (risk-taking, venturesome, and able to redeploy resources) or *survivors* (conservative, traditional, with some diversification and possible 'passive' adoption of agri-environmental programmes).

9.5.1 Niche markets and regional speciality food products

With food becoming a statement of taste, sophistication and fashion in particular niche markets (i.e. sub-groups of mainly middle- and high-income consumers with particular tastes, life-styles and aspirations), there are opportunities for a relocalization and reintegration of agriculture within marginal agricultural regions in developed market economies. Indeed, certain food markets are increasingly being shaped by the demands of particular groups of consumers. A life-style image is related to class and income, and more affluent groups are able to perceive food as not only good to eat but also good to think (Beardsworth and Keil 1992). Thus for the 'new consumer' a relationship emerges between food and personal identity.

The growing demand by such groups for high-quality, wholesome and traceable food raises issues over just what is meant by *quality* and how regional speciality food products can be linked to particular places. Quality is a contested concept, but relates to the satisfaction of consumer needs (Vastoia 1997) and a consistent level of performance, taste and so forth provided by the product (O'Neill and Black 1996). It has connotations of being able to command higher prices and is a positional characteristic – something that is above minimum standards. A quality product is 'differentiated' in some positive manner that is recognized by consumers. It is possible, therefore, to identify a number of indicators of quality (Ilbery and Kneafsey 1998, 2000a). These range from ensuring *specification* of production method and raw materials used, establishing an *association* with a region, local environment or cultural heritage, to generating consumer *attraction* through such attributes as taste, texture and appearance, and achieving some form of official *certification* for the speciality food product (Table 9.2). Thus quality implies a number of 'actors' in the agro-food system, from initial producer to final consumer. While the consumer may perceive

Table 9.2 Indicators of quality in speciality food and drink products

1 Achieving *certification* by a professional organization, government or other external bodies:
  - appellation contrôlée
  - organic symbol
  - quality mark
  - self-regulation
2 Establishing *association*, either geographically within a region or historically with a tradition or culture:
  - regional designation
  - traceability
  - historic/cultural importance
  - local environment
3 Ensuring *specification* of:
  - production method (e.g. small-scale workshops, 'authentic' recipes)
  - raw materials (e.g. local milk, water from a particular spring)
  - ownership
4 Generating *attraction* by tapping into the subliminal wants of consumers:
  - designs
  - texture, flavour and taste
  - freshness/appearance
  - premium price
  - consumer perception

Source: Ilbery and Kneafsey (1998: 331).

quality in terms of food safety, appearance, taste and traceability, producers often see quality as a market device to increase sales, differentiate their product and gain premium
prices.

The linking of regional speciality food products to particular places is a necessary
prerequisite for the relocalization of food production. Bell and Valentine (1997: 155)
argue that there are already strong links between products and places, so that 'almost any
product which has some tie to place, no matter how invented this may be, can be sold as
embodying that place'. The creation of a strong regional identity is thus vital and the
'fixing' of local food products to place can be achieved through marketing and place
promotion (Ilbery and Kneafsey 1998, Ray 1998). Emphasis is placed on *selling places*, but
the image created must be distinctive to the region and not related to all rural areas. So,
marketing becomes particularly important because constructions of quality will be
interpreted, understood and consumed differently by individuals in different places and at
different times.

With the growing demand for regional speciality food products, it is important that the
sector is properly regulated. In the UK, for example, the Food Safety Act (1990) has effectively handed over the regulatory power for consumer protection to corporate retailers.
This requires supermarkets to show 'due diligence' by demanding 'traceability' for all the
products they sell. Not surprisingly, they have taken the initiative on such issues as food

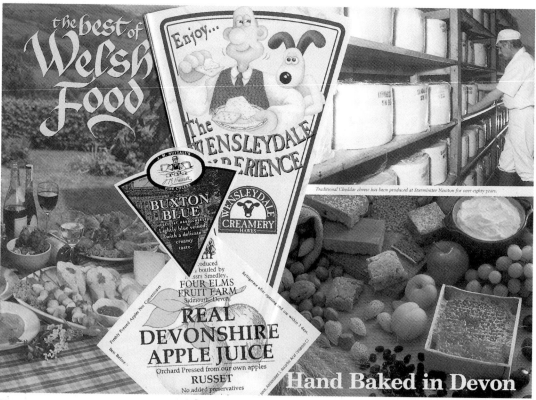

Plate 9.2   Speciality food and drink products in the UK.

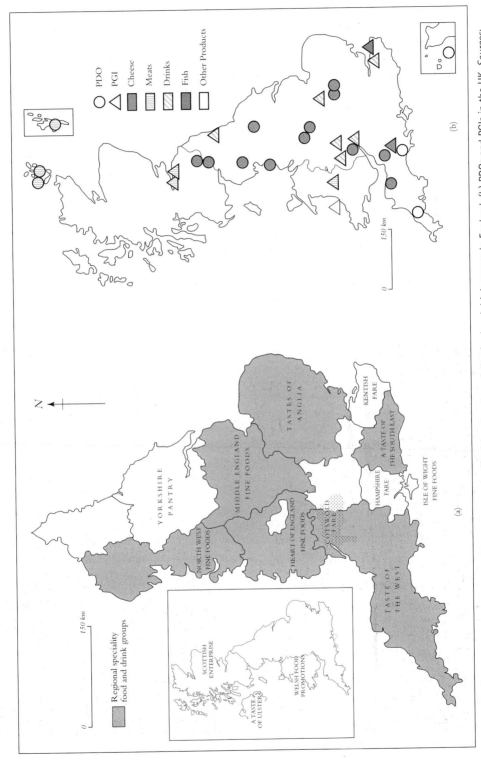

Figure 9.2 Regional speciality food and drink products in the UK: (a) regional speciality food and drink groups in England; (b) PDOs and PGIs in the UK. *Sources:* Ilbery and Kneafsey, 2000a, 2000b

quality, authenticity and convenience. Supermarkets have also seen the potential of using regional imagery and marketing to exploit the quality food market, often creating 'false' images of traditional fare and wholesome food (Gilg and Battershill 1998). Most quality assurance schemes have been initiated by retailers and by supermarkets; very few have been started by speciality food producers (Morris and Young 1997). Traditional retailers are in danger of losing their market share, partly because they cannot compete with the supermarkets and partly because consumers demonstrate less confidence in their claims of quality. Thus, there is a real possibility of agribusiness TNCs and corporate retailers appropriating the speciality food market and integrating it into their industrialized agro-food systems. Buck et al. (1997) have produced early evidence of such a process in relation to organic farming in northern California.

One response to such a process is for the state to regulate regional speciality food products through a formal system of quality 'labels'; another is for small producers to establish regional speciality food groups. As an example of the latter, the government quango 'Food from Britain', with financial support from the Ministry of Agriculture, has overseen the establishment of six regional speciality food and drink groups in England (Figure 9.2a). These regional groups are able to offer a broad range of products to attract buyers, arrange trade fairs and exhibitions, provide training on marketing and food promotion, and gain access to markets and funding. The largest regional group, Taste of the West, has 190 producer members, and survey evidence suggests that they define quality in terms of specification and attraction rather than official certification schemes and association with a regional environment, culture and tradition (Ilbery and Kneafsey 2000a).

It is for this reason that official certification of 'quality' regional speciality food products, where product can be related specifically to place, is required. In this context, the EU has recognized the potential of differentiating speciality food products on a regional basis. Two related regulations were passed in 1992, one of which (Regulation 2081/92) is concerned with speciality food products from specific areas. In an attempt to harmonize and develop Community-wide quality labels for particular food and drink products, producers can apply for either a PDO (Protected Designation of Origin) or a PGI (Protected Geographical Indication). A PDO relates to a product originating in a specific region where the quality of that product is due exclusively to a particular geographical environment with its inherent human and natural factors. In contrast, a PGI relates to a product originating in a particular region which possesses a specific quality, reputation or other characteristics attributable to that geographical origin (but not necessarily due to its natural environment or raw materials). Such quality 'marks' protect producers from attempted copies of their product and act as a marketing device to attract customers. A scheme of this nature aims to replace national quality labels with an international system and thus ensures fair competition between EU producers; it also helps to ease the transfer of locally-produced speciality food products to regional, national and international markets.

The United Kingdom already has over 20 PDO/PGI designations (Figure 9.2b). Specialist cheeses dominate the PDOs, indicating that these are made exclusively from local raw materials (e.g. Stilton cheeses). In contrast, drinks dominate the PGIs. Thus Newcastle Brown (a beer) is a PGI because, although associated with the city of Newcastle, not all of the ingredients come from that region. There are now over 300 PDO and 150 PGI designations in the EU. Some member countries have either very few or none (e.g. Finland,

Sweden, Denmark, the Netherlands), while others have a large number (e.g. France and Italy with over 90 each and Greece and Portugal with over 60 each). These differences suggest that food has more cultural significance in some European societies than others.

9.6 Meeting the challenge of food production

This chapter has examined various dimensions of the globalized agro-food system. As noted earlier, some commentators (e.g. Le Heron 1993, Pritchard 1998) suggest that a third food regime (Box 9.1) is emerging which, whilst being quite flexible in nature, is focused on the supply of fresh fruit and vegetables for 'elite' consumption in the developed world. Unfortunately, such a situation is likely to exaggerate further the huge inequalities which already exist in global patterns of food supply and consumption, thus condemning large numbers of people to continued poverty and hunger.

Such inequalities are, of course, a function of the prevailing capitalist mode of production, in which agribusiness TNCs and corporate retailers have developed considerable power and so control the global food supply system. They have targeted their activities on particular 'hot spots' around the world, where 'productivist' farming methods are practised and integrated into essentially industrialized systems. Indeed, farming is just a small cog in large agro-commodity chains. These hot spot regions are not confined to the developed world and can now be found in a number of 'newly agriculturalizing countries', as demonstrated in the Kenyan case study. A majority of farmers and thus regions, in both developed and developing countries, have been left out of these commodity chains and so become marginalized. Nevertheless, small-scale producers in some LDCs often attempt to supply food for what is perceived to be lucrative export markets rather than focus on the production of much-needed staple products for the indigenous population. Continued liberalization of world trade in agricultural products and the movement towards greater market orientation in agricultural policy can only increase international competition in food production and further strengthen the power of agribusiness TNCs and corporate retailers.

These developments present a major challenge for the future supply of food to the most needy groups in the world, and especially to regions of high population growth in LDCs. Such a challenge cannot be resolved without state regulation. However, such regulation must not be imposed on different regions in a 'top down' fashion. Instead, endogenous 'bottom up' programmes of integrated rural development are required, which are multisectoral, sustainable and based on local needs through the principle of subsidiarity. Thus, local self-sufficiency becomes the first priority, rather than satisfying the demands of 'elite' groups from developed countries. In the case of LDCs, this means a refocusing of attention on the production of basic staple foods for the local population. This requires the dependency culture on export crops to be broken, which in turn necessitates changes in international and national policy. The benefits of relocalized production must also be spread to all sectors of society and not confined to local elite groups. In the marginal agricultural areas of developed countries, farmers need to adjust to the continued globalization of the agro-food system by embarking upon more sustainable and 'alternative' forms of agriculture, based on local knowledge and customs. Indeed, a focus on high-quality, authentic and traceable regional speciality food products for niche markets is one possible pathway. As shown in the second case study of regional speciality food products, this is especially so if they are 'fixed' to territory, so that there is potential for the local area to keep more of the economic benefit and, at the same time, retain some control over the type of economic activity that occurs there (Ray 1998).

In the final instance, global–local interactions are vital when attempting to meet the challenge of feeding the world's growing population. Thus the encouragement of 'bottom up' forms of rural development must be complemented by a much fairer global system of food trade, in which the vast food-growing potential of regions such as Eastern Europe can be utilized to feed those millions of people still experiencing hunger.

Learning outcomes

Having read this chapter, you should understand:

- The different schools of thought concerning the complex relationships which exist between global population growth and food production.
- The continued globalization of the agro-food system in terms of the concept of food regimes.
- The dominance and spatially uneven penetration of the global agro-food system by transnational agribusinesses.
- The adoption of 'productivist' farming techniques in parts of the developing world to satisfy the tastes and life-styles of particular groups of consumers in developed countries.
- The increasing demand for, and renewed significance of, high-quality regional speciality (niche market) food products from specific places within the developed world.
- The main issues facing global patterns of food production and consumption in the twenty-first century.

Further reading

Dyson, T. (1996) *Population and Food: Global Trends and Future Prospects,* Routledge, London. An excellent exposition of the complex relationships between global trends in both population growth and food production.

Goodman, D. and Watts, M. (eds) (1997) *Globalising Food: Agrarian Questions and Global Restructuring,* Routledge, London. A diverse collection of essays on aspects of the globalization of food production and consumption.

Grigg, D. (1985) *The World Food Problem 1950–1980,* Blackwell, Oxford. A scholarly account of hunger, poverty and world agricultural development in a thirty-year period after the Second World War.

Ilbery, B. and Bowler, I. (1998) From agricultural productivism to post-productivism, in Ilbery, B. (ed.) *The Geography of Rural Change,* Longman, London. A review of conceptual and empirical research on the main dimensions of 'productivist' and 'post-productivist' farming systems in the developed world.

Le Heron, R. (1993) *Globalised Agriculture: Political Choice,* Pergamon, Oxford. A theoretical and empirical analysis of some of the major issues surrounding the continued globalization of the agro-food system.

Useful web-sites

www.europa.eu.int   The official site of the European Union. Within this very large site, Directorate General 6 has responsibility for Agriculture. The DG 6 site contains information on the EU's Common Agricultural Policy, key speeches and policy papers, statistics, links to member-state sites and an Information Resource Centre.

www.foa.org   The official site of the United Nations Food and Agriculture Organization. The site contains information on FAO programmes and FAOSTAT, an online multilingual database with information on food production and trade.

www.maff.gov.uk   The site of the UK government's Ministry of Agriculture, Fisheries and Food (similar sites exist for other national governments, for example ***www.usda.gov*** is home of the US Department of Agriculture). The site contains information on UK agricultural policy, statistics and links to other government departments, non-ministerial departments and NGOs.

www.oxfam.org   Home of Oxfam International which is dedicated to addressing the structural causes of poverty and related injustices. The site includes Oxfam International's Position Papers that address some of the issues raised by this, and other, chapters.

www.soilassociation.org   Home to the UK organization dedicated to promoting organic food and farming. The site contains an online library with a variety of information from local food links to policy papers. The site also contains information on regional food groups in England (such as ***www.heff.co.uk*** for the Heart of England Fine Foods Group).

CHAPTER 10

MARCUS POWER

Alternative geographies of global development and inequality

What is the geography of the Third World? Certain common features come to mind: poverty, famine, environmental disaster and degradation, political instability, regional inequalities and so on. A powerful and negative image is created that has coherence, resolution and definition. But behind this tragic stereotype there is an alternative geography, one which demonstrates that the introduction of development into the countries of the Third World has been a protracted, painstaking and fiercely contested process. (Bell 1994: 175)

Topics covered

■ Development statistics and indicators

■ Development as knowledge and power

■ Theories and subjects of development

■ International agencies and non-governmental organizations

Learning to understand and appreciate social or economic differences relative to the area of the world in which we live is a difficult and value-laden process. The media relay images and information about inequality within and between worlds almost every day, but the assessment of global inequalities has been widely disputed. Conditions of poverty clearly vary between different areas as does the way poverty is experienced, so how can a universal definition be agreed upon? Is wealth to be measured in terms of Richard Branson's ability to circumnavigate the globe in an expensive hot-air balloon or in terms of an African family's ability to pay for its children's education? One key measure used to depict patterns of global unevenness is Gross National Product (GNP) per capita (see Box 10.1). The World Bank is fond of monetary criteria like these which categorize the countries of the world according to groups of income, literacy, life expectancy, levels of primary school enrolment or volumes of export trade. But these indicators often provide only a very superficial insight into global patterns of impoverishment and inequality (Watts 1993a, 1993b). World Bank and United Nations development reports regularly carry bland statistical sections which often seem loosely related to 'on the ground' dynamics. A major problem with the maps produced from these statistics is that they have allowed some observers to label whole areas as 'Third World' or 'Global South' as if the same could be said of all its constituents (Wood 1985). As Morag Bell puts it:

> In emphasising what people are deprived of (as is implied by poverty), they [statistics] impose a negative uniformity upon non-western societies. As objects of study poor people become categories and are labelled as a homogeneous group. (Bell 1994: 184)

Box 10.1   The UK and Uganda

A comparison of the UK and Uganda is useful here in allowing us to evaluate the relative value of global indicators of inequality (Figure 10.1). Both countries are roughly the same size but their populations clearly do not have the same levels of access to water, health services or education. The UK can even afford US$ 586 per head on defence expenditure compared with US$ 7 per head in Uganda. These statistics clearly suggest massive inequalities between the two countries such that UK citizens can apparently expect to live 37 years longer than their Ugandan counterparts. In understanding how this is possible and how to respond to this unevenness there must be agreement on the significance of these inequalities and their causes or origins. There must also be an appreciation of history (Uganda was a British colony) and of the fact that not every Ugandan citizen has a life expectancy of 40 years or earns US$ 190. Ultimately simple measures like these tell us very little about wider social and economic differences or about how these variations are produced.

10.1   Geography and 'third world' development

The problem with labelling vast regions as 'low income' or 'Third World' is that a set of negative assumptions are often made and a 'negative uniformity' is imposed. Thus evidence of unevenness and injustice in the contemporary world is not always effectively conveyed in these statistical measures which are taken as indices of what constitutes 'development' (see Box 10.2). There is always a politics of labelling which it is necessary

UGANDA AND THE UK

UGANDA

Area: 236,000 square kilometres

Population: 19 million

Life expectancy at birth: 40.2 years

GNP per head: US$ 190

Population with access to safe water: 38%

Population with access to health services: 49%

Defence expenditure per head: US$ 7

Public expenditure on education as a percentage of GNP: 1.9%

Military expenditure as a percentage of combined education and health expenditure: 18%

THE UK

Area: 245,000 square kilometres

Population: 57.7 million

Life expectancy at birth: 76.7 years

GNP per head: US$ 18,340

Population with access to safe water: 99%

Population with access to health services: 99%

Defence expenditure per head: US$ 586

Public expenditure on education as a percentage of GNP: 5.4%

Military expenditure as a percentage of combined education and health expenditure: 40%

Figure 10.1   A comparison of the UK and Uganda: some key statistics. *Source*: International Broadcasting Trust, 1998

Box 10.2 An overview of global inequalities

Some of the most widely used indicators of global development are drawn from the reports of major international development institutions like the UNDP or the World Bank. Many of these international agencies acknowledge the weaknesses of their data collection methods and include detailed technical definitions and glossaries of the indicators they use. According to the UNDP *Human Development Report* the international standardization of data collection methodologies has proved extremely elusive: 'despite the considerable efforts of international organizations to collect, process and disseminate social and economic statistics and to standardize definitions and data collection methods, limitations remain in the coverage, consistency and comparability of data across time and countries' (UNDP 1997: 143). Similarly, the World Bank warns that 'Differences in timing and reporting practices may cause inconsistencies among data from different sources ... you should be cautious when combining these data' (World Bank 1999: 12). Nonetheless both the UNDP and the World Bank frequently aggregate significant amounts of development indicators into generalized categories and world regions. UNDP reports often aggregate countries into one of three categories (low human development, medium human development and high human development). The World Bank also prefers a tripartite division of the world as low income, middle income and high income. The table below includes a selection of indicators drawn from the World Bank's *World Development Indicators* (World Bank 1999) published in April 1999. Although the grouping of world regions in this way does point to significant patterns of global inequality, these indicators are ultimately meaningless given the massive differences within countries, let alone between these overarching artificial world regions in terms of data collection and availability.

	Girls in secondary school (per cent of total enrolment)	Electricity use per capita (kWh per capita)	Personal computers (per 1,000 people)	Value added in services (millions)
Low income	43	279	2.2	303,160
Middle income	48	1,222	15.8	2,661,945
High income	49	7,941	264.4	14,573,643

Source: World Bank Indicators, World Bank (1999: 17–21)

to understand and be critically aware of. All too often the 'developing world' has been defined as a 'problem' for Western governments which can only be resolved with the intervention of Western 'experts', donors, technology or ideology.

Definitions of the 'Third World' have been contested, as have the origins of the phrase (Auty 1979, Mountjoy 1976, O'Connor 1976, Pletsch 1981, Wolfe-Phillips 1987). Yet the concept of three worlds can hardly be said to convey a precise meaning. The geographical boundaries of the 'Third World' vary enormously according to which statistical indicators are used (Chaliand and Rageau 1985, Griffiths 1994, *New Internationalist* 1999) and the idea has different meanings in different cultures. The idea of 'three worlds' has provided many observers with a convenient shorthand expression and is undoubtedly linked to the formation of 'mental maps' and 'imagined geographies' of inequality (Bell

1994). Given that definitions of development have often been contested, these 'imagined geographies' often overstate a set of (tragic) stereotypical common features of the so-called 'Third World' such as poverty, famine or political instability. In discussing unevenness and inequality therefore a representation or an imagination of the people and places subject to the development process is constructed which is only partial and provisional. An 'alternative geography of the Third World' is proposed here, based on a recognition that development has been a 'protracted, painstaking and fiercely contested process' (Bell 1994: 175). This chapter tries to offer such a critical 'alternative geography' which aims to challenge some of the 'representations of peoples and places upon which grand [development] theories are based' (Bell 1994: 193–4).

It is extremely difficult to summarize the lives of people who are in some way subject to development and the variety of struggles that they encounter. Thus relations between 'worlds' are imagined and simplified by geographical expressions like 'North' and 'South', 'rich' and 'poor', 'developed' and 'underdeveloped'. These terms have to be approached with some caution. Becoming particularly popular after the publication of the Brandt Report in 1980 (Brandt 1980), the idea of 'North' and 'South' is widely seen as preferable to the three worlds scheme. However, no definitions were mentioned in the report and some of the boundaries it delineated between 'North' and 'South' were considered erroneous. Subscribers to this, and the three worlds scheme, have been criticized for the simplicity of these divisions and their failure to recognize diversity, the world does not consist of a series of individual national economies in the way often suggested in United Nations and World Bank reports. Many commentators have argued instead that there is a single global economic system with ever-closer degrees of interdependency (Allen 1995). This global capitalist economy, it is argued, *actively produces* inequality and uneven development:

> A big part of the 'economic development', i.e. the wealth, of the rich countries *is* wealth imported from the poor countries. The world economic system *generates* inequality and it runs on inequality ... It is a fraud to hold up the image of the world's rich as a condition available to all. (Lummis cited in Rau 1991: 46–7, emphasis in original)

If the global economy is a pyramid and everyone is encouraged to stand on top, how is this to be arranged? Development equality – 'catching up' with the wealthy through economic activity – increasingly seems less likely as we enter the twenty-first century and as evidence grows of widening and deepening global inequality. 'Development' has served then as a 'lighthouse' (Sachs 1992) or as a 'lodestar' (Wallerstein 1991) into which several different movements, governments and institutions have invested faith and meaning. Like the term 'Third World', development partly represents a geographical imagining, a representation of a better world, a belief in the idea of correctable inequalities/injustices between nations, states and regions and within the existing global economic orders. Escobar (1995) argues that the idea of the 'Third World' also gives enormous power to Western development institutions to shape popular perceptions of Africa, Asia or Latin America.

In a famous speech in 1949 President Truman spoke of an 'underdeveloped' world and of a 'handicap and threat both to them and the more prosperous areas' (Porter 1995). Born of the ashes of colonialism and of the fires of anti-communism in the United States, 'development' occupied centre-stage as an ideal in the global arena after the Second World War. Truman goes on to explain the need for 'modern, scientific and technical knowledge' and announces the beginning of a 'bold new program' to resolve 'underde-velopment' and poverty in 'backward' areas (Sachs 1992). Since then 'development' has become:

> an amoeba-like concept, shapeless but ineradicable [which] spreads everywhere because it connotes the best of intentions [creating] a common ground in which right and left, elites and grass-roots fight their battles. (Sachs 1992: 4)

Ideological interpretations of development have varied but they have usually always claimed 'the best of intentions'. The idea of development is often discussed in relation to 'developing countries' (itself a value-laden expression) but development has taken numerous forms in numerous locations and is actually still relevant in societies which proclaim themselves to be 'developed'. A sense of development being simul-taneously important in both 'developing' and 'developed' societies was recently for-mulated in a Human Development Index (HDI) (UNDP 1998). Again assuming that 'development' can be depicted in a single measure, the UNDP Index combines data (by country) on life expectancy, literacy, income, environmental quality and political freedom. Development is defined here as a series of numerical stages with a linear progression indicated (statistically) as degrees of departure from Western norms and ideals. Human development is represented as a series of league tables which measure progress with the unquestioned underlying assumption that every country can reach the highest levels of performance set in 'the West' or have access to the 'purchasing power' of the 'upper income' world. As in the case of GNP per capita, the HDI does in many ways (albeit at a very crude level) point to growing gaps between dif-ferent areas of the world. Its significance is undermined by the failure of its 'gener-alized' indicators. However effective statistical measurement of development can become, it is necessary always to be critical of the way in which statistics are collated and definitions formulated and to be careful in the assumptions and prescriptions made as a result.

This chapter is divided into four further sections. The first discusses two of the most important perspectives that have been formulated on the relations between develop-ment and inequality: the *modernization* and *dependency* schools. Although there have been many other different strands of development thinking, both these perspectives have been widely influential and remain relevant to an understanding of theory and practice today. Section 10.3 outlines the need to view development historically and formulate a sense of how it has been redefined through time. The 1960s have been described as the 'high decade of development', as a period of great optimism. Conversely, the 1980s were described as the 'lost decade for development'. So how have historical forces shaped our understanding of the geography of development? The next section focuses attention on people and places and explores how the subjects of development experience developmental interventions. Three major 'brokers of devel-opment' (non-governmental organizations (NGOs), states and international institutions) dominate the literature on development, but how are the needs and aspirations of intended beneficiaries understood by these agents of the development process? The final concluding section of the chapter poses questions about the political, cultural and economic implications of globalization for the future of development theory and prac-tice and examines possible alternatives to development based around local cultures and knowledges.

10.2   Conceptions of development

'Development thinking' (Hettne 1995) or the sum total of ideas about development theory, ideology and strategy, has shown numerous twists and turns in the twentieth century. Since all-encompassing definitions have been contested and controversial, little consensus exists today about what constitutes 'development'. However, some core conceptions have emerged and some have continued relevance in the contemporary world. Although there are many different strands of development thinking to explore, the modernization and dependency approaches have been two of the most influential in the twentieth century. In discussing the different conceptions it is important to think about where and when they have emerged (Slater 1993). Most reflect some of the priorities of development thinking characteristic of their era. The formation of development theories therefore depends on different perceptions of 'development challenges' at different times (see Box 10.3).

10.2.1 The modernization school

Watnick (1952–3) once argued that the task of development was to provide 'an ethos and system of values which can compete successfully with the attraction exercised by

---

Box 10.3   Coca-Cola wars in Uganda

Since the Coca-Cola manufacturing licence for Uganda came under the ownership of the South African Bottling Company in 1996 its market share has risen from 30 per cent to 60 per cent with the volume of cases sold doubling to approximately 72 million in 1998 (*Guardian*, 23 December 1998). A key issue in the successful marketing of Coke in Uganda has been the company's ability to distribute 3,500 cooler boxes to drink vendors, opening four ice-plants. Uganda's sources of power, like many African countries, are subject to frequent disruptions and limited infrastructure makes certain rural areas inaccessible. Few Ugandan companies could hope to compete with this level of organized distribution or the US$ 16 million Coca-Cola invested in a bottling plant in the western city of Mbarara which produces some 96 million bottles a year at capacity. Ugandan soft-drinks companies would also be hard pressed to compete with Coke's marketing capacity which employs seven South Africans and a workforce of 800.

Figures for the consumption of Coke in Africa are slightly paradoxical in that Mozambique, a country with one of the lowest GNP per capita incomes on the continent at US$ 100 (World Bank 1997), consumes an average of 35 drinks per person per year, one of the highest rates in Africa (*Noticias* 1998). Ugandans, however, consume an average of 18 bottles per person despite a much higher GNP per capita income of US$ 250 per person. This suggests that globalization, in which African markets are penetrated by globally available products like Coca-Cola, is itself an unequal and uneven process, affecting different parts of the same world regions in different ways. The example of Coke's emerging domination of soft-drinks markets in Uganda also raises questions about the links between globalization and development, since this 'success' could be identified as a success for development in that it has improved infrastructure and created jobs. Does the widening availability of Coke represent a form of progress?

Plate 10.1 A rally and demonstration of popular discontent with GM Foods in Delhi in March 1999. A variety of women's groups have been at the forefront of the campaigns in India to reduce dependency on Monsanto's technology. (Images of India Picture Agency)

Communism' (Watnick 1952–3: 37). Watnick was calling for the modernization of 'underdeveloped areas' and he paints a picture of 'underdeveloped peoples' (p. 22) confined to 'backwardness' but torn between the appeal of communism and the prospect of Western modernization. This was an essential characteristic of the *modernization school*, which was often dualistic, opposing 'traditional' to 'modern' life-styles, 'indigenous' to 'Westernized' as if no country or citizen could belong to both categories.

Hirschmann (1958), another key proponent of modernization ideas, voiced the optimistic view that the forces of concentration ('polarization') will 'trickle down' from the core to the periphery at national, regional and global scales. Governments were advised not to intervene in this polarization process despite the fact that Hirschmann (1958) was advocating 'a basically unbalanced economic growth strategy' (Potter *et al.* 1999: 46). Gunnar Myrdal (1957) adopted a more pessimistic view, arguing that capitalist development produces deepening regional and social inequalities. This happens when population migration, trade and capital movements come to focus on the key growth points of the economy, such as the largest urban areas. Friedmann's (1966) core–periphery model adopts many of the same assumptions about the polarization of development in 'transitional societies'. This model suggests that a number of stages exist in the national development of urban centres leading to a final stage of an 'interdependent system of cities', which represents the spatial fruition of the development process.

Another modernization theorist who identified stages of growth in the development process was W. W. Rostow (1960). Again the focus was on a top-down 'trickling' of capitalist development from urban-industrial areas to other regions (Stöhr and Taylor 1981). Rostow (1960) predicted that nations would 'take off' into development, having gone through five stages which he likened to the stages an airplane goes through before take off, from taxiing on the runway to mid-flight. Ranging from stage one, 'traditional society', to stage five, the 'age of high mass consumption', the theory takes our faith in the capitalist system for granted since Rostow assumes that all countries will be in a position to 'take off' into development. Further, in common with other modernization approaches, Rostow's model devalues and misinterprets 'traditional societies' which represent the 'lowest' form or stage of development. The advanced state of modernization is always represented as 'Western modernization'; traditional societies seem like distant, poor relations.

Geographies of inequality and development cannot be neatly summarized as a set of prescriptive stages. Urban areas, for example, are themselves subject to uneven development and inequality. Modernization approaches have also largely failed to address the importance of gender, assuming that men and women occupied equal positions in terms of power relations and decision-making. The 'trickle down' effect has often failed to materialize among those who have been the subjects of modernization projects.

10.2.2 The dependency school

Important questions need to be raised about where the 'core' ends and the 'periphery' begins. Radical *dependency approaches* that emerged in the 1960s and 1970s challenged this notion of positive core–periphery relations, identifying instead exploitation between 'satellites' and 'metropoles'. The *Dependencia* school is most commonly associated with Latin America but also emerged in Africa, the Caribbean and the Middle East. Drawing upon Marx's writings about the unevenness of capitalist development, dependency scholars such as Celso Furtado (Furtado 1964) drew attention to the mode of incorporation of each country into the world capitalist system. This is identified as a key cause of exploitation. The dependency school is most commonly associated with the work of André Gunder Frank who published a seminal paper on 'The development of underdevelopment' in 1967. This thesis is relatively uncomplicated in that it views development and underdevelopment as opposite sides of the same coin; the development of one area often necessitates the underdevelopment of another. Dependency on a metropolitan 'core' (e.g. Europe, North America) increases the 'underdevelopment' of satellites on the 'periphery' (e.g. Latin America, Africa). Unlike many modernization approaches, dependency theorists (*dependistas*) sought to view development in historical contexts, and argued, for example, that colonialism helped to put in place a set of dependent relations between core and periphery.

These peripheral satellites, it was argued, were encouraged to produce what they did not consume (e.g. primary products) and consume what they did not produce (e.g. manufactured/industrial goods). Thus, rather than likening the process of economic growth to an aeroplane setting off into a blue sky of urban-based, Western life-styles and consumption, the dependency school was arguing that many 'underdeveloped' areas had been stalled on the runway by unequal relations and a history of colonialism, denying them a chance of ever being airborne, 'modern' or 'industrialized'.

Key criticisms directed at the *dependistas* were that the theory represents a form of 'economic determinism' and overlooks social and cultural variation within developed and underdeveloped regions. Another point of contention was that the dependency theorists seemed to be calling for a de-linking from the world capitalist economy at a time when it is undergoing further globalization and economic integration. Elements of the dependency writings were, however, quite thought-provoking and remain relevant, particularly their contention that the obstacles to development equality were *structural*, arising not from a lack of will or poor weather conditions but from entrenched patterns of global inequality and 'dependent' relationships. A brief comparison of the two schools of development thinking is useful here, particularly given that dependency was primarily a critique of the classical traditions of modernization. According to Bell (1994) both perspectives deal in dualistic either/or scenarios:

> these contrasting schools of thought share a common starting point – poverty is viewed in deprivationist terms while their different views of development have a positive connotation. (Bell 1994: 184)

Concepts of development like the modernization and dependency perspectives are far from being static, uniform or unified, however. In this way it is sometimes easier to refer to modernization and dependency 'schools' since neither represents a singular commonly-agreed approach and both are complex and varied. As we shall see in the following section, it is important to consider theories and practices of development in their specific historical and geographical contexts.

10.3 Development practice (1): history and development

The phrase 'traditional society' was particularly important in the modernization school but historical perspectives were often lacking in early writings about development (Rist 1997). As Crush (1995) points out, development is primarily 'forward looking', imagining a better world, and does not always examine issues of historical and geographical context. The historical process by which 'gaps' began to emerge between 'North' and 'South' has been interpreted in a variety of ways but a key question has been to what extent did European expansion and colonialism 'underdevelop' (Frank 1967) large areas of the world? The impact of European expansion was not uniform; the geographical patterns of expansion varied, as did the motivations for it. Not all countries were affected in the same way by colonialism because of their different histories, cultures and resources. European expansion is usually seen as reaching its zenith before the Second World War but the influence of Europe and 'the West' more generally has endured well beyond 1945. Hall (1992) argues that an important divide was put in place between 'the West' and 'the rest' as a direct result of this imperial expansion, reminding us that 'the West' is much more of an idea than a geographical reality. By accelerating contact between cultures and economies 'the West' was presented as 'the best' and most advanced or 'civilized' of all humanity. Many accounts of the history of European expansion are thus dominated by the presumed supremacy of 'Europe' and 'the West' with only limited references to the complex histories and cultures of the areas that were colonized. Similar criticisms have been made of some development theories (e.g. modernization) which have also often been based around 'Western' philosophies, experiences and histories (Hettne 1995; Slater 1992).

Many recent histories of development have dated its beginnings as an area of theory and state practice to President Truman's speech of 1949. The idea of development is, however, much older than this (Cowen and Shenton 1996) and has much more diverse geographical origins. Development was not a simple 'gift' following contact with Europeans but predates the 'age of discovery' (1400–1550) and the 'age of empire' (1875–1914) (see Section 1). Indigenous peoples in Africa, the Americas, Asia and Australasia had highly developed and sophisticated cultures and technologies prior to colonization (Dickenson *et al.* 1996). Given that the particular nature of colonialism varied and was inspired by often very different motivations, definitions are not easy to formulate. Colonialism has been variously interpreted as an economic process of unequal exchange, as a political process aimed at administration and subordination of indigenous peoples, and as a cultural process of imposing European superiority. All three explanations are clearly important in understanding histories of colonialism, and its legacies are important to any understanding of contemporary development (see Box 10.4). As Clapham (1985: 20) put it:

> There is still no more striking, even shocking, reminder of the impact of colonialism in Africa than to cross an entirely artificial frontier and witness an instant change of language.

Box 10.4   Cuba and the socialist 'Third World'

Although many conceptions of global development have no problem in distinguishing between 'developed' and 'developing' countries, the same has not been true of distinctions between 'capitalist' and 'socialist' or centrally planned economies. The boundaries between second and third worlds are thus often unclear, as are wider definitions of 'socialist development' and 'socialist government'. The geography of socialism has therefore been diverse and is by no means static. There are immense political, economic and social differences within the so-called 'Socialist Third World' (Forbes and Thrift 1987). One useful way of distinguishing between 'socialist' states is the nature of their political structure and the degree of 'democracy' therein. The regime of Fidel Castro in Cuba provides a case in point. After the overthrow of Fulgencio Batista's corrupt regime in 1959 Cuba established a Communist state under Castro in 1962. Despite a thirty-six year US blockade on Cuba and crumbling infrastructure, Cuban socialism brought significant advances in health and education in a social revolution that Castro hoped to export to Africa and the Caribbean. The ideas that lay behind this attempted transformation may have been devalued since the end of the Cold War and the collapse of the former Soviet Union but Castro (now 71) remains in power, outliving his major leadership rival in Cuba Jorge Mas Canosa and surviving refugee crises, internal plots and external assassination attempts. One of Castro's strengths has clearly been his ability to manipulate Cuban senses of national pride and national identity in the face of external aggression. In January 1999 Castro told a conference in Havana that he knew 'almost nothing' about economics but he continues to predict that the global capitalist economy is unsustainable and can only end in crisis (*Observer*, 24 January 1999). The human rights record of Castro's regime has been poor as dissent continues to be brutally suppressed, political prisoners imprisoned and independent newspapers and TV stations banned (*New Internationalist* 1997). The Cuban example, however, disrupts many conceptions of development which assume that as a relatively small, 'satellite' state, Cuba would be largely irrelevant to the overall scheme of development practice. For regimes from Mozambique to Chile, Tanzania to Vietnam, Cuba provided critical logistical and ideological support in the struggle for 'socialist development'.

Box 10.5   Globalization and GM foods

'Food and agriculture is very much a global reality which shapes our lives in profound cultural, ideo-logical and economic ways' (Goodman and Watts 1997: 3).

Since the UN staged the first World Food Conference in Rome in 1975, amidst growing anxiety about food availability and food prices, the world food economy has changed in many different ways. At the level of *production* and *distribution*, there has been a radical restructuring. These areas have been internationalized or become more global in scope and character, particularly since the emerg-ence of transnational corporations (TNCs). By the second World Food Conference in Rome in 1996 UN agencies were becoming increasingly aware of the ways in which TNCs are (re)shaping the world food economy in ever more global ways, fundamentally reorganizing the division of labour in the global agro-food system over the last 25 years (see Chapter 9).

Transnational biotechnology corporations like Monsanto, Hoescht, Pannar and Calgene have claimed that the recent emergence of GM or genetically-modified crops has the potential to eradi-cate global food shortages and malnourishment, offering sustainable food security for the poorest parts of the world. These claims have, however, been widely contested in the 'global south'. In March 1999, leading scientists and activists from 12 countries (including Malaysia, India and Bangladesh) met at a conference in Delhi to discuss the prospect of 'biodevastation' associated with the intro-duction of GM crops and vowed to bring down Monsanto and other biotech TNCs and to build a global mass movement for sustainable and organic agriculture (BBC 1999).

In April 1999 an alternative year-long seed *Satyagraha* was launched in India in direct opposi-tion to Monsanto's GM seeds. Vandana Shiva, Director of the Research Foundation for Science Technology and Ecology that helped to launch *Satyagraha*, has described Monsanto's seeds as 'ter-minator technology' and has promised that '[w]e will drive Monsanto out of this country. It will be destroyed. Free trade in food is a recipe for famine and farmer's suicide' (Shiva 1999: 9). Almost 1,500 organizations, many of them established by women, have come together across India to reject the new GM seeds like Monsanto's Bollgard Cotton which can destroy beneficial species and can create superior pests. Many Indian environmentalists have described the recent patent crisis as a 'neo-colonial' ploy by the West and the World Trade Organization to (re)colonize the 'developing world' (*Asian Age* 1999: 5). Many biotechnology transnationals are now turning to South Africa as a potential test ground for GM organisms since the country currently has no regulations to control modern biotechnology (a Bill to establish a regulatory system is currently making its way through parliament). South Africa is not a signatory to the International Convention on Biodiversity that aims to protect the world's biological resources. Signatories of this convention agree that an international protocol on biosafety is needed to ensure the safe handling and transfer of GM organisms between countries.

In demographic terms colonialism effected profound transformations in relations between ethnic and cultural groups and in relations of production (particularly in agricul-ture where patterns of land ownership were often radically altered). The nation-states that emerged following the granting of independence to many countries between the 1940s and the 1970s often did not correspond to contemporary patterns of ethnicity because the inherited forms of government administration tended to reflect the interests of the depart-ing colonial power. Nevertheless, a key outcome of the 'colonial experience' has been the ability of 'post-colonial' states to engage with the global capitalist economy. Many 'post-colonial' states maintain important political, cultural and economic ties with their former colonial rulers. In order to understand how these and wider connections with the capitalist world economy have emerged, it is necessary briefly to examine the concept of decolonization. This is usually taken to refer to the formal political withdrawal of colonial powers but African writer Ngugi Wa Thiongo (1986) has talked of a wider, on-going process of 'decolonizing the mind'. Colonialism put in place important political and economic relations but the cultural legacies of colonialism bequeathed deep social and cultural divisions in many societies. The contribution of women to development, for example, was devalued in many colonial societies (Coquery-Vidrovitch 1996) which left profoundly unequal social relations between men and women to be resolved after independence.

In the process of decolonization 'development' became an over-arching objective for many nationalist movements and the independent states they tried to form. Although experiments with development were tried in many colonies, the idea of development was invested with the hopes and dreams of many newly emerging independent nations that wanted to address inequalities and divisions in their societies (Rahnema 1997). As we have seen in the preceding section, global inequalities have widened since 1945 and this has complicated many post-colonial development strategies. Several recent shifts in the global economy, and the emergence of what has been termed 'neocolonialism' have undermined attempts to meet promises made in the early post-colonial era. Particularly in the last three

Plate 10.2   The environmental pollution left by Shell's oil refineries in Nigeria has had particular implications for the people of Ogoniland who have seen very little of the profits generated by oil reinvested in their state. Some observers have argued that Shell's relationship with the Ogoni people and the Nigerian state is one of 'neocolonialism'. (Still Pictures/Detlee Pypke)

Box 10.6   **The World Bank**

When the World Bank was set up at the Bretton Woods Conference in 1944, it was seen as a way to bring about shared prosperity in a world recently shaken by war. According to a Brazilian delegate at the Bretton Woods Conference, the desire to establish a central lending source to which governments could contribute was 'inspired by a single ideal – that happiness be distributed throughout the world' (IBT 1998: 3). Fifty-five years on the World Bank has increased its influence in the distribution of global equality and 'happiness'. In 1990 the Bank's then President, Barber Conable, remained confident of alleviating world poverty: 'We know a great deal about who the poor are, where they are and how they live. We understand what keeps them poor and what must be done to improve their lives.'

   Despite this unbridled confidence in the Bank's capacity to resolve global inequalities the organization has been widely criticized for the environmental, social and cultural impacts of its work (Oxfam 1995, Hildyard 1994, 1997). The Bank's World Development Report (WDR), which in 1997 had a production budget of US$ 3 million, is presented as 'cutting edge' research when in reality the Bank has become increasingly estranged from the wider intellectual community. The WDR overlooks far more research than it includes and often neglects indigenous knowledges and perspectives on development. The 1997 WDR, for example, was entitled *The State in a Changing World* (World Bank 1997) which would have been an excellent opportunity to discuss critical new perspectives on the state and to provide an open review of some of the wider cultural and political dilemmas facing contemporary states. Instead the World Bank's promises of 'radical shifts' in thinking about the state are, on closer analysis, remarkably similar to the older 'economistic' view of a minimalist state. According to current President James Wolfensohn the World Bank wants to increase its role as 'provider of leadership in the field of development knowledge' but the organization's ability to understand 'who the poor are' or 'where the poor are' remains limited. Escobar (1995) shows how the Bank has often made people the 'objects', not the 'subjects', of development, pointing to the economistic language of the Bank and their predilection for statistics. 'Leadership' seems a long way off, since it is taking the organization so long to understand and appreciate the way in which gender is relevant to development or why projects often fail to reach intended beneficiaries. With its US$ 20–25 billion annual lending programme (IBT 1998) the World Bank has enormous potential for alleviating poverty, but not without a radical rethinking of its approach to development. The current system of 'one dollar, one vote', where voting capacity depends on a member government's financial contribution, has to be transformed so that all 178 member states can have a say and not just a handful. Individual national governments that choose to adopt World Bank policy also need to consult more widely, inviting greater social participation in this process.

---

decades, a number of factors have served to sustain the linkages between the 'core' and the 'periphery', usually to the detriment of the latter. For example, the emergence of transnational corporations (TNCs) that seek raw materials, markets and profits on a global basis has had wide-ranging impacts on the world food economy (see Box 10.5). Further, since the beginning of the 1970s there has been a sustained and substantial flow of investment funds from Europe and North America to other world regions. Wealthy Northern banks were able to lend vast sums of money to African, Latin American and Asian governments and investment in manufacturing plants increased due to rising costs of labour and environmental protection in the global 'North'. This latter trend has led to what has been called the **New International Division of Labour** (NIDL) but this is by no means the first division of labour between 'the West' and 'the rest'.

The term 'neocolonialism' has often been deployed as a means of pointing to the ways in which the NIDL is often similar to earlier divisions of labour. This is not to suggest that there has been a simple reinstatement of colonial power relations. Continuing relations of 'dependency' have also been related to more recent historical events such as the 'debt crisis' which began in the 1970s and 1980s. As a consequence of heavy borrowing commitments many countries have become increasingly dependent on development assistance from international institutions (Figure 10.2). Between 1985 and 1992 Africa disbursed US$ 1.6 billion in debt payments, diverting government spending in crucial areas like education and health (Oxfam 1993: 3). The severity of the debt crisis gave key financial and regulatory institutions like the IMF and the World Bank (see Box 10.6) a renewed influence in the affairs of national governments.

Between 1980 and 1993 Southern countries lost around US$ 100 billion because of a major decline in the value of commodity prices in international markets (Potter *et al.* 1999). A sharp deterioration in the terms of trade between North and South made it very difficult for many countries to make debt repayments, further increasing dependence on the World Bank and IMF who then imposed **Structural Adjustment Programmes** (SAPs) on these countries. The 'neoliberal' perspective advanced by SAPs urges a withdrawal of state intervention in the national economy, a liberalization of trade relations, the privatization of government institutions and a devaluation of national currency. SAPs have often had a whole variety of negative impacts on the poor, who are affected by rising unemployment, inflation, devaluation, the withdrawal of state subsidies and the end of free education and health care as a national objective (Oxfam 1995). Between 1980 and 1993 sub-Saharan Africa's debt more than tripled to around US$ 183 billion which, although much less than the Latin American total of US$ 446 billion, still means that Africa owes more than it earns.

In September 1996, for the first time, the World Bank and IMF agreed to reduce the debt burden of some of the poorest countries with the Heavily Indebted Poor Countries (HIPC) initiative. There are currently 41 HIPCs and the Bank's President recently described the initiative as 'good news for the world's poor' (IBT 1998: 12) but Japan and Germany have opposed it from the outset and the IMF is a reluctant partner. Richer countries were initially unable to negotiate the sharing of the 'debt burden' even though the sums of money involved (an estimated US$ 5.5–7.7 billion) are relatively small. The 'burden' was nothing compared to that carried every day by millions of poor people who pay the price for their governments' indebtedness and choice of SAP. In any event, the problem is that this initiative barely scratches at the surface of the estimated US$ 245 billion owed by the world's poorest countries. The provision of overseas development assistance or aid by the world's richest economies is a similarly limited gesture. In 1970 the UN urged 'developed countries' to allocate at least 0.7 per cent of their GNP to overseas aid by 1975. In the USA in 1994 alone this would have represented some US$ 47 billion, a fraction of the 2.2 per cent annual increase in that country's GNP between 1980 and 1993 (UNDP 1999). This commitment has since been reiterated at numerous UN development conferences but the response of the world's most prosperous nations has fallen well below such expectations. In May 1997 the UK overseas development budget had fallen to 0.27 per cent of GNP (DFID 1997) and although the New Labour government

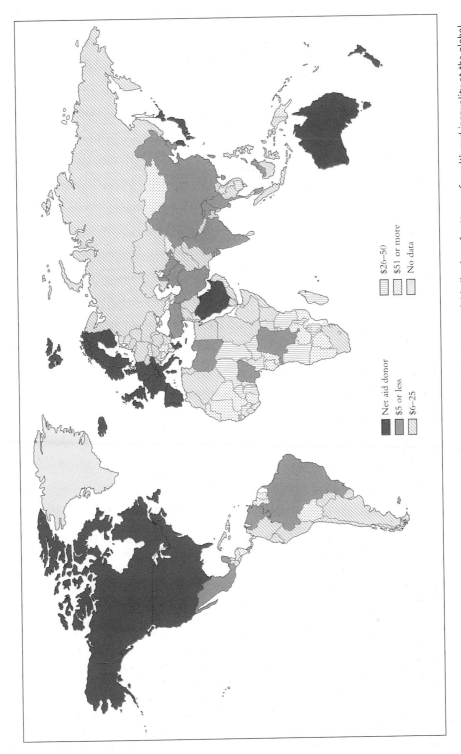

Figure 10.2 Flows of multilateral and bilateral aid between the 'North' and 'South' give a useful indication of patterns of wealth and inequality at the global level. *Source:* World Bank, 1998b

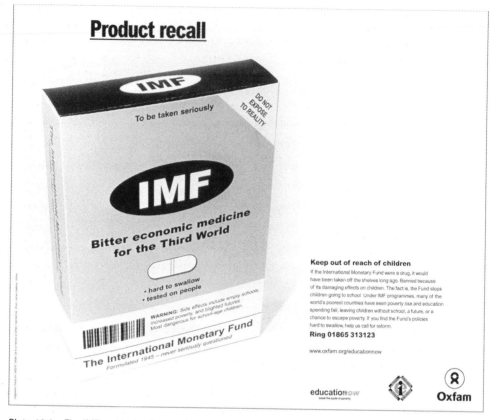

Plate 10.3   The IMF and its hard to swallow 'bitter economic medicine' has had major implications for 'Third World' countries as depicted in this Oxfam campaign. Its side effects include empty schools, increased poverty and 'blighted futures'. (Oxfam)

has pledged to meet UN targets, restrictions on total spending have made this impossible before the year 2000. Like many other forms of foreign development assistance, UK contributions have often been tied to the purchase of British goods and services (German and Randall 1993). US aid was also often closely related to perceived US geopolitical interests (Luke 1991, Ó Tuathail 1994). Since development assistance can take many forms (such as disaster relief, longer-term development aid, food aid or military aid) donor governments can manipulate and condition the receipt of these contributions (Earthscan 1997, Owens 1996).

The environmental consequences of development aid have not always been widely appreciated and indeed early development thinking was very much based on the idea of taming nature and natural forces (Jarosz 1992). Running almost parallel with the growing levels of industrialization led by public and private sector borrowing in the 'global South' in the 1970s and 1980s has been the emergence of environmental consciousness in the arena of development. Key stages in this process were the Brandt Report (1980), the Brundtland Commission on Environment and Development (WCED 1987) and the Rio Earth Summit of 1992 which considered the possibility of direct linkages between poverty and environmental degradation (see Box 8.4). The notion of 'sustainable development' is also relatively recent, although the term is exceptionally vague and appeals to mainstream development institutions that believe in evolutionary rather than revolutionary change. The concept of sustainability has its roots in Northern streams of environmentalism (Redclift 1984, Adams 1990, 1993) and does not always reflect perceptions of environmental crises held by people in the 'global South'. Some commentators have even suggested that 'sustainable development' is incompatible with neoliberal economics which favour the principles of the market, since there is much evidence that such perspectives have aggravated conditions of poverty and other causes of environmental degradation (Watts 1993a).

The history of recent international development conferences since the environment re-emerged on the international stage with the first Earth Summit in Rio (see Boxes 8.3 and 8.4) in 1992 is interesting and tells us something about recent shifts in development thinking (Earthscan 1992). Like many other UN gatherings, Rio was big on rhetoric but weak on actual commitment to change, whilst consensus was difficult to achieve between world regions with very different priorities. An international conference on population and development was staged in Cairo in 1994 and this was followed in 1995 by a conference on social development and human rights held in Copenhagen. One of the most interesting UN conferences followed in Beijing in 1996 on the theme of Women and Development. This last meeting, perhaps more than any of its predecessors, illustrated the complexities of the UN's task in finding consensus amidst the diverse cultures and societies that were represented (Whittaker 1997). The meaning of gender and its relationship to development is socially produced and this varies enormously between the different cultures and societies of the 185 member states. In a way, international institutions like the UN have the potential to play a very powerful role in shaping 'gendered' conceptions of development and in raising awareness of gender issues. With a staff of over 500,000 and a total budget of US$ 1.35 billion, the UN agencies have crucial resources at their disposal and have the potential (if not always the commitment) to bring radical changes in international gender relations.

10.4 Development practice (2): subjects of development

People are central to the development process and an integral element in all development strategies. However, their needs have often been ignored and there has been a general failure to consider the implications of development policies for individuals, households and communities. Development strategies have also been driven by economic objectives, relegating provision for people's basic needs, their empowerment or efforts to raise their living standards to second place (Potter et al. 1999). Development strategies often assume that 'people' and 'communities' are homogeneous and passive rather than differentiated and dynamic. The knowledge and skills of people in particular communities have often been overlooked in favour of broader and less sensitive approaches to 'modernization' which bypass local cultural traditions and histories. Three major agents of development have dominated the literature on development: nation-states, international development institutions and NGOs. It is difficult, however, to view any one of these three 'brokers of development' in isolation since, for example, nation-states are the primary focus of the

Box 10.7   Children and development

One of the clearest examples of global inequalities comes from a comparison of the relative position of children in societies of the global 'North' and 'South'. In wealthier societies children are often well-fed, have clean clothes and access to an education until the age of 16 or older and have good protection against ill-health. Their counterparts in some of the world's poorest countries have few of these assurances. A report from ActionAid (1995) clearly shows that children and children's work have been neglected research areas despite the fact that as Robson (1996: 43, cited in Earthscan 1998) has argued '[b]y the age of 10–12 years, some children may contribute as much to household sustenance as adults'. Again gender is crucial in determining the roles and responsibilities of children with girls often expected to provide assistance with household duties and boys involved in more public forms of labour. Access to education for male and female children is also often unequal with the continuation of female children's education being seen as much less of a priority for many poorer households (Momsen 1991). Children have also often been caught up in the political conflicts that divide many societies and make development a more complicated proposition. Graça Machel's (1996) study on the impacts of armed conflict on children makes clear that 'in the course of displacement, millions of children have been separated from their families, physically abused, exploited and abducted into military groups, or they have perished from hunger and disease'. Surveys cited in Machel's study indicate that mortality rates among displaced children can be as much as 60 per cent higher than rates for conflict-affected, non-displaced children in the same country (Earthscan 1998). A recent study by a human-rights group also revealed that as many as 300,000 children, some as young as 8, are in active combat around the world in government forces or armed rebel groups (*Independent on Sunday*, 6 December 1998). While the UN definition of a minor is anyone under 18, existing international law sets 15 as the minimum age of recruitment into armed forces and participation in hostilities. Uganda has the lowest official recruitment age of 13, but it should be noted that the recruitment of under-18s is not confined to countries of the global South. Thirteen of the 15 EU member-states still recruit under-18s which for Human Rights Watch would seem to be part of the problem: 'If the British army is recruiting at 16 because it needs to catch young men at the peak of their physical fitness then they are relying on the same arguments that we hear from Kabila' (Laurent Kabila, the ruler of the Democratic Republic of the Congo, has a regular army believed to include up to 20,000 child soldiers) .

Source: Independent on Sunday, 6 December 1998.

United Nations while NGOs often try to work at the interface between national governments and international governmental institutions.

Many international development organizations have often made erroneous assumptions about the distribution of power and decision-making within the nations and communities where they operate. Consider the level of the household, for example. It is taken for granted that an equal division of labour exists between men and women within households and communities. As we have already noted, national statistics often conceal significant variations within and between specific regions, between rural and urban areas or between men and women. These aggregations of household labour and resources often fail to capture the changing and important contributions of household members such as children (see Box 10.7). The contributions of women to the productive and reproductive activities of the household are still widely overlooked in national 'official' development statistics (Momsen 1991, Østergaard 1992). Nonetheless, the household has been, and will likely remain the basic living unit in many parts of the global South, the key to controlling and organizing production, consumption and decision-making. Household sizes and structures differ quite considerably from those of the global North, often including an extended family of grand-parents, aunts, uncles and other more distant relatives. Inequalities are ever present. Manifesting themselves at this level of the household means that all these households do not have the same level of access to income, assets (e.g. land) or local and national power struc-tures. It is important for international development organizations to understand the differ-ent roles and responsibilities of men and women within these diverse households. The contribution of women in particular to household income and welfare needs to be much more carefully understood (Kabeer 1994, Moser 1993, Wuyts *et al.* 1992).

In development policy and practice it has often been assumed that the household in the 'global South' corresponds to the Western model of the nuclear household, yet these households are far from homogeneous in their structure and in the amount of control over decison-making and resources that women enjoy. The contribution of women to the development process, of their knowledge, time, labour and capital, has often been under-estimated by development 'experts'. Mohanty (1991) rejects the idea of the stereotypical 'third world woman' as the passive recipient of social services that has often been con-structed by development planners and theorists alike. Similarly, the visibility of women's

Plate 10.4 A gathering of delegates from around the world to debate connections between gender and development in Beijing in 1995. Consensus was hard to achieve across the range of countries represented but the event nonetheless demonstrated the growing involvement of the UN in discussions about gender. (Popperfoto/Reuters)

productive, reproductive and community management roles in the recording of national development statistics has been widely criticized by gender and development (GAD) researchers (Rathgeber 1990, Moser 1993). Women were often not defined as 'productive' contributors to their societies, as it was abstractly assumed that women were largely confined to the private reproductive or domestic sphere (Marchand and Parpart 1995). In agriculture, for example, access to credit, rural extension services, land ownership and farm technology has often been denied to women, despite the fact that women produce more than half the world's food (Koczberski 1998). Development planners have often neglected the importance of women's 'double day' (Moser 1993) or the limitations placed on their time by the need to combine reproductive (e.g. child care) with productive (e.g. farming) activities. In addition to these two important roles, women often participate in the management of their communities. They organize political, social and economic structures and play a crucial role in the management of the environment and natural resources (Sondheimer 1991, Dankelman and Davidson 1988, Shiva and Moser 1995, Sittirak 1998).

Although the importance of women's participation in development planning has increasingly been recognized by international institutions, as we have already seen, the capacity of national governments to respond effectively to the needs of their societies at the 'grass roots' has been severely curtailed by indebtedness, reduced commodity prices, dependence on international donors, recession and protectionism in the global economy. The state has typically been seen as the primary agent of development, particularly since the 1960s when many countries attempted ambitious large-scale modernization projects with the backing of international development institutions and commercial lenders. The formation of new states, particularly in sub-Saharan Africa, followed at a time when development ideals were becoming increasingly popular worldwide. The 'age of great optimism' imbued these emerging states with a sense of the importance of development in the post-colonial era. Again, as with so many other aspects of 'development' there are no universal definitions of the state and states do not behave in any predetermined way. Ideology also has an important bearing on the way the role of the state in development is viewed. For example, reflecting the World Bank's 'neoliberal' framework, the World Development Report (WDR) views an effective state almost exclusively 'through the lens of economic efficiency' with states seen as providing either 'barriers' or 'lubricants' to free-market economic reforms. Not suprisingly, the Bank often 'locks society out' (Hildyard 1997: 43) in its definition of the state (Figure 10.3). Although the WDR calls for greater participation in decision-making in order to bring 'government closer to the people', a number of key policy areas are seen as 'off-limits' since they require 'isolation from political pressure' (World Bank 1997, 117).

The term 'empowerment' is often used by the Bank and other sections of the 'development industry' (Ferguson 1990) to describe their commitment to popular participation in decision-making for development. Facilitating 'people's participation' is now on the agenda of most international development institutions, including the United Nations Development Programme (UNDP) which produced a major review of participation and democracy in 1993 (UNDP 1993). For Rahnema (1997), the increasingly widespread acceptance of the idea of participation suggests that it has been severely diluted and has lost some of its more radical potential:

> governments and development institutions are no longer scared by the outcome of people's participation ... there is little evidence to indicate that the participatory

Figure 10.3   This illustration depicts the idea that World Bank 'experts' view the role of the state in development as comprising a simple formula or recipe, with several distinct 'ingredients'. *Source:* 'The World Bank and the State: a Recipe for Change?' by Nicholas Hildyard, Bretton Woods Project

approach, as it evolved, did, as a rule, succeed in bringing about new forms of people's power. (Rahhema 1997: 118–23)

Escobar (1995) also seems somewhat disillusioned with the development concepts of 'participation' and 'empowerment', pointing out that development planners and politicians have often tried to manipulate experiences of participation to suit their own ends. Munslow and Ekoko (1995) similarly argue that empowerment has been stronger on rhetoric than in reality and that advances in widening political participation have been slow, particularly in Africa, where the 'democracy and development' agenda has been manipulated by political movements across the continent (Baker 1998). Empowerment speaks of making people the agents of their own development while doing little about the causes of inequality, and without transforming the nature of ties with international development institutions and their ideologies.

The idea of empowerment and its increasing acceptance in development circles is partly a consequence of the rising influence of non-governmental organizations (NGOs) which have often been highly localized and focused on particular issues. NGOs have been defined as a 'residual' category that includes a wide variety of formal associations that are not government-led, even though definitions vary enormously and many NGOs do receive finance and support from their home governments. They can operate at a local, national or international level, but there is often little overall co-ordination between different NGOs even though they usually work on similar issues, sometimes in the same country or city. A key issue that has been raised in critiques of NGO activity is the closeness of their contact with local cultures and knowledges. Although many have argued that the work of NGOs is often poorly monitored and sometimes imposes foreign solutions in delicate situations, many observers have argued that they can provide an effective link between state and society, often working in 'very practical ways' (Edwards and Hulme 1992: 14).

The relationship between the state and NGOs is crucial to the prospects of inclusive democratic development. Particularly in situations of political instability or conflict NGOs can provide critical linkages in the absence of effective or disputed state administration. Médecins Sans Frontières (MSF), for example, has a policy of *témoignage* or witnessing

Plate 10.5 Campaigners in London demonstrate the need to 'drop the debt now!' through the Jubilee 2000 campaign which has attempted to raise awareness about the scale of the debt problem and the need for Tony Blair's New Labour government to act. (Format Photographers/Ulrike Preuss)

which includes establishing a presence near people in danger, reporting on their condition and 'engaging in public condemnation when there are massive and repeated violations of human rights' (Cohen 1998: 7). Many international development institutions fail to recognize and understand the causes and conditions of conflict and its wider impact on development. According to modernization theories, ethnic identities were seen as irrelevant and as something which belong to a distant 'tribal' past. However, ethnicity remains a very important, if neglected, feature of the 'development process' and has been at the heart of a number of recent post-colonial conflicts in Africa (e.g. Angola or Rwanda) (see Box 10.8).

One of the world's most acute and growing problems that has arisen out of conflict and instability is the increasing number of internally displaced people, or internal refugees (Black and Robinson 1993). An estimated 20–22 million people were internally displaced by conflict throughout the world in 1998 (Earthscan 1998). Definitions of internally displaced people vary, since people can be forced to flee their homes for reasons of war and conflict (civil or international), disasters (natural or man-made), from development projects themselves (construction of dams or urban clearances) or changes in the economy (e.g. unemployment or inflation). The ethnic and cultural identities of internally displaced peoples are far from being apolitical or static for the people caught up in struggles for power (Black 1998). As Bennett (1998: 13) argues, 'social identity and power relations are reformulated in the process of flight', and so a sense of identity is crucial to the way people experience and cope with displacement and indeed development more generally.

Responsibility for ensuring peace and security around the world lies with the Security Council of the United Nations, of which there are five permanent members – the US,

---

Box 10.8   Rwanda: looking beyond genocide

For those with limited background knowledge of the history and geography of Central Africa, the genocide in Rwanda in 1994 can seem very difficult to interpret and understand. Media representations relying upon the image and myth of Africa as the 'Dark Continent' (Jarosz 1992) explained the tragedy in terms of 'ethnic cleansing' and deployed myths and images of tribal savagery (Prunier 1995). In fact, Hutu and Tutsi cannot really be described as 'ethnic groups' in that both speak the same language, inhabit the same territory, and respect similar traditions and taboos (Destexhe 1995). It is important to view the Rwanda genocide in historical context, looking for example at the specific forms of German and later Belgian colonialism. Successive colonial governments accentuated differences between these supposedly distinct 'ethnic groups' and aggravated tensions between them, fuelling racial hatred. Rwanda became independent in 1962 and only one year later several massacres of Tutsi people began, led by the new Hutu government administration. Appeals to ethnic identity after independence became a way of winning support and in 1972 President Kayibanda imposed a quota on the Tutsi people allowing them only 10 per cent of the places in schools, universities and civil service posts (Prunier 1995). Successive presidents further fractured the country along the lines of identity depending on which region of the country people came from until in 1990 the Tutsi-led Rwanda Patriotic Front (RPF) launched a major offensive against Hutu people, launching an invasion from neighbouring Uganda. In July 1994, the UN made the fateful decision to withdraw its peace-keeping troops from Rwanda and by the end of that year half the population had been killed, been displaced within the country or were in exile.

France, China, Russia and the UK. These countries represented the key powers at the time of the formation of the UN. It seems somewhat contradictory, however, that they are still entrusted with world peace and stability when these permanent members of the Security Council (Buckley 1995: 5) make some 85 per cent of global arms sales. Particularly after the major periods of instability in Rwanda and Somalia, UN peace-keeping activities have been widely criticized as reactive rather than preventative (Fisas 1995). The UN Security Council seems to have responded less effectively to 'low intensity' conflicts on the global periphery but has invested considerably more commitment to conflict resolution in Europe. It had 'no trouble coming up with the US$ 1.5 billion *a day* that was needed to support allied action against Iraq in the Gulf War' (Buckley 1995: 20). In many ways the UN system suffers from a 'distorted sectoral focus' (Righter 1991: 187) and a lack of overall policy co-ordination where several UN agencies duplicate the same activities. Like other international development institutions, however, we must remember that the UN is quite varied 'in terms of the shape and function of its organs, the rules and practices concerning their activities and their power to influence the behaviour of other organizations' (Potter *et al.* 1999).

10.5   Conclusions: uneven globalization

> Radical development geography must adopt a framework that is liberated from the tyranny of dualism and allows for changes in the world economy and variations between states – one that accommodates alternative historical and contemporary geographies, cutting across the North-South divide. (Bell 1994: 188)

In recent decades one of the major trends has been that the world in which we live has become ever more global in character and orientation (globalization is the subject of Section 4). Global corporations and global marketing activities have resulted in the availability of standardized projects and products and communications have significantly improved, creating new regimes of 'interconnectedness' (Allen 1995) (Figure 10.4). In studying globalization and its impacts it is necessary to consider its economic, political and cultural strands (Allen 1995). For Potter *et al.* (1999) two generalized views on the relations between globalization and development have emerged. The first stresses the homogenization of culture that globalization engenders and points to the way in which places are becoming 'if not exactly the same then very similar' (Potter *et al.* 1999: 80). A second 'more realistic view' of the links between globalization and development also:

> stresses the likelihood of social and cultural homogenisation, with key American traits of consumption being exemplified by the 'coca-colonisation', or 'coca-colaisation' and the 'Hollywoodisation', or Miamisation of the Third World, replete with McDonalds golden arches. (Potter *et al.* 1999: 80)

Both these perspectives on globalization fail to recognize that not everybody passively consumes or is affected in the same way by the availability of 'global' products like Coca-Cola, McDonalds or Hollywood. Globalization provides important historical contexts and frameworks through which development has been understood and experienced. Similarly globalization has to be viewed in relation to the conditions of 'unequal development' which characterize the contemporary world. The notion of a 'shrinking world' has been

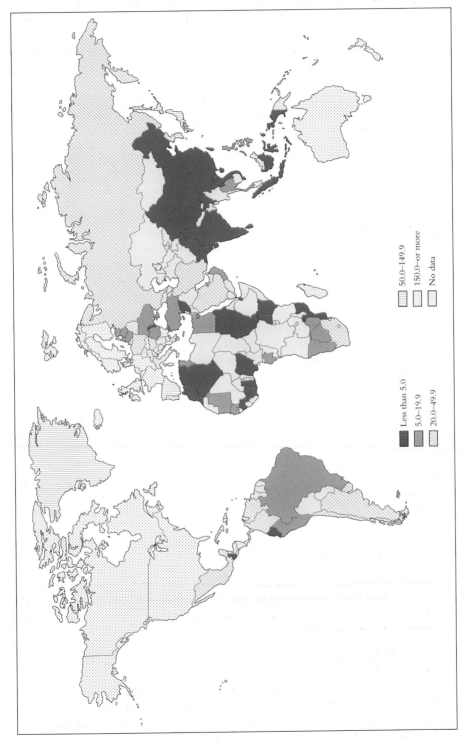

Figure 10.4 Personal computers per 1000 people. The ownership of personal computers varies enormously between different world regions, suggesting that not all countries can afford to travel on the global information superhighway and that globalization is geographically uneven. *Source:* World Bank, 1998b

important in discussions of the impacts of globalization, but patterns of global inequality have not shrunk after fifty years of development theory and practice. Ake (1996) reminds us that for Africa and other 'peripheral world regions', globalization is a highly uneven process which has not affected all areas of the so-called 'Third World' in the same way. Development in Africa, for example, has never failed; it just never got started (Ake 1996). In addition, although development thinking has itself been far from monolithic, many conceptions had the habit of 'robbing people of different cultures of the opportunity to define the terms of their social life' (Esteva 1992: 9). The *Dictionary of Human Geography* (1994) defines development as a process of 'becoming' and as a 'potential state of being', but also mentions how:

> The achievement of a state of development would enable individuals to make their own histories and geographies *under conditions of their own choosing* (*Dictionary of Human Geography* 1990: 103; emphasis added)

Local perspectives and knowledges of development have been often overlooked by international development institutions and state approaches have often been heavily centralized and 'top-down'. Whilst many sections of the 'development industry' have in recent years come to acknowledge the importance of people power and popular participation in development (Friedmann 1992), 'empowerment' has been co-opted by the mainstream of development theory and practice and has been stripped of its radical potential. UN agencies in particular have been criticized for their neglect of participatory approaches and their excessive bureaucracy but prospects for reform look bleak. Many recent 'anti-development' critiques (Esteva 1992, Escobar 1995) appear disillusioned with the future of the development industry and its capacity to understand and alleviate world poverty. Anti-development writers have not sought a new way of theorizing development or development alternatives, but have written about the need to think in terms of 'alternatives to development' (Watts 1993a). Potter *et al.* (1999: 14) are unduly critical of this perspective, dismissing it as a 'Utopia for New Age travellers'. Corbridge (1995) also criticizes anti-development writings, arguing that they romanticize and universalize the life-styles of indigenous peoples. Certainly many writers have emphasized the importance of grass-roots participation and popular social movements, but they have also re-emphasized the importance of local perspectives and cultural identities. Escobar (1995), for example, has argued that development has 'created abnormalities' such as poverty, underdevelopment, backwardness and landlessness and then proceeded to address them in

Learning outcomes

Having read this chapter, you should understand:

- The importance of critical and alternative approaches to the study of global development.
- The historical forces and pressures that shape development theories and practices.
- The role of some of the key brokers of development such as states, United Nations organizations, communities and NGOs.
- The diversity of peoples, places and spaces of development at the beginning of the twenty-first century.

a way which denies value or initiative to local cultures. Escobar (1995) is not offering a new all-encompassing conception of development but calling the whole process of 'development thinking' into question, showing how the globalization of development ideas and categories has been mediated by historical forces and by local peoples and social identities.

Further reading

Bell, M. (1994) Images, myths and alternative geographies of the Third World, in Gregory, D., Martin, R. and Smith, G. (eds) *Human Geography: Society, Space and Social Science*, Macmillan, London, 174–99. Illustrates the stereotypical images associated with the idea of 'three worlds' and contests the myths of backwardness and underdevelopment. Makes a case for an alternative geographical approach to representations of people and places in the South.

Dickenson, J., Gould, B., Clarke, C., Mather, S., Prothero, M., Siddle, D., Smith, C. and Thomas-Hope, E. (1996) *A Geography of the Third World*, Routledge, London. Slightly more traditional than Potter *et al.* (1999) and structured around the notion of the Third World, but provides a useful introduction to geographical approaches.

Potter, R., Binns, T., Elliott, J. A. and Smith, D. (1999) *Geographies of Development*, Longman, London. Although dismissive of the anti-development approach, this book offers a wide-ranging discussion of theories of development, urban/rural spaces and the important institutions of global development.

Radcliffe, S. A. (1999) Re-thinking development, in Cloke, P., Crang, P. and Goodwin, M. (eds) *Introducing Human Geographies*, Arnold, London, 84–92. Presents a critical discussion of the role of state power in development and of the importance of gender relations. Forms part of a larger three-chapter section, 'On Development Geographies' with contributions from P. Routledge and S. Corbridge.

Shiva, V. (1998) *Biopiracy: The Plunder of Nature and Knowledge*, Devon, Green Books. A useful example of some critical contemporary writing about development which raises questions about the plunder of nature and knowledge in the name of development.

Useful web-sites

www.ids.ac.uk/eldis/ The ELDIS site is prepared by the Institute of Development Studies (IDS) at the University of Sussex and offers an indispensable gateway to a range of relevant development resources. The site allows detailed searches (by country or theme) and contains an up-to-date 'what's new' section with the latest findings about development issues.

www.oneworld.org/odi The Overseas Development Institute (ODI) site contains an excellent search option, numerous policy and briefing papers and links to relevant development networks. The ODI describes itself as 'Britain's leading independent think-tank on development issues' and offers a critical and informative approach to studying development.

www.oneworld.org/   'Oneworld.net' (available in five languages) provides useful 'spotlights' on contemporary development issues and a range of links to associated sites and organizations. Information on working in development is available here, as is information about global development campaigns (e.g. debt relief, GM foods or the WTO campaign).

www.wto.org/   The official World Trade Organization web-site offers a variety of information about this controversial global organization and related 'trade topics'. Useful links to other global development institutions are available, as is an archive of speeches by WTO officials and a library of relevant development resources.

www.info.usaid.org/ and ***www.dfid/gov.uk/***   The two sites relating to the US Agency for International Development (USAID) and the British Department for International Development (DFID) with a range of information about British and American government development policy. Contains useful policy statements, case-studies and relevant links.

Production, exchange and consumption

The idea of going 'global' is often associated with the economic aspects of human geography. This section of the book demonstrates how 'the economic' shapes many aspects of human and organizational behaviour and the associated spatial patterns of growth and development. Many economic processes and their geographical outcome are increasingly a function of links and dependencies that extend far beyond immediate local or national environs; they embrace diverse interactions with places and people in many different parts of the world. Many of the symptoms of this 'globalization' are most obvious in the sphere of consumption (Chapter 14). Coca-Cola, McDonalds, Levi jeans, or Microsoft computer operating systems and software are just a few examples. Although many such products and services are often slightly customized to conform with local tastes, cultures or regulations (often reflected in the advertising used in different markets), they essentially offer a uniform experience or convey a common fashion statement wherever they are consumed in the world (see Chapter 14).

Consumption is both a consequence of and a driver of globalization. The shift from production systems largely focused on national markets to production facilities of the same company distributed around the globe has been accelerating. In part it reflects a search for new markets as national markets have become saturated. But it also reflects the demand created by a growing band of global consumers linked to common sources of information via the satellite, cable and Internet services that have metaphorically made the world a much smaller place. During the first quarter of the twentieth century Fordist methods of production helped both to fulfil and to create demand for high-volume, low-cost, standardized products at a time of steadily increasing incomes per capita, at least in the developed economies (Chapter 12). The more recent transition towards post-Fordist production with its emphasis on greater specialization, customization, and even more fragmented divisions of labour and horizontal rather than vertical integration of production has simultaneously confirmed the importance of the local alongside the global. The locality has been reasserting its role as the focus for production in harness with global influences. The massive production plants employing thousands of workers that were symbols of the Fordist era have to some extent been replaced by industrial districts made up of networks of firms, small and large, each contributing particular skills and knowledges. Together they form distinctive nodes of production such as Silicon Valley in California or the City of London.

This is not to say that large firms are no longer important in the globalization of economies. On the contrary, they are even more influential in shaping the geographies of production and consumption or the distribution of spaces that are engaged with the global and those that are increasingly disengaged (Chapter 11). Many of these transnational corporations have more resources at their disposal than many national governments. They therefore exercise significant influence on the decisions relating to the economy, investment in infrastructure or regulation by nation-states and by international organizations such as the Organization for Economic Co-operation and Development (OECD) or the World Trade Organization (WTO).

Consumption and production are mediated or enabled by the circulation of finance capital (Chapter 13). National governments, companies large and small, or individuals require finance to allow them to execute investment strategies in the case of companies or purchasing decisions in the case of individual consumers. Advances in information technology and political shifts have greatly increased the global circulation of finance and encouraged innovation in the financial instruments traded daily on stock, commodity, currency and other exchanges in business centres around the world. Yet again, however, as with the other themes explored in this section of the book, the flows of finance capital and access to it are spatially uneven. This operates to the significant disadvantage of marginalized nations, regions and groups within society while conferring even greater advantages on others than they already possess.

This section demonstrates how the geography of the economy incorporates global as well as local economic perspectives, reflecting changes over time and space. Some of the causes of the resulting uneven economic development are highlighted. The analysis suggests that there are some important issues and questions that economic geographers can usefully address at the start of the twenty-first century.

CHAPTER 11

PETER DANIELS

The geography of the economy

Topics covered

- The changing nature of economic geography

- The rise of the global economy

- The inevitability of globalization?

- The significance of transnational corporations

- Localities and the geography of the economy

- New economies in the twenty-first century

The economy is everywhere. Wherever we happen to live or to work our daily existence is influenced by economic relationships of one kind or another. Yet we tend to take for granted this crucial system of activities and administration through which society uses the resources at its disposal to provide for its material needs in the context of scarcity and to produce wealth. There is rarely a day that passes without references in the news media of most countries to the condition of the economy. The balance of exports over imports, the interpretation of indices such as the unemployment rate or consumer spending, announcements of factory closures, the creation of new jobs, or the performance of, and

Plate 11.1 Economic and business information occupies an important place in daily newspapers and periodicals such as *The Economist* or *Business Week*. (Alan Thomas)

prospects for, the diverse activities that together make up a national economy are all scrutinized regularly. Indeed, in many countries separate sections of the 'quality' daily newspapers are devoted to the economy and to business news.

The 'national economy' is just one of many types of economy. There are many other types that tell us something about the scale or the functions involved. These include: global economy, local economy, market economy, command (or redistributive) economy, informal economy, hidden economy, subsistence economy, peasant economy, Internet economy, digital economy, and space economy. The latter has particular resonance for geographers. It refers to the way in which **geographical space** is organized by the actions and behaviour of the activities that together make up the economy. Every economic activity requires and occupies some form of space. This may range from a one-room (15 sq. m) office used by a self-employed management consultant working from home, to the large sites (hundreds of hectares) occupied by car production plants in Japan or the US, or the very extensive (hundreds of sq. km) areas occupied by cattle ranches in the American West or sheep stations in the Australian Outback. Likewise, the families that provide labour for the production process and are a major source of demand for goods and services require space in which to live. Invariably, the disposition of different kinds of spaces is influenced by the availability of appropriate resources, including the means of production, so that some spaces are better suited to certain activities than others.

11.1 Economic geography

Familiarity with the discipline of economics is helpful for studying the geography of the economy and some economists are interested in spatial analysis (e.g. regional economists or urban economists). Geographers interested in spatial structures with reference to anything from a peasant economy to the global economy comprise the major sub-discipline of **economic geography**. It 'is concerned with the spatial organization and distribution of economic activity, the use of the world's resources, and the distribution and expansion of the world economy' (Stutz and de Souza 1998: 41). In order to measure and to explain such changes it is necessary for economic geographers to understand the levels of, and reasons for, the interaction between activities in an economy. This is usually referred to as **spatial interaction** and/or economic dependence/interdependence. It is manifest as flows of passenger or freight by air, rail, road or sea within and between countries or the daily flows of money between international stock exchanges (see Chapter 13). Telecommunications networks provide a platform for flows of information and knowledge within companies and between them and their suppliers and clients. There are fast growing flows of e-commerce and other transactions arising from the expansion of the Internet and the World Wide Web. Such interactions change over time and vary in spatial intensity. Economic geographers therefore study and try to explain trends in local, regional, national or international growth, or structural changes in the composition and location of activities in different types of economy.

In its early guise economic geography was termed **commercial geography**: a largely factual compilation of the circumstances under which different commodities were produced and exchanged around the world. Gradually, those interested in the geography of commodities changed from being generalists to specialists in the study of particular commodities, such as oil or coal, or manufacturing such as iron and steel. Although largely

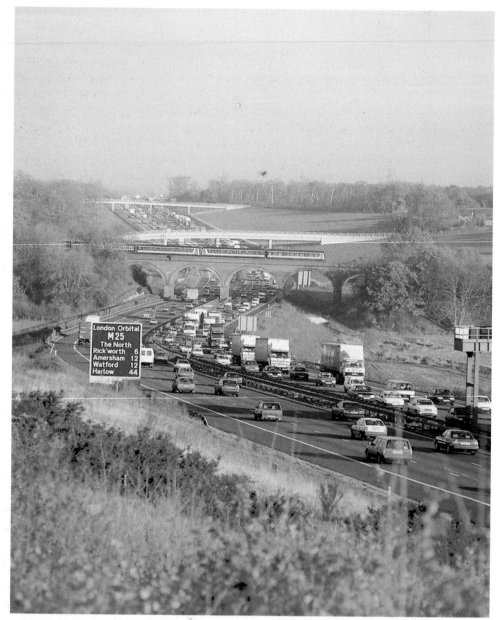

Plate 11.2 Economic growth has been accompanied by a dramatic increase in demand for transport within and between cities, which often results in traffic congestion at peak hours. (Quadrant Picture Library/David Sparkes)

descriptive, some analysis of the factors affecting the location of these activities helped to move the emphasis towards explaining observed spatial variations in the pattern of commodity production and trade. The identity of economic geography was consolidated by the adoption of the principles of neo-classical economics (Box 11.1) whereby the

Box 11.1   Models of economic location

There has been much interest in economic geography in the development of theories or models that help to explain the development of economic activities within a spatial context. One of the most important is *neoclassical economic theory* which was first used to model patterns of agricultural land use by Von Thunen in 1826 (Hall 1966), later applied to industrial location (Weber 1929) and subsequently used to explain the distribution of services and of settlements (Christaller 1966). In each of these and numerous other cases the objective is to develop generalizations about patterns of economic activity. Such is the complexity of the real world, however, that it is necessary to make some assumptions. Decision-makers behave in a rational fashion; they possess perfect knowledge; everyone is attempting to maximize profits; competition is unconstrained; economic activity takes place in a uniform physical environment (this is often referred to as an isotropic surface). The key assumption is that distance is the main influence on decision-making by households and businesses and that the resulting spatial patterns of economic activity can be explained by examining the relationship between distance and transport costs. The outcome of this approach is an ideal or optimal pattern of land use or industrial location. The main criticism is in relation to the unrealistic nature of the assumptions and the failure to account for many other factors that have an impact on the geography of economic activity. These include changes in the technology of production and consumption, the variety of ways in which businesses are organized and make decisions on where to locate, the role of social and cultural factors, and the effects of international and national political and regulatory environment.

    The limitations of neoclassical models encouraged the development of alternatives. Amongst the most significant was the *behavioural model*. While still seeking to arrive at generalizations, the focus is shifted to the role of the individual as the principal source of explanation for spatial patterns of economic activity. Things such as the motives, opinions, preferences and perceptions of the individuals making location decisions are brought into the equation and it is not expected that the outcomes are optimal. Thus, decision-makers do not have perfect information and are not assumed to maximize profits. Examples of the behavioural approach to economic location include studies of office location (Edwards 1983), consumer behaviour (Potter 1979), and location change by manufacturing industries (Hayter and Watts 1983). The behavioural approach also has its limitations. It is too descriptive and only highlights variations in behaviour rather than why an entrepreneur starts a certain business at a particular location. Too much attention is paid to factors specific to the firm or households at the expense of more general processes operating within the wider environment.

    The *structuralist approach* to explaining the location of economic activity and the way it changes represents a more holistic approach than behaviouralism. The premise is that behaviour is shaped or constrained by wider process in the social, political and economic spheres. Notions of culture and of class rather than individual ideas determine the spatial structure of economic activity. Firms work within a capitalist society in which production is essentially a social process structured by the relationship between capital and labour. It is not enough to explain patterns of economic activity by studying the various sectors such as manufacturing for their own sake. Rather, it is necessary to examine the 'hidden' mechanisms or processes that underpin economic patterns. These cannot be measured in the way typical of the neoclassical or behavioural approaches so that a good deal of the structuralist approach is devoted to developing theories about, for example, the way in which changes in economic conditions affect the requirements for production that in turn change the requirements of an economic activity at a given location that might cause it to move elsewhere (see Massey 1981).

operation of modern capitalist society is geared towards optimal resource allocation in terms of maximization of social welfare. This was derived from the influential ideas of Engels and of Marx who promoted the idea that the economic basis of society determines the form of its social institutions (see Box 11.1).

Such economic determinism dominated economic geography for much of the 1970s and the 1980s. During the 1990s it has been challenged by new economic geography. What precisely is 'new' is the subject of debate (see, for example, Bryson *et al.* 1999) but in essence it is about recognizing the importance of the 'cultural' when interpreting the 'economic' (see Chapter 14). Life-styles, beliefs, languages, ideas, imaginations and representations are seen as interacting with the economic to produce culturalization of the economy rather than the economization of culture. Goods and services incorporate cultural attributes in, for example, their design, marketing, packaging and likely benefits to users. Material possession of a good (car, camera, watch) or consumption of a service (tourism, fast food, financial transaction) is only part of the experience (a real as well as imagined event) of using or being seen to use or to possess it. Consumption of goods and services involves beliefs about what it says about us as individuals or groups: social or job status, image or wealth. Indeed, economic geographers have previously recognized cultural factors when using terms such as socialist economy, Chinese family production networks, or the various 'corporate cultures' encountered in transnational corporations (TNCs); only recently have they considered the economic and the cultural to be symbiotic rather than worthy only of separate study.

11.2 The economic problem

Each of us has a variety of needs and wants that, when combined together, make up a society's demand for goods and services. The economic problem (Heilbronner 1972) is how to devise a system that fulfils this demand in a manner that makes effective use of the physical and human resources that must be combined to produce and to distribute the output of goods and services. The possibilities are diverse but can be distilled to just three controlling or co-ordinating mechanisms (Dicken and Lloyd 1990). The first is tradition. Here, the allocation of resources to production is based on a set of 'rules' based on past practice with, for example, the passing on of land from father to eldest son providing a social framework for resolving the economic problem.

The second approach is by recourse to command. In a command economy redistribution of wealth is a key objective and this requires public ownership of the factors of production. A central political authority, elected or self-appointed, uses directives to set production targets for a defined period (say three or five years) for different economic sectors or regions within an economy. Centralized direction of national economic development is combined with control over the allocation of human and physical resources in order to achieve the targets and to distribute the output. Some of the developing countries such as Cuba or China use this approach to solving the economic problem and, until recently, it was also the favoured approach in the former centrally-planned (communist) economies of the Soviet Union and of Eastern Europe (e.g. East Germany, Poland or Czechoslovakia).

Recourse to the market is the third approach to the economic problem. In contrast to a command economy, the market economy operates on the basis of decentralized decisions by consumers and by firms on, for example, how much to produce, what prices to charge and how and when exchange transactions take place. The workings of the modern market system are complex and difficult to generalize about. Witness the diverse interpretations placed by professional economists on the significance for an economy of the latest unemployment figures, a fall in interest rates or a steep rise in the price of crude oil. In such a **capitalist** system, most of the resources and the means of production are controlled by a relatively small proportion of individuals and firms who are seeking to improve their economic well-being through **competition** (Box 11.2). Assuming that there is more than one producer of a given good or service, each is seeking to create the most favourable exchange terms (such as lower price, better quality, superior design or better after-sales support) in order to retain or to enhance their share of the market. Success depends in part on the wage-labour market that transforms the labour provided by individual workers into goods or services according to price signals. Because such workers own neither the means of production nor the goods/services that are produced they are vulnerable to changes in the requirement for their labour. This may result in slower wage growth, demands for an increase in productivity, changes in working practices, the acceptance of short-term rolling contracts or, in the worst case, loss of their jobs.

Although it is theoretically founded on the basis of freedom of action, in practice a market economy is the object of regulation by national governments and international

---

Box 11.2   Perfect competition

According to classical economic analysis, *perfect competition* is a concept common to all markets. It is argued that the price mechanism creates the most satisfactory outcome to the problem of how best to allocate the factors of production and the use of scarce resources in an economy. Perfect competition has a number of requirements. On the *supply side*, these are:

- that many firms compete against each other, each firm supplying such a small proportion of output that it is unable to influence the price by changing its output;

- that factors of production are perfectly divisible and mobile, so that each firm is able to increase output without forcing up, for example, the level of wages or the cost of assembling raw materials;

- that firms have complete freedom to enter or to leave an industry; and

- that each firm has complete knowledge of the behaviour and activities of its rivals.

Almost as a mirror image, the requirements on the *demand side* are:

- that there are many buyers, with no customer able to vary price by the amount that they purchase;

- that each buyer has complete knowledge of the quantities offered for sale and the price involved in all transactions;

- that transport costs are absent, so that no buyer is prevented from taking full advantage of different prices in different parts of the market; and

- that the products of firms are homogeneous, so that the product of firm i in industry x is a perfect substitute for the product of any other firm j in industry x.

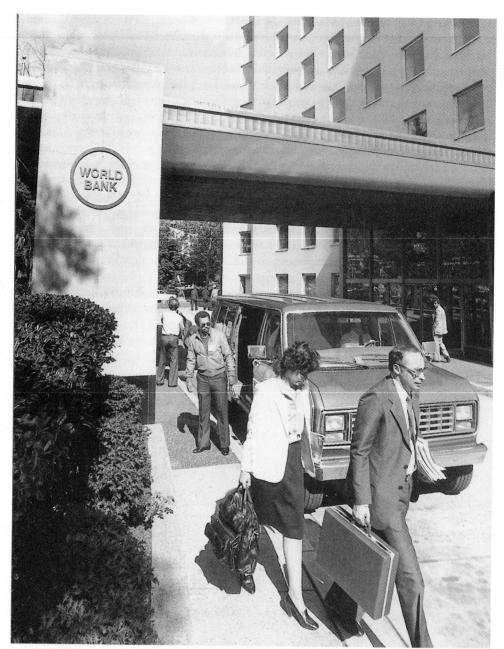

Plate 11.3   World Bank headquarters, Washington, D.C. (Camera Press/Tom Hanley)

institutions. The market is not necessarily the most effective and equitable way to allocate the rewards of production. Society needs to control the behaviour of those who own the means of production as well as to provide mechanisms for sharing the benefits as fairly as possible. Thus, in recent years the geography of the economy has been increasingly influenced by the actions of national governments or groupings of nation-states (or trade blocs)

Box 11.3 Transnational corporations

The geography of the economy, whether global or local, is being shaped by decisions made by transnational corporations (TNCs). You may well encounter a related term: multinational corporation (MNC). The latter suggests a firm that has premises or production plants in several countries while the former describes a firm operating in at least two countries including its home country. TNC is therefore a much more all-embracing term than MNC; there are many more TNCs than there are MNCs and they provide the basis for a more realistic assessment of the scale of international investment.

The definition of a TNC is that:

- It controls establishments/economic activities in at least one other country apart from its home country.

- It possesses an ability as a result of its size or the ownership of particular knowledge or skills to move its operations and its resources quickly between international locations, i.e. it is relatively footloose.

- It can exploit or take advantage of differences between countries, regions or cities around the globe in factor (land, labour, capital) and non-factor (information, knowledge, regulation) endowments.

- It owns and controls overseas activities, although this is not a pre-requisite for TNC status since there are many other ways, such as franchising, licensing or joint ventures, of achieving a presence in markets outside the home country.

The incidence of TNCs varies by industry sector and by country. The majority are small relative to the limited number of very large global TNCs but they generate significant flows of capital, knowledge, information, expertise, products, raw materials or components amongst their own establishments or between countries. There are few parts of the world that are not in some way affected by the activities of TNCs and for some countries it makes the difference between inclusion or exclusion from the economic mainstream.

such as the European Union (EU) or the North American Free Trade Area (NAFTA). International bodies such as the World Trade Organization (WTO), the United Nations Conference on Trade and Development (UNCTAD), the International Monetary Fund (IMF) or the World Bank also exert considerable influence on patterns of economic development. The geography of the economy has also been significantly shaped during the last quarter of the twentieth century by the activities of very large businesses (transnational corporations or TNCs) with markets and operating facilities that extend well beyond their home countries (Box 11.3).

11.3 The economic system

We will return to some of these issues later since they are the source of some of the most interesting questions for economic geographers to address in the twenty-first century. Before doing so it is necessary to elaborate on the notion of an economic system. What is it and how is it structured? This is important if, as economic geographers, we are keen to understand the processes that produce spatial patterns of economic activity. A system is a set of interrelated objects that are connected together as a working unit or a unified whole. In this case the objects are, amongst others, homes, shops, offices, manufacturing plants or oilfields that are connected by the movement of labour, information, raw materials, finance, e-mail messages, knowledge or freight necessary to sustain their particular role in the economy.

The convention is to divide this system into **economic sectors**. The precise details may vary depending on where we look, but in essence there are four main sectors within the economic system. Activities engaged in the exploitation of natural resources, such as agriculture, fishing, mining or oil extraction, form the **primary** sector. Much of the output from the primary sector has limited use and value until it has been transformed in some way to become part of usable goods. Whether this takes place near the source of the primary commodity or at a location some distance away (therefore requiring transfer between point x and point y) it requires a **secondary** sector or manufacturing. The products created by manufacturing processes may be immediately suitable for final use by consumers or may be components of other final products. Either way, they will have to be distributed to the places and markets where they can be consumed or purchased. This is the role of the **tertiary** sector, which includes wholesale and retail trade, transportation, entertainment and personal services.

Plate 11.4 Large-scale distribution facilities at strategic locations with easy access to motorway networks serving extensive regional or national markets are a vital part of modern production. (Quadrant Picture Library/Paul Sherwood)

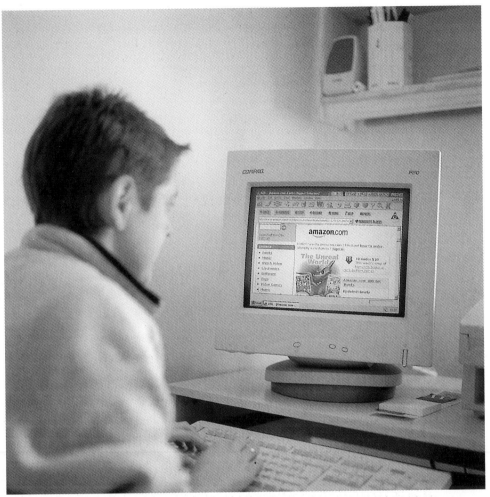

Plate 11.5 Customers can now purchase a wide range of goods and services across international borders from the comfort of their own homes using Internet sites such as Amazon.com. (Alan Thomas)

Improvements in transport and telecommunications and their close integration with the rapid advances in computing since the early 1980s have transformed the secondary and tertiary sectors. Producers have greatly improved access to international markets and consumers' expectations in relation to products or to quality of service have been much enhanced. The growth of international purchasing via the Internet is most symbolic of this change. More carefully designed and refined goods and services that meet the specialist requirements or needs of sub-groups of individuals or firms within the economy or specific types of national or regional market are much more common and have been made possible by the emergence of the **quaternary** sector. This includes banking, finance, business services, the media, insurance and administration. These are the activities that assemble, transmit and process the information and knowledge that is required by the other three sectors to enable them to adjust effectively and efficiently to the changing social, economic, and geographical parameters of doing business in the twenty-first century.

11.4   A dynamic economy

It should be apparent from this very brief outline of the economic problem and the economic system that relationships between the sectors, their internal composition and the balance between them are constantly changing. This takes place at all levels of resolution, from the rural locality to the largest metropolis, from peripheral to core regions, from developing to developed economies. This dynamism is nothing new but during the closing decades of the twentieth century it has been dominated by changes which will set the agenda for the launch of the twenty-first century (see also Chapter 12).

Since the 1970s employment in the secondary sector has been contracting, even in developing economies and the post-socialist economies that have invested heavily in manufacturing as a means of advancing their economic development (see also Chapter 3). Take the case of the United Kingdom. The steel industry, for example, employed 165,000 in 1980; by 1999 the figure had decreased to approximately 40,000. In clothing and textile manufacturing there were some 840,000 jobs in 1980, down to 215,000 in 1999. The decline of employment in the coalmining industry is even more dramatic; just 10,500 were employed by the industry in 1999 compared with more than 190,000 just fifteen years earlier (1984). The primary sector, which is already a very small part of the economy in employment terms in most developed countries, is also contracting as more productive methods of farming are introduced and crop yields enhanced through improving disease resistance or the application of fertilizers (see Chapter 9). This leaves the tertiary and quaternary sectors that have expanded and diversified as information and telecommunications technology has shifted production towards goods and services whose functionality is determined more by the knowledge and expertise incorporated in their design and specifications than anything else (Figure 11.1).

These structural and functional readjustments to the economic system have been taking place at the same time as, and perhaps because of, the **internationalization** of economic activity. As the term implies, this refers to the spread of economic activities beyond national boundaries. This is not a new phenomenon, but the rate at which it has been taking place as well as the types of economic activities involved accelerated enormously during the 1980s and the 1990s. Internationalization should not be confused with **globalization**. The latter is a more inclusive term and is the culmination of a transition from an international to a global economy. In this sense globalization is very much a contemporary phenomenon that refers to the integration across national boundaries of markets, finance, technologies and nation-states to a level that has not been witnessed before. This has enabled nation-states, TNCs and individuals to extend their reach around the globe faster, further, deeper and at lower cost than could ever have been imagined even ten years ago. As Dicken (1992: 1) has put it: 'These two themes are intimately connected in the sense that the pervasive internationalization, and growing globalization, of economic life ensure that changes originating in one part of the world are rapidly diffused to others.'

The global economic system is now the envelope within which all the various other systems mentioned earlier in this chapter must fit. In principle at least, globalization implies convergence in the economic, social and cultural values of the localities, regions and nations that together comprise the world economy. Economic diversity is replaced by economic uniformity. However, as the other contributions to this section of the book

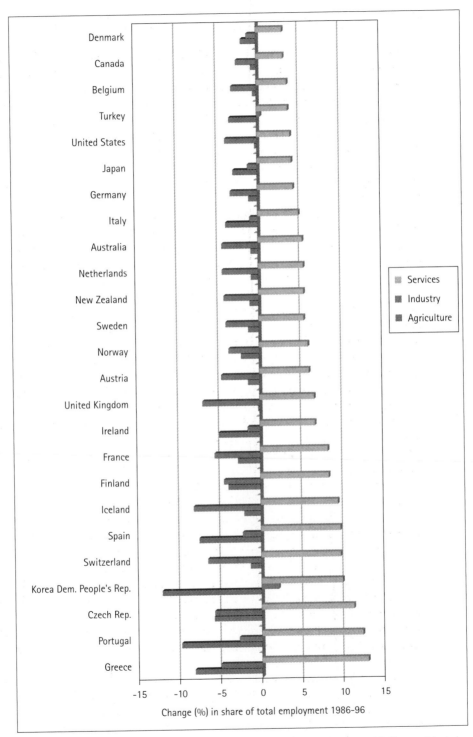

Figure 11.1   Changing structure of employment by sector, selected countries, 1986–96. *Source*: Adapted from OECD, 1998

Figure 11.2   A proportional map of the world, by share of GNP, 1998.

Table 11.1   GNP-based measures for the relative performance of low, middle and high income countries, 1997

Country groups	Population (millions)	GNP ($ billions)	GNP per capita ($)	PPP GNP[1] (per capita $)
Low income	2,036	712	350	1,400
Middle income	2,857	5,411	1,890	4,320
High income	927	24,001	25,890	22,930
World	5,820	30,125	5,180	6,260

Source: Extracted from data in World Bank (1999).

Note: 1 Gross national product converted to international dollars using purchasing power parity rates. An international dollar has the same purchasing power over GNP as a US dollar in the United States. This tends to put a higher value on goods and services that are not traded, resulting in a higher value of GNP for many developing economies.

show, we should beware such sweeping assumptions. They are a convenient shorthand, but at the expense of overlooking the significant variations in the economic health or the level of participation of different parts of the world in the global economic system. A map of the world in which the countries are drawn on a scale proportional to their share of total gross national product (GNP) amply illustrates that convergence is a long way from being achieved (Figure 11.2). Almost a quarter of the world's total GNP is concentrated in the United States. Almost half of the world's population is in low-income countries (Table 11.1) producing less than one thirtieth of world GNP in 1997, equivalent to just US$ 350 per capita. The high-income countries have less than one sixth of the world's population but five sixths of total GNP in 1997 at a rate per capita that is seventy-five times greater than for low-income countries where poverty is often a major problem (Box 11.4). Every country may not feel part of the global[ization] system, but every country is being globalized and shaped by it (Friedman 1999).

11.5   The rise of a global economy

Amin and Thrift (1994: 2–5) have enumerated some of the consequences of the transition to a global economy. First, *there has been a marked increase in the power of finance over production*. Finance capital now takes many forms and moves almost seamlessly and with great speed across the globe, especially between the world's financial markets centring on stock and other exchanges (see Chapter 13). Electronic trading has ensured volatile and fast-moving financial markets that can transform (either positively or negatively) the economic prospects of companies and, more importantly, of national or regional economies overnight.

An illustration is provided by economic events in 1997 when Asia had reached its peak as an engine for worldwide economic growth. This was based on the region's consumption of vast quantities of raw materials used by its flourishing consumer goods conglomerates (especially those from Japan, South Korea and Taiwan) that exported merchandise to every corner of the world. In early December 1997 the government of Thailand announced that it was closing almost all of the country's leading finance houses, mostly private banks, that had become almost instantly bankrupt. This was its response to a free

Box 11.4

Income distribution and poverty

Commonly used poverty measures are based on estimates of the population whose consumption, expenditure or income falls below a prescribed level. This might be set at an international poverty line of US$ 1 per day which is one of the most commonly used measures (see chart below). The proportion of the population that falls below this line can be very large. In nine of the countries able to provide estimates (see below) over half the population were below the poverty line. There are eight countries with more than 50 per cent of the population below the US$ 1 poverty line, including Guinea-Bissau (88.2 per cent), Zambia (84.6 per cent), Uganda (69.3 per cent) and India (52.5 per cent). Poverty is clearly a widespread and serious problem in these and several other countries and it places them at a considerable disadvantage in the search for domestic economic growth or a share in the expansion of the world economy.

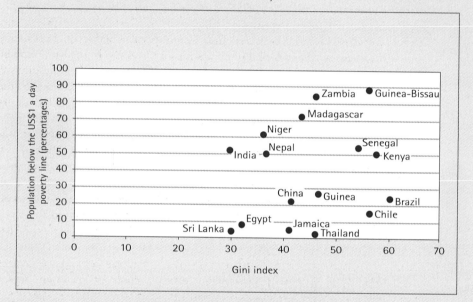

Population below the US$ 1 a day international poverty line and the Gini index

Note: The Gini Index is a convenient and widely used measure of income inequality. It has a value of zero for perfect equality of incomes and 100 for perfect inequality.

Source: International Labour Office 1999

fall in the value of the nation's currency (the Baht) which had been targeted by global speculators who took the view that Thailand's economy was much weaker than the published figures (or speculator sentiment) warranted. With the notable exception of Singapore, this triggered a dramatic loss of confidence in the economies of the other emerging markets in the region. The capital, often in the form of foreign direct investment (FDI) that had flowed into the region on the back of its success, was rapidly moved elsewhere.

The economic contagion did not stop there. Prices for commodities such as crude oil, aluminium and copper began to collapse as demand fell, triggering impacts on the economies that supplied them. Not the least of these was Russia which, since the early 1990s, has been trying to refashion its economy in the image of a market rather than a centrally-planned system. With many of its pre-market economy factories totally inadequate for the task of producing internationally competitive goods there was a heavy dependence on the income from crude oil sales, and foreign loans to fund economic readjustment. The latter were attracted by unrealistically high interest rates on financial instruments such as government bonds (as much as 70 per cent by early 1998). Foreign banks and corporations were persuaded to purchase these in the belief that the IMF would help Russia, thus protecting their loans, if there was a risk of default. But the collapse of crude oil prices triggered by the Asian crisis inevitably had a major impact on Russian government revenues and made it difficult for the interest to be paid on government bonds. By August 1998 Russia found it necessary to devalue its currency (the Rouble) and had defaulted on its government bonds.

The Russian economy is small by world standards but because the financial system is so global, events there had further consequences elsewhere. With some of the world's largest banks and trading firms reporting losses on their Russian investments of several hundred million dollars, it was necessary to compensate by selling assets in countries that were more financially sound. The Brazilian economy had been performing quite well during the mid-1990s but investors lost confidence during 1998. Although the government raised interest rates to 40 per cent in order to retain capital, there was a flight of investment to the safest locations possible such as the United States or some of the European markets such as the United Kingdom or Germany.

A second consequence of the rise of a global economy is the pivotal role of *knowledge as a vital factor of production*. Since knowledge, its creation and exchange, is embodied in people rather than machines, economic welfare relies more on 'producing' an educated and skilled workforce. There is disagreement over whether this is linked to growth of GDP or to more efficient production. Third, *technology has become transnationalized*, especially amongst knowledge-intensive economic activities such as financial services or telecommunications. This does not necessarily mean better access to the factors of production; the increased complexity of the opportunities so created means that only those producers and institutions with the resources to manage technology can really take advantage. Fourth, *globalization of technological change, the mobility of finance capital, and transportation and communication has been accompanied by the emergence of global oligopolies*; production in a particular industry is dominated by a small number of firms. The large share of the world market held by Microsoft for computer operating systems and software is a good example. There is now a sense in which corporate and economic survival requires 'going global'. Fifth, the power of individual nations to regulate their own economic development or to exercise a strong influence on the outcome, for example, of trade agreements has been diluted by *the rise of transnational institutions* that co-ordinate, steer and even regulate aspects of the world economy. Examples include the IMF, the United Nations Conference on Trade and Development (UNCTAD), the World Trade Organization (WTO), or the Organization for Economic Co-operation and Development (OECD). Sixth, *the 'wiring' of the world* has stimulated more extensive cultural flows and has 'deterritorialized signs, meanings and identities' (Amin and Thrift 1994: 4). Last but not least, *these symptoms of globalization present themselves as new [global] economic geographies*. These have been characterized in various ways, including the centralization of economic power in global cities (Sassen 1991), as a 'borderless' global economy (Ohmae 1990), as a 'space of flows' (Castells 1989), or as a new global division of labour (Dicken 1992).

11.6   Is globalization of the economy inevitable?

So the globalization of the economy is well established and will become even more entrenched? Not necessarily. There is lots of evidence suggesting that we have become daz-

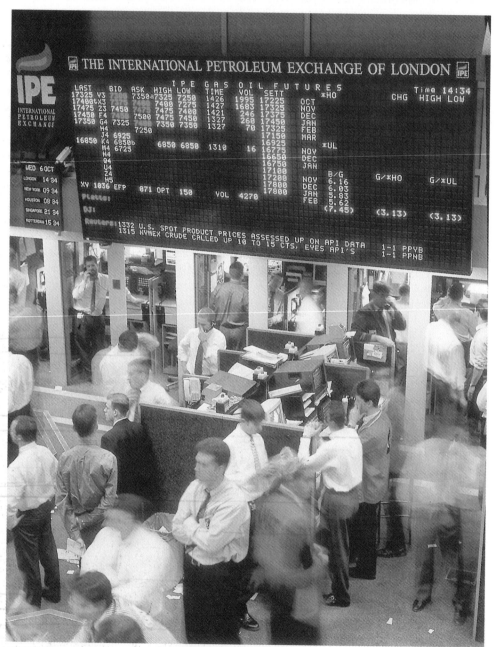

Plate 11.6 Exchanges such as London's Petroleum Exchange exert a powerful influence on the dynamics of the global economy. (Science Photograph Library/Jerome Yeats)

zled by the rhetoric of globalization. Two examples will suffice. First, labour markets have largely continued to function within national boundaries. Of course many millions of people (see Chapter 7) move temporarily or permanently to jobs in countries other than their own. But the numbers involved do not stand comparison with the massive migration of labour from Europe to North America or to Australia and New Zealand in the late nineteenth century and the early part of the twentieth century. In Asia, Latin America or Europe labour markets are essentially national for, amongst others, linguistic, cultural and educational reasons. The European Union (EU) has actively tried to dismantle the administrative and legal barriers to cross-border movements of labour or the ability to practise certain professions within the Single Market. But this has not triggered a massive shift of labour from the less prosperous (Southern) regions to the more prosperous (Northern) European states.

Again, globalizing markets might be expected to become more uniform with respect to price and product specification as a result, for example, of the liberalization of trading rules or the pressure of competition on high-cost producers to lower their prices. On the contrary, price differentials for the same products (cars, cameras, personal computers, washing machines, books, etc.) have remained as wide, or even wider, as in the past. Individual national governments may be responsible because of higher value-added taxes (VAT) on goods or services, less than full compliance with the universal aspects of trade agreements such as equal treatment for all firms or labour whether from the home country or from outside, or the application of import quotas. Companies may also be culpable through the use of inefficient distribution networks, use of less cost-effective transport or by charging premiums for minor product modifications in order to meet environmental or other regulations in particular markets. The possibilities are numerous and as a result product price convergence remains potential rather than real.

Plate 11.7   The world headquarters of Microsoft Corporation, Redmond, WA. (Microsoft)

Thus, a truly integrated world economy may be some way off but it will certainly be a central part of the development agenda for the early decades of the twenty-first century. Technology will continue to set the pace, changing the way in which information and knowledge are transferred and accessed, or modifying the nature of work and skills required. This may change the balance of demand for labour between richer and poorer economies and between and within each of the major economic sectors. Technology in the form of satellite and other surface-based telecommunications media has also made it much more difficult for governments to regulate their economies (using trade barriers as well as internal controls). Levying taxes on transactions of all kinds (financial, currency, retail, entertainment, information, etc.) that can take place electronically without the impediment of national boundaries is also much more difficult. This adds to the potential for further displacement of jobs and growth opportunities for less developed economies and to increased uncertainty as TNCs can relatively easily move production from one country to another.

Ultimately, however, further integration of economies will also depend upon national and international political priorities and the extent to which individual nations are prepared to compromise or to give up some of their sovereignty (see Chapter 16). Greater economic integration in Asia, for example, is dependent on a reduction in cultural differences/values across the region. A new pan-European currency, the Euro (€), introduced in January 1999 has yet to demonstrate that it can become established as a viable alternative to established international currencies such as the US dollar, the Japanese yen or the pound sterling. It remains to be seen whether the majority of the population of France,

Plate 11.8 Globalization has been accompanied by increasing geographical concentration of corporate control as symbolized by the density of office development and the skyline of Manhattan, New York City. (Geophotos/Dr Tony Waltham)

Germany and the other countries that have joined the Euro initiative will switch from their Franc or Deutschmark bank accounts to Euro bank accounts (some 9 million accounts had been opened by May 1999).

11.7 Trade and comparative advantage

One important reason for creating the Euro is to reduce the cost of cross-border trade and investment amongst the participating countries. Such trade is one of the most obvious manifestations of the globalization of the world economy, especially since 1945. Since that time the value of the world's total GDP has increased by a factor of approximately six, but the volume of world trade in goods has increased more than fifteenfold. More recently, trade in services has also been growing rapidly and now accounts for more than 20 per cent of world trade (by value) (Table 11.2). Exports as a percentage of GDP have grown from around 7 per cent 50 years ago to more than 15 per cent today (World Bank 1999). In addition to the more open climate for trade created by organizations such as the General Agreement on Tariffs and Trade (GATT) and its successor, the WTO, expansion has taken place because trade is considered good for economic development.

The export side of trade is considered especially important since it helps to protect existing jobs as well as to create new ones. Imports are viewed, at least by governments, more cautiously because they are perceived to lead to job displacement. However, according to trade theorists as far back as Mill (1821) and Ricardo (1817) the benefits of international trade arise from the goods received rather than those exported. This arises from the principle of **comparative advantage** whereby nations or regions tend to specialize in the production of those goods or services in which they have the greatest ratio

Table 11.2 Trends in merchandise and commercial service exports, world regions, 1985–96

World region	Sector	Exports (Value, $US bn)			Change (%)	
		1985	1990	1996	1985–90	1990–96
North America	Merchandise	310.0	521.8	827.2	68.3	58.5
	Services	72.8	155.1	242.7	113.0	56.5
Latin America	Merchandise	108.5	145.6	252.1	34.2	73.1
	Services	17.6	29.1	47.0	65.3	61.5
Western Europe	Merchandise	776.3	1,637.1	2,295.3	110.9	40.2
	Services	193.7	417.6	593.2	115.6	42.0
Central and Eastern Europe/Baltic States/CIS	Merchandise	156.1	105.4	171.8	−32.5	63.0
	Services	–	–	–	–	–
Africa	Merchandise	80.6	102.1	116.7	26.7	14.3
	Services	11.3	18.6	26.8	64.6	44.1
Middle East	Merchandise	102.3	133.8	162.9	30.8	21.7
	Services	–	–	–	–	–
Asia and Far East	Merchandise	416.4	792.4	1,462.1	90.3	84.5
	Services	61.6	132.4	284.4	114.9	114.8
Total world exports	Merchandise	1,950.2	3,438.2	5,288.1	76.3	53.8
	Services	357.0	752.8	1,194.1	110.9	58.6

Source: WTO, *Annual Yearbook*, 1997.

Box 11.5

Comparative advantage

Comparative advantage explains the tendency for countries (or regions/localities within countries) to specialize in certain goods and/or services even if they have the ability to fulfil their needs from domestic production. As long as countries or regions specialize in those products or services in which they have comparative advantage, they will gain from trade. Advantages can stem from spatial variations in, for example, mineral or land resource endowments, from variations in the educational levels of the labour force, in access to markets, or differences in levels of technology. For this to work effectively it is necessary to assume a system of free trade, hence the significance of the trade liberalization that has been high on the agenda of many countries since the 1980s. There is a nagging concern, however, that comparative advantage is not reflected in actual patterns of world trade. We would expect the biggest flows to be between the countries with the largest cost differences. It seems that consumer tastes and geographical proximity are actually more important than cost differences. This explains the fact that the vast majority of trade is between countries with relatively small cost differences, often involves similar rather than different goods (such as cars, electrical goods of all kinds, certain kinds of business services such as management consulting), and occurs on a 'nearest neighbour' basis. Well over half of EU goods and services trade takes place between the Member States while the US has Canada and Mexico as its major trading partners.

of advantage over other areas (or for which they have the lowest ratio of disadvantage) (Box 11.5). Porter (1990) has suggested the concept of competitive advantage as a more useful way of explaining how individual nations participate in the world economy. It is broader than comparative advantage in that it incorporates differences in values, cultures, histories and institutions as well variations in endowments of the factors of production.

History also plays a part. The City of London proved to be an innovative location for financial and insurance activities during the late eighteenth century. Originally stimulated by the Britain's extensive trade with its colonies and subsequently diversifying as the industrial revolution really got under way during the nineteenth century, London finance houses had the established prestige, experience and knowledge that later entrants such as New York or Paris had yet to accumulate.

The combination of absolute, comparative and historical advantage endows some countries and TNCs with significant 'clout' within the globalized economy. The home countries of TNCs tend to be the developed economies of Europe and North America. It is true that lower wage costs or longer working hours in the developing countries or the newly emerging economies of Eastern Europe now make them an attractive alternative, especially in relation to the ongoing **international division of labour** (Coffey 1996, see also Chapter 12). Production in many manufacturing and services industries is increasingly subdivided into a series of discrete processes undertaken by separate workers or groups and there is scope for companies to seek locations that offer lower wage costs. But factors such as lower productivity, higher training costs and less efficient transportation and telecommunications infrastructure can often outweigh the wage cost advantages of the developing economies. This is not to say that if the circumstances are right, footloose production will not shift. A good example is the rise of India's 'Silicon Valley' (Mortished 1999) in and around Bangalore, southern India (Box 11.6) or the relocation to Jamaica of ticket processing and other high volume, labour-intensive, routine functions by US firms.

" WHITE-FACED MEN SHRIEKED WITH HOARSE VOICES THAT THEY SOLD, EVERYBODY SOLD" (*p.* 53).

Plate 11.9 A busy Exchange in the City of London, late 19th century. (Mary Evans Picture Library)

<table>
<tr><td>**Box 11.6**</td><td>**Silicon Valley in a developing economy: Bangalore, India**</td></tr>
</table>

Bangalore in southern India has experienced massive growth of computer software and related high tech industry since the mid-1990s. Wage differentials provide part of the explanation. An Indian graduate employed as a software developer in Bangalore might earn US$ 3,700 per annum in 1999 compared with US$ 60,000 for the equivalent in California's Silicon Valley. Even though India is a poor country, generally lacking basic infrastructure, its education system produces a large number of highly trained engineers, computer scientists and mathematicians who, in addition to their native dialects, speak the global *lingua franca*, English. Time zone location means that its workers are at their desks when those of the US or Japan are heading home or already in bed. Providing that they can access their markets/clients Indian companies are well placed to supply knowledge-based services to the global market. In just five years Bangalore has emerged as a location for hundreds of computer software companies, including Infosys (first launched on the Bombay stock exchange in 1992 and on the Nasdaq stock exchange in 1999 where it was valued at US$ 2.7 billion), WIPRO, and Tata Consultancy Services. Using two telecommunications satellites, the Bangalore firms provide online update, repair and computer code writing services to multinational clients such as British Airways, Boeing Corporation, Goldman Sachs and Levi Strauss. Offshore Software Development Centres offer staff teams specializing in configuring existing computers to counter the 'millennium bug' or to handle the modification of company computer systems to provide salaries, invoices and other transactions in the new European currency (Euro). A large number of British banks and building societies have contracts for software maintenance with Bangalore firms and British Telecom has outsourced most of its software development to an Indian venture. However, such is the demand that wage inflation at 20–25 per cent per year is threatening to erode, at least in the long term, the cost advantage of Bangalore.

11.8 The rise of transnational corporations

The example of Bangalore will no doubt be repeated elsewhere (such as the Pearl River Delta region, southern China). The success of this region and others like it around the globe depends partly on the actions of TNCs. If we group the top ten TNCs by their assets and according to whether their home countries are developed, developing or transition economies, the scale differences between global household names such as Nestlé, Ford or IBM, as measured by employment, and their opposite numbers is immense (Table 11.3). The leading TNCs from the developed countries are primarily in the automotive and petroleum exploration and distribution sectors. Those from the developing countries, with one exception, are in diversified activities, construction and transportation. The latter is also prominent amongst TNCs based in the transition economies, but all of the TNCs in the top ten are much smaller than almost all their opposite numbers elsewhere. However, size of TNC is not a good indication of transnationality (Table 11.3). While Nestlé has by far the highest index for the developed country TNCs, all the other scores are in the same range as those for the developing country TNCs. A larger number (three) of the transition economy TNCs actually have higher indices of transnationality than any of those in the other two groups apart from Nestlé.

Table 11.3   The top ten TNCs (by assets) in developed and developing economies, emerging economies, 1997,[3] ranked by transnationality index

Rank (by assets)	Corporation	Country	Industry[2]	Employment		T Index[1] (%)
		Developed countries		Foreign	Total	
9	Nestlé SA	Switzerland	Food and beverages	219,442	225,808	93.2
5	Exxon Corporation	US	Petroleum expl/ distribution	–	80,000	65.9
3	Royal Dutch Group	Netherlands/UK	Petroleum expl/ distribution	65,000	105,000	58.9
8	Volkswagen Group	Germany	Automotive	133,906	279,892	56.8
7	IBM	US	Computers	134,815	269,465	53.7
10	Daimler-Benz AG	Germany	Automotive	74,802	300,068	44.1
6	Toyota	Japan	Automotive	–	159,036	40.0
2	Ford Motor Company	US	Automotive	174,105	363,892	36.2
1	General Electric	US	Electronics	111,000	276,000	33.1
4	General Motors	US	Automotive	–	608,000	29.3
		Developing countries				
3	Jardine Matheson	Hong Kong/China	Diversified	–	175,000	75.0
4	First Pacific Company	Hong Kong/China	Electronics	40,400	51,270	74.4
7	Sappi Limited	South Africa	Paper	9,492	23,458	61.9
5	Cerrex SA	Mexico	Construction	10,690	19,174	56.6
2	Daewoo Corporation	Republic of Korea	Diversified	–	–	50.8
1	Petroleos Venezuela SA	Venezuela	Petroleum expl/ distribution	11,849	56,592	44.5
9	China National Chemicals	China	Diversified	625	8,905	43.1
6	Hutchison Whampoa	Hong Kong/China	Diversified	17,013	37,100	37.3
10	LG Electronics Inc	Republic of Korea	Electronics and electrical eqmt	32,532	80,370	30.1
8	China State Construction Engineering Corporation	China	Construction	5,496	258,195	27.3
		Emerging economies[4]				
5	Atlanstka Plovidba	Croatia	Transportation	–	528	96.5
8	Adria Airways	Slovenia	Transportation	–	585	95.0
1	Latvia Shipping Co.	Latvia	Transportation	1,631	2,275	81.5
4	Molokov a.s.	Czech Republic	Trade	576	1,000	64.8
6	Piiva Group	Croatia	Pharmaceuticals	1,616	6,680	37.7
3	Gorerje Group	Slovenia	Domestic appliances	607	6,717	36.0
2	Podravka Group	Croatia	Food, beverages, pharmaceuticals	501	6,898	32.6
7	Skoa Group	Czech Republic	Diversified	1,073	19,830	10.6
9	MOL Hungarian	Hungary	Petroleum and natural gas	628	20,140	5.1
10	VSZ a.s. Kosice	Slovakia	Iron and steel	58	26,719	1.7

Notes: 1 The transnationality index is the average of three ratios: foreign assets to total assets, foreign sales to total sales and foreign employment to total employment. 2 Using US Standard Industrial Classification as used by the US Securities and Exchange Commission. 3 1998 data for Emerging Economy TNCs. 4 Based on survey responses from Croatia, Slovenia, Hungary, Lithuania, Slovakia, Czech Republic, Macedonia (TFYR), Rep. Moldova, Romania and Ukraine. Source: Extracted from UNCTAD (1999), based on UNCTAD/Erasmus University database and an UNCTAD survey of the top TNCs in Central and Eastern Europe.

Box 11.7   Foreign direct investment

For statistical purposes, *foreign direct investment* (FDI) is defined as the acquisition by an individual or enterprise resident in one country of assets located in another. There are numerous difficulties in identifying such flows and collecting reliable information about them. Especially important is the fact that many take place as transactions internal to TNCs and are therefore difficult to enumerate. Where FDI involves financial transactions they are included as part of FDI if they result in a material interest in a foreign enterprise (usually defined as a shareholding of at least 10 per cent).

World FDI inflows exceeded US$ 400 billion in 1997 with developing countries accounting for more than one-third of the total. This represented a doubling of the flows since 1992 (US$ 200 billion). FDI has been growing faster than other indicators such as GDP, exports of services or merchandise, or domestic investment in recent years and is clearly contributing to the increasing internationalization and integration of economies with firms producing more and more outside their home country. For example, in 1997 7 per cent of global GDP was generated by the production of foreign affiliates of TNCs and if parent firms were included the proportion would be much higher. It is also the case that FDI between developed countries is more important than FDI from developing countries. However, some of the biggest inflows in recent years, although still a relatively small proportion of the total, have been recorded by the newly emerging economies of Central and Eastern Europe.

FDI inflows and outflows can be expressed as a percentage of GDP to provide a simple index of economic integration. The value has risen from 1.0 per cent in 1980 to 2.3 per cent in 1996 and for inward and outward stock of FDI from 10 per cent to 21 per cent. The equivalent index for trade flows as a percentage of GDP has remained relatively constant over the same period. This suggests that global economic integration is being facilitated more by FDI than by trade flows.

With many countries making it easier for FDI by liberalizing their controls and regulations, cross-border mergers and acquisitions have become an important part of FDI.

World FDI and related indicators 1996 and 1997

Item[1]	Year 1996	1997	Growth rate (per cent) 1996–7
Inflows of FDI	338	400	18.7
Outflows of FDI	333	424	27.1
Inward stock of FDI	3,065	3,456	12.7
Outward stock of FDI	3.115	3.541	13.7
Cross-border mergers and acquisitions[2]	163	236	45.2

Notes: 1 Billions of US$.
       2 Majority-held investments only.

Source: UNCTAD (1998), based on FDI/TNC database and UNCTAD estimates.

The merits or otherwise of global economic integration are seen by some as closely entwined with the behaviour of TNCs. Depending on the viewpoint, they may be seen as the saviours of otherwise disadvantaged economies because of the technology or the job opportunities that they provide or, conversely, they may be portrayed as ruthless exploiters of low-cost labour and inadequately protected resources in environmentally sensitive areas. Either way they cannot be ignored because, whatever measures are used to illustrate globalization, TNCs invariably figure prominently (Daniels 1999). They are a major source of foreign direct investment (FDI) and sales outside their home countries by affiliates are growing faster than exports and were worth about US$7 trillion in 1995 (Box 11.7). This is not insignificant, since it exceeded the total value of all world exports of goods and services. Multinational companies are also major investors in buildings, equipment and people at locations outside their home country (FDI) although relative to all investment, especially by the developed economies, the proportion is relatively small (6–7 per cent) (see, for example, Daniels 1999).

How can we explain the expansion of TNCs that has been going on for a large part of the twentieth century and will certainly continue well into the twenty-first century? Given that all companies are driven by capitalist principles, they need to sustain growth of profits for stakeholders and for reinvestment in their plant and infrastructure. Saturation of national markets reduces the opportunities to become larger and to produce more and thus limits the ability to make savings through economies of scale (Box 11.8). These are the reduced costs per unit of output as firm or plant size increases. The benefits do not necessarily involve the creation of production facilities in the overseas markets, since it is often now cost-effective to transport finished goods, especially high value-added products, to geographically dispersed locations. Transport developments such as containerization, larger vessels and faster handling and transfer facilities at seaports have been important. This also makes it possible for multinationals to outsource the manufacture of component parts to global suppliers while continuing to assemble the finished product at a small number of key locations. Examples include car assembly, electrical goods such as personal computers or video recorders, and aircraft construction.

Some economies of scale are not material, rather they arise from opportunities for global advertising and marketing in support of a global brand image of the kind

---

Box 11.8 Retailing in a global high street

The world of retailing is shrinking. Growth and improvement in telecommunications and the opening up of new sources of cheap labour in the post-socialist economies of Eastern Europe and China have created new opportunities for retailers with global ambitions. Until quite recently (early 1980s) retailers relied on local manufacturers for their merchandise; now they can easily source it from the developing world. The high street or shopping mall is therefore more cosmopolitan with respect to the variety of sources from which its goods for sale are obtained. As a result prices have been cut in Europe and especially in the US so that retailers need to grow larger to make decent returns. The larger the retail operation the greater the scope for big economies of scale. The buying power of a large over a small retailer enables the former to command lower unit prices for bulk purchases and larger contracts for more work than smaller rivals can offer. As a result an elite group of global retailers is commanding an increasing share of retail sales in many countries. This often involves expansion based on an established retail formula such as McDonalds or Benetton. Most of the successful global retailers are general merchandisers such as Wal-Mart (US), 7-11 Japan or Gap. Several also include food sales because they bring in customers on a more regular basis than sales of non-food items such as white goods (refrigerators, freezers, washing machines), for example.

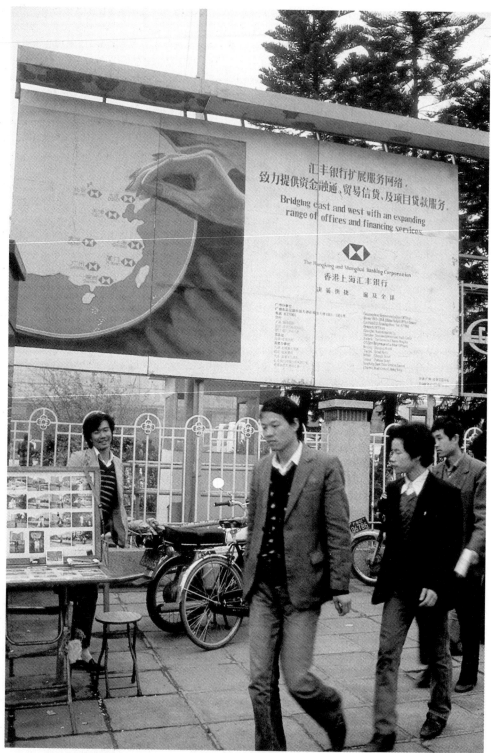

Plate 11.10 Parts of China, such as Shanghai or Guangzhou just across the border from Hong Kong, are experiencing rapid economic transformation and this is expected to continue well into the 21st century. (Format Photographers/Joanne O'Brien)

associated, for example, with some soft drinks or fashionable clothing. The development of TNCs can also be explained as the outcome of vertical integration whereby efficiency gains, improved quality control and protection of proprietary information are controlled by acquisition of the forward (customer) and/or the backward (supplier) linkages in the production chain. In some cases these acquisitions will be outside the home country of a firm, thus transforming it into a TNC. The 'follow the leader' effect in the rise of TNCs should not be underestimated. Some of the manufacturing firms such as IBM, Kodak or Ford that began production at plants outside the US during the 1920s and 1930s were generally successful and it was logical for competitors to follow suit. In addition, some other activities, such as accountancy, legal, financial and advertising firms, that have close links with multinationals and provide important advice on critical aspects of their overseas expansion have also found it necessary to follow their clients to locations outside their home countries. Global networks of offices and specialist consultants allow business and professional service firms to continue providing the level of support and expert advice that clients already receive from the home country headquarters of these firms.

FDI by TNCs is not uniformly distributed. Although developing country TNCs are on the increase, most of the effective foreign investment is by developed country TNCs. Most of their investment (more than 60 per cent in 1998) goes to other developed economies. There are signs that the balance is shifting towards the developing economies and newly emerging markets such as China or Eastern Europe that have significant consumer potential. For those countries able to attract it, FDI helps to secure access to capital, technologies and organizational expertise. The net effect is to facilitate the modernization of infrastructure, to increase industrial capability and to improve the quality and breadth of much-needed financial, business and professional services. China, a vast potential market with more than one-fifth of the world's population, which was closed to inward investment until very recently, is now the leading recipient of investment attention from the world's TNCs. Prior to the collapse of the Asian economies almost two-thirds of the investment flows to developing countries in 1996 and 1997 went to the Asian countries (excluding Japan). Some of the states of Latin America such as Brazil or Chile are also attracting significant FDI. Eastern Europe was the destination for significant flows of investment in the mid-1990s but this fell sharply following the knock-on effects of the Asian financial crisis. Much of the African continent, however, continues to be bypassed for any form of investment, even though many of its countries are resource-rich. Consumer purchasing power is low and average incomes per capita are not anticipated to grow in a way that will raise the consumption of major consumer goods to a level that will justify the attention of most TNCs.

To summarize, globalization of the economy has had the effect of further strengthening the role of market forces while, as a result of advances in telecommunications and transportation technology, easing some of the constraints imposed by geography. As the real cost of international telecommunications has declined during the 1990s, global financial integration has strengthened. Globalization has also created opportunities for new kinds of international trade, especially in services, and allowed easier diffusion of innovation and practices that encourage economic development and participation. The importance of non-government organizations, multinational companies and regional trading blocks such as the EU or NAFTA for shaping the economic geography of globalization has been greatly enhanced and will probably continue to increase.

11.9   Localities and the geography of the economy

While the geography of the economy is increasingly accommodating a global dimension this is not at the expense or to the detriment of individual localities. As Porter (1990: 158) has noted, 'it is the combination of national and intensely local conditions that fosters competitive advantage.' It has been suggested that 'territorial policies . . . should now focus more on mobilizing local resources to build competitive strength and create jobs. This change of emphasis can add impetus to the restructuring of national economies by reinforcing the capacity for self-generating change' (Johnston 1998). Localities are taken to mean subdivisions within nations such as cities or regions or places that often have a particular economic identity because of the kinds of activities that take place there. Some localities even smaller and less complicated than cities, such as rural districts that are the foci for producing very particular kinds of wine (France), dairy products (Denmark) or woollen goods (Wales), for example, are also included. Many are identified with named territorial units such as administrative areas, while others are industrial districts or agglomerations whose identities are derived from specific economic activities, such as Motor Sport Valley (Pinch and Henry 1998) in southern England or Silicon Valley, California (see Chapter 12). While these are often highly specialized and self-contained localities, often with systems of governance and regulation that fit their particular needs, they are inextricably linked with the wider national and global economic system.

These 'new industrial spaces' thrive on the dynamism, innovation and 'untraded interdependencies' (knowledge and information that circulates through the transfer of key workers between firms or via social and other networks) that are made possible by spatial agglomeration. Even TNCs, that are now often portrayed as being 'placeless' because their operations are so extensive and relatively mobile, started from businesses that were initially nurtured in a particular locality with its own economic and other characteristics. There is evidence that the locally-shaped attributes of firms are carried through into their transformation and organization as TNCs (Dunning 1981).

On the other side of the coin is the question of how, once they have become globalized, TNCs leave an imprint on localities where they have manufacturing plants, offices, research and development laboratories, distribution facilities or retail floorspace, for example. Part of the answer is determined by the way in which a TNC operates. Those that rely heavily on in-house production and exchange of goods and services will have a less marked impact on a locality when compared with those that externalize their requirements for component manufacture or services ranging from security and cleaning to marketing or personnel recruitment. Local orientation will also vary by type of corporate function with headquarters having stronger connections to suppliers and services than, say, a manufacturing plant. The framework used to manage corporate operations will also be influential. Some TNCs grant a good deal of autonomy to their subsidiary or branch establishments for their operational and strategic planning. Others exert highly centralized control and direction from their world headquarters or from a small number of regional headquarters covering, for example, Western Europe or the Asia Pacific. The style adopted also tends to vary according to the type of industry but may also vary within similar industries.

It is actually very difficult to generalize about the embeddedness or otherwise of TNCs in localities; even within one organization this may vary according to the style and social or community networks accessible to individual managers or teams (Box 11.9). This reflects the importance of the most visible aspect of corporate local embeddedness: the links with suppliers of goods and services. As the demands of flexible production and 'just-in-time' supply of inputs such as car components have escalated so has the involvement of TNCs with their suppliers become much closer. This is partly driven by the need to maintain the very exacting quality control standards required for the production of sophisticated products for ever-more demanding clients and customers. It is also driven by the need to share the technological expertise, which many small firms cannot expect to be able to acquire very easily if they are to fulfil the expectations of the TNCs. This may extend to TNC staff working at the premises of supplier firms or being involved with the training of their staff. The 'localness' of supplier firms becomes quite important in these circumstances. On the other hand there are numerous examples of TNCs bringing together personnel from around the globe at one training establishment or at a research and development facility in the home country to transfer the necessary information, knowledge and training to their preferred supplier firms. Whatever the model used, such arrangements make it more difficult for local economies not already part of established supplier–TNC networks to become participants. As some global–local relations deepen, others remain as shallow as or shallower than ever before.

Localization as a counterpoint to globalization of the economy is also being sustained by national governments that are decentralizing resources and decision-making to sub-national governments (World Bank 1999). Sub-national spending was 10 per cent of the national total for Mexico in the mid-1980s, rising to more than 30 per cent in the mid-1990s; equivalent figures for South Africa are 21 per cent and 48 per cent, or for China, 45 per cent and 58 per cent. This has allowed localities to exercise greater autonomy over the institutions and instruments that they put in place, using the insights gained from detailed knowledge or local networks to capitalize on economic advantages or to create an economic and social environment that will attract new investment. Such decentralization still

---

Box 11.9   Embeddedness of economic activity

The advantages of local agglomeration for some economic activities have been eroded as more of them, especially TNCs, have a presence that is simultaneously local and global. This hybrid identity is sometimes referred to as *glocalization*. There has therefore been a shift in the literature on industrial agglomeration towards the idea that an institutional atmosphere is important for maintaining identities and competitiveness. This new institutional sociology is encapsulated by the notion of *embeddedness* (Grabher 1993). Economic action is now seen as social action and is affected by a network of relations between 'actors' (firms, institutions, individuals) that may be located within an agglomeration but can also be located in different agglomerations. Instead of trying to understand and explain the continuation of existing industrial agglomerations or the emergence of new ones via economic considerations such as vertical disintegration of changes in product life-cycles, attention has turned to social and cultural explanations. Levels of inter-firm collaboration, social consensus, the nature and intensity of institutional support and networks for local business and the circumstances encouraging skill formation, innovation and the circulation of ideas are the kinds of factors that are considered important (see Amin and Thrift 1994, Storper 1993).

requires some national co-ordination because some localities within a nation, especially cities, are usually better positioned to act as independent competitors for globally mobile investment or employment. Localization has boosted urbanization, not least the concentration of economic activities, while globalization, the rise of the service economy and increasing international competition offer opportunities as well as threats to urban areas (Sassen 1994; European Commission 1999). There are fewer barriers to separate local markets as trade liberalization is limiting the prospects for national intervention in flows of capital and trade. Metropolitan areas are major sources for national economic growth in developed and developing economies. They are the ultimate expression of the benefits of localization such as scale and agglomeration economies, access to diverse and lower cost information, and an environment conducive to innovation. They are also more easily promoted as locations for investment and good factor productivity. The strength of economic growth in urban centres can be used to stimulate vibrant rural industrial sectors such as Guangzhou, north of Hong Kong, or Zhejiang and southern Jiangsu adjacent to Shanghai, all in China (World Bank 1999) (Plate 11.10).

11.10 New 'economies' in the twenty-first century?

If these are the circumstances now, what are the chances that they will change in the face of some of the new types of economy that mark the opening of this century? These are the Internet, e-commerce or digital economies whose presence was already apparent at the end of the 1990s and which are rapidly growing and diversifying. It has been estimated that in 1996 there were fewer than 40 million people around the world connected to the Internet; by the end of 1997 the figure had increased to 100 million (US Department of Commerce 1998) and at turn of the century it was approximately 200 million. The great bulk of commerce and international trade online, including the Internet, consists of the sale of services such as telecommunications, information technology, publishing and the media, travel and tourism, retailing, transportation and professional services like consulting, architecture and engineering. Although consumer sales gain much of the publicity, much of the growth of electronic commerce is taking place between and within business. Online revenue growth has quadrupled since the mid-1990s. In 1998 it totalled some US$ 35 billion inter-company and US$ 15 billion retail, worldwide. By 2003, the value of goods and services traded via the Internet will increase to US$ 1.3 trillion (almost one-sixth of the value of world trade) (Price Waterhouse Coopers 1999, International Data Corporation 1999). The potential for growth in direct cross-border sales to consumers is enormous as Internet access, user confidence, payment systems and Internet/Web security continue to improve.

Apart from numerous legal, regulatory, security and other challenges that it presents, the global digital economy is likely to trigger changes in the geography of the economy. Just as the industrial revolution brought about significant economic and social changes (both positive and negative) in the shape of new jobs, new industries and new industrial regions, so might the digital economy stimulate its own revolution. Perhaps the Internet really will remove the friction of distance on economic interaction once and for all but only if a truly global infrastructure is put in place. One of the first challenges is to 'plug-in' national economies to the Internet. To be excluded will be to widen the gap that already exists between those countries 'inside' and those 'outside' the world economy. In

Figure 11.3(a) International connectivity via the Internet and related telecommunication services, 1991. © 1991 Lary Landweber and the Internet Society

Figure 11.3(b)   International connectivity via the Internet and related telecommunication services, 1997. © 1997 Lary Landweber and the Internet Society

just over five years (Figure 11.3) the connectivity of countries to the Internet has improved dramatically. In late 1991 most of the countries of Africa and large tracts of the former Soviet Union, Asia and the Far East had e-mail services only. By mid-1997 most countries worldwide had Internet and/or Bitnet services. Being connected, however, is only part of the picture; there remain vast differences in the geographical distribution (number, density and processing power, for example) of the computers, telecommunications networks and software required to participate in the digital economy.

A second challenge posed by the digital economy is concerned with human resources. It is likely that more jobs will be created by the digital economy than are lost but these jobs are generally in higher-skilled and better-remunerated occupations. All economies face the big challenge of retraining existing workforces and training future workers to fill the new jobs that will be created. History suggests that the response to this challenge will be uneven, with comparative advantage enabling some economies and some localities to respond more quickly and effectively than others. Perhaps less certain is how this will modify existing spatial patterns of economic development. For example, the digital economy will enable a wide range of purchasing and banking transactions to be undertaken from home via the telephone, personal computers, televisions and even mobile phones that provide access to Internet services. An international door-to-door delivery time of 24 hours or much less is now commonplace, depending on the type of good or service or the location of the supplier. Moreover, cross-border transactions and payments are as easy as those made to the local supermarket or bookstore, although they present a real challenge to governments that are concerned about how to gather and assign any taxes linked to these transactions. With shopping sites on the Internet presenting themselves as 'virtual malls' and encouraging users to place their selections in 'baskets' or 'carts', they are attempting to replicate offline shops. A major shopping survey in the US in late 1998 reported that 39 per cent of those with access to the Internet were spending less time at real shopping malls and local stores because they could easily buy products online (*An Atlas of Cyberspace*). This, combined with trends such as increased working from home or teleworking, whether self-employed or as an employee of a TNC, points to the potential for significant changes.

The challenge for economic geographers will be to understand how the digital economy impacts on the structure and pattern of economic activities at different scales, on the structure of localities and their relationship with the global economy, or on the relative importance of nations and TNCs in regulating or shaping the outcome.

Learning outcomes

Having read this chapter, you should be aware of:

- The nature and changing concerns of economic geography during the last quarter of the twentieth century.

- The economic problem and the alternative approaches that have been adopted to manage the allocation of resources, etc., at country level.

- The importance of economic restructuring and the shift from manufacturing to services in the modern economy.

- The significance of the globalization process in the structure and spatial development of the world economy.

- The major role performed by transnational corporations in trade, foreign direct investment and other indices of globalization.

- Alongside globalization, the continuing importance of localities for the evolving geography of the economy.

- The challenges posed to economic geographers by the rapid growth of the digital economy and its potential for changing established patterns of economic activity.

Further reading

Bryson, J., Henry, N., Keeble, D. and **Martin, R.** (1999) *The Economic Geography Reader: Producing and Consuming Global Capitalism*, John Wiley, Chichester. Charts the revival and expansion of economic geography during the 1990s using selected papers that explore the evolving economic geography of the advanced capitalist economies. Covers globalization, new spaces of production and consumption, restructuring of welfare and new landscapes of work.

Dicken, P. (1998) *Global Shift: Transforming the World Economy*, 3rd edition, Paul Chapman, London. A broad-ranging overview of the evolution of internationalization and the globalization of economic activities. Focuses on the complex interaction between transnational corporations, nation–states and the rapid developments in information and communications technology, drawing on fields outside geography such as political science and economics.

Healey, M. J. and **Ilbery, B. W.** (1990) *Location and Change: Perspectives on Economic Geography*, Oxford University Press, Oxford. Provides a good broad understanding of the evolution of the economic landscape with particular reference to key economic sectors in developed market economies and to a lesser extent developing and planned economies. Includes overviews of trends and theoretical developments in economic geography prior to the onset of globalization.

Knox, P. and **Agnew, J.** (1998) *The Geography of the World Economy: An Introduction to Economic Geography*, 3rd edition, Arnold, London. A synthesis of the factors shaping the development of contemporary economic patterns in market-oriented and centrally-planned economies. Strong emphasis on interdependence of economic development at different spatial scales, from local through national to international.

Useful web-sites

http://www.ilo.org International Labour Organization. Documents and statistics on a wide variety of employment and labour market issues for individual countries and on a comparative basis for countries around the world.

http://www.un.org United Nations. Reflects the wide-ranging responsibilities of the UN but includes a useful section on global economic and social development, including an annual report on trends and issues in the world economy. Links to other web-sites.

http://www.wto.org World Trade Organization. Offers a wide range of documentation and statistical material about global trends in trade in goods and services, regional aspects of trade, electronic commerce and background research and analysis on all aspects of world trade. Links to related web-sites.

http://www.europa.eu.int/comm/index_en.htm The European Commission. Readers of this chapter will find the links to the various Directorates-General (DGs) sites within the Commission (Industry, Transport, etc.) useful starting points for monitoring economic development patterns, problems and policies across the EU.

http://www.oecd.org Organization for Economic Co-operation and Development. Represents mainly the developed economies and provides free documents, summaries of OECD economic surveys and statistics.

http://www.unctad.org United Nations Conference on Trade and Development. Focuses in particular on the interests of less developed countries in relation to international trade and foreign direct investment. Publications, statistics and links to related web-sites.

JOHN BRYSON AND
NICK HENRY

CHAPTER 12

The global production system: from Fordism to post-Fordism

As a world we have never consumed as much as we do today and enjoyed as high a quality of life. Other chapters deal with the environmental implications and inequalities of such life-styles which the world is now, finally, beginning to face up to. This chapter is about the very large, and complex, capitalist economic system producing the goods and services we consume in our everyday lives. We begin by exploring the meaning of a *global* production system and argue that we have, in fact, had different forms of 'global' production system for centuries. The continued development of this system over time (and space) is attributable to economic specialization driven by the continuous extension of the **division of labour**. This key concept is the subject of the second major section of the chapter. The remaining sections examine Fordism, the global system of production which dominated the second half of the twentieth century, its subsequent economic failure (crisis), and the post-Fordism global system which has now taken its place as we enter the twenty-first century.

Post-Fordism is a global system of production like never before. When journalists, politicians and TV reporters mention economic globalization, they are referring to the unprecedented speed, scope and extent of the post-Fordist global production system. The final section outlines the characteristics of this production system and its distinctive global economic geography. The conclusion summarizes the key arguments of the chapter.

12.1   A global production system?

The story of economic geography is one of continued geographical expansion ('globalization'); in this sense, globalization is not just a twentieth- and twenty-first-century phenomenon. There has been a 'global' production system for centuries, although it is difficult to ascribe a date to the origins of this system. However, such a definition of 'global' should be relative; that is, relative to the period of time under discussion.

For example, Massey (1995) describes the example of Chang Ch'ien, ambassador to the Emperor of the Chinese Han Dynasty, sent in 138 BC in search of allies. He made it to Bactria (today part of what is now Afghanistan) which was, in fact, an old outpost of Greek influence previously conquered by Alexander the Great. Other historical empires to name might be Carthage, Muscovy or the Roman Empire. Each of these empires included trade with 'far-flung' and 'distant' places; 'global production systems' of their time. The Roman economy stretched well beyond the borders of its empire, beyond the line of the Rhine and the Danube, and eastwards to the Red Sea and the Indian Ocean. It has been recorded, for example, that the Roman Empire had a trade deficit of 100 million sesterces a year with the 'Far East' and that ancient Roman coins are still being dug up in India today (Braudel 1984).

A similar story for the fourteenth century may be compiled from the more than 150,000 surviving letters of Francesco di Marco Dantini, an Italian merchant (Origo 1988). Origo's superb study of the life, work and thoughts of this merchant reveal the extent of Francesco's 'global' economy. England was very much a part of this economy. Between 1343 and 1376 one of the greatest Florentine (Florence, Italy) banking houses collapsed. The reason was to be found in England. Edward III had defaulted on the repayment of his loans to the bank. The collapse of this bank triggered a domino effect with almost every smaller trading and banking company in Florence following suit (Origo 1988: 78). Despite the date, this story has a familiar, contemporary, ring as it concerns an action or event made in one country that triggers impacts in the economy of another (such as the Asian financial crisis of 1997, see Chapter 11).

Although lengthy travel times made England very distant from Florence, in the fifteenth century money flowed from Italy to England and in return wool flowed from England to continental Europe. Francesco was a participant in this trade. He purchased English wool from three Italian firms with branch offices in London. These firms sent their buyers riding all over England acquiring wool from individual farmers and the great monastic houses (Origo 1988: 71).

The years in which Francesco was buying English wool were also the years in which the English cloth industry was developing. Rather than exporting wool, England began to export cloth and began to compete seriously with the cloth-making industries of Flanders and France (Origo 1988: 72). Even in the fourteenth century areas of specialist production existed, or early forms of commodity production whereby some areas specialized in the production of raw materials and others in the production of finished products.

Francesco's business carried with it high levels of risk; the fourteenth century was an uncertain world of disease and, for merchants, ship losses and robbery. His business activities were driven by information flows and informed by predictions of what would sell at a certain price. The information came via letters sent by special couriers and messengers, as well as by the postal service run by the Venetian Republic between Venice and Bruges, which took only seven days.

The 'global' production system of the fourteenth century was complex and volatile. The difference between Francesco's world (globe) and that of the twenty-first century is one of scale, scope and extent. Most history books would describe Francesco's production system as 'international' rather than 'global'. But, in fact, relative to the period, and Francesco's life, it could be argued that he was part of a global system. The extent of the globe (Francesco's view of the world) was delimited by far-away places with which he traded, such as England. Today, we might view as 'far-away' (from the UK) such places as China or Cameroon. Commercial jet aircraft can circumnavigate the globe several times in the time (seven days) taken for Francesco's messages to be conveyed by post from Venice to Bruges.

It is the fact that individuals, goods and services can travel vast distances in very short periods of time, that has made the terms 'global' and 'globalization' very much late twentieth-century creations. And it *is* important to differentiate between earlier international ('global') production systems and the post-Fordist system of today. Economic globalization is the process by which the economic relationships between places are constantly being reworked and shrunk (Allen and Massey 1995). As a result the economic globe is continually changing shape; it is also becoming smaller as the ties that bind places to the global production system become ever tighter. Whilst news of Edward III's default took days to arrive in Florence, advances in telecommunications meant that the Asian financial crisis began reverberating around the world's financial markets in minutes. It is the scale, scope and extent of the current 'global' production system, and processes of globalization, which allows differentiation from earlier forms of production system and geographical expansion.

It is geographical expansion throughout history that has seen countries, products and people become enmeshed in the 'global' capitalist production system. This is a history of how increasing numbers of people, companies, regions and countries gradually gained specialist economic expertise. The issue is how to explain this process.

12.2 The division of labour

The key to answering this problem is found in two places: in Josiah Wedgwood's eighteenth-century porcelain factory in Stoke-on-Trent (Staffordshire, England, also known as 'the Potteries') and in the work of Adam Smith, the economist.

12.2.1 Wedgwood and 'making machines of men'

In 1759 Josiah Wedgwood established a porcelain factory in Stoke-on-Trent. At the beginning of the eighteenth century most pottery was produced for the local market. By 1750, some potters had begun to export to other regions of England as they had developed expertise in particular types of pottery. When Wedgwood established his workshop most potters sold their products by sending completed pots direct to market or to a merchant. In 1774, Wedgwood decided to alter this practice by producing a catalogue. Catalogue-selling only works when factories can produce goods that are uniform in quality as well as detail. The problem for Wedgwood was that potters liked to show their skill and creativity rather than produce standardized pots. To prevent product variation, he

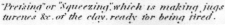
"Pressing" or "Squeezing", which is making jugs turenes &c. of the clay, ready for being fired.

Engraving designs on Copper Plates for producing the much admired 'blue printed pots' &c.

Figure 12.1   The division of labour in pottery production, 1827. *Source*: Forty, 1995: 35

Plate 12.1   A typical specialist craft activity – book binding. (Format Photographers/Jenny Matthews)

therefore introduced enamel-printed transfers that were mass-produced and applied to the pots before firing. The hand painters could no longer stray from the designs printed in the catalogues.

The only other source of variation came from the potters who were still able to introduce variations into the form of a pot. To overcome this problem, in the 1750s Wedgwood divided the work required to manufacture a pot into seven different tasks (Forty 1995: 32) (see Figure 12.1). Each employee undertook a single task, and because no employee was responsible for a single pot, no individual was able to alter the design. This was Wedgwood's attempt to convert his employees into 'such machines of men [*sic*] as cannot err' (Wedgwood, cited in Finer and Savage 1965: 82). At each stage of the production process, however, employees were still able to make small variations to the design. Wedgwood kept dividing the production process into more and more specialist tasks. This increased his control of the process, as well as permitting him to replace skilled labour with less skilled, cheaper labour. Josiah Wedgwood's effort to standardize the production of pots by splitting the production process into smaller and smaller tasks is a process known as 'the division of labour'.

12.2.2 Adam Smith and the division of labour

In 1776 Adam Smith published *The Wealth of Nations*, in which he established the foundations of economic theory. For Smith:

the greatest improvement in the productive powers of labour, and the greatest part of the skill, dexterity, and judgement with which it is anywhere directed, or applied, seem to have been the effects of the division of labour. (Smith 1977: 109)

To illustrate the importance of the division of labour, Smith explores pin-making, suggesting that an individual without knowledge of the production process would be unable to make even one pin in a day. The production process of pin-making can be divided into eighteen distinct operations: from the drawing out of the wire, to making the pin head. If one person performs all of these tasks then they might make 20 pins in a day. If, however, individuals specialize in particular tasks, by introducing a division of labour, Smith shows how ten people could make 48,000 pins in a day.

Smith reasoned that the increase in the quantity of work that results from the division of labour is the product of three factors. First, the dexterity of employees improves and this contributes to an increase in speed. Second, time is lost when an employee has to change from one type of work to another. Tools have to be found, the working area altered or the worker must move to a different part of the shopfloor. Time is also lost as the worker adjusts to a new task. Third, machines permitting cheaper and faster production can replace labour. A division of labour precedes the replacement of a task by a machine. It is through a division of labour that distinct parts of the production process become visible, and as they become visible machines can be developed to replace labour.

12.3 The division of labour: spatial and gendered

12.3.1 Spatial divisions of labour

Adam Smith's division of labour explores specialization within factories. His theory, however, made no explicit reference to geography. The obvious extension of the division of labour was to incorporate geography into the process and to produce what has been termed a spatial division of labour (Massey 1995, Knox and Agnew 1998). Once a production process had been subdivided into its component parts it is a comparatively simple step to develop factories dedicated to a particular part of the production process, such as a steel mill, paper or packaging factory. This spatial division of labour began with manufacturers establishing specialist units that were contiguously located, but soon developed into a more dispersed spatial division of labour. Different parts of the production process were located close to either sources of raw materials, cheap labour or the market place (see Box 12.1). Very quickly, some regions and countries came to specialize in particular types of economic activity. For example, an (international) division of labour soon developed between the industrial countries (core) producing manufactured goods and the non-industrialized countries (semi-/periphery) supplying raw materials and agricultural goods as well as a market for manufactured products.

This spatial division of labour can occur *within* firms. One location (a major advanced city such as London) is used for the headquarters (HQ), another for research and development (R&D), and yet another (a peripheral region, South Wales, or country, Malaysia) for the manufacturing branch plant. Moreover, this process is not confined to manufactuirng companies. In a small (back) office in County Kerry (Ireland), a group of women process the insurance claims of the fifth largest insurance company in the United States. The claim forms are flown across the Atlantic, checked, computerized and downloaded onto the parent company's computer in New Jersey, USA (Allen 1995: 56).

Plate 12.2 An early factory. Matthew Boulton's Soho Factory, Birmingham, 1774. This factory manufactured buckles, buttons and jewellery. (Birmingham City Council Department of Leisure and Community Services)

12.3.2 Gender divisions of labour

That it is a group of women in Ireland doing this work is no coincidence. The division of labour is also gendered (and racialized) producing a **gender** division of labour. In most industrialized countries women are about as likely to be in paid work as men. However, only rarely do women work alongside men, performing the same tasks for the same remuneration. Most jobs are clearly identified as either 'men's work' or 'women's work'. 'Female' jobs usually pay less than 'male' jobs, and are frequently seen as less prestigious (Williams 1993). A gender division of labour occurs as a result of perceived, but rarely actual, differences between women and men.

In the labour market certain jobs (tasks) are labelled on the basis of a set of feminine and masculine characteristics. For example, women are employed on assembly lines in the semiconductor industry as it is considered that, in comparison to men, they are more manually dextrous – they have 'nimble fingers' (what of 'male' professions such as chefs, pianists, electricians?) – and are able to endure long hours of boredom (Wilkins 1983, Elson and Pearson 1981). Another important perceived feminine characteristic is related to motherhood, and the suitability of women for the caring professions such as nursing and teaching (why can't a man be an *au pair*?). Finally, female jobs frequently involve the body as an ornamental object. One of the best examples of this process is in the case of the *female* flight attendant who is required 'to deploy "skills" and abilities which they are deemed to possess simply by virtue of their sexual difference from men' (Tyler and Abbott 1998: 434). Tyler and Abbott sat in on a number of selection days for female flight attendants. They found that applicants were rejected for the following reasons: 'the applicant

Box 12.1   Alfred Weber and the location of industry

Traditionally economic geography was concerned with describing and explaining the location of industrial activity. A series of attempts were made to develop models that would explain as well as predict the best locations for manufacturing activities. One of the most influential theories was developed in 1909 by Alfred Weber, a German economist. Weber's theory relies on four assumptions: raw materials are not ubiquitous and are fixed in location; markets are separate from places of production; transportation costs are the major factor in determining the location of a factory; and companies operate in a competitive market. On the basis of these assumptions Weber suggested that transportation costs are the product of two related factors. First, the weight of the raw materials required to construct a product in relation to the weight of the completed product; second, the distances over which the materials and the product have to be transported. The identification of these two factors allowed Weber to develop an index of cost – known as the *ton-mile*. The best location for a plant is one with the lowest total ton-mileage. Weber developed a simple model, the locational triangle, to illustrate the relationship between these two factors.

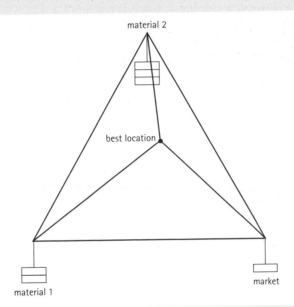

This model can be converted into a simple machine known as the *Varignon frame*, in which distances and weights are simulated by a set of scaled weights and pulleys connected by wires. The point where the wires reach a state of balance is the optimum location. Weber developed this model by introducing complications into the analysis. This model was rediscovered in the 1950s and a number of attempts were made to develop more sophisticated theories of industrial location. All of these theories failed to replicate the reality of real industrial location as they ignored both the behavioural aspects of industrial location as well as the role played by government. It is well known that the preferences of senior managers for locations that provide life-style attractions (countryside, golf courses, access to cultural facilities, etc.) may overrule more economic rationalizations for specific locations.

Box 12.2

Export processing zones and the Bangladesh textile industry

The newly industrializing countries (NICs) like Taiwan, South Korea and China have been especially active in export-driven industrialization. These countries have established export processing zones (EPZs) to attract local and international capital to establish factories for export. EPZs attract foreign direct investment through low tax rates, reduced foreign exchange controls, the provision of factory and warehouse space and exceptions from tariffs and export duties. Countries like Bangladesh that aspire to become NICs have also established EPZs. The textile and clothing industries have played an important role in EPZs and they are also one of the most important employers of female labour globally. Bangladesh's clothing industry was export driven from its establishment in an EPZ as there is little demand for manufacturing clothing within the country. The success of the EPZ has meant that local textile producers are no longer capable of providing the quantity or quality of cloth required for export garments. The EPZ relies on imported textiles and a cheap, predominantly female workforce. Garment manufacture in Bangladesh uses simple inexpensive operator-guided electric machines. However, the technology is gradually changing and computerized attachments are being introduced. These changes are marginalizing women, as the new machines require new skills that are considered to be high-tech male jobs rather than low-skill female jobs. Women are in danger of losing their jobs and of being excluded from operations requiring new skills (Jezkova 1995).

---

was too old, the applicant's skin was blemished, the applicant's hair was too short, too messy or too severe, the applicants nails were too short or bitten, the applicant's posture was poor or the applicant's legs were too chubby ... another applicant ... was rejected on the basis that she was slightly "pear shaped"' (Tyler and Abbott 1998: 441–2). The world of work would be a different world indeed if 'not being pear-shaped' was a selection criterion applied to men!

The gender division of labour also operates spatially; from pools of labour in the 'third' world to clusters of women living in suburban areas or rural regions such as County Kerry. Branch plants, call centres and back offices are located close to supplies of (often cheap) female workers (see Box 12.2). One of the best examples of a spatial gender division of labour occurred in the 1960s and 70s in the United Kingdom with the relocation of government offices to South Wales. The civil service was aware that South Wales possessed a large potential reserve of unemployed women. Offices that were relocated there included the Companies Registration Office (Cardiff) and the Driver and Vehicle Licensing Agency (Swansea). Both offices require low-skilled, poorly paid, malleable office workers to undertake routine administrative functions such as checking that forms have been filled in correctly, data entry and processing and data retrieval (Winckler 1987). The best location may not necessarily be the cheapest but the location with the relatively best-educated or least unionized workforce.

12.4   Fordism

The growth of capitalist production and today's global economy is the story of the continued expansion, both geographical and functional, of the division of labour. The nineteenth century was a period in which successful family firms that were located in the core economies grew by developing new markets and by the acquisition of less successful companies. It was a period also in which family capitalism, or companies mainly run and controlled by families, started to be replaced by forms of organized capitalism. This involved the separation of ownership from control, as management increasingly became a professional occupation rather than the task of the family owner. The professionalization of management is part of the continual escalation in the complexity of the division of labour; from the general manager to specialists in finance, human resource management, public relations and so on. This period is also associated with escalations in the division of labour, the international division of labour and the gender division of labour.

One of the key figures in this process was Henry Ford, the American automobile manufacturer. Ford pioneered the development of mass production based on assembly-line techniques and the introduction of scientific management into the workplace. He also encouraged mass consumption through advertising, higher wages and the supply of relatively inexpensive products. Mass production cannot occur without mass marketing to encourage mass consumption. Both these business techniques were developed and refined in the United States before being exported to other countries. There are two important locations to consider in this story: the cities of Chicago and Detroit.

Plate 12.3   The City of Chicago in the 1930s. (Popperfoto)

12.4.1 Chicago and the meat-packing industry

The history of the Midwest of the United States is the history of the city of Chicago. This history has two sides: a history of Chicago's relationships with the Great West, its hinterland (Cronon 1991), and a history of the development of Chicago and its industries (Miller 1997), especially the meat-processing industry. Meat-processing is an interesting industry as it is a product of an early division of labour: the separation of the production of food from its final consumption. Chicago became the centre of the United States meat-processing industry as a consequence of two related processes.

First, in the 1880s, meat-packing became the first assembly-line industry as it proved to be impossible to mechanize the production process. An assembly line is organized around a division of labour, with the production process divided into a number of parts with each part located adjacent to each other so that tasks are undertaken in a set sequence. Rather than using technology, the division of labour was used to reduce production times and cost. The first assembly line, or 'sort of dis-assembly line' (Lacey 1986: 104), was was divided into thirteen processes; live animals went in at one end and ten minutes latter reappeared as dressed meat products. In 1893 a workforce of 25,000 men, women, boys and girls processed nearly 14 million animals creating products valued at US$200 million (Miller 1997: 199).

Second, as the populations of American cities increased it became impossible for each city to rely on local meat supplies. Alongside this change, cattle production expanded dramatically on the unfenced plains of the Midwest (Cronon 1991: 221). Cattle were sent to Chicago in slatted railway cars, and shipped from there to be slaughtered in the cities on the East Coast. In 1878, the owner of a Chicago slaughterhouse hired an engineer to design a refrigerated railway car that could be used to ship processed meat rather than live animals. The result was dramatic. The trade in transport of live animals was undermined and Chicago developed into the centre of the United States beef-butchering industry. With the refrigerated car, Chicago 'moved from a center for the transfer of live cattle to a center for slaughtering and packing cattle' (Miller 1997: 208). A transportation innovation resulted in a new spatial division of labour, with Chicago developing a specialism in the production and supply of meat products.

12.4.2 Detroit, Henry Ford and Fordism

Most economic geographies of Fordism begin with Henry Ford's development of the Model T motor car, the first successful mass-produced car. Yet the introduction, in 1913, of the assembly line in Ford's Detroit plant relied on the experience and expertise developed in Chicago's meat-packing industry. At the Ford factory knowledge of other assembly lines was obtained from managers who had previously worked in the arms industry, meat-packing and in steel mills. This knowledge along with the ideas of Adam Smith and F.W. Taylor, the founder of scientific management and Taylorism, were developed and refined to such an extent that each job in the Ford factory was carefully timed and subdivided into a number of smaller operations of equal duration. An employee's job was reduced to a single task that usually lasted considerably less than a minute. Ford workers when asked what they did in the factory would, for example, say 'My work was to put on bolt no. 46' (Ling 1990: 145).

Fordism, the production system named after Henry Ford, was a combination of:

1 The extended division of labour.
2 Improvements in the manufacture of parts so that they became increasingly standardized.
3 The realization that machines should not be grouped together by type on the shop floor (e.g. all presses together) but arranged in the sequence required to manufacture a product. This idea was developed in the arms industry, but Ford expanded it by linking each part of the production process together by a constantly moving conveyor belt: the assembly line.

The secret of Ford's success lay in standardization, efficiency and marketing. In the first decade of the last century most car manufacturers sold a small number of cars but at prices substantially above production costs. Ford reversed this logic by maximizing sales of low-priced cars, each of which carried a small profit. As more cars were sold, production costs declined as economies of scale were achieved and Ford was further able to reduce the cost of the Model T. These cost reductions stimulated additional sales as customers found that they could now afford to buy a Ford, but only a Ford. Ford's other secret was that he only manufactured one car – the Model T. This provided him with a cost advantage, as he did not have to manufacture a wide range of different parts. The Ford Model T was originally built like any other motor car; it sat on a cradle while workers swarmed around it. The demand for the car was to change this. In 1909 18,664 cars were sold, but this had increased to 34,528 by 1910, and to 78,440 by 1912 (Lacey 1986: 105). Yet the Model

Plate 12.4 Ford Model T assembly at Highland Park, 1914. (Topham Picturepoint)

T was still expensive as the price reductions enabled by mass production were still to be introduced. In 1908 a new Model T cost US$825 which compared with the average teacher's salary of US$850 (Lacey 1986: 94). By 1917 the price had declined to just US$360 and as low as US$290 in 1934. The demand for the Model T was generated by advanced publicity and the fact that the car contained some genuine innovations, for example it was the first car with an engine protected against the elements by steel casings.

For Henry Ford the success of the Model T was a real problem. Production could not keep up with demand. The solution was the development of the first assembly line in an automotive factory and an increasing division of labour. Production went up, but at the expense of the quality of life of Ford's employees. Staff turnover in the Ford factory in 1913 was running at a staggering 370 per cent (Ling 1990: 147). In the same year, Ford surprised his competitors by raising pay from US$4 to US$5 a day and simultaneously reducing working hours from nine to eight hours a day. This was an attempt to retain staff, but at the same time the existing two nine-hour shifts were replaced with three eight-hour around the clock shifts. More cars were produced, yet each worker was paid more (Lacey 1986: 117). The five-dollar day was widely considered by newspapers and rival car companies to be an error of judgment on Ford's part. However, his critics were unaware of the savings in labour costs created by the new continuously moving assembly lines: in 1912 it took 728 worker-minutes to assemble a car in comparison to only 93 minutes a year later (Lacey 1986: 120).

Plate 12.5   A completed Ford Model T. (Corbis UK/Bettmann)

12.4.3 The rise of the branch plant

The Ford Motor Company's final innovation was the development of branch plants. The Model T left the main Ford plant in Detroit as little more than a chassis, motor and body. Important accessories like headlights were installed by branch agencies established and run by Ford. Branch agencies also undertook car repairs at a time when most cars had to be returned to the factory for repair, but they also meant that Ford had the facilities to sell cars direct to the consumer. It was a short step from a branch agency to a full assembly branch plant and what prompted this step was transportation costs.

The Model T was transported from Detroit by fitting as many vehicles as possible into a rail freight car. Railway companies charged for this service on the basis of a minimum weight of 10,000 pounds per eleven-metre freight car. Such a car held four fully assembled Model Ts but these only weighed 4,800 pounds. Ford was being charged for 5,200 pounds which was not being transported, and on top of this railway rates for carrying completed cars were 10 per cent higher than normal freight rates as cars were considered to be fragile and difficult to manoeuvre. The solution to this problem was the establishment of assembly plants located close to the main markets (a spatial division of labour within the firm). Freight cars packed with enough parts to assemble 26 cars were sent from the main Detroit plant. The development of assembly plants halved Ford's transportation costs, but the savings were not passed on to the consumer. Assembly plants also

Figure 12.2 Map of the US Fordist region: the American manufacturing belt in 1920. *Source:* Knox and Agnew, 1998

Figure 12.3 Ford in Western Europe, 1950–84. *Source*: Chapman and Walker, 1987

embedded the Ford Motor Company into local communities encouraging local newspapers to adopt the company. Free publicity and the impression that Ford was a local company resulted in additional sales of the Model T.

12.5 The geography of Fordism

Fordism has its own distinctive geography. From the 1920s the heartland of US Fordism developed between Boston and Chicago and from Cincinnati to Milwaukee (Figure 12.2). In the United Kingdom the West Midlands consolidated its position as the workshop of the world (Bryson *et al.* 1996) whilst, in Germany, manufacturing production consolidated itself in the Ruhr. These regions became the centres of mass production activities and of reasonably well-paid but monotonous jobs. This does not mean that other regions did not become industrialized, but that the scale and intensity of manufacturing production in these areas was different in scale and scope.

12.5.1 The new international division of labour

The obvious development from branch plants located within the confines of a nation-state is the establishment of branch plants located in overseas countries. An **international division of labour** (a form of spatial division of labour) occurs when the process of production is no longer confined primarily to national economies. This is not a new process, as regions of the world have specialized in particular segments of the production process for a long time. Examples include wool production in England during the fourteenth century or the export of cocoa beans from West Africa from the nineteenth century onwards to support the rapidly growing European chocolate industry.

As mass consumption developed alongside improvements in mass communications (Williams 1998) ever-larger companies developed with the capacity to mass-produce affordable goods. It was a period of consolidation of businesses, of mergers and acquisitions and of the development of monopoly capitalism. By 1957 America's top 135 corporations owned 45 per cent of the country's industrial assets (Baran and Sweezy 1966: 45). The period after the Second World War was one of economic growth, of nearly full employment, of prosperity, of growing car ownership and, in the words of Harold Macmillan, in the United Kingdom it was a time when 'you've never had it so good'. It was also a period in which multinational companies developed through the creation of branch plants (Dicken 1998). American companies like Ford had established branch plants in Europe as early as 1910, but until just after the Second World War such plants were very similar to Ford's American assembly branch plants. From the 1950s large companies reorganized their production processes and began to develop a **New International Division of Labour (NIDL)** with parts being supplied from countries all over the world. Effectively companies were searching for the cheapest locations to manufacture and assemble components. The key industries driving this NIDL were the car and electronics industries. The result of the NIDL in the car industry was a highly complex network of cross-border flows of completed vehicles and component parts. In 1967, Ford consolidated its European plants into a European division and began to develop cars especially for the European market place (see Figure 12.3). A European division of labour was introduced with engines assembled in Germany, transmissions in France and electrics in the United Kingdom. Today, the onset of multinational companies that act globally to exploit the international division of labour (remember the example of the US insurance company in Ireland mentioned earlier) has resulted in competition between nation-states and regions to attract companies that are seeking the best location to undertake parts of their production process (see Box 12.2).

12.6 The crisis of Fordism

The post-war years of economic growth and expansion peaked in the mid-1960s. The 'crisis' hit different companies and countries at different times. In America the heyday of Fordism can be identified with the debut party of Henry Ford II's daughter, Charlotte. This took place on 21 December 1959 and cost US$250,000. At that time in the US, no one had heard of Toyota. Ford and the other American car manufacturers had dominated the market for cars in the United States. Fordism was a sellers' market, in which manufacturers dictated what could be purchased; the wishes and the whims of consumers were secondary. During the heyday of Fordism the quality of products declined, but their cost increased. All of this was to end in the 1960s when European, followed by Japanese, car companies began to export cheap, high-quality cars to the United States.

The crisis of Fordism emerged in the early 1970s drawing to a close the post-war

---

Box 12.3 Fordism in crisis in the West Midlands region, UK

After the Second World War the West Midlands became the industrial heartland of the United Kingdom and the workshop of the world. It was the centre of the United Kingdom's car industry at a time when Britain made more cars than any other country except the United States. Indeed, the West Midlands conurbation accounted for over 8 per cent of UK manufacturing output. Between 1951 and 1961, the region's employment increased by 14 per cent compared to a national average of only 8.5 per cent. This decade of employment growth was driven by five manufacturing sectors: vehicles, metal goods, metal manufacture, mechanical engineering and electrical engineering. In 1961, just under 65 per cent of the labour force was employed in manufacturing compared to a figure of 39 per cent for the United Kingdom. Between 1951 and 1966 the region gained 361,000 jobs. The West Midlands was at the heart of the Fordist boom in the United Kingdom.

The crisis of Fordism emerged in the early 1970s in the West Midlands economy. Between 1971 and 1993, just over half a million manufacturing jobs were lost in the region, a decline of 50 per cent in total manufacturing employment. By the late 1980s the West Midlands had shifted from being one of the most prosperous regions in the national economy to being one of the most depressed 'peripheral' regions. The West Midlands crisis was the result of a number of related factors (see Bryson et al. 1996), but at the heart of the problem was overseas competition and cheap imports which undermined the profitability of the area's companies. By the early 1980s many large West Midlands companies had closed or severely rationalized production. For example, GKN in the early 1970s employed 17,000, but by 1980 this had fallen to only 5,000. The industrial decline of the West Midlands left a legacy of industrial dereliction, empty obsolete factories, and high long-term unemployment. The industrial heartland of the UK economy had become an industrial wasteland.

industrial boom. Economic growth in the industrialized countries fell, unemployment increased and household disposable incomes fell. Unemployment in the advanced economies averaged around 4.5 per cent in the 1960s, but grew to 10 per cent by the mid-1980s (Knox and Agnew 1994: 209). The crisis was caused by the combination of a series of economic pressures including the inflexibility of Fordist industrial production, the high cost of production in the heavily unionized advanced economies and increasing overseas competition. At an international scale, industrial production was again reorganized away from high-cost to low-cost locations and a further spatial division of labour evolved. One of the most pronounced consequences of this crisis was the decline in the industrial manufacturing base of the advanced economies. This decline was most evident in former core industrial zones such as the US Midwest or the West Midlands region in the UK (Box 12.3). De-industrialization affected all aspects of the economies of these regions as spiralling multiplier effects undermined the client base of small subcontracting firms, and rising long-term unemployment and industrial dereliction discouraged new inward investment (Massey 1988).

With hindsight, it is now generally recognized that the crisis of Fordism marked a crossroads in the historical development of capitalism. Just as the industrial revolution or the onset of Fordism marked distinctive phases of capitalism, so it is now believed that with the crisis of Fordism came an epoch-making transformation in the organization of capitalism. The 'post-Fordist' debate concerns the nature and direction of such epoch-making change (Amin 1994). This transformation of the organization of production – for example, of technology, firms, labour, markets (consumption patterns) – has brought with it a further transformation in the geography of the 'global' production system.

12.7 Post-Fordism

There is still a strong debate about the contours of post-Fordism. How do we characterize and explain its defining features? One full-blooded description of post-Fordism is that given by Hall (1988, quoted in Amin 1994: 4):

> A shift to the new 'information technologies'; more flexible, decentralized forms of labour process and work organization; decline of the old manufacturing base and the growth of the 'sunrise', computer-based industries; the hiving off or contracting out of functions and services; a greater emphasis on choice and product differentiation, on marketing, packaging and design, on the 'targetting' of consumers by lifestyle, taste and culture rather than by categories of social class; a decline in the proportion of skilled, male, manual working class, the rise of the service and white-collar classes and the 'feminization' of the work force; an economy dominated by multinationals, with their new international division of labour and their greater autonomy from nation state control; and the 'globalization' of the new financial markets, linked by the communications revolution.

While there is general agreement with this description, there is greater disagreement as to why these changes have come about; disagreement on how to explain these new economic realities through the use of theory. A wide selection of theories have been put forward – including 'informational capitalism', 'flexible accumulation', 'neo-Fordism', 'disorganized capitalism', 'post-industrialism' and 'the fifth Kondratiev' – to account for these major changes in the economy and society.

Four major processes driving economic change are commonly identified:

- Technological change: the rise of information technology; flexible machines in flexible manufacturing systems.
- Labour practice change: the use of flexible, multi-skilled core workers combined with numerically flexible peripheral workers (e.g. temporary, contract, part-time, zero hour).
- (Combining to produce) industrial organization change: the growth of the flexible firm; 'lean and mean' multinationals and small and medium-sized enterprises, concentrating on their 'core' specialisms as part of networks of collaborating firms (both small and large).
- Market demand/consumption change: increasingly differentiated (customized, segmented, niche, life-style) demands for goods and services.

Furthermore, these processes are taking place within the context of 'globalization' – a highly contested term both within, and without, economic geography (see Allen and Massey 1995, Chapter 3; Bryson *et al.* 1999, Part Two; Chapter 11 in this volume). In the contemporary economy, the term is used to denote at least two distinct processes. First, it depicts the quantitative shift of increasing integration of the world economy as more and more parts of the globe are drawn into the networks of capitalist production and markets. Simple examples would be the rise of the NICs (Newly Industrializing Countries) such as Korea, Singapore, Malaysia and Thailand to name but a few (Thompson 1999), the growing presence of China as it liberalizes its economy towards capitalist production (Ogilvy and Schwartz 2000) and, of course, the growing, but highly uneven, insertion of the post-socialist states into the global economy following the fall of the Iron Curtain (Stenning and Bradshaw 1999). It is in this sense that the global production systems of today are 'more global' than those of Francesco di Marco Dantini in the fourteenth century or even those depicted in Fordism's new international division of labour. Second, however, is a qualitative dimension that has driven the use of the term 'globalization' to describe the current economic era. In this sense, the term global economy is used to stress how production and investment crosses the globe with very little regard for national boundaries (Dicken 1998, Allen 1995). Markets are global, not national (for example, the EU or Latin America), and the major players in such an economy talk of global products – Fiat's Tipo 178 world car, for example, or Coca-Cola or Microsoft Software. They question the need to have a 'home-base' in a nation state as against a headquarters located, as with all the other parts of the production process, in the best location to meet its particular requirements (such as access to skilled labour and finance capital).

12.7.1 Post-Fordist production system(s)

If the Ford Motor Company – with its assembly-line mass production in branch plants around the world to meet the demand of a mass market – is taken as the exemplar of the Fordist production system, then today's post-Fordist system is best understood through a number of 'typical' examples. There have been a number of alternative responses to the challenges of the new production era. Currently, the post-Fordist production system comprises a *mix* of different ways of organizing production at a number of spatial scales. Table 12.1 contrasts two of the major alternatives to Fordist mass production.

Table 12.1 Alternative production systems in post-Fordism

Characteristic	Flexible specialization	Fordist mass production	Japanese flexible production
Technology	Simple, flexible tools/machinery; unstandardized components	Complex, rigid single-purpose machinery; standardized components; difficult to switch products	Highly flexible (modular) methods of production; relatively easy to switch products
Labour force	Mostly highly skilled workers	Narrowly skilled professionals 'conceptualize' the product; semi/unskilled workers 'execute' production in simple, repetitive, highly controlled sequences	Multi-skilled, flexible workers, with some responsibilities, operate in teams and switch between tasks
Supplier relationships	Very close contact between customer and supplier; suppliers in physical proximity	Arms-length supplier relationships; stocks held at assembly plant as buffer against disruption of supply	Very close supplier relationships in a tiered system; 'just-in-time' delivery of stocks requires 'close' supplier network
Production volume and product variety	Low volume and wide (customized) variety	Very high volume of standardized products with minor 'tweaks'	Very high volume; total partially attained through production of range of differentiated products

Source: Adapted from Dicken (1998), Table 5.2.

Examples of all *three* forms of production (flexible specialization, mass production and flexible production), and, indeed, further hybrid mixtures of them, can be identified in the post-Fordist global production system. What is common to all, however, is the response to the processes of change identified above: all represent the struggle for competitiveness in a global economy through some form of flexible production. For example, at the heart of flexible specialization is a skilled (craft) worker using flexible machinery to produce small volumes of customized goods. This has led to the resurgence of small, independent, entrepreneurial firms which develop as part of dense networks of interrelated, specialist firms (see Box 12.4). In contrast, Japanese flexible production implies flexible specialization but for the production of extremely high volumes of products and on a global scale (Fujita and Hill 1995) (see Box 12.5). Meanwhile, the Fordist mass production system has been shaken to its roots (see, for example, Box 12.3). Adaptation has seen some companies survive through further global extension in the search for lower costs of production and higher economies of scale through increased production for new markets (sometimes called global Fordism).

The response to change has produced three characteristics of this new post-Fordist global production system.

Benetton

In 1965 Benetton owned a single factory in Venice. In 1968 it purchased a retail store in the Alpine town of Belluno. In 1990, the company had over 5,000 outlets in 60 countries. Today, Benetton comprises a handful of factories and warehouses, its own investment bank and financial service organizations, a Formula One team for advertising purposes (see Box 12.7 on Motor Sport Valley), its own set of high-profile stores and around 7,000 franchised shops in 120 countries around the world. This 'clothing services company' has direct employment estimated at only 2,500 employees, 1,300 of whom are at the company's home-base in Trevino, Venice. Around 90 per cent of its production is undertaken in Europe – the majority in the Veneto region – by a dense network of around 500 small- and medium-sized subcontractors. The company turned over 4,200 billion lire (£1,329m/$2,092m) in 1998.

Benetton is a small firm which has made big but not in the sense of 'big' used in the days of Fordism. The company epitomizes post-Fordist production, combining flexible production with global reach based on a 'network' form of production and selling. Whilst the majority of the European textile industry has moved to cheap labour locations (textiles was a leading industry in the construction of the NIDL; see Box 12.2 on EPZs), Benetton has combined niche marketing, high-quality design and state-of-the-art manufacturing processes to build a global position within textile clothing.

The cash registers of Benetton's shops are linked, by satellite if necessary, to Trevino (its home-base). Headquarters staff know precisely what is selling day-to-day (it keeps a particularly close eye on 'flagship' stores) and set production levels on the basis of orders received from shops. Clothes arrive in the stores within 10 days of production. Benetton undertakes key (high value-added) production tasks – design, cutting, dyeing – within its highly sophisticated and computerized factories whilst 'putting-out' the more labour-intensive tasks to its network of local subcontractors.

http://www.benetton.com

12.7.2 The big are getting even bigger

The move to a global economy has triggered an ever greater number of multinational firms (see also Chapter 11). In 1969, the number of multinationals – companies that produce goods or market their services in more than one country – in the top 14 industrial countries of the world was estimated at 7,000. By the early 1990s, the figure had risen to about 24,000. In 1998, the figure was put at 53,000 multinationals worldwide. Also, multinationals are becoming more multinational; they are involved in more countries and in greater numbers of more varied overseas production relationships.

Multinational production (production by companies overseas) now outweighs exports as the principal route to meeting overseas demand and it is estimated that two-thirds of world trade is accounted for by multinationals. Furthermore, up to a third of world trade is actually cross-border, but *intra-firm*, trade. That is, different parts of the same company, different parts of the corporate spatial division of labour, sending goods and services between their sister factories and offices, and crossing national borders when doing so (see Figure 12.3). Moreover, despite the growth in total numbers of multinational firms, the *share* of the global economy accounted for by the top 100 multinationals has continued to increase. The largest 100 firms account for almost 30 per cent of total world sales by multinationals and employ (directly) 6 million workers worldwide (UNCTAD 1997, quoted in Held *et al.* 1999).

Box 12.5 **Global Toyotaism**

The post-Second World War rise of Japanese manufacturing multinationals to global power is one of the stories of twentieth-century capitalism. Flexible specialization is used in the main to refer to network systems of production among small producers. Global Toyotaism refers to how, in Japan, flexible specialization is more often found in collaborative relations among large parent companies and smaller supplier firms (see Table 12.1; Figure 12.4, *keiretsu* diagram). Economies of scale are gained through the flexible mass production of a variety of products in a system. Craft flexibility and the most advanced information technology are combined – automation with a human mind – in a distinctive organizational system that has produced higher assured quality, flexibility and continuity in production. On the shopfloor, work teams, job rotation and learning-by-doing take precedence in a system which integrates conception and execution throughout the production process.

This integration is continued throughout a supply chain based on JIT (just-in-time) principles. The parent firm, suppliers and subcontractors continuously consult to assure flexible delivery of high-quality, error-free parts to the assembly line to be fitted just-in-time. This system fosters spatial agglomeration in Toyota-style industrial districts. As the Japanese giants have globalized production, they have 'transplanted' similar-style districts comprising a parent plant combined with suppliers (see Section 12.8.3 on the new industrial spaces).

http://www.toyota.com

12.7.3 And Small may be Beautiful once again

On the other hand, some of the most dynamic centres of economic growth under post-Fordism have become so due to the rapid growth of large numbers of new, small and medium-sized enterprises (SMEs) (Birch 1987, Storey 1997). In particular, a connection has been made between the production system of flexible specialization and the dynamic opportunities it has afforded for small firms (see Box 12.4). This contrasts starkly with the Fordist era when the cutting edge of capitalist organization was seen as (i) the very large firm and the very large plant and (ii) the need to be large because of vertical integration: doing for yourself (bringing 'in-house') as much of the production process as possible (if in different places).

12.7.4 'No one can do it all any more': networks of production

The increasing pace of technological change, the need for 'global presence', and the drive for flexibility of production and products (speed in response to rapidly changing market demand) has created a situation where one firm, no matter how large, can do it all on its own. The result is the rise of networks of production (production *dis*integration) such that the global economy is labyrinthine in its interrelationships between firms, large or small and global or local (see Dicken 1998, Chapter 7). These production relationships take a myriad of forms (compare Box 12.6 with Box 12.4). Examples include strategic alliances, (international) subcontracting, franchising, subsidiaries, joint ventures, R & D networks and sponsored spin-offs to name just a few. In 1996, the world's 41 largest car producers shared 244 strategic alliances between them with Peugeot of France involved in 22

Box 12.6   Nike

Nike, the US sports footwear, clothing and equipment company, is one of the leading examples of network production. In contrast to Benetton and Toyota, exploiting the international division of labour through the exhaustive use of international subcontracting has driven Nike's strategy from its very beginnings. Co-founder Philip Knight's founding theory (in 1964) was to produce high-quality running shoes designed in the US, manufactured in Asia, and then sold in America at lower prices than the then popular West German-made running shoes. Today this US$10 billion company, headquartered in Beavertown, Oregon, has a global brand based on high quality design and marketing, employs around 16,000 staff and does not wholly own a single integrated production facility. Throughout its short history, Nike has subcontracted virtually its entire production to a changing array of factories, predominantly in South-East Asia. It has shifted in unison with the changing international division of labour in the search for lower production costs. Whilst the company has exclusive partners for high-end products, its main product base of volume training shoes is produced by volume subcontractors which the company is able 'to turn on and off like a tap' in response to poor performance or changing market demand. Currently, Nike's official 'factory profile' is of the Samho Factory, Qinqdao, in China which employs 6,750 workers, 90 per cent of whom originate from outside the region and live in factory dormitories.

http://www.nike.com

agreements with other car companies (Knox and Agnew 1998: 208). Yet, in 1990, the automotive sector was estimated to account for only 5 per cent of strategic alliances by sector (Dicken 1998: Table 7.3).

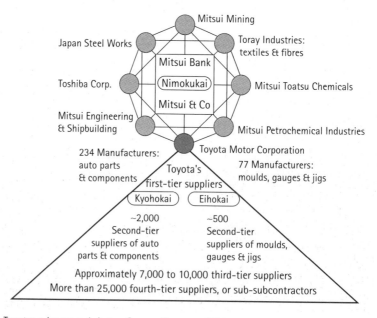

Figure 12.4   Toyota: a Japanese *keiretsu*. *Source*: Harryson, 1998

Some of the most well known forms of networks include the Japanese *keiretsu* and South Korean *chaebol* (see Figure 12.4). Whilst very different in form from the production network of Benetton or Nike, all these examples are successful in producing flexibility and global presence through production networks and indirect labour. Gone are the days of Ford's River Rouge plant in Detroit which, at its height, used 70,000 workers to transform the raw materials going in at one end into over one million black Model T Ford's exiting at the other.

12.8 Post-Fordism: a new global production system

The transition from Fordism to post-Fordism has dramatically redrawn the economic geography of global production. It has produced also one of the paradoxes of modern

Plate 12.6   An aerial photograph of Silicon Valley. (Air Flight Service Mapping Photographers)

times. On the one hand, the global economy – and the largest firms that comprise it – has never been as large, or reached as far, as it does today. Few places and people in the world are not touched by this global spatial division of labour, let alone by the consumption dynamic which creates the demand for this global production system. Yet, on the other hand, the onset of post-Fordist global production has triggered a *reconcentration* of the division of labour. A major characteristic of today's global economy is the 'resurgence of the region' (Sabel 1989) or the reagglomeration of production ('clusters') (Scott 1988; see also Chapter 11). Particular places – cities, towns, regions – are becoming famous as *the place* to produce particular products. Probably the most well-known example is Silicon Valley, the massive agglomeration of semiconductor and computer producers in southern California. Other examples would be the three financial centres – London, New York and Tokyo – which dominate the world's financial and business service industries. These New Industrial Spaces (NIS) are just one of three distinctive characteristics of the global geography of post-Fordist production: global reach *and* global concentration, an increasingly complex international (global) division of labour, and local geographical concentration (agglomeration).

12.8.1 The global triad

The concept of the global triad encapsulates both the global nature of the world economy and, even at this spatial scale, elements of its geographical concentration. The term describes how the world economy is now essentially organized around a tripolar, *macroregional* structure whose three pillars are North America, Europe and east and south–east Asia (see Figure 12.5). By 1994, these three regions' share of total world manufacturing output had increased to 87 per cent and they generated 80 per cent of world merchandise exports. The triad also accounted for 90 per cent of the world stock of foreign direct investment (FDI) (Dicken 1998: 60–1). Of the world's largest 500 multinationals in 1996, 443 (89 per cent) were headquartered in the US, EU or Japan. Switzerland, South Korea and Canada accounted for 33 of the remaining 57 companies.

Nevertheless, increased concentration within the triad hides, for example, the exceptional economic growth of what have become known as the Asian Tiger NICs (see South Korea above), and the opening-up of the former Soviet Union and east central Europe to capitalist production. Meanwhile, Africa has seen its tiny share of world trade and investment decline over the preceding decades. Overall, whilst the tentacles of the global production system continue to spread this is overridden by an increased concentration at the global level within the three macroregions of the triad.

12.8.2 The newer international division of labour

Under Fordism, the new international division of labour (NIDL) was taken to describe the increasing foreign direct investment (often through branch plants) by advanced economy (US, European) multinationals in other advanced economies and, eventually, in the less developed countries (for example, Taiwan and South Korea). Indeed, this increasing trend became part of the crisis of Fordism (deindustrialization) for the advanced economies as home-country plants were closed down and production shifted to new plants in the global periphery (see Box 12.3).

With the transition to post-Fordism, it has been argued that the global economy is

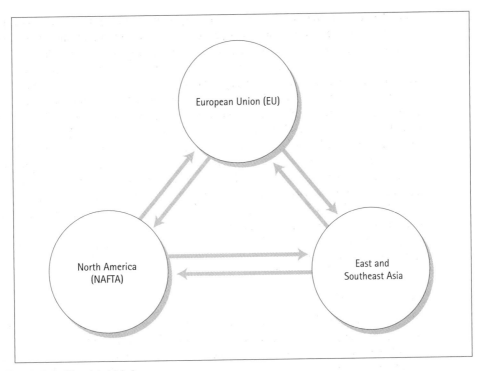

Figure 12.5   The global triad

being shaped by a '*newer* international division of labour' (Coffey 1996). There are several processes at work shaping this global division of labour. First, whilst foreign direct investment remains important, firms from the advanced economies are using other, less direct, ways to exploit the cost advantages of the less developed countries. For example, and drawing on the network approach to production, 'manufacturing' firms in the advanced economies perform less and less of their own physical production. Rather than building manufacturing branch plants in LDCs, they concentrate on the higher value-added elements of design and marketing and engage in international subcontracting with local companies in the LDCs. Nike is a prime example of this strategy (see Box 12.6). Secondly, an increasing share of the activities being relocated is involved in the provision of services. One example is the growth of the Indian computer software industry over the last two decades (Lakha 1994; see also Chapter 11). In 1988, the top five software exporters in India were either fully foreign (US) owned companies or partnerships with overseas (US) companies. Another example, used at the beginning of this chapter, is that of the New York Life Insurance Company's 'back office' in Ireland. This case highlights also the third aspect of the newer international division of labour; namely, that an increasing share of outward FDI by firms in advanced economies is destined for other advanced economies. It is this trend which has contributed to the increased dominance of the global triad mentioned above. However, and fourth, within this triad some of the LDCs which were among the original destinations for FDI by core nations have begun to undertake FDI themselves. The Asian Tigers are a prime example of this. Moreover, there are two

Plate 12.7   Korean investment in South Wales, UK: the LG factory, Newport. (Terence Soames Photography)

distinctive patterns to this FDI. First, as the original NICs have seen their economies develop, so too have their labour costs. One result has been their replication of the search for low-cost labour through investment in their less developed neighbours such as Indonesia or Bangladesh (Box 12.2). Secondly, there has been the growing pattern of 'reverse' FDI (Lall 1983). Whilst Japan's investments in the US and Western Europe – often directed at the deindustrialized regional economies such as South Wales (Cooke 1995) or the US Rust Belt (Florida 1996) – are well established, more recent sources of FDI include the Korean *chaebols* such as Samsung and LG and Malaysian multinationals such as Proton. Whilst 'labour' is only one driving force behind this 'reverse' process of investment, its impact on the international division of labour is highly significant.

12.8.3 The new industrial spaces

The term new industrial spaces (Scott 1988) is used to describe the set of dominant economic regions and spaces which have grown in tandem with the transition to post-Fordist flexible production. It is these economic nodes which encapsulate the shift in the geography of production – in particular, the process of reagglomeration of production – with the move to post-Fordism. They take a number of different, overlapping, forms.

- New industrial districts: Dense agglomerations of, normally, small and medium-sized firms specializing in the high-quality production of a particular good or service. The archetypal example is the 'Third Italy' – home of Benetton – with individual towns and districts renowned for their specialist production of goods such as textiles

(Como, Prato), ceramics (Sassuolo), shoes (Ancona) and special machines (Parma, Bologna, Modena). Other examples include Jutland in Denmark (furniture, shipbuilding), Småland in Sweden (metalworking) and Baden Württemberg in Germany (machine tools, textile machinery).

- High technology (sunbelt) areas: A diverse range of technopoles, corridors and innovative complexes, often in previously unindustrialized areas, which have grown through their capture of the 'sunrise' high technology industries of information technology, electronics, pharmaceuticals, R & D, etc. (Castells and Hall 1994). The most famous example is Silicon Valley in southern California, birthplace of the integrated circuit and the personal computer. The experience of Santa Clara County (as it was known then), a mainly agricultural region, is the growth story of the era. In 1950, it employed around 800 manufacturing workers. Today, Silicon Valley employs nearly 400,000 high technology workers alone. Companies born in the region include Hewlett Packard, Intel, Apple and Sun Microsystems. Other, more modest, examples include Britain's Western Crescent/M4 Corridor, Boston's Route 128, the Ile de France Sud, and Japan's Technopolis programme.

- Flexible production enclaves within old industrial regions: This varied classification attempts to capture the variety of economic development that has taken place in previously industrialized regions. Often rocked by the decline of previously core 'sunset' industries (shipbuilding, coal, steel, heavy manufacturing, etc.), some alternative economic growth has been achieved through firms and sectors applying flexible production techniques. This might occur through in-situ restructuring of firms both large and small or the attraction of new, often foreign, investment. Examples of the former would include the resurgence, largely based on small firms, of the textile and film industries of Los Angeles, or the post-production industries of Soho, London. Examples of the latter include the large concentration of overseas investment from Japan and Korea in South Wales or the resurgence of the US Midwest through Japanese 'transplant' investment (see Box 12.5). This highlights the local and regional impacts of the 'reverse' FDI discussed above.

- Global cities (Sassen 1994): These are the pinnacle of the global urban hierarchy as cities once again become key sites of production with the move to service industries in the advanced economies (see Chapter 11). Cities such as London, Tokyo and New York are the command and control centres of the global economy. This means more than just being the headquarter sites for global companies. In addition, they are the industrial districts of the financial and business services and the centres of the corporate services that provide the expertise for the management of global production. These centres include the 'servicing' industries – the cleaners, security guards, couriers and pizza delivery personnel – who service the corporate elite. They are sites also of the elite cultural and retail industries; the galleries, theatres and fashion houses. A second tier of global cities includes, for example, the 'metaregional' capitals of Paris, Frankfurt, Sydney, Hong Kong, São Paulo and Mexico City.

12.8.4 The global mosaic of post-Fordist production

Even a simplified outline of the post-Fordist geography of production offers, then, a complex, multiscale picture of economic growth. The economy is global like never before and at this scale it is no longer a simple case of allocating countries to the core, semi-periphery and periphery. Yet, continued global reach has gone hand-in-hand with increased geographical concentration (localization). This is no more starkly shown than in Britain's Motor Sport Valley (see Box 12.7) in which some of the largest global companies in the world have beaten a path to semi-rural Oxfordshire – and an array of comparatively tiny firms – to produce their flagship motor-sport products (racing cars). The ever-changing international division of labour is producing increasingly complex, and interdependent, relationships between, and across, metaregions, nations, cities and localities, between firms large and small, and between economies old and new.

What is undeniable, however, is the substantial reorganization of global production with the move to post-Fordism and the new, uneven, geography of winners and losers that this entails. For the 'heavy industry' communities and Fordist regions and cities of old, for example, this has often meant devastating change as new power-house regions and nations such as southern California or South Korea have displaced them on the economic map. For other regions, such as sub-Saharan Africa, their position has barely changed, their economic marginality as evident under post-Fordism as it ever was. For yet others, such as the post-socialist states, the onward march of global capitalism has thus far implied an uncomfortable ride on the roller-coaster path of economic development.

12.9   Conclusion

The history of economic development is one of increased economic specialization moving hand-in-hand with geographical expansion; the combined and uneven extension

---

Box 12.7

The 'global agglomeration' of Motor Sport Valley

'Within an hour's drive, two hours at the most, it is possible to get anything in motor sport done, have anything designed, anything built'

Over 75 per cent of the single-seat racing cars used in more than 80 countries across the world are British-built. This includes the vast majority of Formula 1 cars and those cars racing in the US-based Championship Auto Racing Team (CART) and Indy Racing League championships. Britain is the base for the majority of the world's rallying teams including giants such as Mitsubishi, Ford, Subaru and Hyundai. In 1995, British-built cars won the top five world motor sport championships; over 50 per cent of the supply chain of those cars was British.

The motor sport industry employs around 50,000 people in several hundreds of small and medium-sized firms clustered in a 50-mile radius around Oxfordshire in southern England, in a region known as Motor Sport Valley. It is a truly global industry with finance, sponsorship, drivers, engineers, parts and expertise coming to the UK from all over the world (look at the names of the companies listed above; they include virtually all of the world's largest car producers). Motor Sport Valley has become *the place* in the world to produce your motor sport product – no matter how large the company and the global choice of locations on offer.

(Box continued)

Box 12.7 (Box continued)

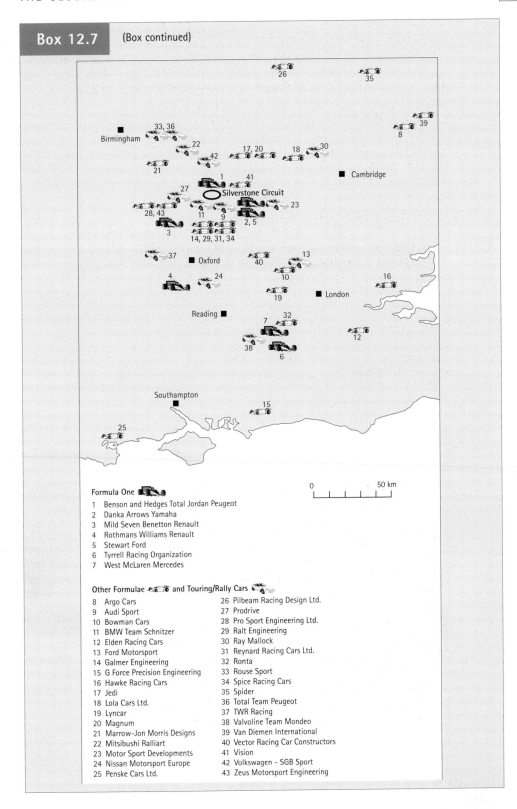

Formula One

1 Benson and Hedges Total Jordan Peugeot
2 Danka Arrows Yamaha
3 Mild Seven Benetton Renault
4 Rothmans Williams Renault
5 Stewart Ford
6 Tyrrell Racing Organization
7 West McLaren Mercedes

Other Formulae and **Touring/Rally Cars**

8 Argo Cars
9 Audi Sport
10 Bowman Cars
11 BMW Team Schnitzer
12 Elden Racing Cars
13 Ford Motorsport
14 Galmer Engineering
15 G Force Precision Engineering
16 Hawke Racing Cars
17 Jedi
18 Lola Cars Ltd.
19 Lyncar
20 Magnum
21 Marrow-Jon Morris Designs
22 Mitsubishi Ralliart
23 Motor Sport Developments
24 Nissan Motorsport Europe
25 Penske Cars Ltd.

26 Pilbeam Racing Design Ltd.
27 Prodrive
28 Pro Sport Engineering Ltd.
29 Ralt Engineering
30 Ray Mallock
31 Reynard Racing Cars Ltd.
32 Ronta
33 Rouse Sport
34 Spice Racing Cars
35 Spider
36 Total Team Peugeot
37 TWR Racing
38 Valvoline Team Mondeo
39 Van Diemen International
40 Vector Racing Car Constructors
41 Vision
42 Volkswagen - SGB Sport
43 Zeus Motorsport Engineering

of the (spatial) division of labour. In different historical periods of production, whether, for example, the Roman Empire, family capitalism, Fordism or post-Fordism, those involved may well have felt themselves to be part of a 'global' production system. From the standpoint of today, the post-Fordist system of flexible production is 'more global' than anything that has gone before. Similarly, globalization is taking place at a scale and intensity greater than previously experienced.

The paradox of this process is not 'the end of geography' but, in contrast, an even greater sensitivity to the geographical choices on offer within the global production system. Indeed, with further global expansion has come greater geographical concentration – at a number of spatial scales – as certain places have become almost the only place to produce particular goods. One thing is certain: today's global production system is more complex and interdependent than ever. It is the role of the economic geographer to both characterize and understand this complexity.

Learning outcomes

Having read this chapter, you should understand:

- Throughout history there have been 'global production systems'. Nevertheless, there is something distinctive about today's post-Fordist global economy.
- Economic production is a story of economic specialization and the continual reworking of the division of labour.
- A spatial division of labour occurs with the geographical separation of production tasks, for example, writing, printing, binding and distributing a book.
- Such a division may be local (different buildings), regional (the UK's North–South divide) or global (the world's North–South divide).
- Different economic geographies are explained by distinctive patterns of spatial divisions of labour.
- Fordism and post-Fordism each have their own characteristic economic geography, their own map of a global production system.
- The emerging economic geography of the twenty-first century is an interlinked mosaic comprising continued global expansion of capitalism hand-in-hand with local geographical concentration.

Further reading

Allen, J. and **Hamnett, C.** (eds) (1995) *A Shrinking World? Global Unevenness and Inequality*, Oxford University Press, Oxford. An accessible edited collection that provides an understanding of the processes that are shrinking the world and the impact of these processes on a range of global economic objects and activities.

Bryson, J., Henry, N., Keeble, D. and **Martin, R.** (eds) (1999) *The Economic Geography Reader: Producing and Consuming Global Capitalism*, Wiley, Chichester. A more advanced collection of edited readings of the 'great and good' in economic geography.

For students wishing to develop their understanding, introductions and references are particularly useful.

Daniels, P.W. and **Lever, W.F.** (eds) (1996) *The Global Economy in Transition*, Longman, Harlow. A good edited collection of essays that explores the restructuring of the world's production systems and its economic geographies. Now slightly dated but a good, well-written introduction to the range of literature.

Dicken, P. (1998) *Global Shift: Transforming the World Economy*, 3rd edition, Paul Chapman, London. The definitive textbook on the global economy, its evolution, structure and geography. The second edition is as useful as the third.

Knox, P. and **Agnew, J.** (1998) *The Geography of the World Economy*, 3rd edition, Edward Arnold, London. A comprehensive and concise 'encyclopaedia' of the history and geography of the world economy. Heavily influenced by a 'world systems theory' approach to understanding economic geography.

Massey, D. (1995) *Spatial Divisions of Labour: Social Structures and the Geography of Production*, 2nd edition, Macmillan, London. This extremely important book in economic geography theorizes, and exemplifies, the concept of spatial divisions of labour. Based on the experience of the UK economy, this is one of the definitive accounts of why economic geography matters.

CHAPTER 13

JANE POLLARD

The global financial system: worlds of monies

Money is simultaneously everything and nothing, everywhere but nowhere in particular.

(David Harvey 1985a: 167)

Money frees you from doing things you dislike. Since I dislike doing nearly everything, money is handy.

(Groucho Marx, quoted in Furnham and Argyle 1998: 177)

Topics covered

- Money in economic geography

- The social, cultural and political significance of money

- Geographies of monies

- Global monies, local monies and monetary networks

- Money, space, and power: geographies of inclusion and exclusion

This chapter explores some of the worlds of money we inhabit, or rather the economic, social, cultural and political networks through which we establish value and manage exchange. Our attitudes to money, how we obtain it and use it, are not simply an economic phenomenon. Geographers and other social scientists are now grappling with the complex social, cultural, political and geographical aspects of money.

As the quote from Harvey (on page 374) suggests, however, money can be a difficult subject. Novelists, film-makers and social theorists alike have grappled with its seemingly contradictory characteristics. Money can be everything and nothing, everywhere and nowhere; money can be the root of all evil and a source of independence and freedom and so the examples go on. Even simple questions like 'what is money?' present some difficulties. Davies (1994: 29) suggests that money is 'anything that is widely used for making payments and accounting for debts and credits'. Historically, a wide range of items, including beads, shells, whales' teeth, cattle, salt, skins, tobacco, beer, gold and silver, to name only a few, have been used as money (Davies 1994). Such variation suggests that our question should be altered. Rather than asking what money is, we should perhaps ask what functions does money perform? This question takes us to the heart of some of the important economic functions that money performs and to the essence of money as a source of social power which, as Groucho Marx observed, can free us from doing things we dislike.

Money performs several functions (Box 13.1). We are used to thinking of money 'oiling the wheels' of commerce and also being a measure of worth or value. Yet these two facets of money can be contradictory. As a store of value, it is desirable for money to be a stable, fixed representation of value. As a medium of circulation and as a form of **capital** – money that is thrown into circulation with the intent of generating more money – it is desirable to have money freely available as credit, divorced from its moorings in the 'real economy'. As Davies (1994: 29) argues, then, the history of the evolution of money

Plate 13.1 The Wall Street Crash 1929: the stock market collapse of October 1929 brought the 1920s boom to an end and led to widespread bank failures, bankruptcies and drastic reduction in the availability of credit. (*Time*, 26 October 1998, p. 74)

Box 13.1

The functions of money

- Money is a *unit of account*: it is the base of economic accounting systems.

- Money is a *measure of value*: it is the commodity against which the values of other commodities can be calibrated.

- Money is a *store of value*, in that you can sell a good for a certain amount of money and then use that money to buy something else at a later date. Money thus allows for the separation of the sale and purchase of commodities over space and time *if* it is a reliable store of value, if money is not devalued over time and space. Preserving this ability of money to store value is one of the reasons why governments are so concerned about the phenomenon of inflation.

- Money is a *medium of exchange and circulation*.

- Money is a *means of payment* and a *standard for deferred payments*.

These different functions of money often come into conflict with each other. As a *store of value*, governments, firms and individuals alike want monies to be relatively stable and maintain their value, and hence purchasing power, between different times and spaces. As a *medium of circulation*, however, it is important for money to be extended as credit.

Banks create credit by extending loans to individuals, firms and governments. In periods of rapid expansion, there is temptation for banks and other lending institutions to create large volumes of credit. Sudden loss of confidence in the quality of credit, as happened in the Wall Street Crash of 1919 or, more recently, in South-East Asia and Russia, can trigger financial panics and devaluation as depositors, investors and financial institutions flee from credit moneys and seek out safer havens in which their moneys will not be devalued (see Figure 13.1).

---

reveals 'unceasing conflict' between borrowers, keen to expand the *quantity* of money in circulation, and lenders, keen to limit the supply of money and, at all costs, preserve the *quality* of monies in circulation. Managing this contradiction, or more accurately, reacting to financial crises resulting from it, has been and still is the mission of various regulatory authorities.

13.1   Money in economic geography

Until relatively recently, it has been standard practice to lament the relative lack of attention paid to money and finance by economic geographers (Leyshon 1995a). While money and finance have received greater attention from geographers since the late 1980s, many economic geography textbooks still devote little, if any, attention to the subject. One rationalization for this state of affairs is simply that money and finance have, historically, appeared well down the list of priorities for economic geographers. As Martin (1999: 3) argues, '[s]ince its inception, economic geography has been pre-occupied with the industrial landscape.' Manufacturing industries were viewed as the 'drivers' of the economy and received more attention than industries concerned with circulation or consumption.

Perhaps a second explanation for the relative neglect of matters financial is that money may not, in some accounts at least, seem very geographical. Many contemporary debates about globalization stress the increasingly integrated, global machinations of the financial system in which 'virtual' monies can be moved around the globe at the touch of a button. As such, O'Brien (1992) has argued that, as far as international finance is concerned, we are witnessing 'the end of geography', that location is of declining significance for firms and consumers in these global times. This view is rather simplistic, however, and overlooks the very particular historical geographies of different forms of money and financial institutions.

A third explanation for the relative neglect of matters financial is that '[m]ost geographers are interested in tangible things, and money ... is very intangible' (Laulajainen 1998: 2). This argument takes us back to Harvey and the difficulty of nailing down the different, contradictory qualities of money. Moreover, the organization of monies into different financial markets can appear dauntingly complex. The 'global financial system' can sometimes seem like something 'out there', outside us, a fast-moving, volatile 'thing', running out of control, increasingly uncoupled from the 'real' economy and operating beyond the regulatory reach of any one nation-state. This global sense of money and finance can be difficult to grasp. We may have a sense that this international financial system is important, but sometimes it is difficult to understand just how its machinations affect us. In what ways did the Barings Bank (a UK merchant bank) crisis of 1995, for example, affect people in Britain? With the south-east Asia financial crisis prominent in the news since November 1998, what has changed for people in Europe, the United States or elsewhere in the world?

For all the difficulties, real and imagined, involved in understanding the workings of international, national and local finance, money and finance are now attracting a great deal of interest from geographers, other academics, governments and the popular press. This is partly explained by a number of recent financial and political crises, including those in Pacific and south-east Asia, Russia and Brazil. These crises have fuelled a growing sense of unease, in popular and academic writings, over the volatility and instability in global financial markets and a concern that regulatory bodies and nation-states are increasingly unable to control financial flows. Also, as Leyshon (1995a) argues, money and finance have assumed a growing social and cultural prominence since the 1980s. This is related to the growth of financial services employment, the ascendance of Thatcherism and Reaganism and the growing media and popular attention devoted to the rhythms and practices of financial services in London, New York and elsewhere.

The growing body of work on money and finance has grappled with a number of challenges including:

- the economic, social and cultural aspects of money
- the links between our everyday experiences of money and the workings of 'the global financial system'
- the changing geographies of money in these increasingly global times
- the creation of progressive, non-exclusionary forms of financial institutions in the contemporary world.

In dealing with these issues, geographers are not talking about 'the global financial system' as though it were a machine, but rather seeking to understand 'global finance' as a social, cultural and political phenomenon. Revealing the *social construction* of 'global finance' is important because the workings of global finance affect the lives and livelihoods of individuals and the predicaments facing governments. The constitution of 'global finance' affects who has access to credit, how, where and how much they pay for it and whether governments can afford welfare, education, healthcare and other services.

In the subsequent sections of this chapter, some of the above issues are explored in more detail. In the next section, it is argued that in addition to its important economic functions, it is important to recognize some of the different social and cultural aspects of money and how it is bound up with individual and group identities. The focus is broadened in the following section which considers the geographies of money that connect individuals and groups in distant places and at different times. By way of conclusion, the final section considers how some groups and regions are included or excluded from particular geographies of money, before exploring some of the ways in which we can think about progressive, inclusive monetary geographies.

13.2   Everyday worlds of money

Although we may be used to thinking of money as primarily an economic phenomenon, geographers and other social scientists are now recognizing that money is saturated with cultural, social and political significance (Leyshon and Thrift 1997, Dodd 1994). In this section, we examine some of the social and cultural aspects of money in different societies. We focus initially on money and power, before moving on to consider money and identity, money as a transformer or leveller and money as a mediator of our experience of social relations.

13.2.1 Money and power

Many of the social and cultural aspects of money are related to its role as 'the very incarnation of social power' (Harvey 1982: 245). Again, however, we confront the contradictory qualities of money. As a means of expressing social power money has both desirable and less desirable qualities. Leyshon and Thrift (1997) talk of two discourses about money, one of suspicion and one of liberation.

To understand the discourse of suspicion, we can turn to the work of Georg Simmel. For Simmel (1978: 277), money is 'the most perfect representation' of the tendency to reduce quality to quantity. As a store of value and unit of account money transforms social relations – qualitatively different commodities and experiences – into an abstract quantity, namely their price. This was what Marx (1973: 205) meant when he talked of money destroying communities and, in so doing, becoming 'the real community'. In contemporary capitalism, money has become *the* mediator and regulator of economic relations between individuals, a measure of wealth and a means of expressing social power. All manner of social, political and cultural issues – from healthcare, housing, education, leisure and sports, the environment, working conditions, and so forth – are debated not in terms of what they are worth, but in terms of *how much* they are worth, in terms of what can be afforded. Money 'affects our very ideals of what is good and beautiful and true' (Mitchell 1937: 371). Money, in this sense, has corroded the importance of other meanings and measures of 'value'.

Yet, although money can be treated with suspicion because it reduces qualities to quantities and introduces 'impersonality and colourlessness' into society (Simmel 1991: 19), there is also a discourse of liberation that accompanies money. First, money provides a degree of individuality, freedom, security and independence for those that hold it. This is especially true in societies like Britain where laws regarding private property buttress ideologies of liberty, equality and freedom. The idea that money is empowering and liberating is central to the desire to possess money, to work for it, to accept it as payment and to save and invest it. Second, as Simmel (1991: 20) argues, money 'creates an extremely strong bond among members of an economic circle'. Simmel argues that people have far more connections to other people in modern, monetized societies than was the case in feudal society. In the **division of labour** in capitalist societies (see Chapter 12, this volume), for example, money ties people together in offices, factories, homes and shopping malls in different parts of the world. Money may in some senses be corrosive of social bonds, but it also creates a community that can forge new connections and hold people together.

13.2.2 Money and identities

Thus far, the discussion about money and power has been rather abstract. One of the ways in which we can bring this discussion down to earth, so to speak, is to think about the role of money in the construction of many aspects of identity. If we think of identity not as something that is fixed or innate but as a 'capacity to keep a particular narrative going' (Giddens 1990: 54), then money can be an extremely important part of telling and maintaining stories about ourselves. In the remainder of this section, we consider four brief examples.

First, money is very obviously tied up with how individuals identify their *class* positions, what type of accommodation they live in, what type of schools they attend, and whether they can afford to stay on and pursue higher education and so forth. Being one of Britain's 'very rich' is associated with social and cultural networking practices which include particular (very expensive) rituals of conspicuous consumption, including balls and dinners and attendance at annual sporting events like Ascot and Henley. Such activities maintain relatively tight-knit networks, often linked by marriage. Within such moneyed networks, children can be sent to exclusive schools to mix with children from other wealthy families and be endowed with the social, if not financial, capital necessary to smooth their journey along their chosen future path.

Second, money stirs passions about nationhood, sovereignty and *national identities*. Examples abound. A glance through the *Oxford Dictionary of Current English* (1985: 737) suggests that 'sterling work' is derived from sterling, of or in British pound, of coin, precious metal, genuine, of standard value or purity (of person or qualities etc.), of solid worth, genuine, reliable. Another example was provided in June 1998, when Germans celebrated the 50th anniversary of the creation of the Deutschemark. Minted in the United States and shipped in crates for distribution in British, French and US occupation zones in 1948, the Deutschemark became synonymous with economic recovery and stability in Germany in the post-war era. It was, moreover, a national symbol of stability and success untarnished by any association with Hitler's National Socialism.

> It is safe to say that no other currency in the world enjoys quite the same sacred status in its country's iconography. The Almighty Dollar may be a pillar of US super-powerdom and the pound sterling is redolent of Empire. But only in Germany is the coin of the realm the country's veritable national symbol, more exalted than either the national flag or the splayed black German eagle' (Bonfante 1998: 25)

Plans for the implementation of the Euro (€) have therefore had to deal not only with concerns about whether it will be an economic success, but also with issues of sovereignty and national identity that are bound up with different national currencies.

Third, just as money can be a powerful symbol of class or national identity, it is also integral to the performance of *gender identities*. By way of example, we can think of how central the role of 'the male breadwinner' is to the construction of many variants of masculinity. The *Guardian* newspaper recently reported the results of some research into changing expectations of the role of fathers. Based on interviews in Rochdale in Greater Manchester, the research found that men, women and children felt that being a 'provider' was the main role for fathers, even in circumstances where women in the same household were working and, in some cases, earning more than the men (Brindle 1999). Feelings of guilt and of not contributing to the family unit were attributed to men who were unemployed or working in a low-paid job. Films like *The Full Monty* and *Roger and Me* illustrate only too well that loss of employment in the car, steel and mining industries is not simply an economic catastrophe for the men involved. For those whose masculinity is rooted in 'industrial work' and the status of a 'breadwinner', the loss of regular employment and earnings can entail wrenching changes in social status, self-esteem and relations with partners and children.

Fourth and finally, another form of identity which is very directly linked to money, and in many societies assuming increasing importance, is the *virtual financial identity* individuals (and indeed countries) are assigned by credit-scoring companies that vet applications for credit. Those who seek credit in the form of credit cards, mortgages and other loans are the subject of 'credit checks' which assess a wide range of factors concerning employment status, incomes, expenditures, outstanding debts, repayment histories and so forth. In the data banks of major credit-scoring companies, virtual financial identities are being constructed to be shared with other financial institutions. It is on the basis of these virtual identities that decisions are taken about an individual's creditworthiness. Credit-scoring companies are not the only groups attempting to construct virtual identities. Many organizations are now in the business of either constructing virtual identities of their customers (for example supermarkets in Britain use 'loyalty cards' as a means of tracking buying habits) or else buying customer information from third-party providers. The aim is to use customer information to extract 'actionable consumer intelligence' (Gandy Jr. 1995), i.e. to inform decisions about targeting marketing initiatives towards those socio-demographic groups which will generate the greatest profits for the lowest marketing related expenses.

13.2.3 Money talks . . .

Money, as an incarnation of social power, can be a language that creates new links between groups of people. As such, money can be a force for making people more tolerant of difference. For example, the importance of Chinese investment in Birmingham, London, Manchester and other British cities is now being recognized and they are seeking to embrace ethnic and racial diversity as part of their economic development strategies. In many British cities, 'Chinese Quarters' and other distinctive neighbourhoods associated with minority ethnicities are taking shape. Similarly, the 'pink pound', the earning, investing and purchasing power of gay consumers, is now being courted by financial services providers, film and music media, travel firms, airlines, health and fitness firms, restaurants and bars. Surveys in Britain, the United States and Australia characterize gay consumers as groups that are relatively affluent, white, active, financially literate, and likely to travel overseas and eat out regularly (Anon. 1999). Although this stereotype is not representative of gay couples on low incomes or with children, the economic significance of pink pounds and dollars – already acknowledged in cities like London, San Francisco, Toronto and Sydney – is being recognized in a growing number of places. Recently, for example, the Business Section of the Johannesburg *Mail and Guardian* (Anon. 1999) investigated the racial and class divisions that characterize 'the pink rand' now sought after by firms in South Africa and elsewhere.

With its ability to forge new associations between individuals and groups, money is also, potentially, a leveller. Money can disrupt traditional cleavages along the lines of class, gender, ethnicity and sexuality. Pop stars, actors, footballers, tennis players and other high

Plate 13.2   Work and identity: mining is not simply a means of earning a living; it is also a source of status and camaraderie for the men involved. (Format Photographers/Sheila Gray)

earners can join the ranks of the financial elite, regardless of their class backgrounds, their ethnicity, where they went to school and so forth. The realization that 'money talks', that it can be transformative, a bridge to a different life (McLuhan 1964) is integral to the appeal of the forms of gambling like lotteries. Research in the USA (which has a long history of lotteries) indicates that spending on lotteries is not distributed evenly throughout the population. Those committing the highest proportions of their incomes on gambling tend to be low-income groups with the lowest levels of educational attainment (Jackson 1994, Brown *et al.* 1992). Lotteries and casino gambling are now very big business in Europe, North and South America, Australia, Hong Kong and, more recently, China. Global lottery revenues rose 50 per cent (to US$100 billion) between 1990 and 1995 (Emerson 1995) and are now a major source of revenue for governments, gaming companies and tourism-related companies. In Taiwan, where lotteries are banned, gangs net an estimated £10 billion a year betting on Hong Kong's lottery numbers (Emerson 1995).

13.2.4 Money as a mediator of social relations

The idea of money being a language that can connect and liberate different individuals and groups suggests that money can, in all sorts of ways, mediate our experience of social relations. By way of example, we can develop our previous discussion of gender identi-

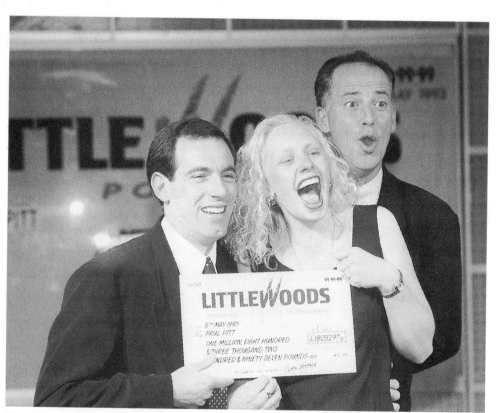

Plate 13.3   Money changing lives? The possibility of a big win, however remote, helps to fuel various forms of gambling. (Topham Picturepoint/AP)

ties and think about how money, its organization and uses are structured in ways that are gendered. Viviana Zelizer (1989), for example, suggests that how monies are exchanged within households is usually subject to a very different, and usually gendered, set of rules than those that govern the use of money in the market place. Zelizer (1989: 351) argues that social structures

> systematically constrain and shape (a) the *uses* of money, earmarking, for instance, certain monies for specified purposes; (b) the *users* of money, designating different people to handle specified monies; (c) the *allocation* system of each particular money; (d) the *control* of different monies; and (e) the *sources* of money, linking sources to specified uses.

Research has shown that men and women in Britain tend to be responsible for different forms of expenditure. Women are primarily responsible for expenditures on clothing, school-related expenses, food and charity donations and men responsible for expenditure on trips and holidays, repairs, redecorating, car maintenance and so forth (Pahl 1989).

Gendered assumptions about earnings and uses of money are also writ large in financial services industries. In Britain, not only are financial spaces like the City of London 'an elitist and masculinist environment' (McDowell 1997: 69), but many financial products target 'the male breadwinner' and are designed around gendered assumptions and stereotypes about work histories and sensitivity to risk. Pension products, for example, are designed to privilege those clients (most usually men) who work full-time and invest premiums regularly, and without interruption, over a forty-year period (Knights and Odih 1995). Similarly, Ravenhall (1999), citing a survey by *Money Magazine*, argues that male financial advisers treat men and women differently, with women being steered towards so-called 'widows and orphans investments', i.e. relatively conservative products with lower risks, and potential returns, than other investments. In Bangladesh, by contrast, it is often women who are the clients of banks, who receive credit and organize themselves and others into village groups to oversee the repayment of debt. Later in this chapter we will examine the establishment of Grameen Bank in Bangladesh. Its founder, Mohammad Yunus, established the bank believing that conventional banks were anti-poor and anti-women and insisted on collateral before making loans. Ninety-four per cent of Grameen Bank's clients are women. Women are regarded as better borrowers because 'they had most to lose, were more careful, had longer vision and were more concerned with the family than men' (Caulkin 1998: 2). In different ways in different societies, then, money can mediate the experience of, in this example, gender relations. Households and financial institutions can operate to reinforce or challenge prevailing assumptions and stereotypes concerning women and money.

In summary, money has cultural, political, social and not just economic significance and this is reflected in the negative and positive discourses about the power of money. Money, as an 'incarnation of social power' is integral to the construction and performance of different identities. It is a 'leveller', a language for linking different individuals and groups, and it mediates our experiences of social relations.

13.3 Geographies of monies

How can we understand the links between some of our everyday experiences of money, described in Section 13.2, and the workings of so-called 'global finance'? This section broadens the focus to consider the geographies of money that connect individuals and groups in distant places and at different times. In these global times, some would argue that geography is relatively less important. Below, we examine some of the arguments concerning the globalization of money and finance, before outlining some of the problems with this view. The section concludes by introducing the concept of *monetary networks* as a way of thinking about money and its 'global' and 'local' qualities.

13.3.1 Global monies?

Many geographers, particularly those working in a **political economy** tradition, have argued that capitalism seems to be speeding up and spreading out (Massey 1994, Harvey 1989). Capitalism is seen to be spreading out in that more and more individuals in different countries and regions are becoming bound up with the logistics of contemporary capitalism. It is also speeding up in that the pace of life seems to be increasing in different parts of the globe. In the second section, we noted the ability of money to transform qualitatively different commodities and experiences into an abstract quantity, a price. This has led many theorists to argue that money is a vehicle for the *homogenization of space*, i.e. a vehicle for making different spaces more similar. While there are ongoing debates about the extent to which capitalism is generating a *global culture* (see Chapter 6), it is difficult to argue with the contention that financial markets in different countries have, since the mid-1970s, become more interconnected (Harvey 1989, Martin 1994). Financial space has become more homogeneous.

When we talk about the *globalization of finance*, however, we are talking about more than just the growth of international financial transactions, or the growing presence of multinational companies in domestic financial markets. Globalization implies

> a strong degree of *integration* between the different national and multinational parts ... the emergence of truly transnational banks and financial companies ... that integrate their activities and transactions across different national markets. And above all, it [globalization] refers to the increasing freedom of movement, transfer and tradability of monies and finance capital across the globe, in effect integrating national markets into a new supranational system. (Martin 1994: 256)

The emergence of this *supranational* system coincides with the increasing difficulties faced by *nationally* based regulatory authorities like the Bank of England or the US Federal Reserve. The 'discourse of suspicion' we mentioned in Section 13.2 rears its head in many popular representations of 'global finance':

> [T]he world is now ruled by a global financial casino staffed by faceless bankers and hedge fund speculators who operate with a herd mentality in the shadowy world of global finance. Each day they move more than two trillion dollars around the world in search of quick profits and safe havens, sending exchange rates and stock markets into wild gyrations wholly unrelated to any underlying economic reality. With abandon they make and break national economies, buy and sell corporations and hold the most powerful politicians hostage to their interests. (Korten 1998: 4)

Whether or not you agree with Korten's view, it is true that the degree of integration of the financial system profoundly shapes and connects the lives of people thousands of miles away from each other. In this sense, Leyshon (1995b) talks of money being able to 'shrink' space and time, bringing some (but not other) parts of the globe relatively closer together through the working of financial markets. A good example of this is considered in Box 13.2, which describes how the fates of property developers in Bangkok (Thailand) ultimately affected the fortunes of a bartender, Graham Jones, in Whitley Bay (UK). The fates of individuals and companies in Britain and Thailand are connected through the workings of foreign exchange markets which link rises and falls in national currencies to the competitiveness of different nation-states.

This example highlights only some of the interconnections between changes in the value of the Thai baht and how these affected a range of businesses across Asia and Europe. We could extend the example by thinking through how redundancies in the north-east of Britain would also have been felt by those not in paid work. For those in households dependent on the wages of someone made redundant from cleaning firms, like Mitie, there would be further belt-tightening or the search for alternative sources of income. For the unemployed, more redundancies in the area might mean more competition for new jobs that are created and so forth.

13.3.2 Forces for globalization

How has this globalization of money come about? There is a range of factors to consider here. First, through the 1980s and 1990s, different governments and international institutions, like the International Monetary Fund (IMF), have pursued neoliberal, 'free market' policies and encouraged the deregulation of financial markets (by eliminating exchange and capital controls) and the liberalization of flows of capital across national borders.

Second, there have been important advances in telecommunications and computing technologies, advances that can be summarized under three headings: computers, chips and satellites (Strange 1999). Computers have transformed payments systems. For many hundreds of years, payments and transfers of money were completed in cash – coins and notes – and written down in ledgers. Coins and notes were superseded by cheques which in turn have now been superseded by electronic monies that can be moved around the world at the speed of light. Semiconductor chips in computers mean that consumers in many countries can use credit and debit cards to pay for goods. More recently, computer chips are being embedded in plastic 'smart' cards to allow the use of 'digital money' to pay for groceries, telephone use and so forth. Finally, there are the systems of communication, using earth-orbiting satellites, that are integral to the operation of computing systems, e-mail, the Internet and other forms of communication that are used by banks, governments and other financial players. These innovations have made 24-hour trading possible as stock markets in different countries are linked by computer. The average amount of currency bought and sold by private currency trades now exceeds US$1 trillion *per day*, more than the assets held by all central banks (Porter and Sheppard 1998).

Third, there have been innovations in tradable financial products that have made it easier and faster to move money around the globe. One example of such an innovation is that of 'derivatives' (Box 13.3).

These, then, are some of the ways in which money has become more global since the 1970s. But what motivates such changes? What motivates the implementation of 'free market' policies, the development of new technologies and financial instruments like

Box 13.2　From Bangkok to Whitley Bay ...

1　Devaluation in Thailand, July 1997

In 1997, several property companies collapsed in Thailand; property prices and the stock market started to fall. Currency speculators, already nervous about slowing growth in the region, started to sell the baht (Thai currency) as they expected the currency to be devalued. In July, the baht was devalued. As a result, Thai exports became cheaper and, to stay competitive, Indonesia, Malaysia, South Korea and the Philippines allowed their currencies to fall sharply. Some Korean firms who needed foreign currency to pay off loans started dumping cheap microchips, forcing microchip prices down from US$10 a unit to US$1.50.

2　Factory closure in North Tyneside, UK, 31 July 1997

Rapidly falling semiconductor prices meant losses of £350 million for German electronics company Siemens. Siemens semiconductors were produced at plants in Tyneside (UK), Taiwan, Germany, France and the US. Managers in Munich announced plans to close their £1.1 billion plant in North Tyneside that employed 1,100 people and had only opened in 1994.

3　The cleaning company

The Siemens factory provided 10 per cent of business in the North East for Mitie, a cleaning company. One third of the workforce of 90 people were facing redundancy and their boss had his salary bonus cut 10 per cent, in line with the loss of work from Siemens.

4　The hotel

Siemens and its contractors, like Mitie, were the single biggest sources of business for the Stakis Hotel. The hotel responded to the closure by switching their market focus. As the flow of German executives and their UK contractors slowed and then ceased, the hotel sought to attract more families.

5　The taxi firm

Foxhunters, a taxi firm, had a contract with Siemens which generated more than a dozen runs a day, usually to the Stakis Hotel, the airport or the university. Since August, Siemens business had dried up and 85 drivers were chasing work for 50. Despite working longer hours, drivers' takings were down by between £100 and £200 a week. Drivers started to economize by bringing in their own lunches and by cutting down their trips to the local pub.

6　The local pub

Takings at Cameron's had fallen by £600 a week since August. The landlord, who blamed the Siemens shutdown for the reduced trade, cut his opening hours. Attempts to drum up more trade by price reductions had little effect. The landlord was hoping for more business around Christmas time. He and his partner had less money for spending on their leisure activities, which included visiting places like Whitley Bay.

7　Graham Jones, bartender in Whitley Bay, October 1998

Graham was fired from his job at a pub in Whitley Bay; there was not enough business to occupy two bartenders in the public bar.

Source: Adapted from Carroll 1998

derivatives? For those working in a *political economy* tradition, like David Harvey (1989), the motivation for these changes is the search for profit. Capitalism is fundamentally about the accumulation of surplus. Competition drives capitalists to seek out new markets, new products and to reduce the *turnover time of capital* (see Box 13.4). In different parts of the globe, money is the language through which the imperatives of capitalism are being communicated.

So, there is a very strong economic rationale for the globalization of money. And it is difficult to argue with the contention that money has become more globalized since the 1970s, that it has, increasingly, connected people in distant places and homogenized financial space. For Richard O'Brien (1992), the growth of the international financial system is tantamount to 'the end of geography', the notion that geography is becoming relatively less important as money, in its different forms, is able to overcome the friction of distance and link distant places together. Others argue that this view is too simple, that geography remains critical to our understanding of global finance. In the next part of this section, we will consider these views in more detail.

13.3.4 Debunking 'global' monies

A rather different departure point for talking about the 'globalization of finance' is to argue that the financial system is really not all that global, that it is more like a *web of connections* between different financial systems, some of which are bound together more tightly than others. This argument develops from noting a number of problems with the notion of a 'global financial system' and global monies.

The first problem stems from the simple observation that 'global finance' is largely the

---

Box 13.3 | **Derivatives**

Derivatives are 'contracts specifying rights and obligations which are based upon, and thus derive their value from, the performance of some underlying instrument, currency, commodity or service index, right or rate' (Cornford, quoted in Strange 1999: 30). Derivatives are essentially financial instruments used to manage risk. They were originally used for *hedging*, i.e. protecting against a change in price. Some of the most common forms of derivatives are 'futures', 'options' and 'swaps'. A futures contract is an agreement to buy/sell a specific commodity or financial instrument at a particular price at a defined location on a stipulated future date.

Derivatives used to be used predominantly in commodities markets. Farmers, for example, would agree to the deferred delivery of crops or other goods and guarantee a particular price. As financial markets have become increasingly volatile, however, derivatives have been extended to more financial instruments and become increasingly sophisticated. Since 1989, trading in derivatives has grown at an average of 40 per cent per year (IMF 1996). As of March 1995, the value of outstanding derivative contacts in 26 countries surveyed by the Bank of International Settlements (BIS) was US$47.5 trillion, twice the value of world economic output (Strange 1999). The complexity and speed of derivatives markets and their relative lack of regulation has worried regulatory authorities. Speculators trade in derivatives contracts and have been involved in some high-profile losses, including those at Barings Bank in London (Tickell 1996).

Box 13.4

The circulation of money as capital

In capitalist societies, the circulation of money as capital is as follows. From left to right in the equation, money (M) is invested by producers to purchase commodities (C), namely labour power (LP) and the means of production (MP), say pieces of wood and wood cutting machinery. Labour power and the means of production are combined in production (P) to make more commodities (C'), in this example, let us say chairs, which are then sold for more money (M') than was originally invested (M).

$$M \rightarrow C \rightarrow \begin{Bmatrix} LP \\ MP \end{Bmatrix} \rightarrow P \rightarrow C' \rightarrow M' \rightarrow \ldots$$

The purpose of production in capitalist societies is to produce profit and accumulate capital. The *turnover time of capital* is the amount of time it takes for money to complete this circuit. The shorter the turnover time, the more often money can be lent out and the more profit can be made. Producers therefore have a very strong incentive to, where possible, reduce the turnover time of capital; time *is* money.

province of North America, Europe and parts of Asia, most notably Japan, Hong Kong and Singapore; it is an idea centred on the experiences of the West and sees financial markets being globalized alongside capitalism (Figure 13.1). Many parts of Africa, Asia and Latin America are not only not included in 'the global financial system', they are being actively excluded as banks refuse to lend money until outstanding debts have been repaid. There is what Massey (1993) describes as a *power geometry* at work then when some commentators describe the financial system as 'global'. For some financial workers in New York, London and Tokyo, international finance may be regarded as 'global' in the sense that all countries and regions of the globe deemed creditworthy and capable of producing profits have been included; sub-Saharan Africa ceases to exist in such a conception of 'the global'.

Second, and resulting from this geographical concentration of the management of 'global finance' in North America, Europe and parts of Asia, some currencies circulate more widely and have a greater spatial reach than others. The yen, the euro and most especially the US dollar are very useful in international markets because they are accepted as forms of payment, unlike, for example, Indian rupees.

Historically, the country that occupies a dominant economic and political position has underwritten the soundness of the international financial system and had its currency accepted internationally as the currency in which commodity prices are quoted and in which payments are made. Before the Second World War, Britain and the pound sterling fulfilled this role; after the Bretton Woods conference in 1944, the US dollar became the key international currency. More recently, as the economic dominance of the US has declined, the yen has become relatively more important in international markets. Other countries that do not use the US dollar, the yen or the euro domestically have to convert their currency to, say, US dollars in order to facilitate international trade.

So, some currencies, like the US dollar, are truly international while some others are national. There are also over 2,000 *local currencies* in operation around the globe, currencies that facilitate exchange only within very specific spatial and social contexts. Local

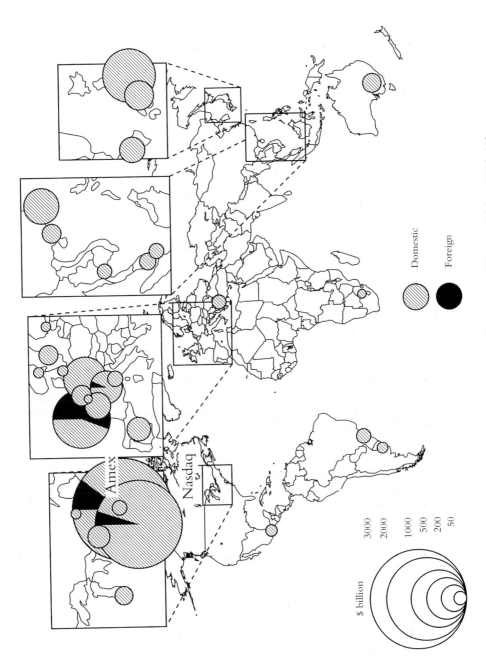

Figure 13.1 Trading values of equities, by major exchange, 1995. *Source:* Adapted from Laulajainen, 1998: 149

Exchange Trading Systems (LETS), for example, are associations whose members list their services needed/offered in a directory. Members of LETS then trade with each other in a local unit of currency, for example, 'bobbins' in Manchester and 'solents' in Southampton (Williams 1996). Since their establishment in Canada in 1983, LETS schemes have spread to Europe, Australia, New Zealand and North America, allowing members not only access to credit, but also the chance to engage in productive activity to earn such credit (Williams 1996). Some local currency schemes are devised in times of hardship to allow local people to trade goods and services when they are unemployed and have little money to spend. Other schemes are motivated by ecological concerns, the desire for community development and social cohesion, and the desire to construct alternative local economic geographies as a form of resistance against the global gaze of capitalism (Lee 1999).

Third, as Martin (1999: 6) argues, for all the talk of 'global finance', there remain different *geographical circuits of money* which form the 'wiring' of an economy, along which 'currents' of wealth, consumption and power are conveyed. Martin (1999) identifies four geographies of money: locational, institutional, regulatory and public. The *locational geography* refers to the location of different financial institutions and markets. Financial institutions and specialized functions (for example, stock markets and foreign exchange markets) tend to be agglomerated in large urban centres, with London, New York and Tokyo sitting at the top of the global hierarchy. Different countries also have different *institutional geographies*, in that they organize their financial institutions and markets in distinct ways. In Britain, for instance, we are used to thinking of the 'big four' retail banks – NatWest, Lloyds-TSB, HSBC-Midland and Barclays – and relatively new banks like the Halifax, Woolwich and others that have converted from building societies. In the USA, by contrast, there were over 9,000 Federally insured commercial banks in 1997 (Pollard 1999). Meanwhile, in parts of Asia and west and south Africa, banks and other financial institutions like Rotating Savings and Credit Associations (ROSCAs) (see Box 13.5) are important in organizing and funding economic activity. ROSCAs take different forms in different regions and vary with the class, gender and ethnicity of their members (Ardener and Burman 1995).

These varied institutional geographies are products, in turn, of contrasting *regulatory geographies*. There are a wide range of supranational, national and regional regulatory spaces that govern the workings of financial institutions and markets. To return to the example of how the Asian financial crisis cost Graham Jones his job in Whitley Bay, a key player in how the crisis unfolded was the International Monetary Fund (IMF). The Chief Executive of Siemens, Heinrich von Pierer, argued that Korea was able to sustain a 'suicidal pricing strategy' for its memory chips (a pricing strategy which led to losses of £350 million for Siemens) because it was receiving financial aid from the IMF (Milner 1998). To give another example, as Martin (1999: 9) observes,

> [M]oney has had a habit of seeking out geographical discontinuities and gaps in these regulatory spaces, escaping to places where the movement of financial assets is less constrained, where official scrutiny into financial dealing and affairs is minimal.

Offshore financial centres like the Bahamas, Cayman Islands, Jersey and Guernsey are attractive because of their low tax rates and minimal regulation. In 1990, an estimated 50 per cent of the world's money stock either resided in or passed through a tax haven (Laulajainen 1998).

Finally, when Martin (1999) discusses *public geographies*, he refers to the role of states in

| Box 13.5 | Rotating Savings and Credit Associations (ROSCAs) |

A ROSCA is 'an association formed upon a core of participants who agree to make regular contributions to a fund which is given, in whole or in part, to each contributor in turn' (Ardener 1995: 1). ROSCAs are known by different names, depending on their form and scale, the social classes of their members and their location. In South India they are known as *kuris*, *chitties* or *chit* funds, in Cameroon as *njangis* or *tontines*.

ROSCAs were well developed in China, India, Vietnam and parts of West Africa by the end of the nineteenth century. Variants of ROSCAs also existed in Scotland and parts of northern England. Members, usually ranging in number from a few to several hundred, make regular contributions, in cash or in kind, to a fund. The fund, or part of it, is then given to each member in turn. The order in which the fund is given out is determined in a variety of ways; lottery, age, kinship seniority, or by rules established by the organizer.

In addition to encouraging regular savings and providing small-scale capital and credit for their members, many ROSCAs have very strong moral and social dimensions. Trust, social solidarity and responsibility are emphasized and members have a strong interest in ensuring that no member defaults on their regular payment. Some ROSCAs are women-only and their potential to empower women, by giving them greater control over income and credit, has attracted the attention of anthropologists, sociologists and feminists (see Ardener and Burman 1995).

distributing monies across regions in the form of goods and services, infrastructure, health, education and so forth, and in transferring monies in the form of various social and welfare programmes.

> What becomes clear from this discussion is that the world remains made up of a patchwork of different financial spaces or systems, and although there has been considerable integration of financial practices and processes across such systems in recent years, we are a long way from the seamless global financial space described by some commentators. (Leyshon 1996: 62–3)

13.4 From 'global monies' to monetary networks

The phrase 'global finance' therefore seems to be something of a misnomer. While it is true that different monies and markets have become more interconnected since the 1970s, what we have been describing as 'the global financial system' seems to comprise some elements that are international in their reach, juxtaposed with other local and national currencies, regulations and institutional geographies. One way of juggling with this complexity is to think about the idea of *monetary networks*.

Recall that in the introduction to this chapter 'the global financial system' was described as a social, cultural and political *network of relations* in which individuals have different capacities to participate. Martin (1999: 11), for example, describes financial markets as 'structured networks of social relations, interactions and dependencies – they are communities of actors and agents with shared interests, values and rules of behaviour, trust, co-operation and competition.'

When we talk about 'structured networks' what do we mean? Dodd (1994: xxiv) argues that monetary networks are, above all, *networks of information* that have five abstract properties (Box 13.6). These networks are formed by the production, interpretation and circulation of information between actors through time and space. Thus we can think of foreign exchange traders in London, Tokyo, New York and elsewhere constituting a network of relations through which foreign exchange trading is organized, regulated, understood and carried out.

How do networks actually work? How are all these different forms of information, that are monetary networks, produced and then moved around? Nigel Thrift (1996, 1998) has argued that the forms of information that constitute monetary networks are always ingrained in *practices*. To give an example of what he means, we can return to the workings of global telecommunications networks that we discussed earlier. Telecommunication networks rely on the workings of thousands of workers and machines to install, maintain, repair and operate them. Similarly, the Internet consists of thousands of miles of fibre-optic cables, most usually laid alongside existing transport infrastructure such as railway tracks, and computing hardware and software linked together, managed and sustained by thousands of programmers, engineers, users and machines (Bray and Kilian 1999). Although the 'global financial system' may sometimes appear to be a rather intangible 'thing', it is really an assemblage of consumers, workers, computers, telephones, office buildings, bits of paper, financial reports and so forth. There is simply no escaping the *materiality* (i.e. the bits of cable, bank branches, computers, telephones, desks, trading floors and so forth) and *practices* (sitting at trading desks using telephones/computers, paying for goods with particular coins, credit cards and so forth) that constitute these networks.

When we talk of 'the global financial system', then, we are really talking about combinations of human beings, technologies and documents which come together to sustain, in particular spaces and times, the flows of information that constitute different monetary net-

---

Box 13.6   Properties of monetary networks

- Networks contain a *standardized accounting system* into which each monetary form within the network is divisible. Thus any goods, priced in terms of this standardized accounting system, are exchangeable.

- Networks rely on *information regarding expectations of the future*; 'money is accepted as payment almost solely on the assumption that it can be re-used later on' (Dodd 1994: xxiv).

- Networks rely on *information regarding their spatial extent*; institutional frameworks will limit the territories in which specific monetary forms may be used.

- Networks are based on *legal information*, usually in the form of rules which govern forms of contractual relations of network members.

- Networks presuppose *knowledges of the behaviour and expectations of others*. In order to trust money, those holding it need to anticipate, and trust, that it can be used and re-used within the network.

Source: Dodd 1994: xxiv

works. Note the use of the plural. Rather than thinking of one 'global financial system' we can think instead of there being a *multitude of monetary networks that overlap and intersect*. This is an important point, because different monetary networks, as we have seen, have very different geographies. Some monetary networks, like those of the foreign exchange markets, have the capacity to connect people and machines in London, New York and Tokyo at the touch of a button. Other monetary networks, like LETS schemes in Manchester, are much more compact and connect people over much shorter distances. Rather than thinking about networks as being either 'global' or 'local', Bruno Latour (1993: 122) argues that we should think of networks as being 'more or less long and more or less connected'.

Monetary networks differ not only in the *length of connections* they are capable of sustaining, but also in the *speed of connections* they are capable of sustaining. We have talked already of the computing networks that facilitate foreign exchange trading and allow banks to move monies around the world in, usually, less than 25 seconds (Warf 1999). One of the attractions of ROSCAs (Box 13.5) for men and women in Ghana and elsewhere is the speed at which news of hardship or an emergency can be spread, and the order of rotation of the fund changed, to help out a member in trouble. ROSCAs may not have quite the speed of connection of foreign exchange trading (assuming, of course, that all the computers are working), but they can react with a speed rarely matched by local banks which are often perceived as distant, impersonal and slow (Ardener 1995).

In summary, financial markets have become increasingly globalized since the 1970s, although truly 'global' markets are concentrated in a select number of major financial centres. There remain distinct regional, national and supranational geographies of money, such that the 'global financial system' can be conceptualized as a web of intersecting networks (Thrift 1996). If we conceive of 'the global financial system' as an intersecting web of networks – some fast, some slow, some long, some short – then we can envisage a whole host of networks that, to differing degrees, connect different geographies of employment, technology, regulation, government policies, leisure trends and so forth. Such a conceptualization provides a way of thinking about how, over a period of fifteen months, the fates of property developers in Thailand became intertwined with foreign exchange markets, semiconductor companies, cleaning companies, hotels, taxi ranks, bars and the loss of a job for Graham Jones in Whitley Bay. This approach also reinforces the point that when we talk about 'global finance', we are not talking about a machine operating 'out there', but a socially, economically, culturally and politically constructed network of relations with which we engage, in some way, shape or form, on a daily basis.

13.5 Conclusion: money, space and power

From the preceding discussion, it should be clear that many monetary networks are very exclusive. Different monetary networks are not open to all who would like to participate and there are some very distinctive *geographies of financial inclusion and exclusion* that have already been mentioned. The benefits of some of the most extensive, fastest monetary networks are, for the most part, extended only to those areas where individuals, companies and governments can produce a profit for the major lenders in North America, Japan and Europe. Moreover access to such networks is carefully screened. Countries, firms, banks and public agencies, like individuals, are credit-rated. On the basis of their credit rating, decisions are made by major banks on whether to offer credit and, if so, for how long, and at what rate of interest. The main international agencies are Moody's Investment Services, Standard and Poor and, in insurance markets, A.M. Best, all US firms (Laulajainen 1998). The Securities and Exchange Commission (SEC) in the US recognized only one non-US rating agency, Ibca, in Great Britain, until 1994 (Laulajainen 1998). Not surprisingly, only the major capitalist economies are assigned high ratings when it comes to government-issued debt in the form of bonds (Figure 13.2). Standard and Poor's so-called 'Investment grade' scales rate types of debt on a scale from 'AAA' ('Superior') through 'A' ('Good') to 'BBB'. Below this are the 'Speculative' grades which range from 'BB' ('Questionable') through to 'D' ('Default').

This is not to say that only the major capitalist economies are able to secure credit. In the late 1970s and early 1980s, total lending to Africa, Asia and most especially Latin America increased sharply. Yet rising interest rates, falling commodity prices and a U-turn in US policy that resulted in rapid appreciation of the value of the US dollar left debtor countries increasingly unable to cope with debt repayments. Affairs came to a head in August 1982, when Mexico suspended its debt repayments, soon to be followed by Brazil, Argentina, Peru, Venezuela, Ecuador and more than twenty other countries by the end of 1983 (Porter and Sheppard 1998). The response of banks to the debt crisis was to drastically reduce their lending to these countries, to effectively exclude them from international bank lending networks.

More recently, investors in North America and Western Europe have been encouraged by fund managers and financial media to invest in so-called 'emerging markets', which include parts of the Russian Federation and former Soviet Republics and other parts of Asia, Latin America and Africa. Again, the lure for investors is the prospect, real or imagined, of higher rates of return than those that can be secured in what are represented to be the more conservative, less volatile, less 'adventurous' confines of Western European and North American equity markets (see Sidaway and Pryke 2000). Some financial markets, as Susan Strange (1996) argued, have become rather like casinos.

The concentration of wealth and power in the major capitalist economies, most especially the US, is such that they hold sway, economically and politically, over the networks through which international lending is conducted and its terms negotiated. Even supranational agencies like the World Bank and the IMF, designed and established after the Second World War to avoid biases in simple bilateral lending, rely on funds from major capitalist countries who exert considerable influence over where the funds go and how they are spent. If South Africa and Russia and other post-socialist economies want to be included in the monetary networks of the World Bank and the IMF, then instituting reform packages based around market-led growth, a reduction in government intervention in the economy and a general mimicking of the political and economic organization of major capitalist economies is the price of admission. Such structural adjustment packages have spread widely through east and west Africa, Latin America and Asia since 1982 (Porter and Sheppard 1998). For those that do not toe the line, life can be made very difficult. On 1 September 1998, in an effort to staunch an outflow of funds, the Malaysian government imposed a freeze on trading its currency, the ringgit, a move not welcomed by currency speculators. Standard and Poor responded by downgrading Malaysia's long-term debt to just above junk status (making it harder and more expensive for Malaysia to obtain credit) and US mutual fund managers pulled out of the country (Greenwald 1998). Money, as we saw in Section 13.2, is a very powerful tool of social and political control.

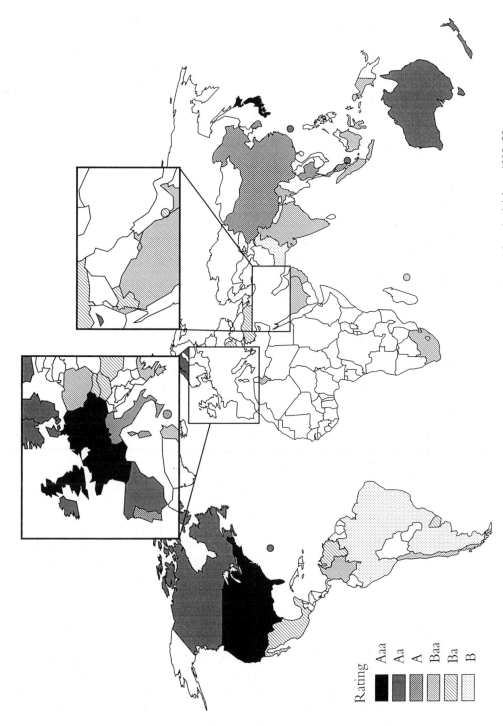

Figure 13.2   Sovereign risk of bonds denominated in a foreign currency, May 1996. *Source:* Adapted from Laulajainen, 1998: 56

Such geographies of inclusion and exclusion operate within countries too as banks, private investors and other lenders look for prosperous growth regions and secure rates of return. Pension fund managers, for example, use savings from all over the UK to invest, disproportionately, in firms located in the prosperous south-east of England (Martin and Minns 1995). In Los Angeles, in patterns repeated in many other US cities, major retail banks have been closing bank branches in low-income neighbourhoods like South Central Los Angeles and opening new branches in prosperous areas like Beverly Hills (Pollard 1996). The financial institutions that tend to move into South Central are so-called 'alternative' financial institutions like pawnbrokers and cheque cashing establishments, which often charge very high fees for basic banking services.

Financial exclusion, whatever its spatial extent, tends to reinforce existing patterns of wealth, poverty and inequality. Those individuals and nations that already have wealth are deemed the most creditworthy and are offered the cheapest forms of credit to expand businesses, pay off debts, buy new goods and services and so forth. For low-income groups and countries still reeling from crippling debt repayment schedules, credit is difficult to find and often very expensive.

13.5.1 Some responses and alternatives to financial exclusion?

There are many, varied responses to forms of financial exclusion. Although some alternative monetary networks, like those of cheque-cashing shops, may be exploitative and expensive to use, there are plenty of examples of other monetary networks that are designed to be inexpensive and inclusive. One very famous example is the Grameen Bank, formed in 1983, which now employs over 14,000 people in 35,000 villages in Bangladesh.

The successes of the Grameen Bank and other micro-credit institutions in Indonesia, Bolivia, Kenya and elsewhere have attracted a great deal of attention and spawned innumerable imitations and different varieties of micro-credit schemes in Africa, Asia, Latin America and elsewhere. Indeed, the prospect of linking access to credit for the poorest with job creation has made micro-finance schemes increasingly popular in Britain, the US and Western Europe, and not just in those regions of the UK and the US that are experiencing financial exclusion.

It is also important to note that some 'alternative' financial networks may be preferred by individuals and households who do not want to use financial institutions like traditional retail banks and instead want to participate in networks that encourage broader social, environmental and other assessments of 'value'. The growing popularity of various micro-credit schemes, like the LETS and ROSCA schemes we discussed earlier, is based not only on their economic successes, but also on their status as 'alternative' and, potentially, progressive schemes. One of the interesting features of micro-credit schemes, like the Grameen Bank, is their often strong emphasis on creating social bonds and a sense of shared responsibility between members; these are schemes in which members recognize other measures of 'value' besides monetary ones. Another feature of the many experiments with microfinance is that it is women who make up the bulk of the membership of many programmes. As Goetz and Gupta (1996) point out, however, access to credit is not necessarily synonymous with control over the money and, more generally, empowerment for women in environments where they frequently face extreme constraints on their lives. As we mentioned in Section 13.2, how monies are exchanged within households is usually subject to a very different set of rules than those that govern the use of money in the market place (Zelizer 1989).

Box 13.7 Grameen Bank

The basic principles of Grameen Bank lending are as follows:

- the bank lends only to the poorest landless villagers in Bangladesh;
- the bank lends, overwhelmingly, to women; 94 per cent of its clients are women;
- loans are extended without borrowers needing collateral;
- the borrower, not the bank, decides what the loan will be used for;
- borrowers pay sufficient interest to keep the bank self-sustaining.

The Grameen Bank has lent out over US$1.5 billion since its establishment in 1983. The average loan is for about US$100. Rather than insisting on collateral or security, women wanting a loan join a village group, attend weekly meetings and assume responsibility for the loans of other group members. Thus, there is strong peer pressure for individual women to maintain payments, otherwise the group is unable to secure further loans. 97 per cent of Grameen Bank loans are repaid.

The success of Grameen Bank has spawned imitations in 58 countries in many parts of Africa, Asia and Latin and North America. Grameen has now branched out into fisheries, mobile telephones, energy and the Internet.

Sources: Bornstein 1996, Caulkin 1998
http://results.action.org/summaryp.html

To conclude, these examples of monetary networks illustrate only too well that money, how it is obtained, how it is used, how it is managed and by whom and what significance it has in society, is an economic, social, political, cultural and geographical phenomenon. Small regional banks, foreign exchange markets, ROSCAs, and a host of other monetary networks, some relatively slow, some relatively fast, make up the 'global financial system' which is, in many ways, becoming more interconnected, linking more people and diverse financial institutions and practices together in distant places. These monetary networks intersect most densely through institutions and practices in North America, Japan and Western Europe. This geopolitical reality structures who has access to

Learning outcomes

After reading this chapter, you should be aware that:

- Money has economic, cultural, social and political significance and mediates our experiences of social relations.
- The 'global financial system' is a socially constructed web of intersecting networks operating at different speeds and over different distances.
- There are distinct regional, national and supranational networks of money.
- There are many geographies of financial inclusion and exclusion.
- There are a range of financial networks which may be more inclusive and politically progressive than others.

credit and on what terms and, in turn, generates the search for all kinds of more inclusionary alternatives.

Further reading

Corbridge, S., Martin, R. and **Thrift, N.** (eds) (1994) *Money, Power and Space*, Blackwell, Oxford. This is one of the early geographical collections of writings on money and a good introduction to a broad range of themes.

Leyshon, A. and **Thrift, N.** (1997) *Money Space: Geographies of Monetary Transformation*, Routledge, London. This is a compilation of Leyshon and Thrift's essays on money.

Martin, R.L. (ed.) (1999) *Money and the Space Economy*, Wiley, Chichester. A wideranging collection of essays covering geographies of banking, financial centres, money and the local economy and money and the state.

McDowell, L. (1997) *Capital Culture: Gender at Work in the City*, Blackwell, Oxford. This is a more specialist book that examines the masculine cultures and practices of parts of the City of London.

Useful web-sites

http://www.ft.com   The *Financial Times* web-site contains useful links to financial and other firm, sector and country data, as well as breaking news stories.

http://cnnfn.cnn.com/   The US-based CNN financial web-site, like the *Financial Times* site, contains numerous links to business, financial and market news, by country, sector and firm.

http://www.worldbank.org/   The World Bank web-site is home to a host of information on World Bank publications, research, data and related organizations.

http://www.cib.org.uk/supersite   This site is maintained by the Chartered Institute of Bankers (CIB), with comprehensive links to a variety of banks, regulatory and professional organizations, country data and information about research, debates and technology in finance.

http://www.grameen-info.org/   The web-site of the Bangladesh-based Grameen Bank. It has information on the mission of Grameen Bank and how it works, together with material on other microcredit schemes.

http://adams.patriot.net/~bernkopf/   A site dedicated to central banking issues, and includes links to Central Banks and their research departments, Ministries of Finance, Bankers Associations and Institutes and Multilaterial financial institutions.

http://www.creativeinvest.com   The site of Creative Investment Research, a US-based firm specializing in 'socially responsible' investment and providing many links on women and minority-owned financial institutions.

CHAPTER 14

PHILIP CRANG

Worlds of consumption

> **Topics covered**
>
> ■ Consumption matters
>
> ■ Global geographies of consumption
>
> ■ Local geographies of consumption
>
> ■ Making connections

14.1   Consumption matters

We geographers rightly pride ourselves on how our subject addresses major global issues, looking for answers to 'big' questions about the character and sustainability of human and non-human lives in the past, present and future. Especially given its inclusion in a book which introduces human geography in terms of 'challenges for the twenty-first century', the subject matter of this chapter may therefore surprise the reader. Much of what we will be looking at under the rubric of 'the world of consumption' is likely to appear, at least at first sight, rather mundane and trivial. Watching television, eating a burger, wearing designer-label clothing, clubbing, drinking a Starbucks cappuccino, and going shopping, are just a few examples. Not only that, but to be blunt, these instances of everyday consumption may also not seem terribly *geographical*. They are, after all, hardly mainstays of school geography curricula. We don't do fieldtrips to Starbucks or the Ministry of Sound (more's the pity). So why would human geographers want to research, teach and learn about such things?

Technically, consumption can be defined as the utilization of the products of human labour. To put that another way, as a topic it covers everything that happens to the things, services and experiences which our systems of production manufacture. So it is about what we do with our food, our housing, our clothing, as well as (for some of us) our Sony Walkmans, our televisions, and other consumer luxuries. In the modern, capitalist world most people now live in, this tends to involve buying many of these goods in exchange for money. However, consumption is not limited to the moment of purchase. Food consumption is about cooking and eating as well as shopping in a supermarket. Consuming

Plate 14.1   The global and local geographies of television: consuming images of the world in the living room. See also Box 14.3. (Popperfoto)

Plate 14.2 Geography going Gucci? See Box 14.4 for an account of some of the unlikely geographies of men's designer fashions. (The Advertising Archives)

clothes is about wearing them, and about having a style of dress, as well as actually hand-
ing the cash or plastic over the counter in the first place. Consumption, then, is fundamentally about money; but it is also, as we shall see, about desire, love, one's sense of self,
and a host of other things besides. Not so trivial perhaps. So let us return, then, to that
question of why consumption matters; and why it matters to human geographers more
particularly.

14.1.1 Consumption matters economically

Let us start with the first part of that question. There are two main sets of reasons why
consumption matters today, and why it provides a major challenge for the future: its economic importance; and its social and cultural importance. As regards the first, until recently
there had been a tendency for academics to neglect the importance of consumption, and
pay far more attention to worlds of production (work and the workplace, the organizational systems that link these together, and the social contracts that frame these (see
Chapter 12)) or the worlds of financial circulation (banking systems, financial markets and
so on (see Chapter 13)). But consumption is far more than the inevitable, final stage of a
one-directional economic production chain, neither is it simply a by-product of monetary circulation. In the form of 'retail capital' consumption in fact plays an increasingly
central role within many economies. It accounts for a growing proportion of paid
employment and often exercises great powers over manufacturing through the organizational power of retailers over their suppliers (for work in this vein see some of the chapters in Parts Two and Three of Wrigley and Lowe 1996).

More generally, every day on news bulletins all over the world 'consumer confidence',
or its lack, is cited as the cause of the 'productive' sectors of an economy booming or contracting. Consumers, or more often 'the consumer' or 'the market', are constantly being
invoked as the reasons why some action must be undertaken: bananas must be a certain
size because consumers want them to be. Hence, Caribbean growers are passed over by
UK food retailers in preference to those in South America. Costs must be cut in banking
as consumers are unwilling to choose institutions that do not offer the best deals, so
branches close and jobs are lost. Supermarkets switch food suppliers due to their concern
with matching the quality standards that 'the consumer' demands. In the words of Danny
Miller, the consumer has become a sort of 'global dictator': 'Today, real power lies
unequivocally with us, that is the waged consumers of the First World' (1995: 10). In
consequence, any squeamishness about studying consumption simply because so many in
the world are not affluent consumers is profoundly misplaced: 'The acknowledgement of
consumption need not detract from the critique of inequality and exploitation, but this
critique is foundering precisely because the enormous consequences and attractions of
consumption are left out of the analysis' (ibid.: 21). A fundamental challenge for the
twenty-first century is to fashion economic forms that fulfil those attractions of consumption, but deal too with its consequences.

14.1.2 Consumption matters culturally

But what about these attractions of consumption? Here we come to its social and cultural
importance. For consumption is rarely, in fact almost never, simply a matter of meeting
material needs. It plays a central role in our wider social and cultural lives. Let us take the

example of food (see Bell and Valentine 1997). We eat because we are hungry and need food to live, yes; but a range of other factors also shape our food habits. There are culturally specific taboos on, and preferences for, certain foods. I do not eat dog or horse meat. Few in Britain do. I do eat other meat, though; many of my friends do not, being vegetarians. Thus consumption can figure as a central component in an understanding of oneself and one's moral place in the world, to the extent that one could barely physically eat certain (in principle edible) things. There are also fashions in food. I now eat pastas or rices most days, but growing up with my parents on a farm in Devon in the 1970s, I never ate them at all. We ate meat and two vegetables for our main meal of the day, had a fried breakfast, and cakes at tea-time. Then, pasta and rice were foreign foods, and did not fit in with the way my mother structured our meals. As a child, I also always had to sit at a table to eat my food. Meals had a clear beginning and end (you had to ask to 'get down from the table'). My dad sat at the head of the table, my mum at the other end, we children round the other sides (there were lots of us). Now, and I am still a little ashamed of this, I skip breakfast, eat a sandwich and chocolate bar for lunch while checking my e-mails at work, and have an evening meal watching television. As you may have noticed, in that eating pattern I pay almost no attention to the dietary quality of what I eat, nor indeed do I regulate quantities. In contrast my partner thinks hard about her nutritional needs, worries about the quality of the food we eat (for example paying extra for organic produce, especially if our daughter is going to eat it) and regulates the amount she eats in accordance with a 'diet'. I too may soon have to adjust the food habits learnt during my twenties. My doctor tells me I am overweight and unfit. He is right.

Eating, then, is socially structured in terms of food choice, and in terms of how those chosen foods are then eaten. It is not a simply a matter of providing fuel for the body. It is bound up with issues of how we understand our bodies (thus, for example, the pattern in Britain is for more women than men to diet, due to the greater pressure put on women to maintain certain body shapes, and to be seen to be restrained in their appetites). It is bound up with issues of how we understand our domestic life, and the ways our households function. It is bound up with our attempts to care not only for ourselves but also for others; my partner buys organic food to both enact and express her maternal care for our daughter; I cook my partner meals to show I love her. It is bound up with issues of social and cultural identity, being shaped by and shaping contours of class and regional and national identity, as well as helping to fashion other life-style groupings. And what is true for food is true across other spheres of consumption. Consumption is about a lot more than meeting clearly defined material needs. In consequence, one cannot explain consumption behaviour purely in material, economic terms.

Just what social and cultural processes do shape our consumption practices, then? To simplify somewhat, one can identify two main sets of ideas on this, both of which emphasize how consumption is not only of enormous economic significance but also of fundamental importance to our social and cultural lives too. On the one hand there are those who emphasize how consumption is socially and culturally produced, how we as individual consumers are, to put it crudely, made to consume by being immersed in a wider **consumer culture**. Here, for example, the philosopher Herbert Marcuse wrote about how in capitalist societies there is a social production of 'false needs' (Marcuse 1964), a sense of requiring far more than one materially needs. More specifically, numerous studies have debated the role of advertising in stimulating consumption, creating desires for products that would otherwise not exist (for a classic study see Packard 1977).

Plate 14.3   'BMW-Alpina: the car with flair for people with flair'. Advertising, commodity auras and the production of false needs; you may want to own a BMW convertible, but do you really need to? (Vintage Ad Gallery)

Plate 14.4 Adidas sportswear was repositioned in the 1980s through its adoption by hip-hop culture, popularized by the likes of Run DMC. (Redferns Ebet Roberts)

In this portrait, then, it is the producers who dominate consumer culture. Indeed consumer culture is a way of ensuring that mass production is accompanied by the appropriate levels of mass consumption. All in all, consumers are portrayed as rather sad figures:

> irrational slave[s] to trivial, materialistic desires who can be manipulated into childish mass conformity by calculating mass producers. This consumer is a cultural dupe or dope, the mug seduced by advertising, the fashion victim ..., yuppies who would sell their birthright for a mass of designer labels. Ostensibly exercising free choice, this consumer actually offends against all the aspirations of modern Western citizens to be free, rational, autonomous and self-defining. (Slater 1997: 33)

On the other hand, some are much more optimistic about consumer culture. Here the emphasis is not on consumers being made to consume, but on how consumption is used by consumers for their own ends. Generally, what are identified here are the more positive social possibilities of consumption, the feelings of agency and self-autonomy it can engender. It is argued that in comparison to many people's worlds of work, worlds of consumption can seem comparatively fertile spaces for creativity, self-fulfilment and, indeed, liberation (Miller 1995). More specifically, some stress the ways in which (many) consumers creatively re-work the products being sold to them, giving them new meanings in the process, and using them as launch pads for their own symbolic and practical creativity (see for example Willis 1990).

Think for example about branded sports shoes. One might see them as classic examples of duping the consumer: vastly overpriced, sold on the basis of their brand status, desired by children who fear social stigmatization without them. On the other hand, the producers are not always in control here, even though they clearly have huge power and benefit enormously in terms of profits. After all it was not they who initially positioned sports shoes as youth fashion wear; it was North American hip-hop culture. Nor, now they actively pursue the youth fashion market, are those sportswear companies able simply to determine whether they stay in fashion; Adidas, Nike and others have all struggled to maintain sales at times when they are deemed to have become unfashionable. So to understand consumption here means understanding certain youth cultural forms and their internal dynamics, as well as understanding the marketing strategies of major transnational businesses. It means recognizing that whilst individual consumers may not control the processes in which they are immersed, they are certainly not passive dupes, but active participants.

More generally, there have been a range of other approaches which have likewise emphasized how consumption involves the use of commodities within at least partially autonomous cultural fields: within existing status structures and patterns of class distinction, so that consumption is regulated by socially structured and inherited tastes (Bourdieu 1984); within the private dream worlds of consumers, so that consumption is about fulfilling our fantasies and desires (Campbell 1987); into domestic relations of love and care, so that consumption is about concretizing our relationships to our nearest and dearest (cooking and sharing a meal, buying treats and gifts, helping someone choose their clothes) (Miller 1998). In all these very different approaches, then, to understand consumption one has to understand the social and cultural relations within which it is incorporated. One cannot simply read off our worlds of consumption from the corporate strategies of producers.

14.1.3 Consumption matters geographically

So far, I have been suggesting that consumption may at one level be mundane and triv-ial, but also, at another, a fundamental part of the economies and cultures of modern soci-eties. Consumption matters. But why is it of particular interest to geographers? Well, for two reasons. First, because geographical questions of space and place are central to the worlds of consumption. Consumption takes place, and its places are central to what it is. In consequence, one can understand modern consumption better through taking a geo-graphical perspective on it. Secondly, consumption designates a crucial set of processes that shape the kinds of places and spaces we live in, the geographies of our modern world. Consumption not only takes place, it makes place. Today's world is shaped by it.

This geographical character of consumption is the main subject matter for the rest of this chapter, and we will be looking at it through three different windows. First, we will be considering arguments over what modern consumption means for geographies of dif-ference, and in particular whether it produces a homogenization of places and ways of life. Here then, our focus will be global, thinking about whether consumption has increasingly corroded culture and its meaningful geographies of place and locality, ripping apart the historical sedimentations that mark places and people as different from each other, and smoothing them over into more homogenized markets that better suit the supra-local operations of capital. Are our worlds of consumption increasingly all the same? Does con-temporary consumption promote the end of geography?

Secondly, we will then shift scales, and think about the more local worlds of con-sumption, the places where consumption happens. Taking the examples of shopping malls and nightclubs, we will be looking at the roles played by such places in the very process of consumption itself, asking just what it is that is going on in them, and what this tells us about the character of consumer cultures more generally. Finally, we will be connecting these local and global geographies of consumption together. After all, consumer worlds are always simultaneously local and global. The food I eat, the television I watch, the music I listen to in my very local, domestic spaces are actually sourced from all over the world. The everyday acts of consumption I undertake in my kitchen and living room are dependent upon huge networks of provision, about which I may know relatively little. The conclusion of this chapter therefore argues that one of the chief characteristics of our world of consumption is its connecting of the local and the global, the immediately pres-ent to the absent, and points to some of the key questions raised by such connections, both intellectually and personally. These, it is suggested, are appropriate challenges for the twenty-first century.

14.2 Global geographies of consumption

14.2.1 Consumption, cultural imperialism and the end of geography

This chapter began by portraying consumption as made up of everyday, rather mundane activities. However, consumption is also at the heart of some of the most significant intel-lectual debates of our age, being seen as central to nothing less than the cultural fate of the world. These debates centre on the extent to which modern consumption both produces and reflects a profound erosion of cultural traditions and differences, homogenizing the world into increasingly similar landscapes, people and social systems as it spreads from its heartlands in North America and Europe. Peet, for example, argues that global capitalism and its consumer culture comprises 'a powerful culture which overwhelms local and regional experience ..., breaking down the old geography of society and culture' (1989: 156). Sack concurs that: '[a]nother modern condition is the trend toward a global economy and culture, which seems to require that places all over the world contain similar or functionally related activities and that geographical differences or variations that interfere with these interactions be reduced' (1992: 96).

More prosaically, it is hard to miss the frequent commentaries on the worldwide spread of products and brands such as McDonalds or Coca-Cola, to the extent that they have entered academic language – in the form of theses of the 'McDonaldization of society' (Ritzer 1996) or of 'Coca-colonization' – as bywords for a consumerist homogenization. Travellers' tales too are replete with mournful stories of how one can trek to the farthest ends of the earth, only to find people doing exactly what the traveller could have seen at home; in the title of Iyer's account from the 'not-so-far East', having 'video nights in Kathmandu' (1989). Some of these issues have already been discussed in Chapter 6. This section will consider them in more detail, focusing upon the issue of consumption.

For many, what modern consumption enacts is a form of 'cultural imperialism'. This thesis takes a number of forms. For some, it is evidenced by the 'dumping' of First World, and especially American, consumer products into Third World markets: 'authentic, traditional and local culture in many parts of the world is being battered out of existence by

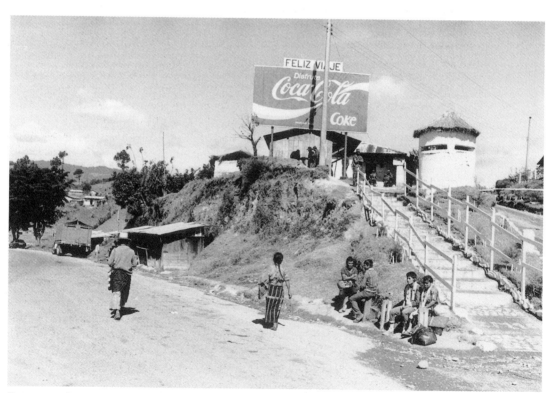

Plate 14.5   Coca-colonization? (Panos Pictures/Julio Etchart)

the indiscriminate dumping of large quantities of slick, commercial and media products, mainly from the United States' (Tunstall 1977: 57). What concerns critics here is both the dominance of First World and American multinational companies (Mattelart 1979) and the predominance of products which promote First World and American values (Dorfman and Mattelart 1975) (for a general review of these emphases see Fejes 1981). Also raised are concerns about the loss of indigenous, 'authentic' local cultures, so that the 'real' culture of an area and a people are lost.

For others, the issue is not just the conquest of some national and regional cultures by others, but the ways in which consumerism destroys all culture. The suggestion here, then, is that what consumption represents is the contemporary dominance of a very peculiar cultural form – consumer culture – which in many respects is barely cultural at all. Consumption, they suggest, is all about the political economy of market shares, niches and profit, and as such fundamentally opposed to the intrinsic values of cultural meaning and communication (Box 14.1). What matters here is partly the loss of cultural diversity – the making of a duller world – but perhaps more importantly the loss of cultural alternatives. Uniformity is doubly problematic, in that it also has elements of conformity. It therefore may result in a form of culture that is inadequate and unfulfilling for many of those it is imposed upon:

> A cultural system which would be adequate for the poorest people in that system would mean a set of instrumental, symbolic and social relations that would help them to survive in meeting such fundamental needs as food, clothing, housing, medical treatment and education. Such needs are not met if they are identified with the consumption of

---

Box 14.1 Consumerism as the end of culture?

'Culture' was one of the crucial terms through which anti-liberal forces counted the cost of modernity. '*Consumer* culture', in this perspective, is merely an ersatz, artificial, mass-manufactured and pretty poor substitute for the world we have lost in post-traditional society. In fact, it is the antithesis and enemy of culture. In it individual choice and desire triumph over abiding social values and obligations; the whims of the present take precedence over the truth embodied in history, tradition and continuity; needs, values and goods are manufactured and calculated in relation to profit rather than arising organically from authentic individual or communal life. Above all, consumerism represents the triumph of economic value over all other kinds and sources of social worth. Everything can be bought and sold. Everything has its price. 'Consumer culture', therefore, is a contradiction in terms for much of modern western thought. (Slater 1997: 63)

The perception here is that everywhere in the world is beginning to look and feel the same. Cities in any part of the world display uniform features determined, for example, by the demands of automobiles; architectural styles become similar; shops display a uniform range of goods; airports – the potent gateway to cultural diversity – have an almost identical 'international' style; Western popular music issues from radios and cassette players from New York to New Delhi ... [I]t is the economic imperatives of multinational capitalism that are behind this cultural convergence. (Tomlinson 1991: 26)

> Kentucky Fried Chicken, Coca-Cola, Aspro, or Peter-Stuyvesant cigarettes. (Hamelink 1983: 15)

What is questioned here, then, are the values that underpin consumer culture, and the ability of consumption to satisfy the 'real' as opposed to the 'false' needs of consumers.

In summary, according to this argument, the world of consumption can be seen as dominated by the concerns and interests of a small number of large multinational companies, mostly based in North America or other parts of 'the West'. They are seen as destroying the distinctiveness and autonomy of local cultures, and as reducing the sense of any alternatives to a particular Western and capitalist way of life. Clearly these are fundamental issues. However, they also rest upon fundamental and debatable assumptions: that the authenticity of a culture is dependent upon its resistance to external forces and artifacts; that non-Western consumers are duped or coerced into having consumption tastes that are inappropriate to their real needs, just as the mass of Western consumers have been too (it seems only the critics of consumer culture know what it is any of us 'really' need); and that consumers passively consume Western goods, and make of them exactly the same the world over. We shall now turn to some of these criticisms of the cultural imperialism approach to consumption, through the lens of work that emphasizes processes not of 'homogenization' but of 'indigenization'.

14.2.2 Consumption and cultural indigenization

Let us start with someone who may seem an unlikely proponent of this idea of indigenization. Marshall McLuhan wrote one of the most famous social science books of the post-war period, as he sought to understand the character of a contemporary society so saturated by the mass media (McLuhan 1964). His argument about the increasing linkages between different places, and phrases expressing this such as the notion of the 'global village', have passed into common parlance, echoed in numerous advertising campaigns promoting the unifying qualities of new communication technologies. However, his argument always went beyond suggesting that this increasing interconnection between people and places – or in the much cited language of the Marxist geographer Harvey (1989), this 'time-space compression' of the world – meant simple homogenization. He elaborates on this in a later paper:

> The question arises, then, whether the same figure, say Coca-Cola, can be considered as 'uniform' when it is set in interplay with totally divergent *grounds* from China to Peru? ... The 'meaning' of anything ... is the way in which it relates to the user ... [For example] it might be suggested that the use of American stereotypes in the United Kingdom yields a widely different experience and 'meaning' from the use of the same stereotypes or *figures* in their American *ground*. The interplay between the *figure* and the *ground* will produce a wide range of different effects for the users. (McLuhan 1975: 44; emphases in original)

McLuhan argues, then, that consumer culture does not float above the world in some abstract space of capital or modernity. It lands and takes root in particular places and times. In so doing it not only impacts on its surroundings; it is also shaped by its locations.

Numerous strands of work on the geographies of consumer culture illustrate this, For instance, and particularly prevalent within anthropology, there are accounts that

Plate 14.6 McDonalds in Beijing: American fast food becomes exotic, stylish Americana. (Popperfoto/Reuters)

emphasize how globally distributed facets of consumer culture – whether particular goods, styles, or cultural orientations – are always locally incorporated. Take, for example, some of the exemplars of homogenizing American consumer culture: Coca-Cola, television programmes, and McDonalds. Each has stimulated accounts of differentiated local incorporations (on Coca-Cola see Miller 1997; on television programmes see Liebes and Katz 1993, Miller 1992; on McDonalds see Watson 1997). Let us start with the rather unlikely terrain of McDonalds.

On an average day over 30 million customers are served at one of 20,000 McDonalds restaurants in more than 100 countries. Over half these restaurants are in the United States, but based on 1995 figures the United Kingdom has over 600 outlets, Brazil over 250, China nearly 200, Thailand nearly 50. Given their datedness, these figures are almost certainly substantial underestimates; a recurrent promotional corporate statistic is that a new McDonalds restaurant opens somewhere in the world every three hours. Not only this, of course, but McDonalds are famed for their uniformity; the same decor, the same basic menu (with very small variations, including the wonderfully named McSpagetti in the Philippines), and the same service style the world over. And yet, McDonalds may not be just the force for cultural homogenization that this suggests. This was certainly Watson's (1997a) conclusion, based on studies of the cultural geographies of McDonalds in East Asia (Box 14.2). These studies found that McDonalds has been localized, indigenized, incorporated into traditional cultural forms and practices, even as it has played a part in the on-going dynamism of those traditions. And exactly how this has happened varies across East Asia. In Beijing, McDonalds has lost its American role as a place of fast

> ### Box 14.2
> ## Cultural geographies of McDonalds in East Asia
>
> In the introduction to a collection of findings on the cultural geographies of McDonalds in East Asia, Watson poses the following questions:
>
> > Does the spread of fast food undermine the integrity of indigenous cuisines? Are food chains helping to create a homogeneous, global culture better suited to the needs of a capitalist world order? ... But isn't another scenario possible? Have people in East Asia conspired to change McDonalds, modifying this seemingly monolithic institution to fit local conditions? ... [Perhaps] the interaction process works both ways. (Watson 1997a: 5–6)
>
> His conclusion is that there is indeed a two-way interaction:
>
> > McDonalds has effected small but influential changes in East Asian dietary patterns. Until the introduction of McDonalds, for example, Japanese consumers rarely, if ever, ate with their hands; ... this is now an acceptable mode of dining ... [On the other hand] East Asian consumers have quietly, and in some cases stubbornly, transformed their neighborhood McDonalds into local institutions. In the United States fast food may indeed imply fast consumption, but this is certainly not the case ... [in] Beijing, Seoul, and Taipei ..., [where] McDonalds restaurants are treated as leisure centers, where people can retreat from the stresses of urban life. (ibid.: 6–7)

and cheap food. Instead, Yan suggests that it has become a middle-class consumption place, somewhere for a special family outing, somewhere 'customers linger ... for hours, relaxing, chatting, reading, enjoying the music' (1997: 72). McDonalds here is seen as American, but Americana means something stylish, exotic and foreign, and as such actually results in the meanings and experiences of McDonalds in Beijing being very un-American! In contrast, in Japan whilst there is a similar leisurely use of McDonalds, it is not a place of exotic social prestige, but a youth hangout, a place where someone in a business suit would be out of place (Ohnuki-Tierney 1997). In Hong Kong McDonalds was likewise marketed to the youth market, and adopted as such initially, but here its American origins are of little contemporary consequence (Watson 1997b). Whilst initially patronized in the 1970s by adolescents seeking to escape products associated both with China and with Hong Kong's perceived provincialism, by the 1990s McDonalds has become routinely local, just another mundane part of the Hong Kong landscape: '[t]oday, McDonalds restaurants in Hong Kong are packed – wall to wall – with people of all ages, few of whom are seeking an American cultural experience. The chain has become a local institution in the sense that it has blended into the urban landscape ... McDonalds is not perceived as an exotic or alien institution ...' (Watson 1997b: 87 and 107).

Hence the meanings and practices of McDonalds – an archetype of global homogenization – vary from place to place. Parallel accounts have been produced through research on a range of other global products from television soap operas to European designer fashions (Boxes 14.3 and 14.4). Three main conclusions emerge from these studies. First, modern consumption may be a global phenomenon, but its forms and meanings are also locally specific. At a broader level the argument here is for a conceptual shift away from seeing the modern world as a singular universal entity, and towards

---

Box 14.3   **Media products and cultural indigenization: the case of *Dallas***

For proponents of the cultural imperialism thesis television programmes are primary instruments for the imposition of Western, American, consumerist values on their global audiences. *Dallas* was the most watched programme in the world in the early 1980s. Set, unsurprisingly, in Dallas, Texas, it was a weekly soap opera following the fortunes, relationships and fabulous squabbles of an extended oil baron family, the Ewings. Among many other places, *Dallas* was very popular in Israel where two media researchers, Liebes and Katz, were investigating what was made of it by a variety of Israeli audiences. What they found was that, far from the programme simply imposing a world view on its audience, different groups of people understood the programme in very different ways and enjoyed it for very different reasons.

> Theorists of cultural imperialism assume that ... [meaning] is prepackaged in Los Angeles, shipped out to the global village, and unwrapped in innocent minds. We wanted to see for ourselves. ... In the environs of Jerusalem, we watched the weekly episode in the homes of Arabs, veteran settlers from Morocco, recent arrivals from Russia and second-generation Israelis in kubbutzim – *Dallas* fans all .... Each cultural group found its own way to 'negotiate' with the program – different types of reading .... We found only very few innocent minds, and a variety of 'villages'. (Liebes and Katz 1993: xi)

The recent immigrants from the Soviet Union denied watching it (even when they did) or dismissed it as American, capitalist propaganda. They saw it in ideological terms. Middle-class kibbutzniks had a more psychological take on the series, concentrating on the personalities of the protagonists and enjoying the task of understanding their motivations. The interviewees from North African and Arabic backgrounds emphasized the convolutions of the plot, framing the programme as an epic family saga. Different people with different backgrounds, and different 'readings' of the same show.

---

recognizing the many 'specific', 'prismatic', 'comparative' or 'global' modernities being forged through consumption practices (see for example Breckenridge 1995 on South Asia; Featherstone, Lash and Robertson 1995 generally; and Miller 1994 on Trinidad). Secondly, rather than necessarily destroying local, indigenous cultures, the consumption of non-local products may be irrelevant to, indeed even helpful to, the on-going production of authentic local cultures. Writing in the Trinidadian context, Miller argues that in a world of globalized consumption 'authenticity has increasingly to be judged a posteriori not a priori, according to the local consequences not local origins' (Miller 1992: 181). It matters less where things come from than what they are used for. Indeed, it may be that the non-local origins of products may be positively beneficial, giving 'the periphery access to a wider cultural inventory' and providing new raw materials for local cultural and economic entrepreneurs (Hannerz 1992: 241). Thirdly, when it does matter where things come from, when their origins are part of what they are valued for, then those origins are understood in locally specific, imaginative ways. America, for example, is mythologized the world over, but in a range of different forms (see, for example, Webster 1988).

> ## Box 14.4
> ### Western goods and cultural indigenization: the case of 'Les Sapeurs'
>
> 'Les Sapeurs' are a group of consumers in what was the People's Republic of the Congo (see Friedman 1994). They are men of the politically marginalized Bakongo ethnicity, drawn especially from the poor in large cities such as Brazzaville and Pointe Noire, who aim to move through a graded and spatialized consumption process of 'Western' fashions. Beginning with the conspicuous wearing of imported ready-to-wear items, Sapeurs look to progress to a stay in Paris ('l'aventure') accumulating *haute couture* from French and Italian design houses whilst usually living in extreme poverty, before returning to Brazzaville's 'Sape clubs' to perform competitively in 'la danse des griffes', a cat-walk display of one's dressed self which is augmented by the sewing of a host of designer labels to jacket lapels. All of which, especially at a first hearing, might be rather too easily dismissed as an extreme case of fashion victimhood, or as a bizarre oddity, thrown up by the maelstrom of global-ized consumer culture as inflected in the Congolese context. However, for Friedman, far from being a cultural oddity, la sape is symptomatic of much more general local–global geographies of con-sumption, in which globally distributed commodities and materials – here Western European designer fashions – are encompassed within local cultural logics of identity practice. In the Congo, understandings of identity are marked by a conception of a cosmic 'life force' which can be tapped into through consumption, and which, especially for the socially marginalized, is located elsewhere, external to the local political and cultural hierarchies. As such, the conspicuous consumption of designer labels becomes something very different than what one might first make of it. It is not any sort of dandyism, or an abandonment of serious cultural projects. Rather it is 'a life-and-death struggle for physical and social survival' (Friedman 1994: 106) that accords to the existing, indige-nous cultural logics of the Bakongo.

14.2.3 Consuming difference

This last point – on the mythologies or 'imaginative geographies' (Driver 1999) attached to the things and places we consume – leads on to a further set of qualifications of the cul-tural imperialist thesis. These concern the assumption that the goal of those promoting consumer culture is simply to eradicate cultural and geographical differences, viewing them as unhelpful encumbrances to the smooth operation of the world of consumption. In fact, one can see cultural and geographical difference being actively embraced, even produced, within consumer culture. There are three different ways in which this is the case.

Firstly, producers may seek to operate across national and regional boundaries, but they also have to attune their products to the tastes of more local markets. Take, for example, global media operations, such as satellite television companies. As Kevin Robins (1995) argues, there is certainly a commercial logic for such companies to operate globally, and to break down the national organization of broadcasting: 'The imperative is to break down the old boundaries and frontiers of national communities ... Audio-visual spaces are thus becoming detached from the symbolic spaces of national culture. ... The free and unimpeded circulation of programmes – television without frontiers – is the great ideal in the new order' (Robins 1995: 250). However, in order to reach beyond the market of business people and tourists looking for a bit of home away from home, programming has

to be localized. This is certainly the view of CNN whose 'news service has been successful in reaching the world's business and political elites, but ... has not significantly penetrated mass markets, where local affiliations and attachments are far stronger' (Robins
1995: 257–8).

Secondly, difference and geography are not just something that those promoting consumption have to deal with and accommodate. Global arrays of cultural and geographical
difference are produced and sold within consumer culture. Exemplary here, it is often
suggested, are the themed environments of consumption spaces such as theme parks and
shopping malls.

In Disneyland and Disney World, for example, the amusements and other consumption opportunities are not just randomly arrayed, but organized into distinct lands (for
example Frontierland themed on the American West, or Adventureland themed on landscapes of European colonialism and imperialism) (see Bryman 1995, Gottdiener 1982,
Wilson 1992). This echoes much earlier consumer festivals and fairs such as the Midway
attractions at the 1893 Chicago exhibition (Rubin Hudson 1979) or Luna Park at Coney
Island, New York (Kasson 1978), and is also a pattern to be found in many large shopping malls, most famously the huge West Edmonton Mall in Canada (Crawford 1992; see
also Goss 1999a). Indeed Gottdiener (1997) goes so far as to posit that consumer culture
has produced a general 'theming of America', so that theme parks become models for
what is happening across the urban landscapes outside them. Sack (1992: 98) also points
to the production of 'pseudo-places' in consumer culture. And Harvey (1989) sees
the same general trend towards highly packaged, and indeed simulated, experiences of

Plate 14.7 Theming the world: Disney World, Florida. (Popperfoto/Reuters)

cultural and geographical difference, such that microcosms of the whole world are encountered in particular consumption spaces, from the television screen, to the High Street, to the mall and theme park.

Rather than eradicating cultural and geographical difference – and thus producing the end of geography – the world of consumption (re)produces geographies, framing certain local places of consumption as global centres. What then becomes the issue is not the end of geography, but what sort of new geographies are taking its place, and the kinds of understandings of people and place that are being circulated through consumer culture.

For instance, these kinds of questions have framed a number of pieces of research on the global arrays of food now being sold and consumed in the United Kingdom, as well as elsewhere (see Cook and Crang 1996). This explicit globalization of food provision is captured in an appreciative review of the diversity of restaurants in London, penned by the listings magazine, *Time Out*:

> The world on a plate. From Afghani ashak to Zimbabwean zaza, London offers an unrivalled selection of foreign flavours and cuisines. Give your tongue a holiday and treat yourself to the best meals in the world – all without setting foot outside our fair capital.
> (*Time Out*, 16 August 1995)

A central concern in research thinking about such developments has been just what sorts of geographical knowledges are being produced through this facet of consumer culture; what conceptualizations of people, place and culture are being mobilized. Take, for example, the following anecdote from the tabloid British newspaper, The *Daily Star*. Under the heading 'korma out for a meal', this item reported on the popularity of 'Indian' restaurants and take-aways amongst British consumers ('curry lovers have sparked an eating out boom in Britain'). At the same time, however, its notion of Britishness was pointedly narrow as, in the very next sentence, it reported that 'just 26 per cent [of those

Plate 14.8 Eating into Britishness: is this 'Indian' curry 'British'? (Alan Thomas)

dining out] chose British restaurants' (*Daily Star*, 5 December 1995). Indian restaurants are a popular British pastime, but not British themselves, it seems. The logic here is all too clear. Indianness is something (non-Indian, white) British people may enjoy consuming, but that does not make it, or the British-Asian (in fact primarily British-Bangladeshi or British-Pakistani, not British-Indian) restaurateurs and waiting staff who embody it, culturally British. Difference is appreciated and enjoyed, at least when it comes to food, but it also distances those with whom it is associated. Potentially, an ethnicized geographical difference is incorporated into a very white British way of seeing the world, in which everything non-white becomes a resource to 'liven up the dull dish that is mainstream white culture' (hooks 1992: 21). See Cook, Crang and Thorpe (1999, 2000) on Britain's culinary multiculturalism and Jackson (1999) on the consumption of ethnicized difference.

14.3 Local geographies of consumption

14.3.1 Let's go shopping

Let us now turn to the local geographies of consumption, the sites where it takes and makes place. In fact, let us go to the shopping centre. Shopping centres (or malls in North America), after all, are routinely cast as the cathedrals of our consumerist age, the symbols of all we value and hold dear. In particular, let us see just what goes on inside a shopping centre, and reflect on what this tells us about the kinds of 'public' space being produced in the world of consumption.

One version of events is that what happens in shopping centres is that we are manipulated into behaving in certain ways, and especially into buying things, by the shopping centre itself. After all, shopping centres are in part designed according to a 'merchandise plan' (Maitland 1985: 8) which tries to maximize the exposure of consumers to goods. In older shopping centres and North American malls this plan tended to be understood in quite mechanical terms. Shopping centres are functional 'machines for shopping'. 'Generator' stores pull us to the shopping centre in the first place, and once we are there 'magnet' stores every 200 metres or so make sure we don't just pop in and out, but explore every part of it. Here, then, 'the public mall [is defined] as essentially the passive outcome of a merchandising plan, a channel for the manipulation of pedestrian flows' (Maitland 1985: 10). There are well known tricks to make us behave as the shopping centre management and its shops want us to: escalators arranged so we have to walk past shop fronts or merchandise when going between floors; hard seats so we don't linger too long without getting up and seeing some more potential purchases; no water fountains so we have to buy expensive drinks.

In more recent shopping centres, however, the manipulation may be more subtle. Increasingly, retail planners and developers see their job as providing spectacular places that people will want to spend time in. This is because research suggests the longer people stay the more on average they spend, and because as competition between centres hots up each one needs more than just the usual shops to attract us in the first place – everything from fountains to funfairs (see for example Goss 1999b). Here, then, the manipulation is less physical and more emotional. The hope is to excite, inspire, relax and please (Box 14.5). Of course, these are promotional intentions rather than shoppers' experiences. As

Plate 14.9   The Bluewater Complex in Kent, the UK's largest shopping mall. (Bluewater Complex Kent)

Chaney (1990) puts it in his review of the MetroCentre in Gateshead, north-east England, the designer's utopian impulses to produce a place of perfection often result in rather underwhelming, 'subtopian' forms. Nonetheless, there is much here to be taken seriously. There is the attempt to produce a fabricated space in which the individual consumer can be made to feel like consuming (see Goss 1999a). There is the emphasis on creating an internal, closed-off, privately owned but partially public environment, divorced from the harsh exterior world, not only climatically but also socially, through the operation of security systems that ensure the absence of any who might threaten this consumer paradise or disrupt the pleasure of the shopper. Thus, the privately owned and managed spaces of malls rework the older public spaces of the street or market:

They reclaim, for the middle-class imagination, 'The Street' – an idealized social space free, by virtue of private property, planning and strict control, from the inconvenience of the weather and the danger and pollution of the automobile, but most important from the terror of crime associated with today's urban environment. (Goss 1993b: 24)

Such shopping centres recognize that there is much more to going shopping than buying things (Glennie and Thrift 1992). Shopping is in part about experiencing an urban space, seeing and being seen by other shoppers. It is a social activity. But they seek to manage this social experience. The street gets re-cast as a purified space of leisurely consumption, cleansed of nuisances that inhabit the 'real' streets, like 'street people'.

However, this fashioning of the mall can be disturbed. Retail spaces can be reclaimed by those who use them. Leading the charge are a mixed assortment of senior citizens and adolescents. Take the case of Bert, 78 years old, and a regular visitor to a shopping mall in New England (US):

I come here at quarter to eleven and I leave at twenty minutes past one . . . I do this five days a week, except I missed one day this year. . . . And I have my lunch here at noontime . . . . I move every half-hour. I go from here down to where the clock is, I go down in front of Woolworth's, then I come back again and go up there and take my bus by the front of JC Penney.' (quoted in Lewis 1990: 126)

For Bert, the mall is not a place to shop, but somewhere sheltered to go outside his house, a chance to meet up with friends. For the mall management there is the option of evicting these low-spending visitors, but ejecting old people forcibly onto the pavement is hardly good public relations. Instead, and as part of wider moves to see malls as the new civic centres, Bert's mall has actually established its own walking club for the elderly, with over 300 members, for whom the doors open at 6 a.m., rather than 9 a.m. when shops open.

Box 14.5	The shopping mall experience: Woodfield Mall, Schaumburg, Illinois

Leaving the comfort of her air-conditioned car, the shopper hurries across the sun-broiled or rain-swept tarmac and thankfully passes through doors into an air-conditioned side mall. Drawn down this relatively dark, single-level tunnel, lined with secondary units, by the pool of light ahead, she reaches one of the squares ahead located at the end of each arm of the main mall system. She discovers this to be a two-storey space. . . . On one side of the cavern is a magnet department store, but before deciding to enter it she turns to see that a second tunnel, much taller than the first with natural light spilling though holes punched in its roof, leads off into the heart of the building. . . . This route now culminates in the central space, Grand Court. . . . Its floor has been quarried and shaped to form an amphitheatre and pools. . . . Emphasis is placed upon the creation of a plausible internal world, full of visual interest and variety, in which one is tempted not simply to pass through, but to stay. (Maitland 1985: 26 and 28)

Adolescents also use the mall as a meeting place, somewhere to hang out. They also spend very little. 'Suburban kids come to malls to look around, meet and make friends, and hang out – because there is nowhere else to go' (Lewis 1990: 130). Or as Ed and Tammy, self-confessed 'mall rat' and 'mall bunny' respectively, put it:

> '[Gesturing around himself] I met all these people here. I've met lots of other people, too. One place where you can always find someone. . . .' (Ed; quoted in Lewis 1990: 130)

> 'I used to come here every Saturday from 11am to 9.30pm and just walk around with my friends, like Gina here, and just walk around and check out the guys.' (Tammy; quoted in Lewis 1990: 130)

Unlike the elderly, the mall is less indulgent of these teenagers. There are constant skirmishes with the security staff over how long ten people can share one Coke at the food court, and some mall rats and bunnies do get banned, getting new hair cuts or clothes in attempts to get back in.

Plate 14.10 'Mall mice', Winchester, UK. Adolescents who hang out at malls are known in the United States as mall rats (boys) and mall bunnies (girls). In the UK, the term 'mall mice' has been adopted to reflect our weaker mall culture. (Alan Thomas)

So what can we make of our trip to the mall? What does it tell us about the local geographies of consumption? Well, it certainly tells us that place matters to consumption. But how? Here the answer is more ambivalent. In part what we see is a pseudo-public space symbolizing and enacting claims for consumption to be all there is to civic life, making citizens consumers in a very concrete way. Yet one also sees the small-scale, trivial resistances that people enact, their refusal to be just consumers in the sense of purchasers, to use this space for their own ends. And that too has a wider symbolism, signalling the possibility that consumption is never just about satisfying those selling to us, but also about fulfilling our own needs for social affiliation and participation.

14.3.2 Are you dancing?

As geographers are increasingly recognizing, shopping centres are not the only sort of consumption space to explore (Crewe and Gregson 1998). How about a night out? How about going clubbing? In London alone there are over 500 club nights on offer every week. In Britain the industry has a turnover of over £2 billion per annum, and it is

Box 14.6 Clubs: spaces of inclusion and exclusion

Clubs vary, of course, in their door policies, but most seek to ensure that the 'right' sort of crowd gets in for the kind of night they are trying to facilitate. In turn, clubbers are not only concerned whether they themselves will get in, but also take an interest in who else gets in as well. Here, as a rather extreme case, is Dwayne, interviewed by Malbon during his research on the geographies of clubbing in London (see Malbon 1997, 1999):

DWAYNE: 'I'm very elitist, I think you have to put a lot of effort in to training your tastes, go to lots of places to do it, and a lot of people for various reasons – because they don't have the time, don't have the resources – just don't do it. I'm very elitist because ... of the way I perceive how taste works. It works strangely. There are social connotations of musical styles and so, umm, if people I didn't like liked the music I like, it would spoil the music ...'

BEN: 'So do you like the people around you to be like you?'

DWAYNE: 'No of course not. I want them to be much cooler than myself. I want to feel "oh God, this is really exciting, this is so brilliant" ... I feel uncomfortable if surrounded by fifteen year olds, but, umm, it's a question of body language, dress, and probably quite a lot of what I fantasise into these people, which in many cases is probably totally away from what they are really like, umm, ... but, I don't talk to them most of the time so I have no way of finding out.'

(Malbon 1999: 55)

For some, the definitions of a cool crowd would be different. They might include those seen as uncool, as mainstream, by Dwayne; they might be defined by their apparent disinterest in such status hierarchies (a funky, friendly crowd as the club listings of *Time Out* magazine would say, chosen more through their insider knowledge than any explicit door policy). But in all these cases it is clear that a club is a place of consumption in which consumers have to make careful judgments about the kinds of places they go to. Where will I get in; where will I feel comfortable; where will I feel like I am where it is at?

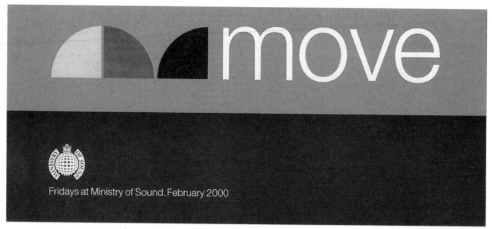

Plate 14.11   The Ministry of Sound: a club, a brand, a place of playful vitality. (Ministry of Sound)

estimated that 42 per cent of the population now visit a club once a year, 43 per cent of 15–24 year olds once a month (Mintel 1996).

As places of consumption, the first thing one might notice about a club is that one has to 'get in' (Box 14.6). But clubs are not just spaces marked by inclusion and exclusion. They are, as Malbon (1997) argues, spaces of play. This should really not need saying. It is patently obvious that people go to clubs to have fun. It is also patently obvious that this fun is achieved through a variety of means. Through enjoyment of the music; through an enjoyment of being part of a crowd of people with whom, temporarily, one can feel at one; through the feelings of competency and bodily expressiveness that dancing can give you; through the sharing of a special place and time with friends; through behaving in ways one cannot in other social situations, for example at work. Yet oddly, having fun is something most academic commentators on consumption say remarkably little about. In fact, to somewhat overstate the case, if someone is caught having fun, the immediate reaction of most academics seems to be to point out why that fun should stop. Clubbers, then, should stop having such a good time and think about the cultural and economic elitism and exclusions upon which their clubbing depends.

However, as Malbon's insightful analysis demonstrates, however pertinent the concerns flagged in this way, just to be a killjoy actually leaves the fun and playfulness of many consumption practices unexplored. It leaves one without any sense of the complicated microspatial practices through which play is enacted; the ways in which one gets to feel, and enjoy being, part of the clubbing crowd; the ways in which one dances so as to lose oneself in the music. It leaves one without any sense of the importance of place to playful consumption; the ways in which one has to be in the right place, with the right music and the right crowd, to have these feelings of exhilaration, communality, and vitality; the way that 'the emphasis is placed on the near and the affectual: that which unites one to a place, a place that is experienced among others' (Maffesoli 1996: 128). And it leaves one without any sense of the wider significance of such playful consumption; of what it says about how contemporary consumer culture is not just made up of atomistic, avaristic individual consumers, but also facilitates temporary, some would say 'neo-tribal' gatherings, through which people come to feel part of something bigger than themselves (see Maffesoli 1996: 72–103).

Clubbing is, then, a place-specific practice of consumption. Whereas in the shopping mall, senses of communal belonging tend to be forged by those occupying the space for purposes beyond those for which it was commercially designed – such as the elderly and the adolescent – in clubs that communal belonging is the product that people are paying to consume. In many ways, clubs are therefore rather unusual places of consumption, characterized by an exceptional emotional and bodily intensity. However, they also point to a wider quality of our local places of consumption; the extent to which we use them, if let in, for experiences of 'public' life and senses of social belonging.

14.4   Making connections

So far this chapter has explored some of the debates over the impacts of modern consumption on both global geographies (of cultural difference) and local geographies (of public space and life). By way of conclusion, I want to think about an emergent concern with reconnecting the local contexts of everyday consumption with the global networks of commodity flows and production that make them possible. This reconnection has emerged as a fundamental intellectual, political and personal challenge in the field of consumption studies. Miller, for example, emphasizes how the 'study of consumption should . . . be increasingly articulated with, and not become an opposition to, the study of the mechanisms by which goods are produced and distributed' (1995: 17). Sack arrives at a similar conclusion on the basis of his determination to develop a moral framework for analysing consumption. Consumers are, he says, profoundly ignorant of the geographies of provision that bring goods to them: 'A shop that sells Colombian coffee does not reveal the social structure that produces the coffee, the economic impact of coffee production on the Colombian economy, or the way coffee growing affects the Colombian environment.' (Sack 1992: 200). Such ignorance, he argues, makes it impossible for consumers to behave with a sense of responsibility to those they are in fact connected to through the webs of commodity flows. In this, modern consumption is deeply amoral.

But what sorts of connections can one make between worlds of consumption and production? One perspective is to see consumers as blinded by a veil that is draped over the things we consume, through a process of what Marx called 'commodity fetishism'. In this portrait the true history or biography of a product, and its basis in the work of others, is obscured to consumers. Instead, one is confronted by false advertising imagery that seeks to associate the product with other fantasy worlds. The task then becomes to puncture these fantasies with some sharp shafts of reality. Hartwick points to the startling incongruities and inequalities that such a puncturing can bring to light:

> Michael Jordan, basketball player extraordinaire, receives US$20 million a year to endorse Nike; Philip Knight, cofounder and chief executive officer of Nike, is worth US$5.4 billion; Indonesian workers are paid US$2.40 a day and Vietnamese workers are paid US$10.00 a week to make the sneakers. . . . The Walt Disney Company is publicly criticized for its treatment of Haitian workers making Pocahontas shirts for 28 cents an hour; Disney's chief executive officer makes US$78,000 a day. (1998: 423)

Smith's account of Starbucks Coffee shops makes similar points. He highlights how such unfairly traded coffee sees only 25 per cent of the retail price returning to the producing country, with even less going to those who actually cultivate the crop (the smallholders, waged workers and seasonal pickers). He contrasts the stylish coffee-shop culture Starbucks epitomizes with ways in which that coffee is sourced. For the poorly paid women who pick the coffee beans, their working day begins somewhere between three and four in the morning. Getting back home at six in the evening, they still have evening meals to prepare, water to fetch from the river, a family to feed and clear up after. Few live in houses with electricity or running water. Along with malnutrition, a major health hazard are the pesticides used on the coffee plants (Smith 1996: 512–13).

So one way to reconnect worlds of consumption and production is to acquaint consumers with the harsh realities of commodity production. It is hard to criticize such an approach. It is a very simple message, but it is always worth getting it across. However, it does have its problems. Rather than articulating worlds of consumption and production together, it can tend to dismiss worlds of consumption, and their positive possibilities, altogether. Consumption just becomes inherently bad. That is neither true, as hopefully the discussions earlier in this chapter have demonstrated, nor terribly attractive in terms of what it suggests one should do. Particularly if one's own worlds of production, as a worker, are thoroughly unrewarding – not usually the case for academic critics of consumption – then giving up all the pleasures of consumption is pretty hard to stomach. Certainly the last two decades in Western politics suggest that failing to recognize the genuine senses of autonomy and creativity that consumption can facilitate, is a sure way to alienate those very same workers in whose name one wishes to re-centre the world of production.

There are, though, more subtle variations on this theme. So-called 'ethical consumerism', in particular, is a highly promising phenomenon. Here, consumption is not disavowed, but it is channelled through concerns with the wider provisioning systems, and their human and non-human participants, of which the consumer is a part. Green consumerism, fair-trade, vegetarianism and other animal welfare concerns are the main strands. Whilst containing some elements of self-denial, and most definitely of hard work, recent research on ethical consumers in London emphasized the positive pleasures gained through such consumer movements (Bedford 1999); the feelings of achievement, the senses of empowerment. However, in its full-blown forms, ethical consumerism is not only hard work and expensive, but requires the adoption of consumption life-styles that are very difficult to maintain unless one is immersed in wider friendship and family networks of like-minded people. Consumption is, after all, thoroughly social. Moreover, the levels of knowledge required to understand the incredibly complex biographies of products are more amenable to academic study than they are to someone who has something else to do. Even the most committed ethical consumers therefore rely on shorthand rationales such as Nestlé is bad, Brooke Bond is good; meat is murder, milk is allowed; and so on.

What this suggests is that rather than simply replacing a superficial and false advertising fetish with the real history of commodities, one can draw people more casually into ethical forms of consumption precisely by extending the fetish or advertising form, and paying attention to the existing pleasures of consumption. Thus, for example, the huge growth in the market for organic produce in Britain and other Western countries at present. It is neatly labelled – with a Soil Association approved designation and logo – and what is more, it can easily be seen as benefiting not only the environment but also those to whom it is being fed (in terms of 'healthiness'). Organic food becomes understood as more generally 'good': good for the environment, good for my family, good for me, good to eat. Of course, not all ethics have such an easy slippage. In particular, the Fair-Trade logo has struggled to make the same market impact. Connecting care for one's nearest and dearest – a very strong existing consumer ethic – to caring for the distant strangers involved in commodity production and distribution has proved trickier. But what would be the point of having a new century, a new millennium indeed, without some challenges still to rise to?

Learning outcomes

After reading this chapter, you should have:

- An understanding of the importance of consumption to modern economies and cultures.
- An understanding of the global geographical impacts of modern consumption.
- A sensitivity to the continuing importance of cultural and geographical difference within contemporary consumer cultures.
- An understanding of the character of public space fashioned within the worlds of consumption.
- An understanding of how one might make connections between worlds of consumption and production.
- An ability to think about the geographies of your own everyday consumption practices.

Further reading

Bell, D. and **Valentine, G.** (1997) *Consuming Geographies: We Are Where We Eat*, Routledge, London. A lively exploration of the geographies at stake in food consumption, at every scale from the body to the home, the city, the nation and the world.

Goss, J. (1993) The magic of the mall: form and function in the retail built environment, *Annals of the Association of American Geographers*, **83**, 18–47. A brilliant, if somewhat sceptical, analysis of shopping malls and the consumer worlds they embody.

Hartwick, E. (1998) Geographies of consumption: a commodity-chain approach, *Environment and Planning D, Society and Space*, **16**, 423–37. A pithy comparison of the advertising imagery used to sell gold and the ways in which that gold is produced. Argues for a connection of consumers and producers.

Jackson, P. (1999) Commodity culture: the traffic in things, *Transactions of the Institute of British Geographers*, **NS24**, 95–108. A review of work on the selling and consumption of cultural difference within consumer culture, by one of the principal geographical writers on consumption.

Mackay, H. (ed.) (1997) *Consumption and Everyday Life*, Sage, London. An excellent collection exploring the role of consumption in people's everyday, cultural lives.

Miller, D. (1992) The young and the restless in Trinidad: a case study of the local and the global in media consumption, in Silverstone, R. and Hirsch, E. (eds) *Consuming*

Technologies: Media and Information in Domestic Spaces, Routledge, London, 163–82. A typically insightful and exuberant analysis from perhaps the best contemporary writer on consumption. Here he analyses the consumption of an American soap opera in Trinidad. His conclusion is that paradoxically this soap, despite its American origins, is authentically Trinidadian.

Geopolitics, states and citizenship

Late in 1998 General Augusto Pinochet, one-time President of Chile (1973–90), paid a visit to London for private meetings and medical treatment. Whilst in hospital he was arrested by British police. The arrest took place on behalf of the Spanish courts who had charged Pinochet with crimes against Spanish citizens in Chile who were murdered by the brutal military regime that Pinochet had headed in the 1970s and 1980s. At once, this reminded the world of Pinochet's politics and the right-wing regime that he had led. It also raised complex questions about sovereignty. Could a former head of state be prosecuted in one country for alleged crimes committed in another? Did this mean that Spain and the UK had the right to violate Chilean 'sovereignty'? Or is 'sovereignty' and the right of the 'nation-state' an excuse for the otherwise inexcusable? Which law should apply? Spanish, Chilean, British, European, International? What about crimes committed abroad or in their own countries by the Spanish and British armed forces in the past? What geographical *scale* of justice should apply? For his supporters in Chile, Pinochet had 'saved' the country from Soviet/Cuban-led communism. He had after all (with decisive support from the USA) overthrown a leftist government, and many of those imprisoned and murdered by his regime were seen as 'communists'. Many other Chileans celebrated Pinochet's arrest. Those who felt themselves or their families and friends to have been victims of Pinochet's regime saw the possibility of some justice. The former British Prime

Minister Margaret Thatcher reminded whoever cared to listen to her that her friend General Pinochet had assisted Britain during its 1982 war with Argentina to recapture the Falkland Islands. So the Chilean story and career of General Pinochet always had *global* dimensions. These incorporate a variety of domains, scales and modes of political analysis, imagination and action; a variety of 'political geographies'.

This section will enrich understandings of what can be meant by 'political geographies'. By the time they finish the section, readers will have an understanding both of some of the ideas that Pinochet drew upon to justify and codify his rule and of powers (and sometimes the fragility) of states, nations, and sovereignty. We aim to whet appetites for the complex topics of political geographies, to encourage critical thinking about often controversial, contested and complex issues and topics.

Those who have reached this section after reading the rest of the chapters (congratulations!) should already be aware that many elements of geography touch on political issues. Such 'politics' may not be confined to those things that are usually designated by the term: governments, elections (or the lack of them), political parties and so on. Rather, readers will have noticed that debates about culture, economic change, history, changing gender relations and many other things described in this text are 'political'. They are political in the sense of being about power, albeit at all kinds of levels and in lots of different ways. So, in an important way, *all* geographies are political geographies. Though focused on geopolitics, nation-states, citizenship and collective actions, the chapters that follow aim to bring that home. The section begins with an examination of the diversity of geopolitical traditions (Chapter 15), showing how the meanings of the term 'geopolitics' have varied and continue to shift and how they articulate local and global visions of geography and power. Chapter 16 introduces the historical and geographical variability of nations and states. It stresses how neither the nation nor the state is to be taken at face value. Or in other words, how nation-states are complex symbolic systems which crucially depend upon particular visions of territory, place and space. In the same vein, Chapter 17 will extend understandings of the contested and uneven formats of citizenship and collective action in an interlinked, but still conflict-ridden world.

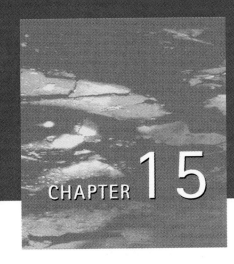

CHAPTER 15

JAMES SIDAWAY

Geopolitical traditions

[US National Security advisors] Robert McNamara and McGeorge Bundy found US policies toward the Latin American military to have been 'effective in the goals set for them', namely 'improving internal security capabilities' and 'establishing predominant US military influence'. That was very important, because, in what they called the 'Latin American cultural environment' – remember, these are serious intellectuals, so they're interested in things like that – it is the task of the military 'to remove government leaders from office whenever, in the judgement of the military, the conduct of these leaders is injurious to the welfare of the nation'.

(Chomsky 1998: 7)

Topics covered

- Origins and history of geopolitics

- Diversity and dissemination of geopolitical discourses

- Changing 'World Orders'

15.1   Introducing the idea of a geopolitical tradition

Whilst the term 'geopolitics' can refer to many things, it is important for students of geography (a subject with which geopolitics has been associated) to understand that geopolitics is commonly associated with a particular way of writing (and thinking) about space, states and the relations between them. Often this takes the form of mapping, emphasizing the strategic importance of particular places.

Geopolitics in this sense sees itself as a tradition. That is, something conscious of its unfolding historical development and with a sense of important founders (and certain key texts written by them). Those who write about this tradition or see themselves as working within or extending it usually trace its origins to the late nineteenth-century writings of a conservative Swedish politician Rudolf Kjellén. Kjellén is reputedly the first person to have used the term geopolitics in published writings (see Holdar 1992). But beyond this idea of a founding moment when the geopolitical 'tradition' begins with the first use of the term by Kjellén, things start to get complicated. The 'tradition' divides, fractures, multiplies and finds itself translated into many languages and cropping up in everything from the writings and speeches of American politicians to texts written by Brazilian generals. All these reinvent and re-work the 'tradition' as they go along. As the introduction to a recently-published collection of 'geopolitical writings' explains:

> the word 'geopolitics' has had a long and varied history in the twentieth century, moving well beyond its original meaning in Kjellén's work. . . . Coming up with a specific definition of geopolitics is notoriously difficult, for the meaning of concepts like geopolitics tends to change as historical periods and structures of world order change. (Ó Tuathail 1998: 1)

In other words, exactly what is meant by the term 'geopolitics' has changed in different historical and geographical contexts. However, a good way to understand something about this idea of a tradition of geopolitics is to look at some of the people, places and ideas associated with one of its most significant appearances, in the ideology of the right-wing dictatorships that ruled many South American countries in the 1960s, 1970s and 1980s. Although today they are nearly all democracies, between the mid-1960s and the mid-1980s many South American countries were dominated by their armies. In most cases they had come to power through military coups and became more concerned with internal security (patrolling the towns and country as a kind of police force and running the wider government) than with fighting or preparing to fight wars with other countries. In virtually all cases the military regimes received a significant degree of support from the United States, which had often trained the military elites in counter-insurgency and provided weapons and logistical support. For in the US, Central and South America and the Caribbean were presented as its legitimate sphere of influence, America's 'backyard'. The US had first declared this to be the case back in 1823 – when it was termed the 'Monroe Doctrine' (so named after the fifth US President James Monroe, 1817–25) (Figure 15.1). However, the full significance of the Monroe Doctrine would only be fulfilled once the USA achieved superpower status after 1945. Alarmed by the presence of pro-Soviet 'communist' forces in Cuba following the successful revolution there in the late 1950s which overthrew a corrupt US-backed regime, the United States was determined to allow no more radical governments to emerge in the region. US political and economic interests would be protected – by whatever means necessary.

Chile is a good example, both of the US role and of the active presence of the geopolitical tradition. On 11 September 1973, the elected left-wing government of Chile (headed by the communist President Salvador Allende) was overthrown in a brutally violent coup led by the head of the armed forces General Augusto Pinochet. The left-wing regime that Pinochet overthrew had already faced American economic and political power and sanctions and the CIA backed the coup. Promising to impose order on a country whose politics had long been democratic but was wracked by left–right divisions, Pinochet and his generals ruled Chile with an iron fist. All opposition was crushed and thousands of people were rounded up by the armed forces. Many 'disappeared' into secret military jails. Hundreds of these prisoners were tortured and murdered. The use of torture became a routine instrument of state. For the next seventeen years, Chile was ruled by Pinochet and his cronies who imposed a new 'neoliberal' economic and political model (of the form that is described in Chapter 10) and murdered any significant opposition. Pinochet finally stepped down as head of state in 1990, when democracy was restored, but not before he had entrenched himself as a senator (in the parliament) for life. He also guaranteed that the armed forces (which he continued to head) would continue to have a significant role in Chile's state structures and that they could not be prosecuted for any crimes committed during the coup and brutal years of military rule.

Figure 15.1 The Monroe doctrine. *Source*: Skidmore and Smith, 1988

Plate 15.1   General Pinochet: geopolitics in action. (Popperfoto/Reuters)

The ideology of the military regime in Chile (like those elsewhere in South America) was intensely nationalist. More importantly it was a particular conservative form of Chilean nationalism. The nation was held to be sacred and the military were rescuing this 'sacred body' from communists, subversives and so on (i.e. anyone who opposed the military vision of ultra-nationalism and order and the neoliberal economic and social model which was now imposed). It is in this nationalism and conception of the nation as a kind of sacred body that geopolitics enters the picture. A few years before the coup, Pinochet had published an army textbook on geopolitics. Intended mainly for use in Chilean military academies, the 1968 textbook indicated that Chile's future dictator took the subject deadly seriously. For Pinochet, geopolitics was a science of the state, a set of knowledges and programmes to perfect the art of statecraft, that is to strengthen the state in a continent and wider world in which it was held to be in competition with others. The starting point and the heart of Pinochet's textbook and of the wider bleak tradition of geopolitics that it presents is an organic theory of the state.

15.2 The organic theory of the state

This idea is at the core of most South American geopolitical writings, and of the wider geopolitical tradition. In summary, it holds that the state or country (and the sense of nation that goes with it), be it Chile, Germany, Turkey or wherever, is best understood as being like a living being. Like any living organism, it therefore needs space to grow and it will be in competition with other living beings. An idea that the state needs 'living space' (a certain amount of resources and land) for the nation to thrive is particularly evident in South American geopolitics. A certain reading of Charles Darwin's idea of a struggle for the 'survival of the fittest' is therefore transferred to the realm of states. Drawing upon conservative German writings of the nineteenth and early twentieth centuries (which had been codified by the German academic Friedrich Ratzel) and with this organic notion of the state in mind, geopolitics claims to identify certain laws which govern state behaviour and which, once identified, can be a guide for those charged with furthering and protecting the 'national interest' (or at least a certain definition of the latter).

It is not difficult to see how this idea could appeal to South American military dictators, for it gives the military, equipped with the supposedly scientific study of geopolitics, a special mission. In other words it legitimizes their rule. Moreover, the organic idea was extended further to legitimize the extermination of those whom the military defined as enemies of the state. People labelled as communist or subversive, indeed all those who oppose the military dictatorship, can be compared with a disease or cancer which threatens the life-blood of the state-organism and is best 'cut out', that is eliminated. The otherwise unthinkable (the murder of thousands of people in the name of the wider welfare of the nation-state) is made to seem natural and a good thing, since it serves the longer-term interest of the health and power of the 'living being' that is the state (see Hepple 1992 for an exploration of this). With the 'cancer' of subversion and disloyalty eliminated the 'state-body' can go on to grow and thrive.

15.3 Brazilian national integration

Aside from Chile, one of the most important expressions of geopolitics in South America has been in Brazil. The military in Brazil overthrew the democratic government in 1964 and stayed in power for the next twenty years. Again the Brazilian military were supported by the USA, whose ambassador in Brazil termed the military coup 'the single most decisive victory of freedom in the mid-twentieth century' (cited in Chomsky 1998: 7). Although never as despotic as the regime led by Pinochet (or similar military governments in Argentina and Uruguay), the long years of military rule gave the Brazilian generals a chance to elaborate and impose a geopolitical vision on the country. One of the most evident aspects of Brazilian geopolitics is the idea that state security requires a measure of national integration. The vast scale of Brazil, the difficulty of travel across its Amazon 'heartland' (the world's largest tropical rain-forest) and the fact that it shares borders with every other South American country except Chile gave the geopolitics of the Brazilian generals a special sense of its national mission and an obsession with the potential for the country to be a great power (known in Portuguese as *Grandeza*). The associated sense of the urgency of the integration of Brazil required the extension of a network of highways

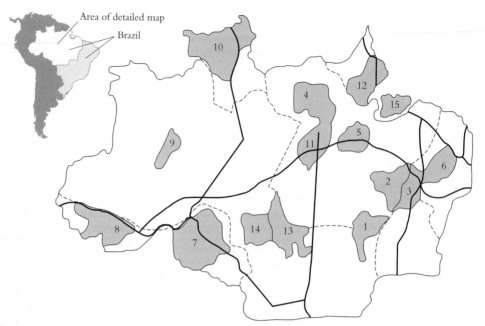

Figure 15.2 National integration in Brazilian geopolitics: Amazon development poles. *Source:* Hecht and Cockburn, 1989: 127

across Amazonia and the settlement of its lands by farmers and ranchers. It is this geopolitical-inspired vision, combined with a highly corrupt system of patronage and favours to those close to the regime, that underlies the enclosure and division of Amazonia into private lands (some larger than European countries such as Belgium) and the accompanying transformation or destruction of the tropical rain-forest (see Hecht and Cockburn 1989) (Figure 15.2).

The consequences of this geopolitically motivated and economically profitable strategy have been disastrous for the indigenous peoples of Amazonia. They have found themselves forced off land that was traditionally theirs, sometimes murdered or attacked by ranchers, the state and settlers and disorientated by the arrival of a frontier culture of violence, destruction and consumption. Poor peasants, particularly from the impoverished north-east of Brazil, have also moved into the Amazon region in search of land and freedom. They are fleeing from the oppression and landlessness they themselves face in the north-east of Brazil, where the vast majority of the land is in the hands of a small elite. Of course, in the geopolitical visions of the Brazilian generals, this population movement is seen as positive, reinforcing the Brazilian population of the Amazon and redistributing what they regard as marginal 'surplus' people in the process. At the same time this would defuse political pressures for land reform in north-east Brazil. In all this we can see how Brazilian geopolitics is implicated in a complex of social and environmental transformations. The destruction of the Amazonian ecosystem is therefore not simply the result of 'population pressure' or abstract forces of development, rather it is to be understood as a social process and therefore as the consequence of particular political choices and (im)balances of power.

15.4 Antarctic obsessions

An interest, at times an obsession, with the last unexploited continent of Antarctica is evident in both Brazilian and Chilean geopolitics. Antarctica is also prominent in geopolitical writings from Argentina and Uruguay and even appears in Peruvian and Ecuadorian geopolitical writings (see Dodds 1997). To understand what this amounts to and what forms it takes necessitates some understanding of Antarctica's exceptional political geography. Antarctica has a unique territorial status, in so far as it has no recognized state on its territory. Everywhere else in the world forms part of the patchwork of states that we learn to be familiar with and take for granted. Antarctica is a stark reminder that there is nothing natural or inevitable about this. The absence of recognized states in Antarctica reflects the inhospitability and remoteness of the continent, the only one without an indigenous human population. Although the plankton-rich waters around Antarctica were exploited for seal and whale hunting in the nineteenth century, it was not until the twentieth century that exploration of the interior began. Even today, Antarctica has only a non-permanent population (at any time) of a few hundred scientists. This presence, plus the relative proximity of the continent to South America and the possibility of exploitable mineral resources, has led to a series of territorial claims. Although Argentinean writers had already represented Antarctica as an extension of the southern Argentine area of Patagonia, the British made the first formal claim in 1908. In turn, parts of this claim were 'granted' to Australia and New Zealand. France and Norway made claims in the 1930s and 1940s, followed shortly by Argentina and Chile. After 1945, the USSR and the USA also established a wide network of bases, although without staking formal claims to territory. The claims made by the Argentine, Chile and the UK overlapped – and foreshadowing the British–Argentine conflict over the Falklands/Malvinas British and Argentine forces exchanged fire in Antarctica in the late 1940s and early 1950s. In the context of this potential for conflict and growing possibility of Cold War confrontation (on the Cold War, see Section 15.8) a United Nations Treaty in 1959 agreed that all claims would be (forgive the pun) 'frozen' for at least thirty years, and the continent reserved for scientific (not commercial or military) use. The Treaty was extended in 1991. But prior to then, it was not clear what the future status of the continent would be. Even today, the Treaty merely defers the issue of claims.

Argentina is the South American country with the largest claim. The Argentine claim has also become central to geopolitical discourse there. As Child (1985: 140–1) explains in a study of geopolitics and conflict in South America:

> For Argentine geopolitical writers, the subject of Antarctica is not only linked to tricontinental Argentina [a power in the South American continent and the South Atlantic], but also to ... national sovereignty, patriotism, and pride. This is a particularly touchy combination after the humiliating defeat of the Malvinas/Falklands conflict. The Argentine National Antarctic Directorate has professors of Antarctic geopolitics on its staff. Through the media, maps, and postage stamps and the centralized educational system, Argentines are constantly taught and reminded that there is an Argentine Antarctic just as much are there are Argentine Malvinas. The need to assert Argentine rights in the Argentine Sea, islands, and Antarctica is linked to dreams and national projects of Argentine greatness.

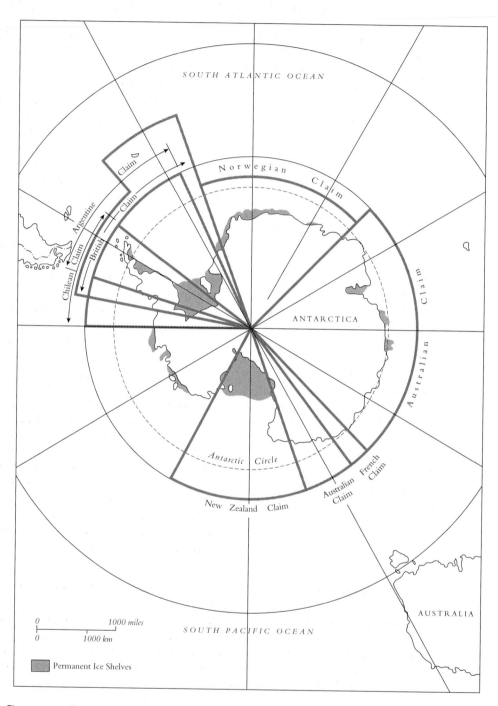

Figure 15.3 Claims on Antarctica. *Source*: Adapted from Glassner, 1993: 498

15.5 Heartland

Although Antarctica and the South Atlantic are significant components of Argentine (along with Brazilian and Chilean) geopolitics, the former also looks beyond other Argentine frontiers. In particular, Argentine geopolitical writers have taken a particular interest in the security of Bolivia, together with some Brazilians and Chileans. They have scripted Bolivia (a relatively impoverished land-locked mountainous country, which has borders with Argentina, Brazil, Chile, Paraguay and Peru) and proximate areas of its neighbours as a key strategic continental 'heartland'. Control of Bolivia, in this vision, would be a vital key to a relative dominance in the South American continent. That Bolivia has a strong revolutionary tradition and was for many years characterized by chronic political instability has reinforced the tendency of the other South American countries to meddle in Bolivian politics. Indeed, during the years (1976–82) when Argentina was last ruled by a geopolitically obsessed military junta, the Argentine armed forces were actively involved in supporting a Bolivian military government. This activity took the form of the kinds of brutal suppression and frequent murder of those (trade union leaders, dissidents, opposition members and leaders) who opposed the military government and its economic and social strategies. The idea of Bolivia as a 'heartland', control of which would be a kind of magic (geopolitical) key to domination of South America, links back to one of the best known genres in classical (European) geopolitical thought. For although the South American countries have seen some of the most significant expressions of geopolitical discourse of modern times, geopolitics originates in Europe and it is to some examples of European geopolitics, including the idea of heartland, that we now turn.

The designation of a heartland was first made by the British geographer (and strongly pro-imperialist conservative politician) Halford Mackinder. In what has since become a widely cited (if less often read) article first published in 1904 following its presentation to the Royal Geographical Society (RGS), Mackinder argued that the age of (European) geographical exploration was drawing to a close. This meant that there were hardly any unknown 'blank' spaces left on European maps of the world. According to Mackinder, the consequence of this closing of the map, this end of the centuries-long task of exploration and discovery, was that political events in one part of the world would invariably affect all others, to a much greater extent than hitherto. There would be no more frontiers for Europeans to explore and conquer. Instead, the great powers would now invariably collide against each other. Mackinder called this end of European exploration 'the post-Columbian age' and the closing of frontiers, the emergence of a 'closed political system':

> From the present time forth, in the post-Columbian age, we shall again have to deal with a closed political system, and none the less that it will be one of world wide scope. Every explosion of social forces, instead of being dissipated in a surrounding circuit of unknown space and barbaric chaos, will be sharply re-echoed from the far side of the globe, and weak elements in the political and economic organism of the world will be shattered in consequence. There is a vast difference of effect in the fall of a shell into an earthwork and its fall amid the closed spaces and rigid structures of a great building or ship. Probably some half-consciousness of this fact is at last diverting much of the attention of statesmen in all parts of the world from territorial expansion to the struggle for relative efficiency. (Mackinder 1904: 422)

Figure 15.4   'The Natural Seats of Power' according to Mackinder (1904).

Given that this was the case, Mackinder claimed to identify the places of greatest world-strategic significance, control of which would give any great power a key to world power. In his 1904 paper, he termed this the 'pivot area'. With Mackinder's address to the RGS and his subsequent article came a series of maps, the most frequently reprinted one of which claims to describe 'The Natural Seats of Power' (Figure 15.4). As Ó Tuathail's (1996: 25) critical account of Mackinder has recently argued:

> Mackinder's January 25 04, address to the Royal Geographical Society, 'The Geographical Pivot of History', is generally considered to be a defining moment in the history of geopolitics, a text to which histories of geopolitics invariably point.

Mackinder himself did not use the term geopolitics. But his 1904 paper and its maps were later appropriated into other geopolitical traditions. Indeed, the map of 'The Natural Seats of Power':

> is perhaps the most famous map in the geopolitical tradition. ... The map is meant to create the illusion of the unveiled geographical determinants of power. It is the global vision, the post-Columbian made possible, the pure panorama produced by the penetrative gaze of the global voyeur. Yet, interestingly, the map is not purely visual but is marked with text. Huge swathes of territory are stamped with a definitive positionality and function: 'Pivot Area', 'Inner or Marginal Crescent', and 'Lands of Outer or Insular Crescent'. These are macrogeographical identities around which Mackinder spatializes history and reduces it to formulaic equations and timeless east/west, land/sea conflicts. 'The social movements of all times', he concluded, 'have played around essentially the same physical features.' (Ó Tuathail 1996: 31–3)

A certain geographical determinism can be detected here, in which the physical environment ('essentially the same physical features' in Mackinder's words) provide the fixed, unchanging backdrop to and therefore something of the explanation of human history. This view, which ignores the social construction of resources (see Chapter 8) and other features of nature, is common to virtually all geopolitics (something which it shares with a great deal of twentieth-century geography). One of the keys formulas in Mackinder's notion of geographical determinism is his identification of a 'pivot-area' or 'heartland zone' in east-central Europe, control of which would be a kind of magic key to world domination. Reworking the ideas of his 1904 article after the First World War, Mackinder refined this notion and penned a formula which encapsulated it:

> Who rules East Europe commands the Heartland.
> Who rules the Heartland commands the World-Island.
> Who rules the World-Island commands the world. (Mackinder 1919: 4)

According to this (simplistic) formula, 'rule' of the eastern portion of Europe offered the strategic path to that of the African-Asian-European continents (which together constitute what Mackinder terms the 'World-Island') and hence a dominant position on the world scene.

In addition to the general backdrop of the emergence of the sense of closed political space, or in Mackinder's terms the completion of the age of exploration (which he called, after its best known figure, Christopher Columbus, the Columbian age), it is important to recognize that Mackinder, whilst claiming to offer a universal and objective model (with 'laws', for example the formula about 'Heartland') is in fact highly subjective. Heartland is another organic metaphor, invoking the idea that territory has a *heart*, like a living being. Moreover, Mackinder's writings reflect not only the moment at which they were penned, but also the place. Mackinder, writing from the vantage point of imperial Britain, is concerned to identify threats and dangers to British power. At the time when Mackinder presented his paper to the RGS, Britain was the pre-eminent world power. It still seemed that way to Mackinder in 1919 when he wrote about the 'Heartland'. But British imperial politicians, like Mackinder, were aware of the growing power of America, Germany and Russia. In fact, potential British imperial competition with the latter in Asia provided a key context to Mackinder's work. As Peter Taylor (1994: 404) explains:

> Behind every general model there is a specific case from which it is derived. For the heartland model this is particularly easy to identify. Throughout the second half of the nineteenth century Britain and Russia had been rivals in much of Asia. While Britain was consolidating its hold on India and the route to India, Russia had been expanding eastwards and southwards producing many zones of potential conflict from Turkey through Persia and Afghanistan to Tibet. But instead of war this became an arena of bluff and counter-bluff, known as the 'Great Game'. . . . Mackinder's presentation to an audience at the Royal Geographical Society would not have seemed so original as it appears to us reading this paper today. . . . Put simply, the heartland model is a codification and globalization of the Great Game: it brings a relatively obscure imperial contest on to centre stage.

Not only does this envisage the world in a particular way, as a 'stage', but it sees only select key actors as the significant figures at play. These are the European powers (plus Russia). Other peoples and places are merely the backdrop for action by White Men. The taken-for-granted racism of Mackinder's model, in which only Europeans make history, is also that of European imperialism and that of the bulk of wider European geographical and historical writings of the time (see Chapter 3).

Yet, although Mackinder's 1904 paper is very much a product of its time and Mackinder's own conservative world view, it has proven durable and has been integrated into rather different contexts, which saved it from the relative obscurity that it deserves as a turn-of the-century imperialist text. The transfer of the discourse of 'Heartland' to Bolivia by South American codifiers of geopolitics has already been noted. In addition, 'Heartland' was appropriated by German geopolitics in the 1930s and 1940s and formed part of the backdrop to Cold War American strategy from the late 1940s through to the last decade of the twentieth century. Taylor (1994: 405) notes:

> First in the inter-war years the heartland theory became an integral part of German geopolitics. It fitted the needs of those who advocated *lebensraum* [living-space], the policy of expanding into eastern Europe [and enslaving and murdering its indigenous populations], coupled with accommodating the U.S.S.R., as a grand continental policy for making Germany a great power again. Second with the onset of the Cold war after World War II, the heartland theory got another lease of life as the geostrategic basis of nuclear deterrence theory. The west's nuclear arsenal was originally justified in part as compensation for the U.S.S.R.'s 'natural' strategic advantage as the heartland power.

The next two sub-sections of this chapter will examine aspects of Nazi, Fascist and Cold War geopolitics in greater detail.

15.6   Nazi and Fascist geopolitics

The formal tradition of writing about space and power under the title of 'geopolitics' also found fertile contexts in Italy, Portugal, Spain and Japan. Influenced and supported by Nazi Germany, all these countries (together with Hungary and Romania) saw the rise and victory of ultra-nationalist or Fascist governments (often through violent struggle or full-scale civil war with democratic or communist forces). In each case geopolitical debates were crucially negotiated through other cultural and political debates about race, nationalism, the colonial pasts and futures, supposed national 'missions' and destinies and the European and global political contexts. The German case has become the best known. In Germany, organic notions of the state had already been popularized by conservative nineteenth-century academics. Moreover, Germany was characterized by extreme political and economic turbulence in the decades following its defeat in the 1914–18 World War. This combination provided a fertile environment for the elaboration and circulation of a distinctive geopolitical tradition. In Ó Tuathail's (1996: 141) words:

> After the shock of military defeat and the humiliation of the dictated peace of Versailles, the Weimer Republic proved to be fertile ground for the growth of a distinct German geopolitics. Geopolitical writings, in the words of one critic, 'shot up like mushrooms after a summer rain'.

The main features of these writings (which they shared with a wider German nationalism, later codified in Nazism) were a critique of the established 'World Order', and of the injustices imposed on Germany by the victors. German claims were often presented

Figure 15.5 'Pan-Regions' as envisaged in Nazi geopolitics. *Source*: O'Loughlin and Van der Wusten, 1990, 'Political geography and the pan regions' Geographical Review 80, adapted by permission of the American Geographical Society

graphically in maps that were widely circulated (see Herb 1989: 97). Like the variants of the geopolitical tradition that were developed amongst right-wing and military circles in Italy, Portugal, Spain and Japan, German geopolitics also asserted an imperial destiny. Indeed, as Agnew and Corbridge (1995: 58–9) explain: 'The Nazi geopoliticians of the 1930s came up with formalized schemes for combining imperial and colonized peoples within what they called "Pan-Regions".'

In this vision, notions of racial hierarchy were blended with conceptions of state 'vitality' to justify territorial expansion of the Axis powers (see O'Loughlin and van der Wusten 1990) (Figure 15.5). In Europe, related conceptions of the need for an expanded German living-space were used as justification for the mass murder of occupied peoples and those who did not fit into the grotesque plans of 'racial/territorial' purity. The practical expression of these was the construction of a system of racial 'purification' and mass extermination. At least 6 million Jewish people were murdered in concentration camps together with millions of others: disabled people, gays and lesbians, gypsies and political opponents. Historical debates about the role and relative significance of German geopolitics within the Holocaust and within broader Nazi ideology and strategy continue (see Bassin 1987, Heske 1986: 87, Ó Tuathail 1996). In the United States in particular, German geopolitics became the subject of lurid tales and depictions, cropping up in media, military and government 'explanations' of Nazi danger. Popular magazines, such as *Readers Digest,* would inform Americans of the 'scientists behind Hitler' describing German geopolitics as *the* key to Nazi strategy. And whilst the relative significance of the German geopolitical tradition in the wider genocidal ultra-nationalism of German Fascism was certainly overstated in such accounts, we should see geopolitics as a particular

Plate 15.2   Nazi concentration camps: racial/territorial 'purification'. (Popperfoto)

expression of *wider* academic and intellectual involvement and complicity in authoritarian state power, war-making and genocide. Much more widely, beyond the geopolitical tradition *per se*, academic geography was deeply implicated in these activities.

An example of this wider complicity was the way that the models of 'Central Place Theory' (an abstract model of the ideal spatial distribution of towns), which were developed in Germany by Walter Christaller in the 1930s, were elaborated with the express purpose of providing planners with a model of German settlement to impose on conquered territories of Eastern Europe, once the indigenous Jewish and Gypsy people had been gassed to death and the majority Slavic population enslaved (see Rossler 1989).

The defeat of the Axis powers in 1945 (culminating in the use of American atomic weapons against the civilian populations of Hiroshima and Nagasaki) and the lurid wartime depictions of Nazi geopolitics in the United States dealt something of a blow to the formal tradition of geopolitics. However, it continued in subsequent decades, both in Spain and Portugal (see Sidaway 1999, 2000), where Fascist regimes remained in power to the mid-1970s and, as we have seen, in Latin America. These regimes (and plenty of others) were integrated into the US-led anti-communist network of allies. In a policy that became known as containment, the US aimed to encircle and block the potential expansion of Soviet power and influence beyond the immediate borders of the USSR and the pro-Soviet states installed by the USSR in the Eastern European territories that it had occupied during its Second World War battle for survival against Nazi Germany. In the allied countries (and as an expression of this strategy of containment), the functions of geopolitics, in particular its 'strategic vision' and claim of 'scientific' objectivity, continued

Dr. K. Haushofer

THE WORLD OF
GENERAL HAUSHOFER

Geopolitics in Action

BY ANDREAS DORPALEN

With an Introduction by
Colonel Herman Beukema, U.S.A.

FARRAR & RINEHART, INC.
New York • Toronto

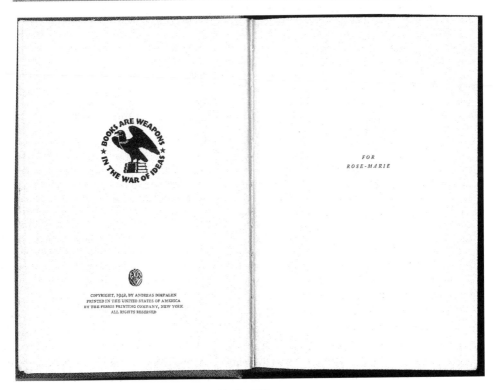

BOOKS ARE WEAPONS
★ IN THE WAR OF IDEAS ★

*FOR
ROSE-MARIE*

Plate 15.3   German geopolitics: as represented in the USA

to operate or were displaced into other disciplines and branches of knowledge, including geography and the expanding subject of International Relations. This displacement of strategic knowledge was particularly evident in the United States, which by 1945 was the greatest power that the world had ever seen, constituting over half of the world economy and (until the recovery of the Soviet Union and its own development of atomic weapons in 1949) possessing a virtually unrivalled military capacity.

15.7   Cold War geopolitics and the logics of containment

> Like a fast-moving glacier, the Cold War cut across the global terrain in the 1940s, scraping and grinding solid bases, carving and levelling valleys, changing the topography. For nearly half a century the Cold War competition between the United States and the Soviet Union dominated the international system. . . . Like all retreating glaciers, the receding Cold War left deposits of debris and exposed an altered, scarred landscape that leaders in the 1990s struggled to measure and fathom. (Paterson 1992: 2)

The Cold War is used as a shorthand description of the conflict (which to many seemed to be a conflict for world domination) between the 'communist' East, led by the USSR and the 'capitalist' West, led by the USA between around 1949 to around 1989. This was not simply a rift between two great powers, but a complex ideological conflict, which often appeared to be about different ways of life and contrasting social systems (the bureaucratic 'command economies' based on predominant state ownership of the means of production, with a ruling class drawn from the bureaucracy, and the capitalist economies based on predominantly private ownership with a ruling class drawn from the owners of capital). It was also a conflict in which *direct* military confrontation between the two great powers was avoided. Hence the metaphor of Cold, i.e. 'poor, frozen or frosty' relationships, but short of all-out 'hot' war, involving direct exchanges of missiles, bombs and so on. In part this avoidance of direct conflict arose because both powers were conscious of the enormous stakes, armed with nuclear, chemical and biological weapons as well as vast arsenals of increasingly high-tech 'conventional' armaments. By the 1960s, they were each capable of destroying virtually all human life several times over. Moreover, neither could attack the other without being sure that the enemy would still possess enough undamaged nuclear weapons to retaliate, a 'balance' of terror known, appropriately enough, as MAD (Mutually Assured Destruction). Instead, conflict took the form of the continual preparation for war, plus proxy wars in what became known as the (originally not pro-West, not pro-East, but contested) 'Third World' and many other political, economic and cultural forms of competition. That is, it was a conflict conducted by *every* means *except* direct military confrontation between the superpowers.

Of course there was no shortage of indirect confrontation and preparation for war. Europe was divided into two armed camps and split down the middle between Soviet and American zones of influence by a fortified military and ideological frontier that became known as the 'Iron Curtain'. Whilst Europe was characterized by an uneasy stability and balance of terror, by the 1970s the term geopolitics had been revived (or rediscovered) by US national security advisors Henry Kissinger and Zbigniew Brzezinski (see Hepple 1986, Sidaway 1998) to refer to the strategic vision deemed necessary to circumvent the sense of growing Soviet power, particularly in the **Third World**, where a wave of successful revolutions had brought left-wing, pro-Soviet governments to power. As has already been noted, one of these was in Cuba, but by the end of the 1970s, there were also pro-Soviet governments in the countries of Indochina (Vietnam, Laos and Kampuchea) where the US had been unable to defeat communist insurgencies despite a massive military effort, in the African countries of Angola and Mozambique, in the Arabian country of South Yemen and in Afghanistan as well as in the small Central American state of Nicaragua. Communist China and North Korea added to American concerns. In this world of super-power competition, geopolitics again found its moment and expression in the visions of the national security advisors and American generals.

But to focus on formal geopolitics would be too narrow. For the relative absence of explicit references to geopolitics in American strategic culture during the 1950s and 1960s did not mean that the kind of mapping of places as pacified/dangerous, ours/theirs, threatening/stable, strategic/secure, and the intersection of a geographical imagination with a military/strategic one that characterize the 'common-sense' understanding of the geopolitical tradition was absent. As has been noted, such understandings proliferated into other academic realms, practices, knowledges and disciplines. As well as cropping up in the subject of International Relations, this proliferation of quasi-geopolitics could be found in academic geography where, under the title of 'political geography', American geographers would proclaim their difference from the 'disreputable' tradition of geopolitics, and then proceed to engage in more or less the same kind of strategic mapping and policy advocacy (see Smith 1984). More widely, American leaders and policy-makers declared that the Soviet Union had to be *encircled* and *contained*. The metaphor of disease (containment) was mixed with that of 'dominoes' – if one country (say Vietnam or Cuba) had 'fallen' to Soviet control or influence, then it could (like a chain reaction) 'infect' other proximate ones. The irony of this (which is equally present in the formal geopolitical tradition) is that in the name of security and strategy, the real complexity of *human* geographies in the places (such as Vietnam or Chile, or Nicaragua or the Persian Gulf) that are deemed

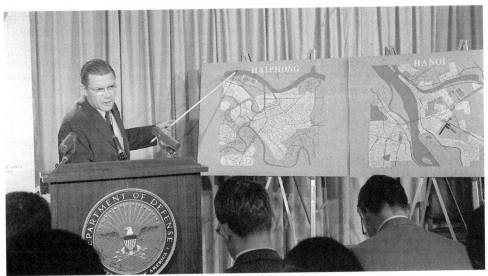

Plate 15.4 'Containment' explained: US Secretary of State for Defense briefing journalists during the Vietnam War. (Corbis UK/Bettmann)

strategic is obscured or erased. Forget about the people, the details of culture and society, what matters is the 'strategic value' of the place. At times this could take quite extreme forms, as with US National Security Adviser Robert McNamara (previously head of Ford, and later head of the World Bank):

> Robert McNamara was, of course, the leading specimen of *homo mathematicus* – i.e., men who behave and believe other men [and, we might suppose, women too] behave primarily in response to 'hard data', usually numbers (infiltration rates, 'kill ratios', bomb tonnage). Like the classic private eye on television, they are always looking for 'the facts', but usually the wrong ones. They miss reality, for they never get close enough or related enough to another society to do more than count things in it. If you relate to a country as a military target you do not need to know anything about it except details as are easily supplied by reconnaissance satellites, spy ships, secret agents, etcetera. You need never know who the victims of your attack were. Your task is merely to assess the results of what you have done and this is done by counting bodies, destroyed factories, enemy soldiers. (Barnet 1973: 119)

Such logic, which has found more recent expression in the 1990–1 Gulf War (see Sidaway 1994: 98, Ó Tuathail 1993), in US actions against what they describe as 'terrorist bases' in Sudan and Afghanistan, and in the 1999 war with Yugoslavia, is seemingly far removed from the prominence of maps in the formal geopolitical tradition. Yet we can see that, in *both* the kind of thought present in McNamara's mind and in classic geopolitics, are moments when the myriad complexity of the world is reduced to a simple black and white representation, be it numerical or cartographic:

> By gathering, codifying, and disciplining the heterogeneity of the world's geography into the categories of Western thought, a decidable, measured and homogeneous world of geographical objects, attributes, and patterns is made visible, produced. (Ó Tuathail 1996: 53)

These logics are both *colonial* and *colonizing*. They are colonial, in so far as they are rooted in the fact-gathering epoch of colonial exploration and share the arrogance of its classifications of peoples and places, which are then turned into a hierarchy of civilized to uncivilized, with the former (who are seen in racist terms) as the 'natural' leaders, or most developed, free or 'advanced'. They are also colonizing, in so far as they expand into and affect other seemingly distant fields of knowledge and social or cultural activities. Everything from scientific debates to art and cinema becomes caught up in geopolitics, and registers its spread, its dissemination.

15.8 Cold War geopolitics in art and culture

The art world and, in particular, the course of modern art in the second half of the twentieth century became intertwined with Cold War geopolitics. More generally, art and politics have frequently been intertwined. The Nazis (and their leader Adolf Hitler, who was a failed artist of minimal talents) detested modern art, condemning it as decadent. For them a gross parody of classical art, sculpture and architecture was all that was fit for the German Reich. Other dictators, such as Joseph Stalin, the leader of the Soviet Union between the mid-1920s and his death in 1953, were also wary of modern art and favoured a garish realist style, depicting heroic smiling workers and peasants. Meanwhile, and especially in the years between the two World Wars, Paris became the centre of production of modern art. But this was disrupted by the devastation of the Second World War and the rise of American economic, cultural and political powers. As Sylvester (1996: 405) explains:

> In the wake of World War II and the catapulting of the United States into superpowerdom and the Cold War, there was parallel activity in US art circles. Media figures, such as Howard Devree and Alden Jewell, advocated a position for US art in the world that would mirror the international ascendancy of the US political economy. Devree maintained in the *New York Times* that the United States was the most powerful country and therefore needed to create art that was strong and virile to replace the art of Paris.

But this was not only words from influential figures. The American Central Intelligence Agency (CIA) stepped in to fund and promote American modern art, bankrolling major exhibitions and tours, to demonstrate the creative and in Devree's words, the 'virile' artistic culture of the United States. This assertion of 'virility' reveals too the masculinity of much American Cold War language and policy, differentiating the thrusting, powerful American society (later personified by people such as President John F. Kennedy) from a Europe portrayed as weak and effeminate and from the more brutal, less cultured men in charge of the Soviet Union. This strategy was more subtle than the vulgar anti-communism that pervaded Hollywood at the time. It reinforced more general arguments about the US as a bastion of 'free expression' and seemed to tally with the relative dynamism of the post-war American economy and the sense that the leadership of the Western World (or, as American commentators preferred to call it, the 'Free World') had decisively passed

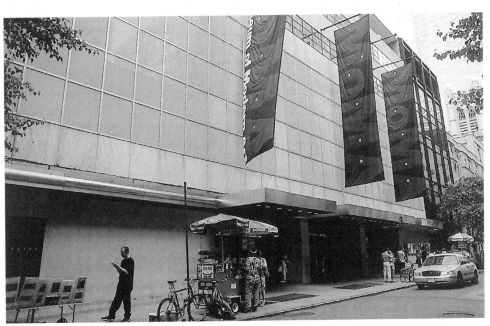

Plate 15.5 American modern art on show: Cold War artifacts? (Trip/M. Barlow)

in military, political, economic and cultural terms to the USA. As Cockcroft (1974: 129) explains:

> CIA and MOMA [the New York based Museum of Modern Art, which was funded by the multi-millionaire Rockefeller family] cultural projects could provide the well-funded and more persuasive arguments and exhibits needed to sell the rest of the world on the benefits of life and art under capitalism.

Of course, the Cold War was waged too in everything from science fiction comics and action movies and thrillers (in which good Americans, like Rambo, and the occasional heroic Brit, like James Bond, faced and defeated totalitarian, alien and frequently communist enemies) to Olympic sports (in which the Soviet, Chinese, East Germans and Bulgarians invested enormous resources in training and drug-enhanced muscles, to 'prove' the superiority of the communist system in the Olympic stadiums).

The important thing to understand here is that whilst the sense of a formal geopolitical tradition may be seen as a self-conscious and fairly direct expression of states, space, power and the relations between them, the latter also take a vast variety of other forms. At the same time, the backdrop of a geopolitical conflict (such as the Cold War) can provide the basis for the making of a coherent sense of a particular way of life, for bonding a country, emphasizing a shared conformist identity and a common enemy (see Campbell 1992; Tyler May 1989).

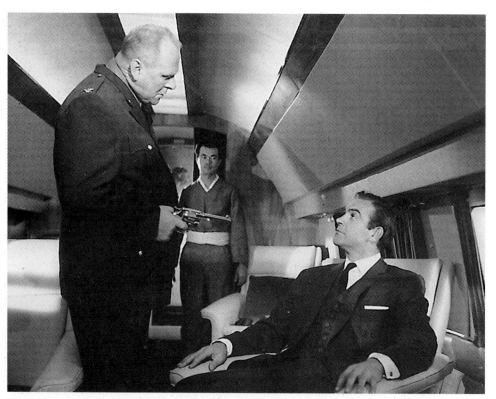

Plate 15.6   Cold War in cinema: James Bond (007) saves the West from communists and criminals. (Topham Picturepoint)

15.10 Conclusions: A new world order and postmodern geopolitics?

The notion of the Cold War faded with the collapse of the USSR and associated communist allies between 1989 and the early 1990s, leaving the United States as the sole effective global military superpower. Yet this Western and American 'victory' has also produced a certain sense of disorientation. This had been evident before, as the so-called 'bipolar' world of the early years of Cold War confrontation in which Washington and Moscow were the main political, economic and ideological points of orientation became progressively more 'multipolar' from the 1960s onwards. That is, as Western Europe and Japan recovered from wartime devastation and as Communist China split on ideological grounds with the USSR, the sense that the world was divided into just two superpower points of orientation lessened. But the sensation of geopolitical complexity was to grow precipitously once one of the established 'poles' (Soviet communism and pro-Soviet regimes in Central and Eastern Europe) collapsed at the start of the 1990s. Having a clearly demarcated and identifiable enemy – as both sides had in the Cold War – offered contrasting but apparently solid ideas of identity, purpose and common cause that has faded in a post-Cold War world. If the Cold War embodied a relatively coherent geopolitical map (East against West, with the allegiance of the Third World as one of the prizes), the post-Cold War world is characterized by a diversity of maps and scripts. A variety of interpretations and geopolitical 'models' now attempt to offer explanation and impose meaning on contemporary events.

Amongst the most influential of these has been the discourse of 'New World Order', articulated by US President George Bush in the early 1990s. In Bush's vision, what is called the 'international community' (the member states of the UN, led by the USA, but with financial and military support from key allies such as Germany, the UK and Japan) would act as a kind of global police force, intervening where and when they felt necessary to maintain or restore 'order'. Both the 1990–1 Gulf War and the 1999 conflict with Yugoslavia were justified and conducted (in part) in the name of building such a 'New World Order'. Critics have pointed out that, as Booth (1999: 49) puts it, the New World Order 'means the New World (gives out the) Orders'. That is, the United States seeks global military hegemony and with the Soviet resistance out of the way, there is little, save its own 'public opinion', to stop it. The 'Monroe Doctrine' mentioned in Section 15.1 is extended to virtually the entire world. Critics point out that this suits Western and especially American commercial interests.

Others have stressed that such American military superiority counts for little and that what is evident in the post-Cold War world is enhanced economic competition between blocs or great powers. In this vision, American power is contested by an integrating European Union and by Pacific powers such as Japan and China. But the terms of the 'contest' are less in conventional military and strategic terms than in competition for markets, productivity and profit. A world of 'geopolitical' competition is, it is argued, being displaced by 'geoeconomic' competition (for an example, see Luttwak 1990). In the US, this has also been interpreted as an impending 'supposed clash of civilizations' (e.g. Huntington 1993), between that of the 'West' and, for example, societies orientated to alternative belief systems such as (Chinese) 'Confucianism' or 'Islam'. Although superficially quite different, notions of geoeconomic competition and a clash of civilizations both betray a sense that the relative military dominance of the USA at the end of the Cold

Box 15.1

The continuing Cold War?

The Cold War in Asia was always a very complex mixture of ideological, territorial and national disputes. An extra dimension was added by fact that China became a nuclear power in the early 1960s at the same time that the Chinese communists split with the USSR on the grounds of ideological differences.

Plate 15.7   The Berlin Wall (symbol of Cold War division) falls in 1989. (Format Photographers/Melanie Friend)

Moreover, whilst it is now a decade since proclamations of the 'end of the Cold War' began to multiply in Europe, in the 'Asia-Pacific' region remnants of the global Cold War confrontation remain in place. Despite substantial changes and accommodations, China, North Korea and Vietnam are still one-party 'communist' political systems. Moreover a series of divisions and territorial disputes whose history is deeply rooted in the Cold War remain unresolved. Most notably, the Korean peninsula remains divided between mutually antagonistic regimes and both the People's Republic of China and the offshore island of Taiwan claim to be the legitimate government of the whole of China.

Arguably, therefore, there is a sense in which the narratives about 'the end of the Cold War' reflect the Euro-Atlantic experience rather than being globally valid. The 'end' may have happened in Europe and the Soviet Union may no longer exist, but Cold War legacies and logics are active in East Asia (see Hara 1999).

War may not in itself be enough to preserve America's sense of leadership, power and world-historical destiny in the context of a complex 'multipolar' world.

Indeed despite early post-Cold War proclamations about the great prospects for a more peaceful and secure New World Order, the world now appears to American strategists as more 'disordered' and unpredictable. 'Enemies' (such as Iraq, Libya, North Korea, Sudan and Yugoslavia) are targeted as the United States continues to declare that certain regimes must be contained or punished. At the same time, high-tech weapons spread, disseminating greater destructive capacity into areas of local confrontation and conflict, such as South and East Asia and the Middle East. Manufacturers in the West, China and the former communist countries have found plenty of lucrative markets for weapons of all kinds, from fighter planes to machine guns. Such exports very often end up in relatively impoverished societies, including many in Africa, the Balkans and Asia that are awash with the destructive debris of the Cold War. And in East Asia a number of disputes and differences rooted in the Cold War endure (see Box 15.1).

Moreover, for the Western powers and for the USA in particular, no longer is the main 'enemy headquarters' so clearly demarcated and fixed behind a stable 'Iron Curtain' – as the USSR and its communist allies were. Although there is no absence of traditional forms of geopolitical discourse, which define a clear boundary of 'our' space against 'them' (this has been evident, for example, in the former Yugoslavia), some have argued that today Western strategy operates in the frame of a new kind of 'postmodern geopolitics'. For example, Gearóid Ó Tuathail (1998: 31–2) examines recent US foreign policy debates and notes how they argue that:

In contrast to the transcendent containment imperative and fixed posture of Cold War strategy, what is required [according to US strategists] in response to persistent territorial concerns and proliferating deterritorialized threats is a geo-strategic doctrine premised on flexibility and speed. The 1995 National Military Strategy of the United States is subtitled 'a strategy of flexible and selective containment'. . . . Threats are described as widespread and uncertain, the possibility of conflicts probable, but their geographic sites are too often unpredictable. This document describes the current strategic landscape as characterized by four principal dangers that the US military must address: 'regional instability; the proliferation of weapons of mass destruction; transnational dangers such as drug trafficking and terrorism; and the dangers to democracy and reform in the former Soviet Union, Eastern Europe and elsewhere'. . . . What is remarkable about these threats is that none has a fixed spatial location; regional instability refers not only to the Middle East but also to Europe and Africa; proliferation and transnational dangers are global; even dangers to reform, which is the only danger explicitly linked to certain places, are potentially ubiquitous, as the 'elsewhere' indicates. Military strategy still has to negotiate territory and place but it has also, in interesting ways, become untethered from place and territory. Anywhere on the globe is now a potential battlefield. The document concedes as much, noting that 'global interdependence and transparency coupled with our worldwide security interests, make it difficult to ignore troubling developments almost anywhere on earth'.

If it began with the modern age of geopolitics, announced by Kjellén and Mackinder, it might be argued that the twentieth century ended in the age of a 'postmodern geopolitics'. Gone is the apparent simplicity of the great power and Cold War confrontations, let alone demarcated and fixed 'heartlands', replaced by the sense of a world of proliferating uncertainty and 'threat'. At least with all this, and the preceding points in mind, we might well argue that a critical examination of 'geopolitics' should not confine itself to the tradition that goes under that name (also see Sharp 1999). To do so would be to miss much wider issues about how broader sets of ideas about space, power, order, society and the international scene are combined into common sense notions about 'us' and 'them' and the whole geographical order of what is presented as the good life.

Learning outcomes

Having read this chapter , you should understand:

- The term 'geopolitics' has been associated with a wide variety of texts and contexts.
- However, geopolitics is frequently associated with the sense of a self-conscious tradition of writing about and mapping the relations between states, geography and power.
- This is usually traced to the late nineteenth-century writings of Rudolf Kjellén.
- One of the key features of this tradition is the idea that the state resembles a living organism which requires living space and will compete with others (the organic theory of the state).
- The oppressive military regimes in South America (notably Argentina, Brazil and Chile) during the 1960s and 1970s elaborated and applied a variety of the geopolitical tradition.
- However, it also proliferated in a number of European countries and Japan, particularly between the two World Wars.
- Since this proliferation of geopolitics was most evident in Fascist states, the defeat of the Axis powers in 1945 signified its relative decline as a formal tradition, although it continued in Fascist Spain and Portugal and in South American military circles.
- Beyond the self-conscious tradition of geopolitics, similar forms of thinking about territory, states and power have proliferated in many disciplines.
- This has been particularly evident during the Cold War, when broader geopolitical discourses suffused many aspects of art, science, culture and daily life.

Further reading

Agnew, J. (1998) *Geopolitics: Revisioning World Politics*, Routledge, London and New York. An historical and contemporary guide which places geopolitics in comparative historical perspective.

Agnew, J. and Corbridge, S. (1995) *Mastering Space. Hegemony, Territory and International Political Economy*, Routledge, London and New York. Part II (Chapters 5, 6 and 7) is useful on aspects of the 'New World Order'.

Barton, J.R. (1997) *A Political Geography of Latin America*, Routledge, London and New York. Amongst other things, this critiques South American geopolitical traditions.

Campbell, D. (1992) *Writing Security: United States Foreign Policy and the Politics of Identity*, University of Minnesota Press, Minneapolis. A challenging but rewarding account of how foreign policy and the Cold War have been productive of America's sense of self-identity.

Dodds, K. (2000) *Geopolitics in a Changing World*, Prentice Hall, Harlow. An accessible and clear introduction.

Heffernan, M. (1998) *The Meaning of Europe: Geography and Geopolitics*, Arnold, London. A very readable account of European geopolitics and much more.

O'Loughlin, J. (ed.) (1994) *Dictionary of Geopolitics*, Greenwood Press, Westport, CT. A useful source.

Ó Tuathail, G. and **Dalby, S.** (eds) (1998) *Rethinking Geopolitics*, Routledge, London and New York. Useful examples of how geopolitics is being critically rethought at a more advanced level.

Ó Tuathail, G., Dalby, S. and **Routledge, P.** (eds) (1998) *The Geopolitics Reader*, Routledge, London and New York. A collection of short texts (readings) on different aspects of geopolitical discourse, including extracts from Mackinder, Haushofer and Kissinger and debates about the New World Order. Good clear editorial introductions. Undoubtedly, the best place to begin a course of wider reading.

CHAPTER 16

JAMES SIDAWAY

The place of the nation-state

The territory of a nation is not just a profane part of the earth's surface. It is a constitutive element of nationhood which generates plenty of other concepts and practices directly related to it: for example, the concept of integrity and sovereignty; border control, conflict, invasion and war. It defines and has some control over many other national affairs, such as the national economy, products, industry, trade, education, administration, culture and so on. Unarguably, the territory of a nation is the most concrete feature of a nation for the management of nationhood as a whole. For a theoretical geographer, it is the territoriality of a nation. . . . For people of a nation, it is a part of SELF, a collective self. It is a nation's *geo-body* . . . Geographically speaking, the geo-body of a nation occupies a certain portion of the earth's surface which can be easily identified. It seems to be concrete to the eyes and having a long history as if it were natural, and independent from technology or any cultural and social

construction. Unfortunately, that is not the case ... the geo-body of a nation is merely the effect of a modern geographical knowledge and it technology of representation, a map. The geo-body, the territoriality of a nation as well as its attributes such as sovereignty and boundary, are not only political but also cultural constructs.

(Winichakul 1996: 67)

16.1   Historical and geographical variability of states

Although it is also other things, notably a provider of services, a system of regulations, ideologies, legal regulations and police powers ('law and order'), flows of capital (budgets, taxation and government spending) backed up by the threat of discipline and violence (for example, armed forces), the state can be interpreted as a form of *community*. As earlier chapters (in particular Chapters 1, 2 and 3) have shown, forms of human community have been extremely historically and geographically variable. This variation is enormous, from the claims of spokespersons for a state (it could almost be any state) that it is a natural embodiment of a 'nation' with the right to a select segment of the world (a territory with boundaries) and an ancient history (so even archaeological remains of past societies somehow come to be classified as 'national monuments', like Stonehenge in England, the Acropolis in Greece, the Pyramids in Egypt and so on) to nomadic communities or complex tribal or dynastic ones with no or little sense of themselves as belonging to or requiring a national state, to those who would reject national and state identity in the name of what are seen as higher or more worthy causes, for example a religious identification, local community or world revolution.

The contemporary system of states, in which all of the land surface of the earth (with the partial exception of Antarctica as detailed in Chapter 15) is divided into state units, whose outlines become familiar to us from maps and globes is after all fairly new. Earlier in the twentieth century, the borders between many of today's states were only vaguely defined, and more recently large areas of the world were ruled by colonial empires or dynastic realms (the Austro-Hungarian in Central Europe, Ottoman in the Balkans and

Plate 16.1   The state as a system of organized violence: armed forces. (Trip/G. Grieves)

Eastern Mediterranean and Ch'ing empire in China, for example) and their inhabitants deemed unfit (not 'civilized' enough) for statehood. Alternatively, the concept of national-statehood was simply not in the political vocabulary. In the latter cases, ethnically, linguistically or religious specific groups owed allegiance to the imperial order rather than to a defined nation. That is, loyalty would be *foremost* to the empire and any sense of national or ethnic identity would be a local or 'private' matter.

Such imperial visions are no longer dominant and today few formal colonies and no large-scale dynastic empires remain. Therefore the territories that were once ruled as part of, for example, a Japanese, British, French, Portuguese or Russian empire are today mostly divided into self-avowed and recognized *sovereign* states. They possess the same apparatus of statehood (leaders, flags, capital cities, administrations, postage stamps, seats at the United Nations and so on) that the former imperial powers have. Even in the relatively short life-history of the author of this chapter, the global map of states has changed. Some have disappeared (like the former East Germany) or split into component parts (like the former USSR and Yugoslavia). All this reinforces the point that states, like other communities, particularly the 'nations' with which they are associated, are not to be taken at face value. The claims made for and on behalf of them deserve critical examination. Having said that, this is not an easy task. For, as Benjamin Akzin (1964) pointed out over 35 years ago, to discuss nations, states and nationalism is to enter a terminological maze in which one easily and soon becomes lost. This chapter will indicate some pathways into the maze. It begins with an account of 'nations' and 'nationalism' before returning to the relationship of these to states. This will also help to pave the way for the discussions of citizenship and political action in Chapter 17.

16.2 Nations as 'imagined communities'

One of the most influential and suggestive critical studies of nations and nationalism is a book by Benedict Anderson (1983) entitled *Imagined Communities: Reflections on the Origin and Spread of Nationalism*. He begins with a reminder of the ubiquity of nations and nationalism:

> Almost every year the United Nations admits new members. And many 'old nations' once thought fully consolidated find themselves challenged by 'sub'-nationalisms within their borders – nationalisms which naturally dream of shedding this subness one happy day. The reality is quite plain: the 'end of the era of nationalism', so long prophesied, is not remotely in sight. Indeed, nationness is the most universally legitimate value in the political life of our time. (Anderson 1983: 3)

Although enormously variable between, say, Nepali, Israeli, Nicaraguan, Vietnamese, Eritrean, American, Greek, Turkish and Irish versions, the ideology of nationalism holds that everyone will have a primary identity with a particular 'nation'. Such communities should be able to express themselves in a state; that is, they should enjoy what is called 'sovereignty' within certain geographical boundaries. 'Sovereignty', which is a term of long vintage and was previously associated with royal dynasties (the sovereign monarch), shifts to the 'people' of a 'nation', and even if a royal figurehead is retained, she or he will have to become in some way a 'national' symbol.

Occasionally nationalisms recognize that the state itself may be multi-national (as in

---

Box 16.1   **Nations without states: the Kurdish case**

Gerard Chaliand (1993: 4) notes that: 'the Kurdish people have the unfortunate distinction of being probably the only community of over 15 million persons which has not achieved some form of national statehood, despite a struggle extending back several decades.'

The lands predominantly inhabited by Kurds are in fact divided between four states (see Figure 16.1): Iran, Iraq, Turkey and Syria (there are also Kurdish communities in the former Soviet Union and migrant populations in Western Europe, especially in Germany).

Figure 16.1   Regions historically inhabited by the Kurds. *Source:* After Chaliand, 1993: ix

Despite early twentieth-century promises from the then imperial powers (including Britain and France), in the end it suited no great power to establish an independent Kurdistan. As in much of Asia (and beyond), the borders were determined in large part by foreign imperialists, taking little or no account of the wishes and rights of local peoples. The rise of Turkish, Iranian and Arab nationalisms – themselves all partly responses to imperial domination – left little space for the Kurds.

Yet frequent attempts on the part of the states in the region to eradicate Kurdish identity and nationalism have not been able to quash either the widespread sense of Kurdishness or the often violent struggle in pursuit of a Kurdish nation-state.

British nationalism, which contains English, Scottish and other affiliations), but in so doing, the wish is usually expressed that somehow a more inclusive national identity will evolve or has evolved which coincides with the boundaries of the state. Britain, Switzerland, South Africa, Nigeria and the United States are all cases where different versions of the claim and goal of an inclusive 'national identity' that supposedly unites disparate 'sub-nations' or communities have been asserted. Similarly a number of 'multi-national' communist countries, notably Yugoslavia and the Soviet Union, sought to regulate nationalism by assigning citizens one of a number of constituent national identities. In the latter case this sometimes involved inventing nationalities to rationalize and simplify more complex tribal and religious identities, whilst asserting that these should be subservient to an overarching sense of Soviet or Yugoslav identity which coincided with the boundaries of the USSR or of Yugoslavia. So, for example, in the USSR, people could be declared on their identity documents as having one of a number of officially recognized nationalities (Uzbek, Latvian or Russian, for example), but they would also and supposedly above all be citizens of the USSR. In due course, it was in part because of local nationalist challenges to wider Soviet and Yugoslav affiliations that the USSR and Yugoslavia collapsed in the early 1990s.

For nationalists, those self-identified nations without their own state (the Kurds are an example, see Box 16.1) or, at the least, an allocated autonomous region within a broader (multi-national) state have a right to one. And frequently, 'nations' within a multi-national state (such as Scotland in the UK or Quebec in Canada) become the basis for claims that they should enjoy full statehood. In many cases too, either the central state or some other community with another affiliation resists. There are numerous examples of this, including the complex case of Northern Ireland, where most Catholics (Republicans) would wish to see the province united with the rest of the Irish Republic (which itself successfully broke away from the British colonial empire earlier in the twentieth century). Most Protestants (Unionists), who are descendants of settler populations from England and Scotland, wish to remain part of a 'United Kingdom'.

The Irish case is just one of dozens of situations where conflicting nationalist and confessional (religious-cultural) logics collide, often with violent consequences. There is clearly something very powerful going on, whereby nationalist visions are linked with particular territories and conceptions of state. Yet, as Anderson (1983: 3) recognizes:

> But if the facts [of the existence of many and sometimes conflicting nationalisms] are clear, their explanation remains a matter of long-standing dispute. Nation, nationality, nationalism – all have proved notoriously difficult to define, let alone analyse. In contrast to the immense influence that nationalism has exerted on the modern world, plausible theory about it is conspicuously meagre. Hugh Seton-Watson, author of by far the best and most comprehensive English-language text on nationalism, and heir to a vast tradition of liberal historiography [theories of history writing] and social science, sadly observes: 'Thus I am *driven* to the conclusion that no "scientific definition has existed and exists."'

Readers may wish to 'prove' Seton-Watson's observation for themselves, by trying to come up with a universally valid definition of a nation. Faced with this task, students (and their teachers) will often work through a long list of characteristics ascribed to nationality. But it seems that exceptions can always be found. Language is a favoured criterion, but then many languages are spoken by more than one 'nation' (English is an example) and some 'nation-states' (Switzerland or Mozambique, for example) are divided into communities

Plate 16.2 'Nation' imagined as 'race': supporters of the far right in the UK. (Format Photographers/Joanne O'Brien)

speaking different languages. Religion is sometimes chosen, but the same objections apply. Ethnicity and race turn out to be dubious criteria, favoured by racists of all stripes, and often part of the basis for national identities, but susceptible to the obvious and undeniable point that everywhere is much too mixed up historically and genetically for such categorizations to be water-tight. Besides, some nationalisms have come to celebrate multi-racialism, as in the 'melting pot' United States or the 'rainbow nation' of South Africa. 'Culture' usually crops up as a criterion, but foolish indeed would be the person who bases a theory or claim on the idea of a homogeneous national culture. For as Chapter 6 has indicated, cultures are always (though of course to varying degrees) heterogeneous and contested. Think, for example, of age and class variations and the coexistence of multiple sub-cultures which characterize every supposedly 'national society'.

All this leads Anderson to declare that nations are in a sense 'imagined communities'. This imaginary status is not to deny that they are not in a sense *real* to those who feel they belong to them. Indeed, Anderson (1983: 6) feels that:

> In fact, all communities larger than primordial villages of face-to-face contact (and perhaps even those) are imagined. Communities are to be distinguished, not by their falsity/genuineness, but by the style in which they are imagined.

The style in which nations are imagined (and therefore the way that they become to seem

Plate 16.3 Banal representations of the nation: postage stamps as national icons. (Alan Thomas)

real by those who belong to them) is, according to Anderson (1983: 7) as 'limited', 'sovereign' and as 'communities':

> The nation is imagined as *limited* because even the largest of them, encompassing perhaps a billion living human beings, has finite, if elastic boundaries, beyond which lie other nations. ... The most messianic nationalists do not dream of a day when all the members of the human race will join their nation in the way that it was possible, in certain epochs for, say, Christians to dream of a wholly Christian planet. ...
>
> It is imagined as *sovereign* because the concept was born in an age in which Enlightenment and Revolution were destroying the legitimacy of the divinely-ordained, hierarchical dynastic realm ... nations dream of being free. ... The gage and emblem of this freedom is the sovereign state.
>
> Finally, it is imagined as *community*, because regardless of the actual inequality and exploitation that may prevail in each, the nation is always conceived as a deep, horizontal comradeship. Ultimately it is this fraternity that makes it possible, over the past two centuries, for so many millions of people, not so much to kill, as willingly to die for such limited imaginings.

So here is the link again between the nation and the state. The state claims to be a sovereign expression of the nation – bound to it and to a particular *place*. That is, it has territorial limits. The nation-state has a *geography*, which is charted, demarcated, mapped and represented to the 'national population' in their school atlas and geography lessons. Such a system of 'national' geographical representation is always combined with an historical vision, a grand narrative of 'national history', often assuming that the nation is ancient, even primeval.

Yet Anderson and other critical accounts of nationalism stress that it is very much a modern ideology. For what makes mass nationalism possible are certain socio-economic and technical transformations, notably the arrival of media and national educational systems. Schooling, newspapers, and later radio and television all help to promote and popularize the idea that people belong to and share in the nation. Whilst others have emphasized the longer historical roots of many nationalisms in pre-modern ethnic affiliations (for example, Smith 1986), for Anderson and most other critical observers, what is striking is nationalism's relative modernity. Even where nationalists imagine the nation as ancient, such an imagination is itself more often than not predominantly a nineteenth- and twentieth-century phenomenon. At a pinch, nationalism might be traced to the seventeenth century. But the ideology of the nation-state was not anywhere very evident before then. So whilst someone might talk about a thousand years of, for example, English or British national history, they conveniently forget that British or English nationalisms are modern concepts, much less than a thousand years old.

A more persuasive critique of Anderson comes from Partha Chatterjee (1993), who argues that, in India at least, a variety of nationalist currents (or voices) have coexisted, for example those of elites, of women, of different castes and so on. Although Chatterjee's arguments are highly specific to the South Asian case, the related point about nationalism and constructions of nation-statehood invoking different conceptions of 'manhood' and 'womanhood' is valid more widely. Although there are immense historical and geographical variations of this, there is a general tendency for women to be seen as particularly important transmitters of 'national culture' (for example, in the idea of a 'mother-tongue' language) (Yuval-Davis 1997) and as somehow the embodiment of the nation. McClintock (1994: 352) thus notes that:

> Nations are contested systems of cultural representation that limit and legitimize people's access to the resources of the nation-state, but despite many nationalists' ideological investment in the idea of popular *unity*, nations have historically amounted to the sanctioned institutionalization of gender *difference*. No nation in the world grants women and men the same access to the rights and resources of the nation-state'.

Moreover, particular ideas of the 'family' are central to most imagined communities:

> Nations are frequently figured through the iconography of familial and domestic space. The term nation derives from [the Latin] *natio*: to be born. We speak of nations as 'motherlands' and 'fatherlands'. Foreigners 'adopt' countries that are not their native homes and are naturalized into the national 'family'. We talk of the 'Family of Nations', of 'homelands' and 'native' lands. In Britain, immigration matters are dealt with at the Home Office; in the United States, the president and his wife are called the First Family. Winnie Mandela was, until her recent fall from grace, honoured as South Africa's 'Mother of the Nation'. In this way, despite their myriad differences, nations are symbolically figured as domestic genealogies.

In several European nationalisms (including 'British') one significance of this is that the sexist notion of women (usually grouped with children) supposedly as *naturally* inferior to men, or as requiring the protection of *naturally* more powerful men, frequently provided the prior backdrop to depicting hierarchies (between dominating and dominated classes, for example) *within* the nation as natural, like those of a family. A similar conception provided part of the racist justification for overseas colonies. Peoples subject to colonial domination were often presented as being equivalent to 'families' of black children ruled by a 'benevolent' white father.

16.3 Constructing boundaries

The extract from Winichakul (1996) quoted at the start of this chapter noted how the 'territoriality' of the modern nation-state is of a bounded space: 'a certain portion of the earth's surface which can be easily identified'. Premodern dynastic realms and empires could make do with loose boundaries. But modern national-states have felt it necessary to *demarcate* their boundaries, to iron out perceived irrationalities and anomalies – sometimes by going to war. In short, they have sought to nationalize and unify 'their' space and 'their' (national) populations and render them into known, surveyed and defensible sovereign territory and communities. As Chapter 15 has indicated, the rise of geopolitics in the late nineteenth and early twentieth centuries represents a particular expression of just such a rationality – but it has wider forms and deeper roots. The nineteenth-century invention and refinement of statistics (*state-istics* = the science of the state) and demography (*demos* = people, *graphy* = writing) and the establishment of geography (*geo* = earth, *graphy* = writing) as a university discipline and field of research are all part of this regime of knowledge and power where people are surveyed and made into subjects of nation-states. For states with land boundaries the occupation and official demarcation of the frontiers is an important part of this process.

Consider the example of the border which is often recognized as the *oldest*, more or less, stable, still-existing border in the world, that between Spain and Portugal. The mutual recognition of the border between Spain and Portugal is usually traced in modern Spanish and (particularly) Portuguese history to the Treaty of Alcañices signed on 12 September 1297 (in fact, the date on the Treaty is 12 September 1335, but amongst other things, our conventions for counting years have altered since). The Treaty has acquired something of the status of a foundational text. In other words, it is interpreted as a kind of proof of the ancient historical basis of the nations concerned.

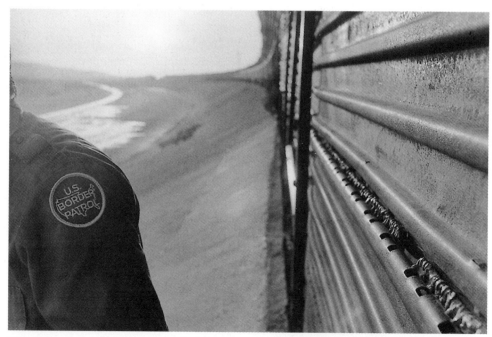

Plate 16.4 Representing the limits of the 'nation-state': boundary patrols. (Format Photographers/Michelle Gienow)

Yet the treaty was *not* signed in the name of states or 'nations'. Named after the Templar castle in which it was sealed and witnessed, the text of the treaty of Alcañices begins with the words 'In the name of God. amen.' and is signed in the presence of the Templars and other holy orders by those who describe themselves as:

> by the grace of God the King of Portugal and the Algarve and by the grace of God the King of Castille and Leon,

and has as its subject:

> towns, castles and lands, town boundaries, divisions and orderings ... [disputes over which] have caused many wars, homicides and excesses, as a consequence of which the lands of both [kingdoms] have been looted, burnt and ruined, weighing heavily on God ... [and] because of our sins, risking the danger of losing them and them falling into the hands of our enemies in the faith and most gravely [causing] the violation of God's will and injury to the holy church of Rome and Christendom. (cited in Martínez 1997: 15; my translation)

Moreover, the actual demarcation of the Portuguese–Spanish border-line had to wait until the nineteenth and twentieth centuries. Two formal delimitation treaties set out an agreed frontier and established mechanisms for its physical demarcation on the ground. But gone are 'amens' and references to 'the will of God'. Instead, the treaties (the first of 1864 for the northern half of the border and the second of 1926 for the southern half of the border) make reference to the need to impose order and well-defined demarcations, to 'eliminate the anomalous situation in which, in the shadow of ancient feudal traditions' (Tratado de Limites, 1866, page 1, my translation), some frontier areas (including a number of villages) were recognized as *shared* or common lands with usage rights by communities of both states/'nationalities'. Following the treaties, such areas were divided and the border, where not demarcated by a river, was marked on the ground by rectangular boundary stones every few hundred metres (although it took a couple of decades to put in place all the boundary markers and resolve local conflicts and differences over the 'fine-scale' division of lands). In other words, the Portuguese–Spanish case indicates how the epoch of modern nationalism (i.e. the nineteenth and twentieth centuries) saw what James Anderson (1995: 71) calls:

> a territorialisation of politics, with a sharpening of differences at the borders of states and of nations between 'internal' and 'external', 'belonging' and 'not belonging', 'us' and 'them'.

By the same logic, or an extreme extension of it, minorities who do not 'fit in' sometimes have to be moved, murdered or deported – or, in the euphemisms of our times, 'cleansed' (see Sibley 1995). This has been associated with, for example, Nazi racial territorial visions of a Europe cleaned of Jews and Gypsies and an expanded racially pure Germany. The Nazi Holocaust represents an extreme case. But 'cleansing' or 'purification' of national space, amounting to the expulsion or murder of people defined as 'foreign' or 'alien', has been widespread in the twentieth century. There are examples from all continents: the creation of modern Turkey whose Armenian population were slaughtered in the second decade of the twentieth century (and which still refuses to admit the rights of a Kurdish minority); the creation of Israel in the 1940s, which was accompanied by the flight of millions of indigenous Palestinian Arabs from territories deemed part of the new Jewish

> **Box 16.2**
>
> ## Irredentism and Secessionism
>
> Irredentism is defined by Mark Blacksell (1994: 299) as: 'The assertion by the government of a country that a minority living across the border in a neighbouring country belongs to it historically and culturally, and the mounting of a propaganda campaign, or even a declaration of war, to effect that claim.' Whilst Blacksell focuses on official irredentism, sometimes irredentist parties and organizations promote the claim without it necessarily becoming their government policy. Often the irredentist claim includes areas where populations are mixed, making for potentially violent conflict. The collapse of Yugoslavia and the Soviet Union have been accompanied by irredentist claims. But there are dozens of other examples around the world.
>
> Often closely related to irredentism is the phenomenon of secessionism, whereby a minority in one state seeks the transfer of a section of the state's territory to another new or existing state. Although it has also had its own local cultural and political dynamics, the conflict in and over Northern Ireland exemplifies irredentism and secessionism. The Irish Republic was born of a secession from the British Empire (whilst at the same time being seen as the first of many national liberations for colonized peoples) in the 1920s. The Irish constitution long claimed sovereignty over the whole of Ireland, thus making an irredentist claim on that part which remained under British jurisdiction, whilst nationalists in what they often called the north of Ireland (refusing the term Northern Ireland which signified a separate jurisdiction) sought secession and a unified Ireland.
>
> Current schemes for the peaceful resolution of the conflict include visions for a measure of joint sovereignty over what (to be as neutral as possible) is called 'the province' (which could be a province of the United Kingdom and/or of the Irish Republic and/or Ulster; a place apart).
>
> Meanwhile, James Anderson (1997: 216) suggests that: 'Northern Ireland is the residue of failures in nation- and state-building, whether viewed from either a British or an Irish perspective. The disputed labels for territory – repeated at a local level in, for example, Derry/Londonderry (or, ironically, "Stroke City") – reflect the rival national identities and suggest that the ideal of the nation-state is unachievable.'

national state; the system of reservations for native Americans in the USA and Canada; the violent exchange of Muslim, Hindu and Sikh populations which accompanied the 1940s partition of India into self-avowed national 'homelands' (Pakistan and Bangladesh) for South Asian Muslims; or the wars of the 1990s in the former Yugoslavia. These logics have also produced calls for the modification of boundaries, sometimes resulting in violence and war (see Box 16.2). No continent has been without examples of such conflict and violence and it is probably impossible to find a state which has not at some time expelled people or murdered them in the name of some nation or other. Yet each nationalism must also be unique, establishing itself as different from others (even if the idioms of flags and so on are similar), constructing a sense of self against others who are defined as outside the imagined community.

The tendency has been to see all this as derived from American and European prototypes. Yet, it is mistaken to regard, say, African or Asian variants of nationalism or constructions of statehood as simply colourful (or failed) replications of a European–American model. To be sure, imported idioms and motifs from Europe and America have been important; for example, the official state language in a fair number of former colonies is that of the old colonial power. And European (and less often United States) imperialism has provided an important part of the backdrop to the trajectory of nationalisms in Africa and Asia. That is, Asian and African nationalisms often took form and arose in response to foreign colonial domination, developing the aim of national liberation and independence from the colonial powers. Yet whilst 'globalizing' capitalism and especially imperialism are part of the picture, African and Asian nationalisms are also (like those of Europe and the Americas) rooted in *local* historical trajectories. In many cases these are of much longer vintage, lending themselves to nationalisms which appropriate (or invent) a vision of a pre-colonial history.

16.4 Nation-states as symbolic systems

As we have seen, the state is the bureaucratic expression of nationalism. This is not to say that a widespread sense of nationalism inevitably precedes the state. Indeed, what is often termed the state apparatus (everything from regulations governing schools, to tax inspectors and politicians) seeks to foster national subjectivities out of the frequently ambivalent and disparate array of identities contained by its boundaries. If, as has been argued by many, nationalism is akin to religion, then the state becomes its symbolic structure. When examining the superficially quite different cases of Australia and Sri Lanka, Bruce Kapferer (1988) claims that nationalism is itself a religion, owing to the fact that, as with most religious-like beliefs, nationalism demands the recognition (his word is 'reification', which means something more than this) of an immanent and all-encompassing entity (the nation). The nation fulfils the role of a sacred cause, something greater than any individual and something which may be worth dying for. The nation is the God or deity of the religion of nationalism and the state is its theology or temple. Moreover, as Herzfeld (1992: 37) notes:

> Every bureaucratic action affirms the theology of the state. Just as nationalism can be viewed as religion, bureaucratic actions are its most commonplace rituals. There are other such everyday rituals: Hegel saw the reading of the morning newspaper as the secular replacement of prayer.

There are indeed many public rituals of nationality and statehood: coronations and remembrance days, military parades, national holidays, national prowess (or the lack of it) at football or the Olympics, swearing in of governments, state funerals. (Football in the UK is also revealing of the complexity of British nationalisms and the survival of separate Scottish, English, Welsh and Northern Irish squads is a testimony to the limits to 'Britishness' and the survival (or 'revival') of other national affiliations.) In all these rituals, the nation is reaffirmed and the state performed and made to seem omnipresent, historical and real. Dramatic cultural or political events covered by the media greatly reinforce senses of national community. It is said, for example, that virtually every adult *American* alive at the time can recall where they were on hearing the news of the assassination of US President John F. Kennedy in 1963. Whilst his funeral, like that of Diana Spencer, became a global media spectacle, it was represented and felt most acutely as a kind of national loss. 'Goodbye *England's* Rose ...' as Elton John put it at Diana's funeral. (It should be noted that there was some controversy over the lyrics of Elton John's tribute song, in so far as Diana was not only the Princess of *Wales*, but a former member of the avowedly *British* monarchy. Rather like the case of football, this bears out the complexities and unresolved tensions of British nationality.)

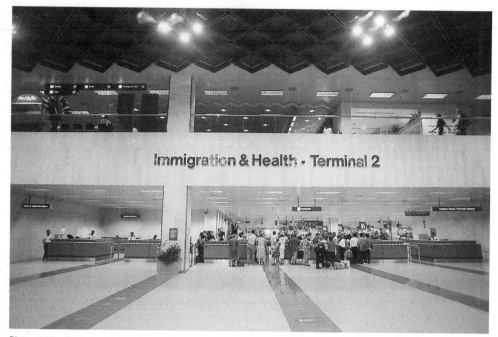

Plate 16.5 Declaring the nation, performing the state: international arrivals at Changi Airport, Singapore. (Robert Harding Picture Library)

But the nation-state also demands (like all effective religions) personal commitment and more minor ritualistic acts, many of which rest on a geographical imagination of inside and outside, belonging and otherness (see Taussig 1997). The anthropologist Michael Herzfeld (1992: 109) examines this theme and is worth quoting at some length:

> Nationalist ideologies are systems of classification. Most of them are very clear about what it takes to be an insider. That, at least, is the theory. In practice, however, divergent interpretations give the lie to such essentialist claims, as to take one prominent and current example, in the debate currently waging in Israel about the definition of a Jew. Such taxonomic [naming] exercises ... are central to the very existence of the nation-state. All other bureaucratic classifications are ultimately calibrated to the state's ability to distinguish between insiders and outsiders. Thus ... one can see in bureaucratic encounters a ritualistic enactment of the fundamental principles upon which the very apparatus of state rests. Seen in these terms, arguments about the number on a lost driver's license or an applicant's entitlement to social security do not simply challenge or reinforce the power of particular functionaries of state. They rehearse the logic of the state itself.

Not only that, but the whole exercise of state power gets taken for granted as the natural order of things. Only when many of those activities which are ascribed to the state are no longer carried out (in situations of war, for example) or when a person finds themselves on the wrong side of a state-sanctioned category (the wrong side of the boundary, the

COME ON IN

Mail investigation reveals how scandalously simple it is for illegal migrants to enter Britain

FRENCH authorities are easing the way for the torrent of illegal immigrants getting into Britain.

A Daily Mail investigation has exposed how France is blatantly passing the problem across the Channel.

Migrants from Europe and Asia are making their way to Calais, where indifferent security means they have the best chance of smuggling themselves onto ferries bound for England. In

The good life - Pages 4-5

three days Mail reporters witnessed dozens of illegal immigrants brazenly clambering over fences to get into supposedly secure lorry parks at the French port.

They watched as men sneaked unchallenged into the backs of lorries waiting to drive onto ferries. Others clung to the axles.

Home Secretary Jack Straw has promised to crack down on the tide of illegal immigrants. But the problem is getting worse. Last year 46,000 asylum

seekers cost the British taxpayer an estimated £600million. This year nearly 60,000 are expected to try their luck. More than half will come via Calais.

Last night the French Interior Ministry admitted that 10,000 illegal immigrants were deported each year from French cities -- but hardly any from Calais. A spokesman said the situation there was 'a unique problem that needed a unique solution'.

Ferry companies are fined for every asylum seeker who arrives on their boats. A source at Dover-based P&O Stena Line, one of three firms operating on the Channel route, said it was

paying £60,000 a week in fines. Last night Tory immigration spokesman David Liddington said the French attitude was an 'utter scandal and disgrace'. 'The British taxpayer and our ferry companies are footing the bill for this inaction by the French and it has got to stop,' he said.

Last night a Home Office spokesman insisted the French were doing well in their efforts to stem the tide. 'The French frontier police have been very active in identifying people trying to get into Britain without proper authority and those who seek to trade in asylum seeking. We welcome this,' he said.

INSIDE: Weather 2, Saturday Essay 12 & 13, Simon Heffer 15, Travel 51-57, Coffee Break 58-62, City 65-67, Sport 84-96

Plate 16.6 Declaring the nation, performing the state: tabliod press in the UK accuses France of failing to impose immigration controls. (John Frost Historical Newspaper Collection) wrong side of the law) is the power (a power over life and death and thousands of lesser things) and the universality of the symbolic order of 'the state' revealed. The state claims the monopoly over these things. And for others to exercise judgment and to punish or to kill is 'to take the law into their own hands'. These things are reserved for the state.

16.5   Conclusions: The place of the nation-state?

There have never been so many states as there are today. In the nineteenth century a wave of new states (countries such as Argentina, Mexico, Bolivia) emerged in the Americas in revolt against Spanish and Portuguese empire. These American prototypes and the USA itself (which dates from 1776) provided an example to nationalists elsewhere in the colonial world. But it was not until the twentieth century that most of Africa and Asia could escape direct colonial domination. This often required violent 'national-liberation' struggles

Plate 16.7   Declaring the nation, performing the state: the Atatürk memorial, Ankara, Turkey. (Camera Press/David Lomax)

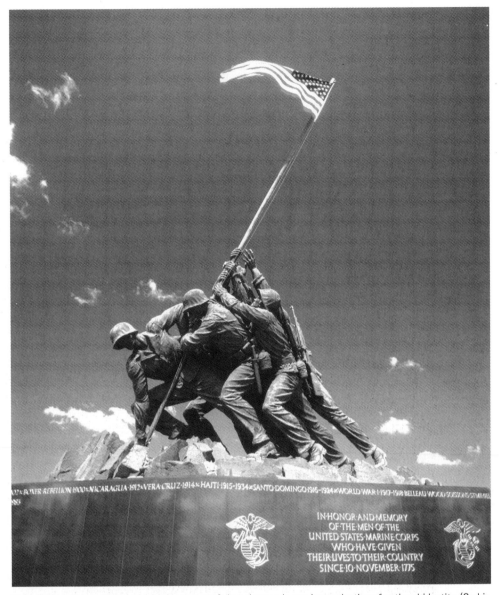

Plate 16.8 War-making and with it a sense of shared struggle can be productive of national identity. (Corbis UK/Bill Ross)

against entrenched resistance from the colonial powers and white settler populations. In the decades between about 1945 and 1975, dozens of new 'sovereign' states emerged in place of the old colonial map. And if, for example, the Tamil nationalists of Sri Lanka, the Palestinian and Kurdish nationalists of the 'Middle East' or the Basque, Scottish and Corsican nationalists in Western Europe get their way, there will be even more.

Yet, for many years the *demise* or decline of the state has been discerned or predicted – and such claims have of late become even more common in discourses about

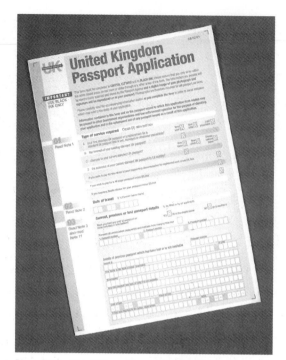

Plate 16.9   Identifying citizenship. (Alan Thomas)

'globalization'. These usually argue that the growing scale and power of transnational flows, particularly of capital (but also of people, ideas and religious affiliations, technologies and so on) is *subverting* the capacity of the state and *weakening* national identities. The nation-state is often described as being 'hollowed-out' or 'eroded'. In this view, the state no longer has the power to command, for example, the society and economy inside its boundaries that was once attributed to it. And such a 'hollowing out' is sometimes seen to point the way to a post-national world (or some kind of *shared* global culture in which national cultures are replaced by a more hybrid, but common global mixture).

However, others object that transnational forces of 'globalization' are really nothing new and that capitalism in particular has shown itself not only able to coexist with states and nations but in symbiotic relationship with them. In other words, they reinforce each other. The same goes for many organized religious movements that coexist or support nation-states. (This is not to say that many movements with religious inspirations do not see themselves as anti-statist [even radically and violently so, as in the case of some fundamentalist Christian militias in the USA]. But in so doing of course they ascribe to the state a certain power and centrality – thereby making it more real. A parallel to help readers understand is the atheist statement: 'There is no God', which cites something called God in order to deny its existence!) Moreover, 'globalization' is not only massively uneven (as we have seen in earlier chapters) but as likely to produce local backlashes as a universal culture of, say, the *same* fast food, Coca-Cola and soap operas and political orientations everywhere in the world. That is, the technologies of capitalism (particularly media) provide the preconditions for strengthening, rather than undermining imagined communities of nationalism, whilst states still act to fine tune the regulatory frameworks for continued capital accumulation (resorting where needed to force to suppress opposition). States still enact laws about business, trade unions, property rights and so on. In most countries, for example, buying a property or land or setting up a business require some kind of registration with the state. No amount of globalization has ended this. However, instead of further entering the large and complex debates about the impacts of 'globalization' on nations and states, for which Anderson (1995) is a good introductory guide, let us return to the conception of the state as a symbolic system, and as a complex of *representations*. The historian-philosopher and activist Michael Foucault (1979: 29) argued in one of his most famous essays that:

> We all know the fascination which the love, or horror, of the state exercises today; we know how much attention is paid to the genesis of the state, its history, its advance, its power and abuses, etc. . . . But the state, no more probably today than at any other time in its history, does not have this unity, this individuality, this rigorous functionality, nor, to speak frankly, this importance; maybe after all the state is no more than a composite reality and a mythicised abstraction, whose importance is a lot more limited than many of us think.

Hence, to recognize the state as, in part at least, a symbolic system is also to recognize that, as Rose and Miller (1992, 172) argue: '"the state" itself emerges as an historically variable linguistic device for conceptualising and articulating ways of ruling'. In other words, things like the 'nation' and the 'state' are *made* real mainly in certain words and deeds, that is, in language and action. Think again of the staging of those *national* sporting, political and cultural occasions, which bring the nation home, and the more mundane bureaucratic acts of state, like showing a passport or filling in a form with your *national* (insurance, registration or identity) number. So perhaps nation-states exist above all as systems of actions and beliefs – an 'imagi*nation*' if you like – which must be continually re-enacted, re-narrated and re-imagined in order to seem important and real to us. Just as they have been here.

Learning outcomes

Having read this chapter you should understand:

- Nations, nationalism and states are complex historically and geographically variable phenomena.

- Nationalism is an ideology (a system of beliefs) which holds that people have a primary identity to a particular nation and that such communities should be able to express themselves in a *geographically* defined state.

- Nations can be understood as a kind of imagined community. They are imagined because not all members of a national community can know each other.

- Such national imaginaries contain a geography, a mental map of national space and its boundaries.

- National imaginaries are also gendered, for example in the idea of mothers as key transmitters of 'national culture' to the next generation.

- States can be understood as complex symbolic systems.

Further reading

Anderson, B. (1983) *Imagined Communities. Reflections on the Origin and Spread of Nationalism*, Verso, London. An expanded second edition from 1991 is available. Likely to be challenging, but worthwhile.

Anderson, J., Brook, C. and **Cochrane, A.** (eds) (1995) *A Global World? Re-ordering Political Space*, Oxford University Press and the Open University, Oxford. Chapter 2 on 'The exaggerated death of the nation-state' is a good place to follow up debates about the impact of globalization mentioned above in Section 16.5.

McDowell, L. and **Sharp, J.P.** (eds) (1997) *Space, Gender and Knowledge: Feminist Readings*, Arnold, London. Section seven on 'Gender, Nation and International Relations' contains extracts of some challenging articles and lots of suggestions for further reading.

Seton-Watson, H. (1977) *Nations and States: An Enquiry into the Origins of Nations and the Politics of Nationalism*, Westview Press, Boulder, Colorado. There are hundreds of case studies of particular nationalisms. Select a nation and you will usually find many to choose from. Seton-Watson is always a good starting point.

Taylor, P. J. and **Flint, C.** (2000) *Political Geography: World-Economy, Nation State and Locality*. Prentice Hall, Harlow. Now in its fourth edition, this text may be consulted for further ideas on most political geography topics.

Yuval-Davis, N. (1997) *Gender and Nations*, Sage, London. An accessible and rewarding survey of its theme.

Finally, two very different studies of European state boundaries and national identities around them:

Berdahl, D. (1998) *Where the World Ended: Re-unification and Identity in the German Borderland*, University of California Press, Berkeley.

Paasi, A. (1996) *Territories, Boundaries and Consciousness: The Changing Geographies of the Finnish-Russian Frontier*, Wiley, Chichester.

CHAPTER 17

MURRAY LOW

States, citizenship and collective action

Topics covered

- The history and geography of citizenship

- The relationships between political parties, states and social movements

17.1   Inside the border

From birth, we find ourselves thrown into a world already divided into mutually exclusive 'geopolitical' compartments each with their own state organizations. A huge number of consequences flow from the apparently arbitrary fact of our being born inside one of these rather than another. Inside each compartment, people find themselves confronted with states which can differ markedly from others in their capacity to intervene in and shape society, their openness to popular needs and demands, and the degree to which they rely on violence to control their populations. They may or may not have access to state services, be able to exercise political rights such as voting or running for office, or be liable to conscription by the government in periods of warfare. These inherited state structures are extremely difficult to replace, or even to dramatically reshape. Revolutionary episodes, in which mass pressure can lead to overall changes in the way states are organized, and where new combinations of social interests can come to control state organizations, are rare in the contemporary world. Furthermore, because of modern immigration law it is not easy for people to move to a different geopolitical compartment in search of political institutions which are less oppressive or more capable of responding to their needs or interests.

Yet much of the time, we take our membership of particular political communities for granted. We also, within limits, usually accept, or take for granted the right of the political institutions responsible for our own geopolitical compartment to make laws, raise taxes, regulate economic and social institutions such as firms and families, and deal with other states on our behalf. This is usually taken to mean that most people accord 'their' states a certain amount of *legitimacy*, meaning that they accept that the political organizations which govern them have the right to do so, and that, on the whole, their laws and policies should be accepted as valid by the populations under their jurisdiction.

In reality, the legitimacy of states and their activities is ambiguous and contested. For example, many people in the world would rather live under alternative geopolitical arrangements because established borders and state organizations, while perhaps legitimate in the eyes of some, do not allow for the expression of their national identities. Virtually everywhere, particular fields of government activity are subject to fierce contestation by groups calling the legitimacy of policies into question, from contemporary 'pro-democracy' movements in, for example, Burma, China or Cuba to 'militia' movements in the United States who dispute the right of the Federal government to levy taxes, interfere in local matters, and to strive for a monopoly on the use of effective organized force (e.g. via control of gun ownership). Furthermore, the absence of noisy or visible opposition to states and their activities does not mean the latter have firm foundations in popular consent. Many people are of necessity too concerned with economic survival to register much interest, positive or negative, in what 'their' states are doing. Most people have limited knowledge of state activities and the 'taken for granted' status of existing arrangements, often central to an appearance of legitimacy, can be quickly undermined with the public revelation of hitherto unknown information about the workings of government or the effects of particular policies.

There are, of course, institutional and ideological mechanisms by means of which modern states can derive some popular legitimacy, however unstable this may prove to be in particular circumstances. Nationalism, an ideology which, amongst other things, tells us which political communities we ought to belong to, has played a primary part in making states and their borders seem justifiable to most of their inhabitants and was considered in Chapter 15. The state organizations dividing up the modern political map have also come to be accepted because of the rights and material benefits they claim to guarantee to their inhabitants including, in some cases, the right to influence state policy through democratic processes. This chapter begins by considering *citizenship*, a status as a member of a particular (geographically demarcated) community which allows claims to be made on its state organizations. Citizenship has developed in socially and geographically uneven ways and, like nationalism, it seems necessarily to have exclusionary implications for those who do not fit dominant ideas about who belongs inside a particular 'geopolitical' unit. Nevertheless, citizenship has helped force states to have a wide range of domestic as well as international policies, and has facilitated the political mobilization of their populations in pursuit of policy change.

Geographers – and other social scientists – have often characterized these domestic processes as involving relations between states and various types of social actors. The second section of this chapter discusses two important types of collective actor shaping conflicts over domestic policies, political parties and social movements, stressing the interdependence between these two channels of citizen influence on states. In conclusion, it considers some of the limits to inherited pictures of political geographies, raising the possibility that the institutions and compromises underpinning our taken-for-granted view of politics inside states may not survive indefinitely. At times this chapter will take readers away from geographical literatures into debates about power, community and politics. However, it will be shown how these need to take space and territory into account.

17.2 Citizenship: an unevenly-developed institution

Citizenship is a central institution underpinning the legitimacy claimed by modern states. It has two dimensions which make this possible:

- Citizenship makes the inhabitants of geopolitical compartments 'members' of political communities. It gives them certain rights of influence over the personnel and policies of state organizations. In return, citizens have certain responsibilities to obey laws, to pay taxes to fund state activity and to mobilize in episodes of warfare.
- Citizenship, partly because of the political pressure it allows citizens to exert, helps generate and guarantee the provision by states of benefits and services which can give men and women a material 'stake' in established political arrangements.

The first of these dimensions shows that the appeal of citizenship as an institution has a lot to do with its connections with democracy, and relates to a kind of 'bargain' where, in return for political rights, citizens are obliged to give their states material and moral support. In this sense, citizenship has a long sense of its history (at least in the West). But it has undergone significant transformations. In the city-states of classical Athens or renaissance Florence, 'membership' of the city could mean an obligation to participate directly in political decision-making. Many people today find this participatory model of citizenship appealing. Historically, however, such participation was restricted to minorities of usually relatively well-to-do males: being a citizen was a privilege marking out some individuals as politically more equal than others.

Plate 17.1   Modern representative democracy does not generally involve direct law-making by citizens, but it does generate political participation of other kinds. This free-trade meeting in mid nineteenth-century London illustrates this kind of direct involvement of citizens in meetings, organizations and activities aimed at influencing representatives, states and other political institutions. (Mary Evans Picture Library)

The crucial innovation in modern models of citizenship was to make it possible to imagine citizenship as an equalizing status, creating political equality even if people were unequal in other respects. In the late eighteenth century, the writings of 'Enlightenment' philosophers such as Rousseau and Kant, and the experiences of political and social revolution in America and France, connected this sense of citizens as political equals with the idea that citizens are ideally self-governing, or *autonomous*. This meant that they should be able to imagine themselves as the authors of what states do. So, in these models, 'membership' of a modern state should permit everyone, as equal citizens, to mobilize politically and influence state policy, although the greater scale at which modern states are organized has usually been said to make the direct participation of citizens in decision-making awkward, and to have led to the development and diffusion of elections and representative democracy.

This vision of citizens as equals in the political sphere, governing themselves through representative political institutions, is at the heart of the ideals of liberal democracy, and political *liberalism* more generally. Liberal political theory places a great deal of emphasis on the *rights* regulating the relationship between individual citizens and their states. Some of these citizenship rights, such as the right to vote, or to free speech and assembly, connect states and their citizens in ways designed to make politics legitimate by freeing channels of communication and influence between states and societies. Others are designed to create barriers between citizens and states by specifying what states cannot do to their citizens, such as seize their property, invade their homes or arrest them without proper legal justification. Modern citizenship rights thus form part of a liberal model of

Box 17.1 The uneven introduction of women's suffrage

Political citizenship rights have tended to develop in ways that reflect and help reinforce existing gender inequalities. Table 17.1 shows how, in many 'first world' contexts, the right of women to vote was only achieved a long time after that for men. In other parts of the globe, some countries, such as those in South America that gained independence in the nineteenth century, show a similar pattern. In many African and Asian countries which were decolonized during the twentieth century, universal suffrage for men and women was introduced simultaneously.

In some contexts, the issue of women's voting rights has been complicated by the existence of other social divisions. In Peru, for example, although women were granted the right to vote in 1955, literacy tests disenfranchised most of the country's peasant population, men and women (about a third of the total) until 1979. It has proved possible for the right to vote to be granted in a regionally unequal manner within a state. In Nigeria, women in the south of the country were eligible to vote from 1957, while women in the predominantly Islamic north were enfranchised only in 1978.

Table 17.1 shows only the first dates from which all men and all women were entitled to vote for selected countries. In some countries, subsequent political history has meant that these rights have been suspended, or rendered meaningless by the absence of other pre-requisites for democracy, such as freedom of speech and association, or by the suppression of political opposition.

Table 17.1 Dates of introduction of universal voting rights for men and women

Country	Votes for all men	Votes for all women
Belgium	1893	1948
Finland	1906	1906
France	1848	1944
Ireland	1918	1918
Spain	1869	1976
Switzerland	1919	1971[1]
UK	1918	1928
USA	1870[2]	1920[2]
New Zealand	1879	1893
Mexico	1917	1953
Uruguay	1916	1932
India	1950	1950
Japan	1925[3]	1945
South Korea	1948	1948
Morocco	1959	1959
Gambia	1961	1961
Kenya	1963	1963

Notes: 1 Switzerland has a Federal system, in which the individual Cantons play an important role in people's lives. It was only in 1990 that women acquired the right to vote on local or cantonal issues in the Canton of Appenzell!

2 These are the dates at which formally universal suffrage was complete. But through the use of poll taxes, literacy tests and intimidation, most African-Americans of both sexes were unable to vote in the southern states until after the passage of the Voting Rights Act of 1965.

3 Men of 25 years and over only.

Sources: Derbyshire and Derbyshire (1989); Krieger (1993); Lane *et al.* (1991); Nelson and Chowdhury (1994).

"Won't you let me help you John?"

Plate 17.2   This British poster from 1910, arguing for the right of women to vote, emphasizes how the granting of formal political citizenship to excluded groups can lead to pressures for changes in state policies. Not surprisingly, established political forces often resist the inclusion of such groups because they anticipate such changes. (Mary Evans Picture Library)

politics which depends on or assumes a division between the state and the private sphere of the economy and households. In the modern world, this model suggests, the ongoing economic inequality of individuals and families in this private sphere is compatible with an historically unprecedented emphasis on the equality of individuals in politics.

Part of the current hype about 'globalization' is about the supposed generalization of

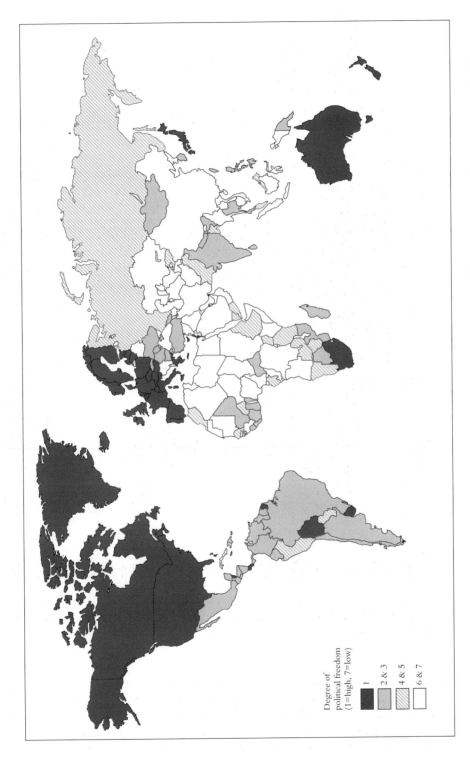

Degree of
political freedom
(1=high, 7=low)

1
2 & 3
4 & 5
6 & 7

Figure 17.1 The uneven development of political citizenship, 1999. Political rights remain highly variable across the globe. This map is based on 'political freedom'. Scores produced by Freedom House, an American organization that monitors political and civil rights (for details visit http:/freedomhouse,org/). Measuring political rights is difficult and controversial, you should interpret the map with caution.

this political model around the world. When everyone lives under it, we have famously been told by Fukuyama (1992), we will have arrived at 'the end of history'. Later in this section, a series of questions will be raised about how far the model, and the citizenship rights it rests on, can really can be seen as 'universal', applicable *in principle* to everyone, everywhere. Here, we should simply note that it has always been criticized for being in many respects out of step with the *facts* of political power. Although the idea that everyone, however unequal in other respects, has the same political weight has been crucial in campaigns to eliminate particular political inequalities, in practice women and the poor have been excluded from democratic arrangements in most countries until well into the twentieth century (see Box 17.1). The underlying reason is obviously that other forms of social and economic inequality affect the degree to which political equality can be institutionalized. The innovations necessary to create liberal citizenship rights are often only partially implemented, can be reversed, and in many political-geographical contexts do not exist at all. Most notably perhaps, the proportion of states in the world which can in some minimal, liberal sense be called 'democratic' has remained, despite some ups and downs, fairly consistently low throughout the last century (see Figure 17.1). In many states, public criticism of, and protest against, government actions remain to be effectively de-criminalized, even in contexts such as the United Kingdom, where recent legislation has imposed new and significant restrictions on public demonstrations. Court systems with powers to block government activity which contravenes citizen rights, and which are deliberately insulated from overt forms of governmental pressure, have emerged to try and regulate state actions and to preserve the abilities of citizens to act politically on their own behalf. Yet such systems seem to merely introduce new forms of minority rule in which judges and processes of litigation which are not resolved by democratic rules increasingly decide important issues. To the extent that modern citizens govern themselves at all, they only do so 'at a distance', through elected representatives. Liberal democracy on this model is therefore often thought to be at best an approximation to popular rule. There have always been critics who have argued for more direct or participatory decision-making in politics to reduce the influence of necessarily unreliable representatives and to bring the scale of politics closer to the scale of people's lives. Frequently such critics draw on the examples of the lost (but reimagined) ancient and medieval city-states or cite more recent tribal and nomadic communities as 'models' of participation.

The second dimension of modern citizenship, the provision of benefits and services by states, reminds us that states have to *do* things to become important to their inhabitants, and for their powers and activities to enjoy some level of popular legitimacy. Historically, states have been organizations which have specialized in violence, and even now at a very basic level being a citizen of a modern state means huddling together with others under the political roof it provides for protection from internal and external threats. Tilly (1985) has described the violent processes through which the modern map of European states took shape, suggesting that states are like rival criminal 'protection rackets', exacting payments in the form of taxes or military service from those who find themselves on their territory. In return, they are able to provide a degree of security to the latter because of their specialization in violence. He underlines the fact that more 'developed' or diverse forms of citizenship have been fairly late in arriving. European states as recently as the 1780s have plausibly been described as 'little more than elaborated networks of drill sergeants, recruiting officers, impressment gangs and attendant tax officials' (Mann 1993:

> **Box 17.2** A simple model of the development of modern citizenship
>
> It has been tempting to characterize the expansion of citizenship through simple evolutionary models. The best known of these is that of T. H. Marshall (1973) in which the development of citizenship can be broken down into several stages, each of which can be thought of as building on rights already established:
>
> - Civil citizenship: legal and civil rights, such as rights of access to the courts, security of property and freedom from arbitrary arrest. These emerged in the seventeenth and eighteenth centuries for selected segments of the population in a few countries.
> - Political citizenship: rights to free speech (including the public dissemination of views on political questions in various media), of assembly and protest, and the right to vote. The expansion of political citizenship is largely a nineteenth-century phenomenon.
> - Social citizenship: rights to state guarantees of socio-economic well-being, such as rights to employment, healthcare, pensions or family benefits. These have been characteristic of the uneven expansions of citizenship in the twentieth century.
>
> Marshall mapped the stages of his model onto specific centuries because it was primarily intended as background to a discussion of the role and future of the British welfare state. Nevertheless, critics have been quick to point out how hard it is to generalize his optimistic scheme to the rest of Europe, let alone the rest of the world (e.g. Mann 1988). In some contexts, such as Germany, social rights have appeared well ahead of meaningful political rights. In others certain 'stages' (notably that of political citizenship) have failed to materialize.
>
> The expansion of citizenship has, moreover, rarely had the linear and progressive character which Marshall envisaged. Even at the level of political citizenship, democratization processes have led to fragile results and have commonly been reversed, giving rise to patterns of oscillation over time between more and less empowering bundles of citizenship rights for much of the world's population.
>
> Marshall's model is also rather descriptive: it is not very useful in *explaining* how citizenship rights emerge and become institutionalized, that is stable over time as a result of their acceptance and enforcement by political and legal institutions. There is much disagreement on this last point among researchers. Some view citizenship rights as the product of struggle by subordinate groups such as the working class and their political organizations (e.g. Korpi 1983). Other, more sceptical scholars see them as innovations sponsored by the economically or politically powerful which are really designed to make the task of governing easier by 'buying off' popular opposition to normal politics (e.g. Piven and Cloward 1971).

371). Put this way, it is easy to see why the legitimacy enjoyed by most states in history has been rather grudgingly given and fragile.

The story of the development of modern citizenship, as it is usually told, involves the expansion of political membership rights which have drawn many states into new fields of activity inside their borders. This story, whatever its flaws (see Box 17.2), draws attention to the fact that citizenship has never entirely been about formal rights to political influence, protection from government, and *political* equality. Citizenship has unleashed a dynamic through which formal political equality is used to make an issue of inequalities in the private sphere and creates pressures to modify them. As a result of this dynamic, being a citizen has become much more central to people's life-chances than in the past, and in the twentieth century there was an uneven but nonetheless massive expansion of state activities in the fields of economic and social policy in most parts of the globe. Large-scale efforts at poverty relief, the provision of income for unemployed people, minimum wages for workers, and health services and education at a range of levels are largely twentieth-century phenomena.

As with the first dimension, links with democracy are central to citizenship's capacity to legitimate modern states. Ideas about popular rule are insufficient to create pressures for democratization: people and movements around the world have struggled for democratic institutions because of the improvements in life-chances more inclusive political regimes seem to promise. In turn, democratic regimes acquire popular support because of their results in terms of socially and economically interventionist policy initiatives. It is not a coincidence that the minimal forms of democracy which have been established in some places in the last hundred years have coincided with the development of welfare states and 'mixed' economies, with states intervening to shape social and economic developments in the interests of their populations. One does not really seem possible in the long run without the other.

Like those characterizing the first dimension of modern citizenship, the policy innovations here have been uneven. The establishment of certain social and economic rights, especially those entailing increased government expenditure, is dependent on a series of economic and political conditions which vary in space and time. In many Third World countries, for example, attempts to create indigenous versions of European welfare states or 'socialism' have been economically difficult and have also challenged the economic and social status of established elites. Both economic development and the changing balance of power and numbers between dominant and subordinate socio-economic groups therefore remain central to explaining the uneven development of citizenship benefits (see Rueschemeyer et al. 1992).

Like other citizenship rights, those associated with 'social' citizenship have developed in highly *gendered* ways. Access to many benefits, such as unemployment benefits and pensions, has often been conditional on participation in the formal labour force. Many women have therefore been excluded from important programmes altogether or have only been able to access them through their husbands. The kinds of benefits available have often reflected a model of society in which male 'breadwinners' are imagined as deserving of state aid, as the chronic inadequacy of childcare programmes in most countries highlights, and women have often been central in images of welfare 'dependency' used to attack the principle of social citizenship, as in the attempted demonization of 'single mothers' in the US or UK in recent years. In discussing this gendering of citizenship, Walby (1994) concluded that although social citizenship rights have been important as a means through which many women have achieved some level of freedom from private domination in the household by men, welfare states have generally been anything but gender-neutral.

Most models of modern citizenship imply that, however spatially and historically uneven its development may be, the rights to which the modern citizen is entitled are in principle applicable to everyone, wherever they happen to be, and whatever cultures or social categories they belong to. Much discussion and criticism of political regimes in Third World or state socialist countries which have not introduced, or have eliminated, many of the rights discussed above, have drawn on a *universal* set of standards based on the experiences of liberal democracies in developed countries. In recent years, the very *idea* of 'universal' citizenship, in which sets of rights can be imagined as in principle applicable to all the individuals inhabiting states across the globe, has come under increasing criticism. These criticisms centre on the *abstraction* from important differences between social groups and societies in such universal models:

1 Contemporary *communitarians* view the emphasis in these models on citizen rights, as opposed to responsibilities, as undermining the sense of obligation citizens should have to the communities and societies in which they live. They suggest that the expansion of state activities and interventions this century, fuelled by abstract ideas about the citizenship rights of individuals, has undermined the role of moral and political civic action rooted in geographically and historically specific communities (e.g. Etzioni 1996). On this sort of account, the question of what states should and should not do has to be answered on a case-to-case basis, according to variations in what community norms suggest. The widespread attempt to reduce state intervention in economic and social life characteristic of many countries in the late twentieth century has been accompanied by escalating criticism of the language of citizenship rights and of the high expectations people have of what states can do for them. This, in turn, has led to communitarian ideas becoming more important in policy discussions. So, in developed countries, far greater emphasis has been placed on the role of voluntary sector organizations at the local level in coping with problems ranging from poverty to crime prevention. Elsewhere, community-based solutions to 'development' issues have become increasingly important in recent years, in part because they better reflect local goals and cultures than policies developed by national governments and international organizations, but also because of the recent emphasis by international institutions and Western governments on reducing the role of the state in Third World societies.

2 From early-modern contention over the regulation of food prices to twentieth-century struggles over social policy, women have reshaped citizenship in important ways (Thompson 1994, Skocpol 1992). Yet in practice citizenship rights such as the rights over property and person, the right to vote, or access to welfare benefits, have been granted in highly unequal ways to men and women. This inequality of access, which has already been alluded to, is no accident, feminist scholars have argued. 'Universal' citizens have, in the Western liberal tradition, been imagined on the model of *male* individuals leaving the clearly-defined 'private' spaces of the home to vote, work and fight (see Prokhovnik 1998). The liberal citizenship model can only pretend to be a universal one, therefore, by systematically concealing relevant gender differences from view, gender differences which might call into question the idea that one model can fit everybody. In this sense, universal citizenship may not only be weak as a solution to gender inequalities, but may also be a source of gender oppression by blocking forms of political representation better able to advocate and decide issues concerning women. Although the issues involved are often very different, this sort of criticism is also useful in considering the relevance of ideas of universal citizenship to political conflicts involving class, ethnicity, 'race' and sexuality (see Young 1989 for a general discussion).

3 The emphasis on universal ideas about citizenship can also be criticized on the

grounds that it is 'eurocentric', meaning that differences between political and cultural traditions in the developed countries of the West and those elsewhere are ignored in the assertion by Western governments and international institutions of catalogues of 'human' rights deriving from European Enlightenment models. So, models of political 'membership' and of state–people relationships stressing religious criteria (such as those in *some* predominately Islamic countries) seem incompatible with universal lists of rights which have a predominantly secular bias and which, for example, may depend on the separation of political and religious institutions. In recent years there has been much concern in Western countries about child labour in Third World countries, which has been expressed in terms of universal ideas about the rights of children. These concerns, in turn, have been criticized for assuming that a Western model of childhood in which work can be replaced by education and labour can be imposed in places with radically different economic conditions and family practices elsewhere in the world.

4 The internal geographical structure of states, particularly the division of powers between central and local governments, poses some deep problems for universal models of the citizen. The varying activities of 'local' states are rooted in the regionally differentiated problems experienced by their local citizens (see, for example, Duncan and Goodwin 1988). By the same token, geographically uneven problems and local state financial resources spell uneven access to the benefits of supposedly 'universal' citizenship rights. In this century, many central state

Plate 17.3 The *centralization* of policy has often been a powerful means of addressing inequalities and injustices. This photo shows black protestors in Alabama in 1965, being confronted by State police. In the USA, decentralized government allowed white people in the South to control State affairs and effectively exclude African Americans from public life. It took massive mobilization by the civil rights movement and Federal (national) government intervention to achieve change. (Topham Picturepoint)

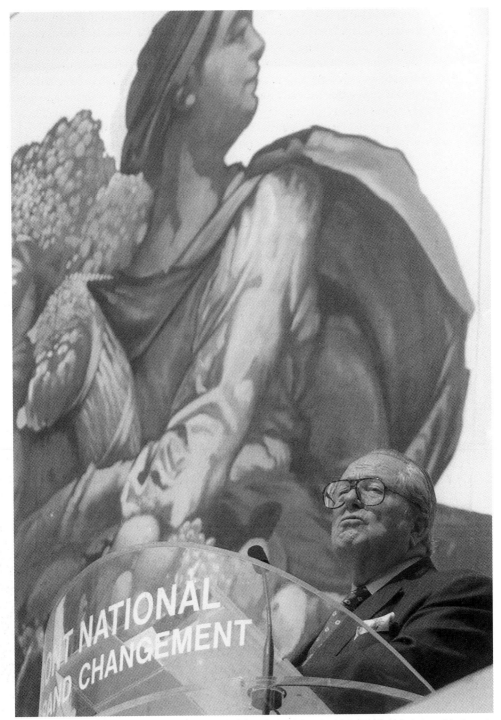

Plate 17.4 Jean-Marie Le Pen, leader of the French *Front National*, the most consistently successful of many European political parties seeking to gain electoral support by defining citizenship in ways which exclude immigrants and resident populations viewed as culturally or 'racially' different. (Associated Press/Michael Euler)

organizations have achieved greater importance than local governments in domestic policy through their role in overcoming such inequalities through inter-regional resource redistribution and the standardization of services. Yet disparities remain, as do tensions between national and local priorities, particularly in contexts such as the United States where decentralized patterns of government and decision-making are formalized within federal arrangements. This appears to be another more or less empirical question about the uneven development of citizenship, or the resources necessary to sustain it, but it also relates to important questions of principle about local versus larger-scale democracy. It is currently fashionable to argue for greater power and responsibilities for local governments, and the enhancement of local and regional forms of citizenship, on the grounds that decisions taken closer to the people have greater democratic legitimacy than those taken by central governments (see Borja and Castells 1997). Yet, greater decentralization of policy-making to regional and local governments in both Europe and North America is likely to make these problems of geographical inequality more acute, and place our inherited concepts of universal citizenship under even more strain (Staehli *et al.* 1997).

5   Most of the above issues are about challenges to dominant Western ideas about universal citizenship from groups and regions who do not necessarily 'fit' the universal pattern. Yet, even inside the dominant institutions of 'the West', the criteria regulating political membership are not really 'universal' at all. Immigration policy reveals very quickly the limits to abstract, universal ideals about what citizenship should be. As modern citizenship has 'deepened', guaranteeing citizens more and more enforceable rights and often costly services, there has been in many countries a 'narrowing' of access to its benefits. Since the nineteenth century, the problem for states of deciding who is or who is not one of their citizens has become more acute, and they have engaged far more actively in regulating admission to the communities within their borders. In many countries there are sizeable populations of persons, resident for work purposes, who do not enjoy many basic citizenship rights. The formal and informal criteria states use to exclude migrants from citizenship rights (birth, nationality, kinship relations with existing citizens and so on) show how citizenship is in reality entwined with questions of nationality and 'race' (Brubaker 1990). In Germany, for example, recent proposals to allow citizenship to those who are not 'blood relations' of Germans have been hotly contested, and in most countries a casual nationalism whereby people 'like us' can be citizens and others generally cannot is very much the order of the day. Thus, at the end of the day, citizenship cannot really be thought of as being fundamentally different from other kinds of belonging. It has in practice proved just as exclusionary as national or racial ideas of community for those people who are seen as not fitting into a given geopolitical space.

This section has reviewed the relationship between citizenship and questions about the legitimacy of states, emphasizing the difficulties involved in assuming that standard modern models of citizenship rights can provide solutions to problems of legitimacy across different historical or geographical contexts or across social divisions such as gender or race. Questions about citizenship are central to political geography because of regional differences within states or between states in what citizenship means, and the differences it makes to people's lives. They are also central because they raise important questions about the difficulties, or even the desirability, of coming up with global patterns for how people and states ought to relate to each other.

17.3 Collective action and policy: citizens and states

States have become subject to enormous pressures from the societies within their borders to transform their actions in different ways. States become sites of contention as soon as their activities impinge on the everyday lives of their subjects. The central interventions characteristic of states over much of their history – particularly their taxing and conscripting activities – have been struggled over intensively and often violently. Many of the citizenship rights discussed in the previous section owe their origins and shape to protracted struggles between different social groups over state policy and the expanded social, economic, and environmental activities of twentieth-century states have in turn generally promoted high levels of contention over a much wider range of policies than in the past.

Some of these struggles are largely *distributional*, concerned with state provision of resources and services or with the distribution of burdens such as taxes or pollution. Other conflicts have more to do with *regulatory* matters such as labour relations, workplace health and safety, environmental protection and the possible risks associated with biotechnology. Others primarily have to do with the recognition and protection of *identity*, from that of the national or ethnic minority groups already considered to the sexual identities of gay and lesbian people (see Brown 1997, who links the politics of sexuality and identity to ideas about citizenship). Geographers have long been interested in the relationships between states at various scales and social groups or social divisions, so as to be better able to explain state activities. To this end, they have often drawn on one of a number of 'theories of the state'. For example, 'pluralist' theories of the state explain policies as the outcomes of conflict between relevant interest groups. Marxist theories of the state, by contrast, view the relationships and conflicts between different social classes as central to understanding why states act the way they do. The use of such theories to understand geographical and historical variations in, for example, the economic policies of states or the power of organized business or landowning elites, is beyond the scope of this chapter (see Block 1987, Judge et al.1995). But all of these approaches, although they do not always emphasize it, depend on understanding how citizens (or firms or other relevant participants in policy conflicts) are able to act collectively in pursuit of their political goals. The second part of this chapter is about such *collective action*, and in keeping with the general theme of citizenship, deals with popular, rather than elite, mobilization.

Such organized mobilization can take many forms. Voting is a form of collective action organized in such a way, and on a large enough scale, that participants cannot really co-ordinate their action effectively. In most contemporary societies, political parties are the organizational forms by means of which of this important but diffuse form of collective action is co-ordinated around sets of contrasting policy aims. *Political parties* are usually general in their appeals. They mobilize coalitions which are usually socially heterogeneous during rather limited and infrequent electoral episodes. Many *social movements*, on the other hand, involve the direct participation of highly motivated groups in political action aimed at achieving specific results.

17.3.1 Political parties and the geography of electoral coalitions

In many countries with political systems open to popular influence, a set of institutions (e.g. parties, elections, representative governments) and practices (e.g. periodic voting, ongoing public political debate) channel political conflict through regulated organizations and routines. Marshall (Box 17.2) related these institutions and practices to the idea of political citizenship rights. Whether the existence of these institutions and practices is enough to label a country 'democratic' is a crucial question. States are subject to all kinds of pressures and influences from economically powerful individuals and business organizations, as well as from movements and pressure groups of a more popular character, which operate through channels other than elections. In many countries elections seem as much influenced by the money channelled to candidates and parties as they are by popular preferences. Worse, the connections between the detail of what governments do and the preferences of electorates are often obscure, leading to the suspicion that electoral democracy is merely a device used by the powerful to help make often unpopular government activities seem legitimate and to channel popular discontent into easily managed and infrequent episodes. All of these worries are, to varying degrees, on target. They also help explain the apparent unease with established political mechanisms felt by many citizens in countries with declining voter turnout in elections. Such problems and qualifications are part of what makes electoral democracy a matter of great concern. It is, on the other hand,

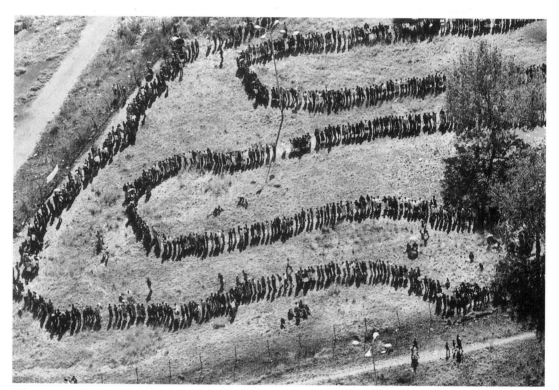

Plate 17.5 The powerful symbolic as well as practical significance of the right to vote is brought home forcefully by this image of voters queuing in the South African elections in 1994, when the majority black population were first able to participate. (Associated Press/Denis Farrell)

hard to deny that the mass participation in elections characteristic of twentieth-century politics in some countries has been crucial in making states pay rather more attention to popular life-chances than they had in the past.

In contemporary 'liberal democracies', political parties are the main organizations responsible for persuading people to vote, for structuring the choices available to them when they do, and for developing sets of appeals and policy positions which, if they gain sufficient support, form the basis of governmental programmes. They are thus central to explaining the ways in which people vote, and crucial actors shaping the relationships between states and their citizens. The practices of representation in which parties engage form the central political mechanism in electoral democracies. As mentioned earlier, it is usually viewed as the crucial political innovation permitting some measure of popular influence over the fairly large states characteristic of the modern world (Dahl 1989, Manin 1997). What do the parties represent? Different parties do represent different social interests, but we should not take this to mean that voters come neatly pre-divided into clearly-identifiable 'voting blocs' based, for example, on class, ethnicity, occupation or gender. If this were the case, we would know what the outcomes of elections would be in advance, simply by looking up census tables! Citizens often have positions on political issues which make sense in terms of their social background, and some parties do become strongly identified with particular social groups. But most citizens have multiple social positions and identities which can push their political identities and preferences in different directions, and make electoral politics less predictable both for citizens and for politicians aspiring to governmental power.

Moreover, political parties, even those such as 'labour' parties or 'nationalist' parties which seem to appeal to a single group, can rarely afford to reflect simply one particular interest. In most contemporary societies there is not an identifiable social group of sufficient numerical strength and internal coherence on which parties can rely to win elections and legitimate their actions in government. A good example of this problem is the situation faced by many socialist parties in European countries. The 'core' interest these parties were founded to represent, the industrial working class, has been steadily declining as a proportion of most electorates, and has rarely formed an electoral majority in any case (Przeworski 1986). Even allowing for definitions of the working class which would include workers in the expanding tertiary and quaternary sectors of the economy, it is clear that building winning electoral coalitions has necessarily involved these parties attempting to attract a variety of social groups simultaneously. These are likely to see themselves as having diverse interests and policy preferences. Likewise, those parties which are strongly identified with business interests, such as the conservative or Christian Democratic parties of Western Europe, cannot aspire to governmental power without finding ways of linking their programmes to large non-business constituencies.

In consequence, much of the most interesting work of political parties goes into actively shaping and reshaping partisan political identities in attempts to weld together heterogeneous electoral coalitions, through face-to-face contact, mobilization at rallies and other public events, and the extensive use of print and electronic media. The creation of new broad electoral coalitions offering otherwise different groups a shared political identity, and suppressing their differences, has often been the crucial factor in mobilizing sufficient support for reorganizations and expansions of domestic state activity. In the twentieth century, the development of some of the most radical democratic sets of policies and institutions designed to protect ordinary people's interests and redistribute social wealth, in Scandinavia, the United States, and in many Third World contexts, have been

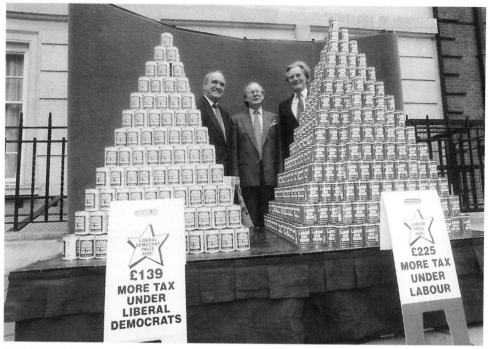

Plate 17.6 British Conservative politicians in 1996 illustrating their claims to tax less by translating their estimates of the tax costs of voting for their rivals into stacks of cans of baked beans. Taxing less is viewed by many politicians as a crucial strategy to gain the support of middle-class voters, who see themselves as less dependent on state expenditures than other social groups. (Topham Picturepoint)

backed up by electoral coalitions uniting rural small farmers and urban workers which hitherto had not seen themselves as falling into the same political category (see Esping-Anderson 1985). To compensate for declining working-class populations and rural support, the parties concerned have, in recent decades, become more and more reliant on adding voters who define themselves as 'middle-class' to their electoral coalitions. In contemporary Europe, the competition for middle-class support has therefore become in many ways the key factor sustaining or undermining established patterns of state intervention. The dilemma for left-of-centre parties is, of course, that increasing reliance on this support often involves diluting their commitments to the very policies they needed middle-class votes to sustain. We can see here how the ability of electoral democracy to deliver material benefits to modern citizens depends not only on the presence of elections and political citizenship rights, but also on the skill and will of parties in creating supportive shared political identities in sometimes rapidly changing circumstances.

This alliance-building activity is, in turn, likely to involve holding together different regional bases of support. In the case of the 'farmer–labour coalitions' just mentioned, this is easy to see. They involve constructing electoral bridges between urban and rural, or central and peripheral, regions of the space-economy. The United States example is especially clear. The creation of the modern Federal (central) government, with its array of institutions and policies in the fields of industrial regulation, social policy and agricultural support was propelled by a striking change in the regional electoral appeal of the Democratic party during the Great Depression of the 1930s (Figure 17.2). The 'New Deal

Presidential Election of 1896

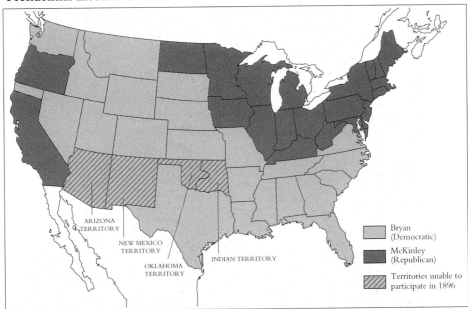

Presidential Election of 1932

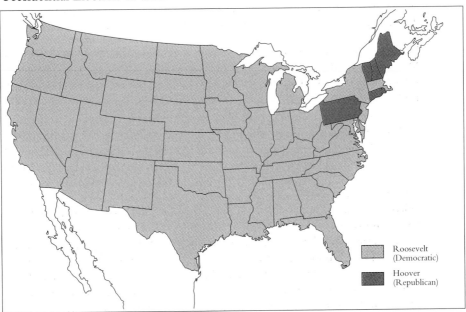

Figure 17.2 The formation of a cross-regional coalition: the United States in the 1930s. Bryan, the Democratic Party presidential candidate, lost the important 1896 election despite winning the support of the southern and western agricultural periphery. Franklin Roosevelt's 1932 victory, and subsequent 'New Deal' reforms, were made possible by linking the Democrats' rural base with the votes of northeastern industrial workers. *Source*: Based on Nash *et al.*, 1990: 670, 823

coalition' which resulted embraced both the traditional southern rural base of the party and working- and middle-class support in the core industrial areas of the north-east and mid-west, a formula which provided unprecedented legitimacy for popular reforms (Bensel 1986). In many Third World contexts, where industrial and service workers form a small and spatially-concentrated proportion of the electorate, the building of farmer–labour coalitions has been central to efforts to create legitimacy for inclusive, non-military, governments (Collier and Collier 1991). This sort of coalition is extremely difficult to construct, and has usually been held together by promises of redistributive social policies and land reform. It has, therefore, often been viewed as a threat by urban and land-holding elites. In several cases, such as the regimes of Juan Perón in 1940s and 1950s Argentina, of Salvador Allende in early 1970s Chile, or of Getúlio Vargas and João Goulart in 1950s–1960s Brazil, attempts to govern on the basis of such cross-regional 'populist' alliances have been an important factor provoking elite-supported military coups.

It is possible for parties to govern on the basis of coalitions which marginalize particular regions on a long-term basis, but this is likely to provoke controversy and, in countries with strong regionalist or nationalist traditions, may even threaten national integrity. Where country boundaries have been created by foreign imperial administrations, for example, in the Balkans, the Middle East, or Indonesia, there often exist several regionally-concentrated linguistic and cultural groups (see Chapter 15). This situation raises prospects of minority nationalism and the formation of party systems organized at the regional rather than the national level. Concerns about national integration have been important in sustaining one-party regimes in several such countries, and efforts at sustaining democracy usually have to cope with the threat of one-region rule. In Nigeria, for example, attempts have been made to legally bar the development of parties with concentrated regional bases of support, and one of the thorniest issues for post-apartheid South Africa has been the degree to which politics should be decentralized, allowing regionally-based organizations like the Zulu Inkatha Freedom Party to 'rule their own roosts' (see Tordoff 1997: 120). In the United Kingdom in the 1980s and 1990s, there was much commentary about the Conservative government's apparent reliance on a 'two-nation' political strategy involving the use of relatively affluent and populous southern England as an electoral base (Gamble 1994, Jessop 1989). With three parties (the Conservatives, Labour and the Liberal Social Democrat Alliance) dividing the vote, it was possible for the Conservatives to retain national power with less than an overall majority, and crucially with low levels of electoral support elsewhere in the country (see Johnson et al. 1988 for details). The possibility of such regionally-polarized rule has been an important factor behind the recent devolution of power to Scotland and Wales and is one factor behind pressures for the UK to adopt a different, less geographically-based, electoral system.

High levels of regional differentiation and inequality can make the construction and maintenance of nation-wide coalitions within parties a difficult task. In Italy, a country with deep inter-regional economic and social inequalities, the Christian Democrats became the dominant party over much of the post-war period by forging links and compromises between northern business and southern agrarian interests, a strategy which was successful in keeping the Communist party out of government, and which involved the support of both the Vatican and the United States government (see Ginsborg 1990). But during a wave of political scandals in the 1990s it became only too clear that a range of illegitimate techniques had become routine parts of holding this alliance together, from the corrupt funnelling of government contracts and finance to linkages with southern Italian organized crime (Stille 1997). As these techniques were progressively uncovered, the Christian Democrat-based inter-regional alliance was 'de-legitimated'. The party collapsed and the entire party system became transformed, along with the 'rules of the game' of the Italian political system. One symptom of the problem faced by the Christian Democrats was the rise of regionalist political forces such as the Northern League seeking to undermine the established geographical organization of the system and sever some of the links built up between North and South. This may seem an extreme case, but the coalition-building in which parties have to engage creates, in general, all sorts of interesting tensions, regional as well as social, within party organizations.

The heterogeneity of partisan coalitions, and the policy fudges and vacillations this helps produce, is one reason why parties in electoral democracies are rarely viewed admiringly by the public or democratic theorists. Another reason is, of course, that parties are not simply sustained by popular votes, but also by financial support from business and other elite sources. Inevitably, then, parties have to be seen as organizations effecting compromises of several kinds, both between different social and geographical interests and between elites and citizen majorities. Although this seems unromantic, it may be that the stitching together of unstable combinations of social forces in which parties specialize has been a key precondition for some degree of democratic government in the past hundred years or so, and for the resulting re-orientation of state domestic activities around some popular concerns. The formation of competing partisan identities may even be an important, and underemphasized, antidote to more monolithic forms of political identity, such as nationalism. In any case, parties are never the only vehicles of political collective action in modern societies. When we view them as part of a range of forms of citizen mobilization which also includes social movements, potential correctives to their compromising character are not hard to find.

17.3.2 Social movements: linking formal and informal politics

Modern 'liberal democracy' is usually thought of as a form of politics characterized by the institutionalization of parties and voting, but it has never taken a purely electoral form. Social movements routinely involve people more directly in collective action to influence state policy or to exert pressure on other organizations, such as firms or parties. The 'formal' political sector laid out in constitutional arrangements and laws regulating elections and political parties has always been accompanied by a lively informal sector, in which citizens march, demonstrate, protest, occupy public and private spaces or more quietly fund lobbying and direct action of various sorts by specialized social movement organizations. Social movements are often written about as if they are a more 'progressive', or at least modern, form of political mobilization than parties. One of the themes of this section will be that, exciting as they are, we can only begin to understand movements if we let go of the temptation to make them seem necessarily more innovative or radical than other forms of political action.

A social movement is a specialized group formed outside the state and party system in an attempt to promote or prevent social change. Social movements may act to influence government policy, as in the case of groups currently campaigning to ban fox-hunting in

Plate 17.7   Members of the Falun Gong movement in China, in April 1999, occupying public space to put pressure on the government to recognize their religion. The image is a reminder that common images of social movements as 'modern', 'progressive' or even 'left-wing' are seriously misleading. This movement eludes conventional Western political categories altogether. (Associated Press/Greg Baker)

the United Kingdom, or to establish greater press freedom in Iran. They may also act without going through political channels to achieve their aims or make symbolic points. This sort of strategy has recently been made famous by Greenpeace interposing its ship, the *Rainbow Warrior*, between whaling vessels and their prey, by Operation Rescue blocking the entrances to abortion clinics in the United States and earlier by 1960s African-American civil rights activists occupying 'whites-only' spaces on buses, at lunch-counters and in department stores. The blockading of ports by animal rights activists seeking to prevent the shipment of livestock, or the dumping of imported meat into the sea by farmers, are two recent examples from the British context.

It is hard to see how contemporary politics could work without such non-institutionalized and often unruly practices. By themselves, votes are a notoriously (and deliberately) blunt way of communicating opinion to those with influence or potential influence over what states do. Participating in social movements is one way in which citizens can more directly 'voice' their desires and beliefs, and express their political and social identities. Moreover, because social movements are by nature more specialized in their concerns and constituencies than political parties, they are a means by which pressure can be exerted on the powerful by groups concerned with specific policy issues. In this sense, social movements are related to 'interest groups' such as business organizations, which seek to influence the details of policy-making outside electoral channels. In electoral-democratic systems characterized by majority decision-making, the views of non-elite minorities are often badly served. Some of the most important recent social movements, such as the American civil rights movement, gay and lesbian organizations in many countries, or the movements for national autonomy or independence within the former Soviet Union, have formed around the protection of, and expansion of opportunities for, minority groups. For all these reasons, and despite the contentiousness and disorder which are often intrinsic to their activities, social movements can be useful to established parties and government policy-makers. They furnish potentially valuable channels of communication between those in authority and specific groups in civil society as well as indicate the intensity of popular concern about specific policy issues.

Social movement activity usually requires much greater sacrifices of time, energy and resources than voting. Looked at from an economic point of view, collective action of this sort can seem irrational: an individual will almost always see his or her contribution to such activity as so marginal that it will hardly seem worth the effort (Olson 1965). Even if individuals are very concerned about a particular social or political issue they are, the story goes, usually happy to rely on others to do all the work to bring about the changes they want. This is one reason why political parties, which represent rather than directly involve the vast majority of the people in their coalitions, might be thought to be predictably dominant when it comes to linking citizens to policy. Yet levels of popular involvement in voluntary group activities, including politically-oriented social movements, can be higher than turnout in elections: a comprehensive recent study (Verba *et al.* 1995) has suggested, for example, that this is indeed the case in the United States, a country with often remarkably low levels of voter turnout in recent years.

How do people and groups get involved in social movements? It is true that feeling strongly about something does not necessarily lead to doing anything about it. Unlike the situations imagined in most economic models, however, movement mobilization does not take place in a vacuum where independent individuals cautiously weigh up the costs and benefits of their potential contributions. Most social movements grow out of established networks of social relationships where cultures of involvement in political action have taken shape, and where people feel a strong sense of obligation to co-operate in activities organized by those they either know or respect. The American civil rights movement, which in some ways became the model for many other social movements in the US and elsewhere in the last quarter of the twentieth century, grew out of pre-established social networks among urban middle-class African-Americans, churchgoers and students in segregated institutions of higher education (McAdam 1983). Successive waves of activism around educational, nuclear power, and environmental issues in Germany have similarly developed out of established subcultural networks involving young people and especially students (Markovits and Gorski 1993).

This 'rootedness' of social movements means that in many cases there are strong links between movement development and particular places or localities. The pre-existing cultures and social networks from which movements draw strength are likely, in many cases, to be highly localized. But it is important to set against this the fact that social movements mobilize at a variety of geographical scales. National mobilizations involving demonstrations, often in capital cities and in the full glare of the media, are perhaps the most obvious, although such movements typically develop from smaller-scale mobilizations around more local issues. As Tilly (1984, 1986) shows in the case of France, social movements have become 'nationalized' in step with the development of centralized state power and policy-making. Much attention in recent years has been given to issues of 'transnational' movement mobilization, around issues concerning the environment in particular (Keck and Sikkink 1998). The effects of such movement activity can be

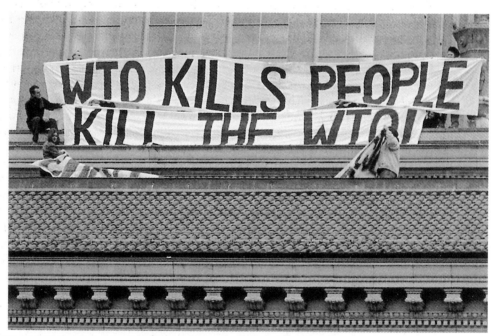

Plate 17.8 Protestors against the World Trade Organization (WTO) in Geneva in November 1999. The movement highlighting the role, power, and lack of democratic accountability of organizations such as the WTO and the International Monetary Fund has a definite trans-national character. (Popperfoto/Reuters)

exaggerated, but clearly some organizations such as Greenpeace and many non-governmental organizations in the development field have a definite transnational character and in recent years there has been much co-ordinated pressure on governments and international organizations by activists and organizations: the mobilizations around French nuclear weapons testing in the South Pacific in the mid-1990s, and the ongoing campaign to cancel Third World debt, may well be harbingers of greater international networking and pressures by the social movements of the future.

Are we living in a period in which social movements are replacing parties as the main form of popular political mobilization? In the context of a declining working-class base for popular challenges to the state, with the accompanying dilution of socialist party politics and increasing popular antipathy to party politics as such, so-called *new social movements* have often been viewed as an emerging alternative form of radical politics (Touraine 1981, Dalton *et al.* 1990). The new social movements which have sprung up since the late 1960s, such as the women's, environmental, anti-nuclear, gay and lesbian and student movements, can be seen as representing a 'new politics' in which direct action and specialized struggles form an alternative to established democratic mechanisms.

New movements are said to differ from previous social movements, particularly the labour movement, in several ways. 'Old' movements were concerned with the advancement of the material interests of their members, as shown by the achievements in economic policy and in the expansion of social citizenship rights of nineteenth- and early twentieth-century worker mobilization. New movements are said to be more concerned with identity and life-style issues. The post-1968 feminist movement, for example, has often emphasized consciousness-raising among women and the celebration of gender difference as much as or more than established concerns over equality of opportunity in politics and labour markets. Environmental activism is difficult to relate to material concerns: so much so that the label 'post-material' politics has been coined to capture the sense that increasing affluence has enabled publics in developed countries to focus more on quality of life issues than they have in the past (Inglehart 1990). New movements are also said to operate more independently of established political channels than old movements and, as a result, to be far more 'radical' in their tactics, employing a variety of direct-action techniques to achieve their goals.

There has been an important 'wave' of social movement activity in most developed countries dating roughly from the mid-1960s. Recent movements have staked out new areas of policy contention and often shown a remarkable capacity for innovation in the tactics they have employed. Moreover, as Dalton and Cotten (1996) show in the case of South Korea, where labour militancy has been replaced by more diverse movement mobilizations around issues of citizenship, women's rights, environmentalism and poverty, such movements have not been confined to Western industrialized contexts (see also Hipscher 1998). Nevertheless, two important critical points need to be made, which suggest that there is a difference between noting the emergence and proliferation of collective action around 'new' issues, and arguing that this marks a radical break from previous forms of movement and from established politics in general.

Firstly, much of the hype about new movements has rested on a series of flawed comparisons with labour movements. Nineteenth-century labour movements were as concerned with the establishment and assertion of working-class identities as they were with 'bread and butter' material issues. Similarly, non-economic life-style- and identity-asserting movements are not recent phenomena. Calhoun (1995) discusses a series of 'counter-cultural' religious and utopian community movements in pre-twentieth-century America which fit the new social movement pattern rather well. But the key counter-example here is probably nationalism which is, among other things, a far from new social movement that is very much concerned with 'non-material', or at any rate non-economic, issues of language, identity, life-style and collective expression.

Secondly, the 'formal' political sector of elections and parties and the 'informal' sector of movement activism are related in quite fundamental ways. Part of the interest in new social movements was stimulated by the thought that political parties were perhaps things of the past. Most commentators would now suggest that parties have been transforming rather than declining (see Katz and Mair 1994, Maisel 1994). More attention is correspondingly being paid to the connections and interactions between movements, parties and national 'systems' of parties (Kitschelt 1990, Müller-Rommel 1990). Some parties grow out of movements. Green parties in Germany, Belgium and elsewhere have emerged out of pre-existing networks connecting environmental and anti-nuclear activism and 'new left' politics (Markovits and Gorski 1993, Kitschelt 1989). The right-wing anti-immigration parties dotting the present-day European political landscape have mostly developed out of background webs of racist organizations, which in some cases are the remnants of earlier attempts at something like party organization. Such party–movement linkages are not new. Earlier waves of party formation in developed countries had been fuelled by workers' and farmers' movements. Many important political parties in African and Asian societies, such as the Indian Congress party, derive from anti-colonial liberation movements.

Even if they originated in other circumstances, established parties often draw on movement support. The Democratic party in the United States has established often fruitful relationships with many recent social movements, including the women's movement and civil rights groups. The British Conservative party has in recent years been pursuing a relationship of apparent mutual advantage with movements which have emerged to protect selected rural traditions, such as hunting with hounds and the defence of private property rights against ramblers. This sort of relationship can be a dangerous game for established party structures and elites, as there is sometimes a fine line between party–movement interactions and the invasion and 'capture' of party organizations by social movement organizations. This can lead to parties being transformed in new directions, and is often a key factor in creating new political possibilities from old political moulds. The periodic capture of some levels of the Democratic party by farmer and labour organizations was a major force for political change in America in the first part of the twentieth century, just as the influence of the religious right within the Republican party has had important repercussions in its closing decades.

Movements, more specialized vehicles for collective action than parties, are among the most exciting 'domestic' political phenomena, and are under-researched in political geography. As some of the debates about new social movements suggest, the contrast with parties should not be taken too far. The common romantic image of movements as popular champions and vehicles of direct action is in many cases well-deserved but they, like parties, are prone to bureaucratization and can become as reliant on financial donations to carry out their activities as many other political organizations. Major environmentalist organizations are often large and professionally-staffed, like the business organizations they oppose (McCann 1986). Others, such as Greenpeace, fund direct action by essentially full-time employees by accumulating donations from their mostly passive memberships.

Indeed, like parties, social movements often claim to be acting on behalf of constituencies who are not directly taking part in their activities. None of these tendencies are inevitable: one of the most successful mass movements, the African National Congress in South Africa, has become a ruling party, but there is as yet little evidence that this has substantially lessened movement mobilization in that country (Klandermans *et al.* 1998). Nevertheless, social movements, like parties and other political collective actors, need to be critically evaluated from the point of view of both how they really operate and how representative they are of the wider groups their activities affect. They also have to be considered in terms of their interactions with, rather than simply their opposition to, other forms of political mobilization.

17.4 Conclusions: limits of modern domestic politics

This chapter has discussed two dimensions of the relationship between modern states and the people inside their territories. Modern citizenship, in conjunction with nationalism, has fostered the development of a certain degree of consent by domestic populations to be ruled by particular state organizations in particular territorial areas. Political parties and social movements have formed links between popular demands and state activity which have enhanced states' legitimacy by subjecting their policies to varying degrees of citizen control. It would be wrong to suggest that any of these mechanisms driving modern domestic politics are the best that could be imagined or achieved, that they exhaust our aspirations for what 'democracy' could be, or that they are operative in all parts of the globe. But it would be equally wrong to ignore the improvement of life circumstances that this recent, albeit geographically uneven, 'domestication' of generally rather predatory organizations has brought about for many people.

Given its recent emergence, it seems appropriate in closing to ask how stable this picture of domestic political activity is likely to be. One of the themes in the discussion of political parties was the importance of broad coalitions, such as those between industrial workers and other relatively underprivileged constituencies such as small farmers, in creating pressures for states to shift their activities in directions favourable to popular interests. Yet, as we have seen, this sort of reformist alliance has become harder and harder to assemble in industrialized countries in recent decades, as economic changes have eroded its social basis. Deindustrialization, social fragmentation and middle-class concern about taxation have put pressure on established patterns of social and economic intervention and the idea of ever-expanding citizenship rights has come under fire, not only from conservatives, but also from the inheritors of socialist traditions. Inclusive democracy as it has been institutionalized in developed countries may depend on the delivery of real improvements to ordinary people's circumstances. It is an open question how far governments can go in withdrawing or modifying various citizenship benefits before real legitimacy problems re-emerge.

Furthermore, the picture of citizenship and citizen mobilization presented here assumes a fairly clear distinction between states and their societies. Another reason why citizenship and established patterns of citizen mobilization may experience dramatic transformations lies in the possibility that traditional government, theoretically responsive to popular influence through parties and movements, is being replaced in many contexts by new forms of governance. In these, states devolve power to non-governmental organizations, or to 'partnerships' between parts of the state and 'private' actors such as business organizations or voluntary organizations. While these new arrangements, which are often temporary in character and highly focused in their aims, may be effective in tackling particular problems, they are rarely subject to direct democratic control. Furthermore, the restructuring and 'down-sizing' of states which results may, in the long term, make it harder and harder for them to rely on the kinds of popular legitimacy which clearly-identifiable *state* benefits and policies make possible. There is an important aspect of this problem which has to do with geographical scale. As the previous chapter noted, the established 'national' states which have formed the main subject of this chapter may cease to be the key organizers of political life, as a result of a general redistribution of power resulting from 'globalization'. If this is true, important consequences for citizenship, and 'national' vehicles of popular mobilization, would follow, as well as for the imagined communities to which citizens see themselves belonging.

There are a whole series of important theoretical and practical challenges ahead in this shifting geopolitical context. What might 'post-national' citizenship look like? Could it be more effective than its forerunners in bridging important social divisions and cultures, or is it necessary to imagine multiple forms of citizenship, each with a certain validity in relation to particular regions or societies, or offering different sets of rights and/or benefits to social groups experiencing different forms of oppression or having different needs? How can governance, as opposed to government, be influenced democratically? How can international organizations be democratized? Should there be international elections, for example, or are conventional diplomatic channels or pressures from social movements or non-governmental organizations enough to give the activities of the United Nations, the World Bank or the World Trade Organization some minimal level of legitimacy? It has been argued (e.g. Linklater 1998) that, after the Second World War and with the development of the United Nations Charter of Human Rights and the concept of 'crimes against humanity', a form of global civil citizenship has been 'under construction'. Is it possible to construct some form of *political* citizenship across borders, languages and cultures? How meaningful could these new citizenships be without forms of non-national governmental activity providing more than formal rights and protections and making some real positive difference to everyday lives?

Modern states are supremely ambiguous social institutions. They are capable of large-scale action to shape societies and the life-chances of their inhabitants through the provision of health care and education, through economic management and the regulation of risks and external threats. They are also, because of their capacity for organized violence, the most dangerous organizations in human history. For a small part of the last millennium, states in some parts of the world have reached accommodations with their populations which have assured their legitimacy in return for higher levels of responsiveness to popular needs. It may be that these accommodations have depended on certain social and geographical conditions which will not prevail far into the next thousand years. Whether this is a pessimistic conclusion or not will depend on the political imaginations and struggles of those shaping the political geographies which replace them.

Learning outcomes

After reading this chapter you should know that:

- To be a citizen of a state is to be a 'member' of a particular spatially-bounded political community.

- The membership benefits associated with modern citizenship are usually said to comprise certain civil and political rights, as well as state benefits and services. The development of access to these rights and benefits has varied greatly across social divisions and geographical space.

- The idea that we can come up with a list of universal citizenship rights, which we think should be available to everyone and used to evaluate the actions of governments across the world, has been increasingly criticized. The ideal of universal citizenship, it is argued, cannot deal with problems of cultural differences between places and communities, or with social divisions such as gender, ethnicity or 'race'.

- Political parties are important organizations through which citizens are mobilized to influence state activities. They are capable of constructing very large coalitions of people for electoral purposes, but these coalitions are heterogeneous and often unstable, because of the social and regional tensions they contain.

- Social movements are specialized forms of popular collective action, generally aiming to promote change in a particular policy area. Social movements and political parties are interrelated in important ways, and it is unlikely that social movements can replace parties as the main form of citizen mobilization.

- The institutions organizing domestic political geographies in the last century or so may be under threat from government restructuring and a general spatial reorganization of world politics ('globalization'). This may lead to the development of post-national citizenships.

Further reading

Mouffe, C. (1993) 'Citizenship' in Joel Krieger (ed.), *The Oxford Companion to Politics of the World,* Oxford University Press, Oxford. A clear and concise discussion of the concept of citizenship.

Painter, J. (1995) *Political Geography: A Critical Perspective,* Arnold, London. A very good introduction to many of the topics considered in this chapter. Chapter 3 is on liberal democracy and the welfare state, Chapter 6 is on social movements.

Taylor, P. and **Flint, C.** (2000) *Political Geography: World-Economy, Nation State and Locality* (5th edition), Prentice Hall, Harlow. This book has always been wonderful at connecting local and national 'domestic' political topics to a global framework. See Chapter 6 for an interesting discussion of the geographies of elections and political parties.

Walby, S. (1994) 'Is citizenship gendered?' *Sociology* 28(2), 379–95. A clear feminist analysis of citizenship, emphasizing inequalities between men and women in terms of access to modern citizenship rights.

Young, I.M. (1989) 'Polity and group difference: a critique of the ideal of universal citizenship', *Ethics* 9, 250–74, reprinted in Anne Phillips (ed.) (1998) *Feminist Political Thought*, Oxford University Press, Oxford. This is a very influential article (and slightly more challenging reading!) which seeks to question the very idea that citizenship rights can be generalized across different social groups. It does not focus on geography, but it is not hard to think of ways in which the arguments might apply to ideas about citizenship which are meant to be valid across different places, countries and regions.

PETER DANIELS•MICHAEL
BRADSHAW•DENIS SHAW•
JAMES SIDAWAY

CONCLUSIONS

Challenges and promises

The root of the word geography combines *geo* (earth) and *graphy* (writing). To engage in geography is to write about the earth. Of course, many other branches of knowledge such as history, sociology, politics, geology (from which geography draws) are also in some sense or other about the world. Traditionally, what has been distinctive about human geography is that it puts an *emphasis* on places, spaces and spatial connections (including those between society and nature), on relations, discourses and imaginations of proximity, distance and territory. Today, human geography embraces issues such as identity, consumption and sexuality which are vital parts of our interconnected, contradictory, complex, conflict-ridden and fantastically diverse planet. We would argue that they are no less 'real world', nor necessarily more or less 'abstract' topics than some of those more supposedly 'staid' issues of, for example, resources, population, cities or the distribution of economic activity. Indeed, we would argue that some of the most interesting, but tricky, challenges for human geography arise from its ambition to take account of the full range of human experiences, achievements and transformations. In other words, human geography is not just about 'out there', it is also about 'in here'. Thus, you all create your own personal geographies derived, for example, from where you live, where you work or when you travel.

Such complexity and challenges are evident when we consider 'globalization'. This is something that is considered a defining characteristic of the times in which we live. As editors we have been conscious of, and have regularly argued amongst ourselves about, what the term takes for granted, what kind of stances we should encourage the contributors to take towards it, and how far we could agree once we saw the arguments of the chapters and sub-sections emerging. As several of the chapters have spelled out, there is certainly a significant sense in which a combination of technical, political, cultural and economic transformations throughout the twentieth century had enhanced the sense of global interconnection and hybridity. Take the case of this book. The example that you are reading may well have been printed in Singapore by a multinational publishing company, whose ownership and 'home' location are difficult to discern. Its production involved dozens of floppy disks, desktop and laptop computers (themselves usually manufactured in Asia), and software written and distributed by the world's largest (by value) transnational company (Microsoft). Numerous drafts of the chapters and sections were e-mailed between authors and editors, sometimes between offices next door to each other, or sometimes across Europe. Just two decades ago the writing and production process would have been very different. There would have been much less dependence on electronic media, more extensive use of surface or airmail to exchange pieces of the manuscript and the production of the book, copy-editing, proof-reading, printing largely completed locally. To describe the contemporary technological, economic, cultural or political tendencies as 'globalization' is, we agreed, to invoke a certain *geographical imagination*, a vision of the growing significance of a global scale of action, of the world as a single place. Yet, we know that much of the world still remains excluded from this hypermobile way of living and working.

This exclusion is made real and meaningful in language. Without the term 'globalization', we might use other terms: 'imperialism', or 'capitalism', for example. Each carries particular connotations. In this way, 'globalization' is a kind of imaginary construct and one on which a certain geographical perception is foregrounded. Yet just because something is imaginary does not make it any less real to those caught up in it. Globalization is seen by some as a positive liberalizing force, breaking down barriers, making capitalism more efficient, and spreading its benefits throughout the world. For others it is a negative force enabling another round of exploitation, often with the further destruction of local cultures and identities. Obviously, globalization is not simply a question of good or bad. Judge for yourself, but do not make the mistake of assuming that everyone everywhere shares your experiences of the world.

Again, consider where you were at the moment of the new millennium. The particular construction of space and time that saw the new year 2000 sweep across the time zones of the earth (measured with reference to a Royal Observatory at Greenwich in the capital of what was once a great colonial power) was also based on a specific geographical and historical imagination. However, the coming of the millennium was represented to us (and lived and made by us all) as an enormous global collective experience, likened by some to the moment when the Apollo 8 spacecraft in 1968 captured that first image of the earth as a whole from space.

Moreover, the idea of a twenty-first century, which frames the title of this text, is, as was noted in our Introduction, global in scope, but rooted in a Western vision. According to the calendars of China, Islam and Judaism this is not the twenty-first century. The calendar that produced 2000 as a millennium year was established first in the regions that we now refer to as Europe and the Middle East and then refined and disseminated over what we now call 'centuries'. In England the calendar now used was not established as official until 1751. And recent controversy over whether 2000 or 2001 marks a 'new century' is in part rooted in the fact that Western mathematics had no concept of zero, until it was imported, via the Arab–Islamic civilization, from India to Europe. So whilst the year 2000 is a Western construct, it turns out to rely on the idea of zero whose origins are elsewhere. One of the lessons here is that the West owes much of its identity and knowledge to ideas, people and artefacts originating outside what have been considered its geographical limits.

As was also noted in the Introduction, our geographies (particularly as taught in Western Europe and North America) also still tend to be rooted in a Western vision, often taking this for granted and assuming that what is the norm is the white Western (often male) 'developed country' experience. In fact only a minority, albeit a relatively power-ful one, of the world's population belongs to such a 'category'. And like everything else, the category itself has a complex history and geography produced out of interchange, interactions and power. Geography is only at the beginning of what promises to be a dif-ficult but promising process of moving beyond such assumptions and outgrowing its modern roots as a Western science tied up with colonial knowledges of the 'rest' of the world.

So, where do you go from here? As a geography student (or any other kind of student for that matter), you will be faced with choices, for example between human and physical geography or between more specialized subjects within one of those traditions. It is in the nature of academic disciplines to sub-divide and to compartmentalize knowledge. Geography is unusual in that it sits at the intersections between the humanities, social sciences and natural sciences. This means that it internalizes both the disadvantages of disciplinarity and the potential advantages of multidisciplinarity. Much of the promise of geography lies on the margins between the sub-disciplines, between human and physical geography or, for example, between economic and cultural geography. In your studies you should make the most of this promise, seek out the space on the margins and look for the connections between the human and natural environments, the economic, the social and the cultural. Many of the challenges that face humankind at the turn of the century also occupy these margins. The emphasis on places and spaces and how they are interconnected, imagined and represented, the hallmark of human geography, is itself promising. We hope that the chapters and arguments brought together in this book will (when supplemented by further reading, critical thought, writing and conversation) encourage our readers to engage with this promise.

GLOSSARY

A

Actor-network   human actors in interaction through non-human intermediaries and power relations.

Agglomeration   the concentration of productive activities in a particular region.

Agribusiness   large-scale, capital-intensive, agricultural businesses incorporating supply, production and processing capacities.

Agro-food system   the highly integrated system of agricultural production in the developed world which involves both upstream (e.g. suppliers) and downstream (e.g. processing) industries.

Appropriationism   the replacement of agricultural inputs with industrial alternatives. It forms a key process in the industrialization of the agro-food system.

B

Balance of payments (BOP)   the difference in value between a country's inward and outward payments for goods, services and other transactions.

Balkanization   progressive subdivision of a region into small political units.

Bands   societies of hunter gatherers typically numbering up to 500 individuals. See Box 1.1.

Billion   thousand million.

Biomass   plant and animal residue burnt to produce heat.

Bottom-up development   economic and social changes brought about by activities of individuals and social groups in society rather than by the state and its agents.

Bounded   applied to the behaviour of decision-makers whose access to information, for example, is constrained by financial resources or time. They also have a limited capacity to process the information that they are able to obtain and they will also be constrained (bounded) by the environment within which their behaviour is taking place.

C

CAP   Common Agricultural Policy.

Capital   money put into circulation or invested with the intent of generating more money. See Box 13.4.

Capitalism   an historically specific economic system in which production and distribution are designed to accumulate capital and create profit. The system is characterized by the separation of those who own the means of production and those who work for them.

Chiefdom   social formation based on societies characterized by the internal and unequal differentiation of both power and wealth and organized around the principle of kinship. See Box 1.3.

City-states   urban regions having political jurisdiction and control over a specific territory. A form of social organization associated with some of the earliest states.

Civilization   refers to an advanced form of social development characterized by such things as urban life, commercial activity, writing systems and philosophical thought.

Cold War   a period extending from the end of the 1950s to the late 1980s during which two ideologically opposed blocs emerged within the world system, both with nuclear capabilities. The first was headed principally by the USA while the other was dominated by the Soviet Union.

Command economy   an economic system characterized by the state-led central planning of economic activity combined with the simultaneous suppression of market-type relations.

Commodity   a good or service manufactured through the use of waged labour and sold in exchange for money.

Commodity fetishism   the process whereby the material and social origins of commodities are obscured from consumers.

Communism   a political theory attributed to the works of Karl Marx and Friedrich Engels. Communism is characterized by the common ownership of the means of production. The foundations for full communism were supposed to be laid during a transitional period, known as socialism. The Soviet Union was the first state to be ruled by a communist government, from 1917 to 1991.

Communitarianism   a moral and political philosophy based on the extension and preservation of community. It is the opposite of liberalism which is regarded as insufficiently sensitive to the social/community sources of self-worth.

Comparative advantage   in economic geography, refers to an advantage held by a nation or region in the production of particular set of goods or services. See Box 11.5.

Conditional reserve   those (mineral or ore) deposits that have already been discovered but that are not economic to work at present-day price levels with currently available extraction and production technologies.

Contact zone   the space in which transculturation takes place – where two different cultures meet and inform each other, often in highly asymmetrical ways.

Containment   the Western strategy of encircling the Soviet Union and its allies during the Cold War.

Core   according to Wallerstein, the core refers to those regions of the capitalist world economy characterized by the predominance of core processes associated with relatively high wages, advanced technology and diversified production. See Box 2.3.

Counterurbanization   population increases in rural areas beyond the commuting range of major urban areas.

Crude birth rate   number of births in one year per 1,000 population.

Crude death rate   number of deaths in one year per 1,000 population.

D

Deindustrialization   refers to a relative decline in industrial employment. It may also refer to an absolute decline in industrial output as well as employment.

Demographic transition model   traces the shift from high birth and death rates to low ones. See Box 7.3.

Depopulation   the reduction of population in an area through out-migration or a reduction in the birth rate below the death rate.

Diaspora   literally, the scattering of a population; originally used to refer to the dispersal of Jews in

AD70, now used to refer to other population dispersals, voluntary and non-voluntary. Evokes a sense of exile and homelessness.

Diaspora space   the spaces inhabited not only by those who have migrated and their descendants, but also by those who are conceptualized as indigenous or 'native'. Similar to contact zone – a meeting point of different cultures.

Disorganized capitalism   a form of capitalism identifiable since the mid-1970s. It is characterized by a relative decline in the importance of extractive and manufacturing industries, together with a relative increase in the importance of services. There is also an increasing tendency for the production process to be dominated by small-scale and flexible forms of organization. See Box 3.5.

Displaced persons   refugees who have no obvious homeland.

Division of labour   the separation of tasks in the production process and their allocation to different groups of workers.

Doubling time   time taken for a population to double in size. See Box 7.1.

E

Edge city (exopolis)   a term referring to an area with city-like functions usually arising on the edge of an already urbanized area or conurbation and heavily dependent on fast communications systems. Edge cities are regarded as symptomatic of the most recent phase of urbanization.

Emerging markets   refers to those markets which are perceived to have a substantial growth potential. The term is often used in relation to countries of the former Soviet Union, Asia, Latin America and Africa.

Empire   an extended territorial political unit or state built up, often by force, under a supreme authority. Empires usually involve rule over alien or subject peoples.

Energy mix   the balance between various sources of energy in primary energy consumption.

Energy ratio   the relationship between energy consumption and economic growth in an economy.

$$\text{Energy ratio} = \frac{\text{Rate of change in energy consumption}}{\text{Rate of change in economic growth (GDP)}}$$

A value greater than 1 indicates that the amount of energy required to create an additional unit of GDP is increasing; a value less than 1 suggests the reverse.

Enlightenment   a philosophy or movement that emerged during the eighteenth century and based on the idea of social progress through the application of reason.

Epidemiological transition   switch in the predominant causes of death noted in the movement from a pre-transition to a post-transition society, primarily from infections to degenerative diseases.

Ethnic groups   groups of people defined according to a common ancestry and culture.

EU   European Union.

Exchange   the process of interchange of goods and services between individuals, groups and/or organizations, whether involving money or not. Can also refer more broadly to social interactions.

Export processing zone   a small closely defined area which possesses favourable trading and investment conditions created by a government to attract export-orientated industries. See Box 12.2.

F

Factors of production   refers to those elements necessary for the effective functioning of the production process and includes typically land, labour and capital.

FAO   Food and Agriculture Organization.

Fascism   a term used particularly to describe the nationalistic and totalitarian regimes of Benito Mussolini (Italy, 1922–45), Adolf Hitler (Germany, 1933–45) and Francisco Franco (Spain, 1939–78).

Feudalism   a hierarchical social and political system common in Europe during the medieval period.

The majority of the population were engaged in subsistence agriculture while simultaneously having an obligation to fulfil certain duties for the landholder. At the same time the landholder owed various obligations (fealty) to his overlord. See Box 1.7.

Financial exclusion   the processes that prevent disadvantaged social groups from gaining access to the financial system.

Flexible specialization   new methods of production such as multi-purpose machines and small-batch production. The term is frequently associated with agglomerations of firms, or industrial districts.

Food regimes   distinct relationships discerned between patterns of international food production and consumption and the developing capitalist system. See Box 9.1.

Fordism   a regime of accumulation involving mass production and consumption. Named after Henry Ford (1863–1947), Fordism is known for a differentiated division of labour, assembly-line production and affordable mass-produced consumer goods.

Fourth World   according to Castells (1998) a term referring to those regions effectively excluded from participation in the 'new' world economy. Importantly, it is of relevance not only to regions and countries of the recognized Third World but also to regions within the developed world. Also sometimes used to refer to indigenous or native peoples.

G

GATT   General Agreement on Tariffs and Trade. A United Nations agency founded in 1948 with one of its main purposes being to facilitate free trade through the gradual reduction of tariff levels.

Gemeinschaft   a form of community said to be common in traditional societies (as distinct from industrial societies) and associated with notions of stability and informal personal contact.

Gender division of labour   a division of labour constructed around gender in which particular tasks and occupations are deemed to be male or female activities.

Geographies of exclusion   the spatial processes by which a powerful grouping consciously seeks to distance itself from other less powerful groupings.

Geopolitics   the term which has been used to refer to many things, including a tradition of representing space, states and the relations between them, emphasizing the strategic importance of particular places.

Gesellschaft   a form of association common in urban-based industrial societies (as distinct from traditional societies) and associated with non-permanent and utilitarian social relationships.

Ghetto   a residential district within an urban region dominated by a specific ethnic group. The term is now commonly associated with notions of deprivation, unemployment and social exclusion.

Global cities   those cities performing a dominant role in the world economy and characterized by specialized service-type functions such as financial markets.

Global production system   refers to the dominant form of production identifiable at a global scale. It has evolved over many centuries into its current form, referred to as post-Fordism.

Global triad   this term describes how the world economy is essentially organized around a tripolar, macroregional structure whose three pillars are North America, Europe, and East and Southeast Asia.

Global warming   an increase in the temperature of the Earth's surface caused by trapping infrared radiation in carbon dioxide, increased amounts of which are produced by burning fossil fuels.

Globalization   a contested term relating to transformation of spatial relations that involves a change in the relationship between space, economy and society.

Glocalization   the interaction between the particular character of places or regions and the more general processes of change represented by globalization.

Green revolution   most commonly refers to the period during the late 1950s and 1960s when new high-yielding forms of wheat and rice were introduced to significant parts of the developing world in an attempt to boost output levels.

Gross domestic product (GDP)   the value left after removing the profits from overseas investments and those profits from the economy that go to foreign investors.

Gross national product (GNP)   a broad measure of an economy's performance; it is the value of the final output of goods and services produced by the residents of an economy plus primary income from non-residential sources.

H

Heartland   identified by the British geographer Halford Mackinder (1904) as the zone in East-Central Europe, control of which would be a key to world domination. The term has since been appropriated by Latin American geopolitics.

High-value foods   includes such foods as fruit, vegetables, poultry and shellfish. World trade in high-value foods has increased markedly during the last two decades.

Hybridity   refers to groups as a mixture of local and non-local influences; their character and cultural attributes are a product of contact with the world beyond a local place. See Box 6.5.

Hypothetical resources   those resources that might be expected to be found in the future in areas that have only been partially surveyed and developed.

I

IMF (International Monetary Fund)   international financial institution that originated from the 1944 Bretton Woods Conference. Its main roles include regulating international monetary exchange and controlling fluctuations in exchange rates in order to alleviate balance of payments problems.

Imperialism   a relationship of political, and/or economic, and/or cultural domination and subordination between geographical areas.

Indigenous peoples   those peoples native to a particular territory that was later colonized, particularly by Europeans.

Industrial revolution   a term that is often taken to refer to the marked transformation of productive forces, initially within the British economic system between the mid-eighteenth and mid-nineteenth centuries. It resulted in the movement of Britain from a largely rural-based economy to one that was dominated by manufacturing and industrial production. Such a transformation resulted in substantial social, political as well as economic changes.

Infant mortality rate (IMR)   number of deaths of persons aged under 1 per 1,000 live births.

International division of labour   a term referring to the tendency for particular countries and regions of the globe to specialize in particular types of economic activity.

Islam   a monotheistic religion founded by the prophet Muhammad in the seventh century AD. Today there are two predominant divisions within Islam, the Sunni and Shi'i.

K

Kondratieff cycle   a term used to describe the cycles of boom and bust evident within the capitalist system since the mid-eighteenth century. Named after the Russian scholar N.D. Kondratieff. See Table 2.2.

L

Legitimacy   with regard to nation-states, a term meaning that the majority of people accept the rule of law of the governing political organizations.

Less developed countries (LDCs)   countries at a disadvantage in today's global competitive environment because their comparative advantage in cheap labour or natural resource endowments have become subordinated to knowledge-based factors. For example, knowledge is the key to innovations in production that ultimately make products more competitive. LDCs suffer from poor productive capacities and competitiveness related, for example, to weak linkages between manufacturing, services and infrastructural sectors such as telecommunications; insufficiently developed human resources; deficiencies in physical infrastructure; and an inability to generate adequate resources to invest in reducing economic and social problems in ways that would help to improve productive capacity. Examples are Indonesia, Nigeria, Ethiopia.

Liberalism   a political philosophy distinguished by the importance it attaches to the civil and political rights of individuals. May also refer to belief in the importance of the free market.

Life expectancy (at birth)   average number of years of life expected on the basis of age-specific mortality schedules for the specified year.

Locality   a place or region of sub-national spatial scale.

M

Market-based states   modern states where the market is the dominant means by which land, labour, capital and goods are exchanged and has a major influence over social and political organization.

Marxism   a form of socialism derived from the teachings of Karl Marx (1818–83). Marxism regards capitalism as an inherently unjust system with the capitalists (those who own the means of production) exploiting the proletariat (those who must sell their labour in order to live). It aims to replace capitalism with a fairer system, socialism maturing into communism. See Box 3.2.

Merchant capitalism   refers to an early phase of capitalist industrial development dominant in the larger urban regions of Europe from the late fifteenth century. Merchants were the principal actors engaged in both the provision of capital and the movement and trade of goods (predominantly bulky staples such as grain and manufactured goods). See Box 2.4.

Modernism   a term typically associated with the twentieth-century reaction against realism and romanticism within the arts. More generally, it is often used to refer to a twentieth-century belief in the virtues of science, technology and the planned management of social change.

Modernity   refers to a period extending from the late sixteenth and early seventeenth centuries (in the case of Europe) to the mid to late twentieth century characterized by the growth and strengthening of a specific set of social practices and ways of doing things. It is often associated with capitalism and notions such as progress.

More developed countries (MDCs)   countries with significant competitive advantages in today's globalizing economy. They have well-developed, increasingly knowledge-based and strongly interconnected manufacturing and service sectors that provide a significant proportion of employment and contribute to significant national and individual wealth. Indices such as literacy levels, incomes and quality of life are high and these countries exercise considerable political influence at the global scale. Examples are the UK, US, Germany, France.

Multiculturalism   refers to a belief or policy which endorses the principle of cultural diversity and supports the right of different cultural and ethnic groups to retain distinctive cultural identities. It has often been criticized for being too symbolic and not politically radical enough in challenging racism.

N

Nationalism   the ideology and sentiment of belonging to a 'nation' and the claim that the 'nation' should be expressed in a 'state'.

Nation-state and **State**   a symbolic system of institutions claiming sovereignty over a bounded territory.

Natural increase   surplus of births over deaths in a population.

Neoliberalism   an economic doctrine promoting market-led growth, deregulation and the privatization of state-owned enterprises.

New economic geography a response to the economic determinism of the 1970s and 1980s. The theory recognizes the importance of culture as an influence on economic processes and outcomes. In this way it draws attention to the culturalization of the economy in contrast to the economization of culture.

New industrial districts (NIDs) areas which specialize in a particular industry because of external economies which result from specialization.

New international division of labour (NIDL) occurs when the process of production is no longer constructed primarily around national economies.

New World Order the notion of a Western-led post-Cold War structure of global power, which sanctions intervention.

Newly agriculturizing countries (NACs) refers to the handful of low–middle income countries which have come to specialize in the export of high-value foods. These include Brazil, Mexico, China, Argentina and Kenya.

Newly industrialized countries (NICs) countries where there has been a recent and significant shift away from primary activities towards manufacturing production. In some cases the proportion of manufacturing production is similar to that of the UK or the US. Examples are South Korea and Mexico.

Niche markets refers to the existence of consumer groups with identifiable tastes and life-styles.

Non-governmental organization (NGO) an organization formed by members of the public and one that has no government connections.

Non-renewable (stock) resources those resources, mainly mineral, that have taken millions of years to form. Their availability is therefore finite as there is no possibility of their stock being replenished on a time scale of relevance to human society.

O

Organized capitalism a form of capitalism that reached its zenith during the period 1945 to 1973. It was characterized by the dominance of extractive and manufacturing industries, large-scale manufacture, mass-production and a significant level of state involvement. See Box 3.1.

P

Palaeolithic the stage in the development of human society when people obtained their food by hunting, fishing and gathering wild plants, as opposed to engaging in settled agriculture. Also referred to as the Early Stone Age.

Pastoral nomadism a form of social organization which is based on livestock husbandry for largely subsistence purposes. Pastoral nomads are characterized by a high level of mobility which allows them continually to search for new pastures in order to maintain their herds of animals.

Pathways of farm business development refers to the different types of agricultural land uses to be initiated by the post-productivist transition. See Box 9.2.

Peasant the term has had a number of different interpretations over the years and in different parts of the world. In general it usually refers to those individuals whose livelihood is largely dependent on the land and on rural subsistence-type activities. The term peasant is usually reserved for those living in organized states, thus distinguishing them from band and tribal members, etc.

Periphery according to Wallerstein, refers to those regions of the world capitalist economy characterized by low wages, simple technology and limited production. See Box 2.3.

Political economy an approach to social study which emphasizes the political/social construction and consequences of economic activity.

Post-Fordism production system comprises a mix of different ways of organizing production at a number of spatial scales. Common to all these types of production is some form of flexible production.

Postmodernism a philosophy which holds that the traits associated with twentieth-century modernism, such as belief in the possibility of managing social change according to sets of agreed principles, are now in retreat in the face of increasing individualism, pluralism and eclecticism.

Post-productivist transition   a term used to describe the movement away from productivist agricultural systems. This new phase of agricultural production is characterized by extensive and diversified patterns of farming and the growing importance of non-agricultural activities in the countryside, for example recreation. See Box 9.2.

Power geometry   the ways in which different social groups/individuals are placed in relation to the forces of globalization, enabling some to benefit and others to be disadvantaged.

Prehistoric societies   societies which have left no written records.

Primary energy   the energy in the basic fuels or energy sources used, e.g. the energy in the fuel fed into conventional power stations.

Productivist agriculture   strongly associated with North America. It is characterized by the development of agro-industrial complexes based on grain-fed livestock. It was at its most influential during the period extending from the 1920s to the 1970s. See Box 9.1.

Proto-industrialization   refers to the early phase of capitalist industrial development in Europe. Characteristics of this period include a significant level of rural-based industrial activity and a low level of technological application in the production process. See Box 2.6.

Proven (proved) reserves   those deposits of a resource that have already been discovered and are known to be economically extractable under current demand, price and technological conditions.

Q

Quality assurance schemes   a series of checks for ensuring the origin and quality of (usually) agricultural foodstuffs typically associated with retailers and supermarkets.

R

Race   refers to the division of human beings into supposedly recognizable groups based predominantly on physical characteristics.

Ranked society   society in which there is an unequal division of status and power between its members, where such divisions are based primarily on such factors as family and inherited social position.

Reciprocity   a system of mutual exchange common in primitive societies. Often involved the mutual exchange of gifts and tokens in order to maintain good external relations or internal social relations.

Redistribution   in pre-capitalist societies a system whereby accumulated wealth and goods were taken and divided amongst members of society by a central ruler or organization, or kept by the centre for its own purposes.

Regulated states   social formations organized on the basis of territory. The regulated, or pre-modern, state was not dominated by the market system with political, social and religious considerations usually being of greater importance. See Box 1.4.

Renewable energy   energy sources such as winds, waves and tides that are naturally replenished and cannot be used up.

Renewable (flow) resources   those resources that are naturally renewed within a sufficiently short time span to be of use to human society. The continued availability of such resources is increasingly dependent upon effective management.

Replacement rate fertility   number of babies that an average woman should have to replace her generation, given prevailing levels of mortality (2.1 per woman in post-transition societies).

Reserves/production ratio (R/P)   if the reserves remaining at the end of any year are divided by the production in that year, the result is the length of time those remaining reserves would last if production were to continue at that level.

Resource   a substance in the physical environment that has value or usefulness to human beings and is economically feasible and socially acceptable to use.

Rural services   social goods that comprise the fabric of non-productive life in rural areas (e.g. education, health, transport, shops).

Rurality   functionalist, critical political economy and social representation approach for distinguishing between rural and urban economy and society.

Rust Belt   a region of the north-eastern USA which suffered substantial industrial decline, especially after the Second World War.

S

Second industrial revolution   a term sometimes used to account for the profound technological and accompanying social changes which affected industrial capitalism from the late nineteenth century. During this period Britain's industrial might was challenged by Germany and the USA.

Semi-periphery   according to Wallerstein, refers to those regions of the world capitalist economy which, while exploiting the periphery, are themselves exploited by the core countries. Furthermore, they are characterized by the importance of both core and peripheral processes. See Box 2.3.

Sense of place   the feelings, emotions and attachments to a locality by residents (past or present), which may be articulated in art, literature, music, histories, or may become part of individual or group memory.

Simulacrum   a copy without an original. An example might be Disneyland's 'Main Street', which represents an ideal American high street, but is not modelled on an original.

Social exclusion   the exercise of systematic economic and political power in a society so as to exclude individuals from life-styles open to the majority.

Social movement   comprised of individuals, groups and organizations united by a common purpose or goal.

Social representation   lay and academic discourses that describe an event or context.

Spatial interaction   a term used to indicate interdependence between geographical areas. Covers the movement of people, goods, information and money between places.

Speculative resources   those resources that might be found in unexplored areas which are thought to have favourable geological conditions.

State   a political unit having recognizable control (claiming supreme power) over a given territory. Unlike earlier social formations like bands and tribes, states have always based their power on their ability to control a specified territory and its inhabitants.

Stratified society   society within which there is an unequal division of material wealth between its members.

Structural adjustment   loans designed to foster structural adjustment in LDCs by promoting market-led growth and a reduction in government intervention in the economy.

Sub-culture   a subdivision of a dominant culture or an enclave within it with a distinct integrated network of behaviour, beliefs and attitudes.

Substitutionism   refers to the increased use of non-agricultural raw materials characterized by the replacement of food and fibre with industrial alternatives. It forms a key process in the industrialization of the agro-food system.

Suburbanization   refers to the movement of middle- and skilled working-class people into residential areas located some distance away from their paid employment.

Sun Belt   refers to the major growth areas of the southern and western parts of the USA during recent years in contrast to the contracting and declining industrial base of the north-east. Can also be used in other parts of the developed world to describe fast-developing regions, e.g. the M4 corridor in England.

Sustainable development a vague yet highly influential term popularized by the publication of the Brundtland Report in 1987. At its simplest it requires contemporary societal development to be undertaken bearing in mind the needs and aspirations of future generations. It has been criticized for failing to represent any real challenge to the prevailing Western development model.

T

Territoriality a term used to describe an expression of ownership and control by an individual, group or state over a particular area of land.

Territory a recognizable region (area of land or sea) occupied and controlled by an individual, group or state.

Third World a rather vague term used to describe those regions of the world in which levels of development, as understood by such measures as GDP, are significantly below those of the economically more advanced regions. The term is increasingly seen as an inadequate description of the prevailing world situation since it disguises a significant amount of internal differentiation.

TNCs transnational corporations. See Box 11.3.

Top-down development economic and social changes brought about by activities of the state and its agencies.

Total fertility rate (TFR) in general terms equal to the average number of live births per woman within a population.

Traceability in economic geography refers to the ability to determine the origin of agricultural foodstuffs.

Trade bloc a group of nation-states that act together in order to further their influence over world trade, e.g. the European Union (EU) or the North American Free Trade Area (NAFTA).

Transculturation the ways in which subordinated or marginal groups select and invent from dominant cultures; although such groups cannot control what emanates from the dominant culture, they do determine to varying extents what they absorb into their own and what they use it for.

Tribe a type of social formation usually said to be stimulated by the development of agriculture. Tribes tend to have a higher population density than bands and are also characterized by an ideology of common descent or ancestry. See Box 1.2.

U

Urban village a residential sector of a city (often inner-city) believed to possess social characteristics typical of a village-type community.

Urbanization refers to the increasing importance of the urban relative to the total population. It is initially stimulated by the movement of people from rural to urban areas.

V

Vital trends changes in fertility and mortality.

W

World Bank international financial institution established in 1944. Its main role is to provide development funds to LDCs in the form of loans and technical assistance.

World systems theory Immanual Wallerstein's conceptualization of the changing nature of the world socio-economic system into three distinct historical categories. See Box 2.3.

BIBLIOGRAPHY

A

ActionAid (1995) *Listening to Smaller Voices: Children in an Environment of Change*, ActionAid, Chard.

Adams, W.M. (1990) *Green Development: Environment and Sustainability in the Third World*, Routledge, London.

Adams, W.M. (1993) Sustainable development and the greening of development theory, in Schurmann, F. (ed.) *Beyond the Impasse: New Directions in Development Theory*, Zed, London.

Aghajanian, A. (1991) Population change in Iran, 1966–86: a stalled demographic transition? *Population and Development Review*, **17**, 703–15.

Agnew, J. (1988) *Geopolitics: Revisioning World Politics*, Routledge, London and New York.

Agnew, J. and Corbridge, S. (1995) *Mastering Space: Hegemony, Territory and International Political Economy*, Routledge, London.

Agnew, J., Livingstone, D. and Rogers, A. (eds) (1996) *Human Geography: An Essential Anthology*, Blackwell, Oxford.

Ake, C. (1996) *Democracy and Development in Africa*, Brooks, Washington.

Akzin, B. (1964) *State and Nation*, Hutchinson, London.

Allen, J. (1995) Global worlds, in Allen, J. and Massey, D. (eds) *Geographical Worlds*, Oxford University Press and Open University, Oxford, 105–44.

Allen, J. (1995) Crossing borders: footloose multinationals?, in Allen, J. and Hamnett, C. (eds) *A Shrinking World? Global Unevenness and Inequality*, Oxford University Press, Oxford, 55–102.

Allen, J. (1999) Cities of power and influence: settled formations, in Allen, J., Massey, D. and Pryke, M. (eds) *Unsettling Cities*, Routledge, London, 182–227.

Allen, J. and Hamnett, C. (1995) *A Shrinking World? Global Unevenness and Inequality*, Oxford University Press, Oxford.

Allen, J. and Massey, D. (1995) *Geographical Worlds*, Oxford University Press, Oxford.

Allen, J., Massey, D. and Pryke, M. (eds) (1999) *Unsettling Cities*, Routledge, London.

Amin, A. (ed.) (1994) *Post-Fordism: A Reader*, Blackwell, Oxford.

Amin, A. and Graham, S. (1999) Cities of connection and disconnection, in Allen, J., Massey, D. and Pryke, M. (eds) *Unsettling Cities*, Routledge, London, 7–47.

Amin, A. and Thrift, N. (1994) Living in the global, in Amin, A. and Thrift, N. (eds) *Globalization, Institutions and Regional Development in Europe*, Oxford University Press, Oxford, 1–22.

Amin, A. and Thrift, N.J. (eds) (1994) *Globalization, Institutions and Regional Development in Europe*, Oxford University Press, Oxford.

Anderson, B. (1983) *Imagined Communities: Reflections on the Origin and Spread of Nationalism*, Verso, London.

Anderson, J. (1995) The exaggerated death of the nation-state, in Anderson, J., Brook, C. and Cochrane, A. (eds) (1995) *A Global World? Re-ordering Political Space*, Oxford University Press and the Open University, Oxford, 65–112.

Anderson, J. (1997) Territorial sovereignty and political identity: national problems, transnational solutions?, in Graham, C. (ed.) *In Search of Ireland: A Cultural Geography*, Routledge, London, 215–36.

Anderson, J., Brook, C. and Cochrane, A. (eds) (1995) *A Global World*, Oxford University Press, Oxford.

Anonymous (1999) South Africa's middle-class pink currency, *Johannesburg Mail and Guardian*, 18 June. http://www.globalarchive.ft.com.

Appadurai, A. (1990) Disjuncture and difference in the global cultural economy, *Theory, Culture and Society*, **7**, 295–310.

Ardener, S. (1995) Women making money go round: ROSCAs revisited, in Ardener, S. and Burman, S. (eds) *Money Go-Rounds: The Importance of Rotating Savings and Credit Associations for Women*, Berg, Oxford.

Ardener, S. and Burman, S. (eds) (1995) *Money Go-Rounds: The Importance of Rotating Savings and Credit Associations for Women*, Berg, Oxford.

Asian Age (1999) We will drive Monsanto out of the country, 6 March, 9.

Askew, I. (1983) The location of service facilities in rural areas, *Regional Studies*, **17**, 305–14.

Atkinson, D. and Cosgrove, D. (1998) Urban rhetoric and embodied identities: city, nation and empire at the Vittorio Emanuele II monument in Rome, 1870–1945, *Annals of the Association of American Geographers*, **88**, 28–49.

An Atlas of Cyberspace – *www.cybergeography.org/atlas/census.html*

Auty, R. (1979) World within worlds, *Area*, **11**, 232–5.

B

Baker, B. (1998) The class of 1990: how have the autocratic leaders of sub-Saharan Africa fared under democratisation, *Third World Quarterly*, **19:1**, 115–27.

Bakhtin, M. (1984) *Rabelais and his World* (trans. Helen Iswolsky), Indiana University Press, Bloomington, IN.

Banister, D. and Norton, F. (1988) The role of the voluntary sector in the provision of rural services: the case of transport, *Journal of Rural Studies*, **4**, 57–71.

Baran, P.A. and Sweezy, P.M. (1966) *Monopoly Capitalism*, Pelican, Harmondsworth.

Barke, M. and Newton, M. (1997) The EU LEADER initiative and endogenous rural development, *Journal of Rural Studies*, **13**, 319–41.

Barnet, R.J. (1973) *Roots of War: The Men and Institutions behind US Foreign Policy*, Penguin, Harmondsworth.

Barrett, H. and Browne, A. (1996) Export horticultural production in sub-Saharan Africa: the incorporation of The Gambia, *Geography*, **81**, 47–56.

Barrett, H., Ilbery, B., Browne, A. and Binns, T. (1999) Globalisation and the changing networks of food supply: the importation of fresh horticultural produce from Kenya into the UK, *Transactions of the Institute of British Geographers*, **24**, 159–74.

Barthes, R. (1972) *Mythologies*, Paladin, London.

Barton, J.R. (1997) *A Political Geography of Latin America*, Routledge, London and New York.

Bassin, M. (1987) Race contra space: the conflict between German Geopolitik and National Socialism, *Political Geography Quarterly*, **6**, 115–34.

Bater, J.H. (1986) Some recent perspectives on the Soviet city, *Urban Geography*, **7**, 93–102.

Bayley, C. (1989) *Atlas of the British Empire*, Hamlyn, London.

BBC (1999) *http://www.bbc.co.uk/science* GM Foods: An Evaluation, accessed 12 December.

Beardsworth, A. and Keil, T. (1992) Putting the menu on the agenda, *Sociological Review*, **40**, 253–93.

Bedford, T. (1999) Ethical Consumerism, Consumption, Identity and Ethics, unpublished PhD thesis, University of London, London.

Bell, D. and Valentine, G. (1997) *Consuming Geographies: We are Where we Eat*, Routledge, London.

Bell, M. (1994) Images, myths and alternative geographies of the Third World, in Gregory, D., Martin, R. and Smith, G. (eds) *Human Geography: Society, Space and Social Science*, Macmillan, London, 174–99.

Bennett, J. (1998) Internal displacement: protecting the dispossessed, in *World Aid '96*, The Winchester Group, London.

Bensel, R.F. (1986) *Sectionalism and American Political Development 1880–1980*, University of Wisconsin Press, Madison, WI.

Berdahl, D. (1998) *Where the World Ended: Re-unification and Identity in the German Borderland*, University of California Press, Berkeley.

Beresford, M.W. (1988) *East End, West End: The Face of Leeds during Urbanisation 1684–1842*, Thoresby Society, Leeds.

Bergeson, A. and Schoenberg, R. (1980) Long waves of colonial expansion and contraction, 1415–1969, in Bergeson, A. (ed.) *Studies of the Modern World System*, Academic Press, New York.

Berman, M. (1983) *All That is Solid Melts into Air: The Experience of Modernity*, Verso, London.

Bernstein, H., Crow, B. and Johnson, H. (1992) *Rural Livelihoods: Crises and Responses*, Oxford University Press, Oxford.

Berry, B.J. (1976) *Urbanisation and Counterurbanisation*, Sage, Beverly Hills, CA.

Betjeman, J. (1937) *Continual Dew*, John Murray, London.

Bhabha, H.K. (1994) *The Location of Culture*, Routledge, London.

Bianchini, F. and Parkinson, M. (eds) (1993) *Cultural Policy and Urban Regeneration. The West European Experience*, Manchester University Press, Manchester.

Birch, D.L. (1987) *Job Creation in America,* Free Press, New York.

Black, R. (1998) *Refugees, Environment and Development*, Longman, London.

Black, R. and Robinson, V. (1993) *Geography and Refugees*, Belhaven, London.

Blacksell, M. (1994) Irredentism, in Johnson, R.J., Gregory, D. and Smith, D.M. (eds) *The Dictionary of Human Geography,* 3rd edition, Blackwell, Oxford, 299.

Blaut, J.M. (1993) *The Colonizer's Model of the World: Geographical Diffusionism and Eurocentric History*, Guilford Press, New York and London.

Block, F. (1987) *Revising State Theory*, Temple University Press, Philadelphia.

Blunden, J. (1995) Sustainable resources? in Sarre, P. and Blunden, J. (eds*) An Overcrowded World: Population, Resources and Environment*, Oxford University Press and Open University, Oxford, 161–213.

Bobek, H. (1962) The main stages in socio-economic evolution from a geographical point of view, in Wagner, P.L. and Mikesell, M.W. (eds) *Readings in Cultural Geography*, University of Chicago Press, Chicago, 218–47.

Bonfante, J. (1998) A German requiem, *The Economist*, 6 July, 25.

Bongaarts, J. (1996) Global trends in AIDS mortality, *Population and Development Review*, **22**, 21–46.

Booth, K. (1999) Cold wars of the mind, in Booth, K. (ed.) *Statecraft and Security: The Cold War and Beyond*, Cambridge University Press, Cambridge, 29–55.

Borgelt, C., Ganssauge, K. and Keckstein, V. (1987) *Mietshaus im Wandel, Wohnungen der behutsamen Stadterneuerung*, S.T.E.R.N., Berlin.

Borja, J. and Castells, M. (1997) *Local and Global: Managing Cities in the Information Age*, Earthscan, London.

Bornstein, D. (1996) *The Price of a Dream: The Story of the Grameen Bank and the Idea that is Helping the Poor to Change their Lives*, University of Chicago Press, Chicago.

Boserup, E. (1965) *The Conditions of Agricultural Growth: The Economics of Agrarian Change under Population Pressure*, Aldine Press, Chicago.

Bourdieu, P. (1984) *Distinction*, Routledge & Kegan Paul, London.

Bowler, I. (1992) Sustainable agriculture as an alternative path of farm business development, in Bowler, I., Bryant, C. and Nellis, D. (eds) *Rural Systems in Transition: Agriculture and Environment*, CAB International, Wallingford, 237–53.

Bowler, I. and Ilbery, B. (1997) The regional consequences for agriculture of changes to the Common Agricultural Policy, in Laurent, C. and Bowler, I. (eds) *CAP and the Regions*, INRA, Versailles, 105–15.

Bowler, I., Clark, G., Crockett, A., Ilbery, B. and Shaw, A. (1996) The development of alternative farm enterprises: a study of family labour farms in the north Pennines of England, *Journal of Rural Studies*, **12**, 285–95.

BPAmoco (1999) *BPAmoco Statistical Review of World Energy 1999*, BPAmoco, London.

BPAmoco (2000) *BPAmoco Statistical Review of World Energy 2000*, BPAmoco, London.

Brah, A. (1996) *Cartographies of Diaspora: Contesting Identities*, Routledge, London.

Brandt, W. (1980) *North-South: A Programme for Survival*, Pan, London.

Braudel, F. (1984) *Civilization and Capitalism 15th–18th Century V III: The Perspective of the World,* English translation, Collins/Fontana Press, London.

Bray, K. and Kilian, T. (1999) God does not pay for it: the political economy of cyberspace, paper presented at the 95th Annual Meeting of the Association of American Geographers, Honolulu.

Breckenridge, C.A. (ed.) (1995) *Consuming Modernity: Public Culture in a South Asian World*, University of Minnesota Press, Minneapolis.

Brindle, D. (1999) Fathers under pressure to become superdads, *Guardian*, 16 June, 9.

Broome, R. (1994) *Aboriginal Australians: Black Responses to White Dominance, 1788–1994*, 2nd edition, Allen & Unwin, St Leonards, NSW.

Brown, D.J., Kaldenberg, D.O. and Browne, B.A. (1992) Socio-economic status and playing the lotteries, *Sociology and Social Research*, **76**, 161–7.

Brown, M. (1997) *Replacing Citizenship: AIDS Activism and Radical Democracy*, Guilford Press, New York.

Brubaker, R. (1990) *Citizenship and Nationhood in France and Germany*, Harvard University Press, Cambridge, MA.

Bryant, C. and Johnston, T. (1992) *Agriculture in the City's Countryside*, University of Toronto Press, Toronto.

Bryden, J.M. (ed.) (1994) *Towards Sustainable Rural Communities*, School of Rural Planning and Development, University of Guelph, Ontario.

Bryman, A. (1995) *Disney and his Worlds,* Routledge, London.

Bryson, J., Daniels, P.W. and Henry, N. (1996) From widgets to where? A region in economic transition, in Gerrard, A.J. and Slater, T.R. (eds) *Managing a Conurbation: Birmingham and its Region,* Brewin Books, Birmingham, 56–168.

Bryson, J.R., Henry, N., Keeble, D. and Martin, R. (eds) (1999) *The Economic Geography Reader: Producing and Consuming Global Capitalism,* John Wiley, Chichester.

Buck, D., Getz, C. and Guthman, J. (1997) From farm to table: the organic vegetable commodity chain of northern California, *Sociologia Ruralis,* **37**, 3–19.

Buckley, R. (1995) *The United Nations: Overseeing the New World Order*, Understanding Global Issues 96/6, Understanding Global Issues Ltd., Cheltenham.

Bunce, M. (1998) Thirty years of farmland preservation in North America: discourses and ideologies of a movement, *Journal of Rural Studies,* **14**, 233–47.

Burnham, W.D. (1981) The system of 1896: an analysis, in Kleppner, P. *et al.* (eds) *The Evolution of American Electoral Systems*, Greenwood Press, Westport, CT.

Butlin, R.A. (1993) *Historical Geography, Through the Gates of Time and Space,* Edward Arnold, London.

C

Caldeira, T. (1996) Building up walls: the new pattern of social segregation in São Paulo, *International Social Science Journal,* **147**, 55–65.

Calhoun, C. (1995) New social movements of the early nineteenth century, in Traugott, M. (ed.) *Repertoires and Cycles of Collective Action*, Duke University Press, Durham, NC.

Calvocoressi, P. (1991) *World Politics since 1945*, Longman, London.

The Cambridge History of Islam (1970), Cambridge University Press, London.

Campbell, C. (1987) *The Romantic Ethic and the Spirit of Modern Consumerism*, Blackwell, Oxford.

Campbell, D. (1992) *Writing Security: United States Foreign Policy and the Politics of Identity*, University of Minnesota Press, Minneapolis, MN.

Carneiro, A. de M. (1990) *Descrição da Fortaleza de Sofala e das mais da Índia*, Fundação Oriente, Lisbon.

Carroll, R. (1998) The chill east wind at your doorstep, *Guardian*, 28 October, 2–3.

Castells, M. (1989) *The Informational City: Information Technology, Economic Restructuring and the Urban and Regional Process*, Blackwell, Oxford.

Castells, M. (1996) *The Rise of the Network Society*, volume 1 of *The Information Age: Economy, Society and Culture*, Blackwell, Oxford.

Castells, M. (1998) *End of Millennium*, volume 3 of *The Information Age: Economy, Society and Culture*, Blackwell, Oxford.

Castells, M. and Hall, P. (1994) *Technopoles of the World,* Routledge, London.

Castles, S. and Miller, M.J. (1993) *The Age of Migration: International Population Movements in the Modern World*, Guilford Press, New York.

Caulkin, S. (1998) Credit where credit's overdue, *Observer*, 15 November, 2.

Cavazzani, A. and Fuller, A. (1982) International perspectives on part-time farming, *GeoJournal*, **6**, 383–90.

Chaliand, G. (1993) Introduction, in Chaliand, G. (ed.) *A People without a Country: The Kurds and Kurdistan*, 2nd edition, Zed, London, 1–10.

Chaliand, G. and Rageau, P. (1985) *Strategic Atlas – World Geopolitics*, Penguin, London.

Champion, A.G. (1994) International migration and demographic change in the Developed World, *Urban Studies,* **31**, 653–77.

Champion, A.G. (ed.) (1988) *Counterurbanisation: The Changing Pace and Nature of Population Deconcentration*, Edward Arnold, Sevenoaks.

Chandler, J. (ed.) (1993) *John Leland's Itinerary*, Sutton, Stroud.

Chaney, D. (1990) Subtopia in Gateshead, the MetroCentre as a cultural form, *Theory, Culture and Society*, **7**, 49–68.

Chaney, D. (1996) *Lifestyles*, Routledge, London.

Chapman, K. and Walker, D. (1987) *Industrial Location: Principles and Policies*, Blackwell, Oxford.

Chatterjee, P. (1993) *The Nation and its Fragments: Colonial and Postcolonial Histories*, Princeton University Press, Princeton, NJ.

Child, J.C. (1985) *Geopolitics and Conflict in South America: Quarrels among Neighbours*, Praeger, New York.

Chomsky, N. (1998) Power in the global arena, *New Left Review*, **230**, 3–27.

Christaller, W. (1966) *Central Places in Southern Germany* (trans. by C.W. Baskin), Prentice Hall, Englewood Cliffs, NJ.

Chung, R. (1970) Space-time diffusion of the demographic transition model: the twentieth century patterns, in Demko, G.J., Rose, H.M. and Schnell, G.A. (eds) *Population Geography: A Reader*, McGraw-Hill, New York.

Clapham, C. (1985) *Third World Politics*, Croom Helm, London.

Clifford, J. (1992) Travelling cultures, in Grossberg, L., Nelson, C. and Treichler, P. (eds) *Cultural Studies*, Routledge, London.

Cliquet, R.L. (1991) *The Second Demographic Transition: Fact or Fiction?*, Council of Europe, Strasbourg.

Cloke, P. (1985) Counterurbanisation: a rural perspective, *Geography*, **70**, 13–23.

Cloke, P. (1996) Rural life-styles: material opportunity, cultural experience, and how theory can undermine policy, *Economic Geography*, **72**, 433–49.

Cloke, P., Crang, P. and Goodwin, M. (1999) *Introducing Human Geographies*, Arnold, London.

Cloke, P. and Edwards, G. (1986) Rurality in England and Wales 1981: a replication of the 1971 index, *Regional Studies*, **20**, 289–306.

Cloke, P. and Little, J. (eds) (1997) *Contested Countryside Cultures: Otherness, Marginalisation and Rurality*, Routledge, London.

Coakley, J. (1998) *Sport in Society: Issues and Controversies*, McGraw-Hill, New York.

Cochrane, A. (1995) Global worlds and worlds of difference, in Anderson, J., Brook, C. and Cochrane, A. (eds) *A Global World*, Oxford University Press, Oxford, 249–80.

Cochrane, A. (1999a) Administered cities, in Pile, S., Brook, C. and Mooney, G. (eds) *Unruly Cities? Order/Disorder*, Routledge, London, 299–344.

Cochrane, A. (1999b forthcoming) Redefining urban politics for the twenty first century, in Jonas, A. and Wilson, D. (eds) *The Urban Growth Machine: Critical Perspectives Two Decades Later*, State University of New York Press, Albany, NY.

Cochrane, A. and Jonas, A. (1999) Reimagining Berlin: world city, national capital or ordinary place?, *European Urban and Regional Studies*, **6**, 145–65.

Cochrane, A., Peck, J. and Tickell, A. (1996) Manchester plays games: the local politics of globalization, *Urban Studies*, **33**, 1317–34.

Cockcroft, E. (1974) Abstract expressionism, weapon of the Cold War, *Artform*, **12**, 39–41.

Coffey, W.J. (1996) The 'newer' international division of labour, in Daniels, P.W. and Lever, W.F. (eds) *The Global Economy in Transition*, Longman, Harlow, 40–61.

Cohen, R. (1998) Recent trends in protection and assistance for IDPs, in Earthscan (1998) *Internally Displaced People: A Global Survey*, Earthscan, Guildford, 3–9.

Coleman, D. (1996) New patterns and trends in European fertility: international and sub-national comparisons, in Coleman, D. (ed.) *Europe's Population in the 1990s*, Oxford University Press, Oxford.

Collier, R.B. and Collier, D. (1991) *Shaping the Political Arena: Critical Junctures, the Labour Movement, and Regime Dynamics in Latin America*, Princeton University Press, Princeton, NJ.

Conzen, M.P. (1990) *The Making of the American Landscape*, Unwin Hyman, Boston.

Cook, I. and Crang, P. (1996) The world on a plate, culinary culture, displacement and geographical knowledges, *Journal of Material Culture*, 1, 131–53.

Cook, I., Crang, P. and Thorpe, M. (1999) Eating into Britishness, multicultural imaginaries and the identity politics of food, in Roseneil, S. and Seymour, J. (eds) *Practising Identities*, Macmillan, London, 223–48.

Cook, I., Crang, P. and Thorpe, M. (2000) Regions to be cheerful, culinary authenticity and its geographies, in Cook, I., Crouch, D., Naylor, S. and Ryan, J. (eds) *Cultural Turns/Geographical Turns*, Longman, London, 109–39.

Cooke, P. (ed.) (1995) *The Rise of the Rust Belt*, UCL Press, London.

Coquery-Vidrovitch, C. (1996) *African Women: A Modern History*, Westview Press, Oxford.

Corbridge, S. (ed.) (1995) *Development Studies: A Reader*, Edward Arnold, London.

Corbridge, S., Martin, R. and Thrift, N. (eds) (1994) *Money, Power and Space*, Blackwell, Oxford.

Corner, J. and Harvey, S. (eds) (1991) *Enterprise and Heritage: Crosscurrents of National Culture*, Routledge, London.

Council for the Protection of Rural England (1993) *The Regional Lost Land*, CPRE, London.

Council for the Protection of Rural England (1995) *Measuring the Unmeasurable: Twenty Indicators for the Countryside*, CPRE, London.

Cowen, M.P. and Shenton, R.W. (1996) *Doctrines of Development*, Routledge, London.

Cox, K. and Mair, A. (1991) From localised social structures to localities as agents, *Environment and Planning A*, 23, 155–308.

Crang, M. (1998) *Cultural Geography*, Routledge, London.

Crang, P. (1997) Cultural turns and the (re)constitution of economic geography, in Lee, R. and Wills, J. (eds) *Geographies of Economies*, Arnold, London, 3–15.

Crang, P. and Jackson, P. (2000) Consuming geographies, in Morley, D. and Robins, K. (eds) *British Cultural Studies*, Oxford University Press, Oxford.

Crawford, M. (1992) The world in a shopping mall, in Sorkin, M. (ed.) *Variations on a Theme Park: The New American City and the End of Public Space*, Noonday, New York.

Crewe, L. and Gregson, N. (1998) Tales of the unexpected, exploring car boot sales as marginal spaces of consumption, *Transactions of the Institute of British Geographers*, 23, 39–53.

Crone, P. (1986) The tribe and the state, in Hall, J.A. (ed.) *States in History*, Basil Blackwell, Oxford, 48–77.

Crone, P. (1989) *Pre-Industrial Societies*, Blackwell, Oxford.

Cronon, W. (1991) *Nature's Metropolis: Chicago and the Great West*, W.W. Norton, New York.

Cross, M. and Waldinger, R. (1992) Migrants, minorities and the ethnic division of labour, in Fainstein, S., Gordon, I. and Harloe, M. (eds) *Divided Cities: New York and London in the Contemporary World*, Blackwell, Oxford.

Crouch, D.P., Garr, D.J. and Mundigo, A.I. (1982) *Spanish City Planning in North America*, MIT Press, Cambridge, MA.

Crush, J. (1995) Imagining development, in Crush, J. (ed.) *Power of Development*, Routledge, London, 1–26.

D

Dahl, R. (1989) *Democracy and its Critics*, Yale University Press, New Haven, CT.

Dalton, B. and Cotton, J. (1996) New social movements and the changing nature of political opposition in South Korea, in Rodan, G. (ed.) *Political Oppositions in Industrialising Asia*, Routledge, London.

Dalton, R.J., Kuechler, M. and Bürklin, W. (1990) The challenge of new movements, in Dalton, R.J. and Kuechler, M. (eds) *Challenging the Political Order: New Social and Political Movements in Western Democracies*, Oxford University Press, Oxford.

Daniels, P.W. (1975) *Office Location: an Urban and Regional Study*, G. Bell, London.

Daniels, P.W. (1999) Overseas investment by US service enterprises, in Taylor, P. and Slater, D. (eds) *The American Century: Consensus and Coercion in the Projection of American Power,* Blackwell, Oxford, 67–83.

Daniels, P.W. and Lever, W.F. (eds) (1996) *The Global Economy in Transition*, Longman, Harlow.

Daniels, S. and Lee, R. (eds) (1996) *Exploring Human Geography: A Reader*, Arnold, London.

Dankelman, I. and Davidson, J. (1988) *Women and Environment in the Third World*, Earthscan, London.

Daugherty, H.G. and Kammeyer, K.C.W. (1995) *An Introduction to Population,* 2nd edition, Guilford Press, New York.

Davies, G. (1994) *A History of Money: From Ancient Times to the Present Day*, University of Wales Press, Cardiff.

Davis, J. and Goldberg, R. (1957) *A Concept of Agribusiness*, Harvard Business School, Boston, MA.

Davis, M. (1990) *City of Quartz: Excavating the Future in Los Angeles*, Verso, London.

Davis, M. (1992) *Beyond Blade Runner. Urban Control: The Ecology of Fear*, Magazine Pamphlet Series, Pamphlet No. 23, Open Media, Westfield, NJ.

Dennis, R. (1984) *English Industrial Cities of the Nineteenth Century*, Cambridge University Press, Cambridge.

Department for International Development (DFID) (1997) *White Paper on Eliminating World Poverty: A Challenge for the Twenty-first Century*, HMSO, London.

Derbyshire, J.D. and Derbyshire, I. (1989) *Political Systems of the World*, Chambers, London.

Desforges, L. (1999) Travel and tourism, in Cloke, P., Crang, P. and Goodwin, M. (eds) *Introducing Human Geographies,* Arnold, London, 296–304.

Destexhe, A. (1995) *Rwanda and Genocide in the Twentieth Century*, Pluto, London.

Dicken, P. (1992) *Global Shift: The Internationalization of Economic Activity,* Paul Chapman, London.

Dicken, P. (1998) *Global Shift: Transforming the World Economy,* 3rd edition, Paul Chapman, London.

Dicken, P. and Lloyd, P.E. (1990) *Location in Space: Theoretical Perspectives in Economic Geography*, Harper & Row, New York.

Dickenson, J., Gould, B., Clarke, C., Mather, S., Prothero, M., Siddle, D., Smith, C. and Thomas-Hope, E. (1996) *A Geography of the Third World*, Routledge, London.

Dictionary of Human Geography (1994) Johnston, R.J., Gregory, D. and Smith, D.M. (eds), 3rd edition, Blackwell, Oxford.

Dobb, M. (1946; 1963) *Studies in the Development of Capitalism*, Routledge & Kegan Paul, London.

Dodd, N. (1994) *The Sociology of Money: Economics, Reason and Contemporary Society*, Continuum Publishing Company, New York.

Dodds, K. (1997) *Geopolitics in Antarctica: Views from the Southern Ocean Rim*, Wiley, Chichester.

Dodds, K. (2000) *Geopolitics in a Changing World*, Prentice Hall, Harlow.

Dodgshon, R.A. (1987) *The European Past: Social Evolution and Spatial Order*, Macmillan, Basingstoke.

Dodgshon, R.A. (1998) *Society in Time and Space. A Geographical Perspective on Change*, Cambridge University Press, Cambridge.

Dodgshon, R.A. and Butlin, R.A. (eds) (1990) *An Historical Geography of England and Wales*, 2nd edition, Academic Press, London.

Dolukhanov, P.M. (1996) *The Early Slavs*, Longman, London.

Donald, J. (1997) This, here, now: imagining the modern city, in Westwood, S. and Williams, J. (eds) *Imagining Cities: Scripts, Signs, Memory*, British Sociological Society and Routledge, London.

Dorfman, A. and Mattelart, A. (1975) *How to Read Donald Duck: Imperialist Ideology in the Disney Comic*, International General Editions, New York.

Dorpalen, A. (1942) *The World of General Haushofer: Geopolitics in Action*, Farrar & Rinehart, New York.

Driver, F. (2000) Imaginative geographies, in Cloke, P., Crang, P. and Goodwin, M. (eds) *Introducing Human Geographies*, Arnold, London, 209–16.

Dudrah, R.K. (2001) Drum N Dhol: British Bhangra, *European Journal of Cultural Studies* (forthcoming).

Duncan, S. and Goodwin, M. (1988) *The Local State and Uneven Development*, Polity Press, Cambridge.

Dunning, J.H. (1981) *International Production and the Multinational Enterprise*, Allen & Unwin, London.

Dyson, T. (1996) *Population and Food: Global Trends and Future Prospects*, Routledge, London.

E

Earthscan (1992) *Facing the Future: Beyond the Earth Summit*, Earthscan, Guildford.

Earthscan (1997) *The Reality of Aid 1997–1998: An Independent Review of Development Cooperation*, Earthscan, Guildford.

Earthscan (1998) *Internally Displaced People: A Global Survey*, Earthscan, Guildford.

Edwards, L.E. (1983) Towards a process model of office-location decision making, *Environment and Planning A*, **15**, 1327–42.

Edwards, M. and Hulme, D. (1992) *Making a Difference: NGOs and Development in a Changing World*, Earthscan, Guildford.

Ehrlich, P., Ehrlich, A. and Daily, G. (1993) Food security, population and the environment, *Population and Development Review*, **19**, 1–32.

Elson, D. and Pearson, R. (1981) Paying for nimble fingers: the exploitation of women in the third world, *Feminist Review*, **7**, 87–107.

Emerson, T. (1995) Fool's gold, *Newsweek*, October, 32–7.

Engels, F. (1845) *The Condition of the Working Class in England*, 1987 edition, Penguin, London and New York.

Escobar, A. (1995) *Encountering Development*, Princeton University Press, Princeton, NJ.

Esping-Anderson, G. (1985) *Politics and Markets*, Princeton University Press, Princeton, NJ.

Espiritu, C. (1986) Transnational agribusinesses in the Third World, in Dembo, D., Dias, C., Morehouse, W. and Paul, J. (eds) *The International Context of Rural Poverty in the Third World*, Council of International and Public Affairs, New York.

Esposito, J.L. (ed.) (1995) *The Oxford Encyclopedia of Islam in the Modern World*, Oxford University Press, New York.

Esteva, G. (1992) Development, in Sachs, W. (ed.) *The Development Dictionary*, Zed, London.

Etzioni, A. (1996) *The New Golden Rule*, Touchstone Books, New York.

European Commission (1999) *Sustainable Urban Development in the European Union: A Framework for Action*, Office for Official Publications of the European Communities, Luxembourg.

F

Featherstone, M. (1995) *Undoing Culture: Globalization, Postmodernism and Identity*, Sage, London.

Featherstone, M., Lash, S. and Robertson, R. (eds) (1995) *Global Modernities*, Sage, London.

Fejes, F. (1981) Media imperialism, an assessment, *Media, Culture and Society*, **3**, 281–9.

Ferguson, J. (1990) *The Anti-Politics Machine: 'Development', Depoliticization and Bureaucratic Power in Lesotho*, Cambridge University Press, Cambridge.

Fieldhouse, D.K. (1999) *The West and the Third World*, Blackwell, Oxford.

Fielding, A.J. (1989) Migration and urbanisation in Western Europe since 1950, *Geographical Journal*, **155**, 60–9.

Finer, A. and Savage, G. (eds) (1965) *The Selected Letters of Josiah Wedgwood*, Cory, Adams & MacKay, London.

Fisas, V. (1995) *Blue Geopolitics: The UN Reform and the Future of the Blue Helmets*, Pluto Press, London.

FitzSimmons, M. (1986) The new industrial agriculture: the regional differentiation of speciality crop production, *Economic Geography*, **62**, 334–53.

FitzSimmons, M. (1997) Restructuring, industry and regional dynamics, in Goodman, D. and Watts, M. (eds) *Globalising Food*, Routledge, London, 158–68.

Florida, R. (1996) Regional creative destruction: production organisation, globalization, and the economic transformation of the Midwest, *Economic Geography*, **72:3**, 314–34.

Flowerdew, R. and Martin, D. (1997) *Methods in Human Geography: A Guide for Students Doing a Research Project*, Longman, Harlow.

Foley, R. (1995) *Humans Before Humanity: An Evolutionary Perspective*, Blackwell, Oxford.

Forbes, D.K. (1984) *The Geography of Underdevelopment*, Johns Hopkins University Press, Baltimore, MD.

Forbes, D. and Thrift, N. (1987) *The Socialist Third World: Urban Development and Territorial Planning*, Blackwell, Oxford.

Forty, A. (1995) *Objects of Desire: Design and Society Since 1750*, Thames & Hudson, London.

Foster, R.F. (ed.) (1989) *The Oxford History of Ireland*, Oxford University Press, Oxford.

Fothergill, S. and Gudgin, G. (1982) *Unequal Growth: Urban and Regional Employment Change in the UK*, Heinemann, London.

Foucault, M. (1979) On governmentality, *Ideology and Consciousness*, **6**, 5–29.

Frank, A.G. (1967) *Capitalism and Underdevelopment in Latin America*, Monthly Review Press, New York.

Freeman, M. (1986) Transport, in Langton J. and Morris, R.J. (eds) *Atlas of Industrializing Britain, 1780–1914*, Methuen, London and New York.

Friedland, W. (1984) The new globalisation: the case of fresh produce, in Bonnano, A., Busch, L., Friedland, B., Gouveia, L. and Mingione, E. (eds) *From Columbus to Conagra: The Globalisation of Agriculture and Food*, University of Kansas Press, Lawrence, 210–32.

Friedman, J. (1994) Globalization and localization, in Friedman, J. (ed.) *Cultural Identity and Global Process*, Sage, London, 102–16.

Friedman, J. and Rowlands, M.J. (1977) Notes towards an epigenetic model of the evolution of civilization, in Friedman, J. and Rowlands, M.J. (eds) *The Evolution of Social Systems*, Duckworth, London, 201–76.

Friedman, T. (1999) It's a small world, *Sunday Times* 28 March, Section 5.

Friedmann, H. (1982) The political economy of food: rise and fall of the postwar international food order, *American Journal of Sociology*, **88**, Supplement, 246–86.

Friedmann, H. (1993) The political economy of food, *New Left Review*, **197**, 29–57.

Friedmann, H. and McMichael, P. (1989) Agriculture and the state system: the rise and decline of national agricultures, 1870 to the present, *Sociologia Ruralis*, **29**, 93–117.

Friedmann, J. (1966) *Regional Development Policy: A Case Study of Venezuela*, MIT Press, Cambridge, MA.

Friedmann, J. (1992) *Empowerment: The Politics of Alternative Development*, Blackwell, Oxford.

Friedmann, J. (1997) Global crises, the struggle for cultural identity and intellectual porkbarrelling: cosmopolitans versus locals, ethnics and nationals in an era of de-hegemonisation, in Werbner, P. and Modood, T. (eds) *Debating Cultural Hybridity. Multi-Cultural Identities and the Politics of Anti-Racism*, Zed, London, 70–89.

Fujita, K. and Child Hill, R. (1995) Global Toyotaism and local development, *International Journal of Urban and Regional Research*, **19**, 7–22.

Fukuyama, F. (1992) *The End of History and the Last Man*, Penguin, London.

Furnham, A. and Argyle, M. (1998) *The Psychology of Money*, Routledge, London.

Furtado, C. (1964) *Development and Underdevelopment*, University of California Press, Berkeley, CA.

Furuseth, O. (1992) Countryside in revolt: rural response to a proposed hazardous waste facility, in Gilg, A.W. (ed). *Restructuring the Countryside*, Avebury, Aldershot, 159–77.

G

Gad, G. and Holdsworth, D.W. (1987) Corporate capitalism and the emergence of the high-rise office building, *Urban Geography*, **8**, 212–31.

Gamble, A. (1994) *A Free Economy and a Strong State: The Politics of Thatcherism,* 2nd edition, Macmillan, Basingstoke.

Gandy Jr, O.H. (1995) It's discrimination, stupid!, in Brook, J. and Boal, I.A. (eds) *Resisting the Virtual Life*, City Lights Books, San Francisco.

Gans, H. (1962) *The Urban Villagers*, Free Press, New York.

Gans, H. (1968) *People and Problems: Essays on Urban Problems and Solutions*, Basic Books, New York.

Garmise, S., Morgan, K. and Rees, G. (1996) *Networks and Local Economic Development*, Papers in Planning Research 160, Department of City and Regional Planning, University of Wales, Cardiff.

Garreau, J. (1991) *Edge City: Life on the New Frontier*, Doubleday, New York.

Gbadegesin, A. (1991) Farming in the urban environment of a developing nation: a case study from Ibadan metropolis in Nigeria, *Environmentalist*, **11**, 105–11.

Gelbard, A., Haub, C. and Kent, M.M. (1999) World population beyond six billion, *Population Bulletin*, **54**, 1–44.

German, T. and Randall, J. (eds) (1993) *The Reality of Aid*, Action Aid, London.

Geyer, H. (1996) Expanding the theoretical foundation of differential urbanization, *Tijdschrift voor Economische en Sociale Geografie*, **87**, 44–59.

Giddens, A. (1990) *The Consequences of Modernity*, Polity Press, Cambridge.

Giles, C. and Goodall, I.H. (1992) *Yorkshire Textile Mills, 1770–1930*, HMSO, London.

Gilg, A. and Battershill, M. (1998) Quality farm food in Europe: a possible alternative to the industrialised food market and to current agri-environmental policies: lessons from France, *Food Policy*, **23**, 25–40.

Gilroy, P. (1993) *The Black Atlantic: Modernity and Double Consciousness*, Verso, London.

Ginsborg, P. (1990) *A History of Contemporary Italy*, Penguin, Harmondsworth.

Glassner, M.I. (1993) *Political Geography*, John Wiley, New York.

Glennie, P. and Thrift, N. (1992) Modernity, urbanism and modern consumption, *Environment and Planning D, Society and Space,* 10, 423–43.

Godlewska, A. and Smith, N. (eds) (1994) *Geography and Empire,* Blackwell, Oxford.

Goetz, A.M. and Gupta, R.S. (1996) Who takes the credit? Gender, power, and control over loan use in rural credit programs in Bangladesh, *World Development,* 24, 45–63.

Goodman, D. and Watts, M. (eds) (1997) *Globalising Food,* Routledge, London.

Goodman, D., Sorj, B. and Wilkinson, J. (1987) *From Farming to Biotechnology: A Theory of Agro-Industrial Development,* Blackwell, Oxford.

Goss, J. (1993a) Placing the market and marketing place, tourist advertising of the Hawaiian Islands,1972–92, *Environment and Planning D, Society and Space,* 11, 663–88.

Goss, J. (1993b) The magic of the mall, form and function in the retail built environment, *Annals of the Association of American Geographers,* 83, 18–47.

Goss, J. (1999a) Consumption, in Cloke, P., Crang, P. and Goodwin, M. (eds) *Introducing Human Geographies,* Arnold, London, 114–21.

Goss, J. (1999b) Once upon a time in the commodity world, an unofficial guide to the Mall of America, *Annals of the Association of American Geographers,* 89, 45–75.

Gottdiener, M. (1982) Disneyland, a Utopian urban space, *Urban Life,* 11, 139–62.

Gottdiener, M. (1997) *The Theming of America: Dreams, Visions and Commercial Spaces,* Westview Press, Boulder, CO.

Grabher, G. (1993) *The Embedded Firm: On the Socioeconomics of Industrial Networks,* Routledge, London.

Green, F. and Sutcliffe, B. (1987) *The Profit System: The Economics of Capitalism,* Penguin, London.

Greenwald, J. (1998) Sticky currency, *Time,* 28 September, 82–3.

Gregory, D. (1982) *Regional Transformation and Industrial Revolution: A Geography of the West Yorkshire Woollen Industry,* Macmillan, London.

Gregory, D. (1994) *Geographical Imaginations,* Blackwell, Oxford.

Gregory, D., Martin, R. and Smith, G. (eds) (1994) *Human Geography: Society, Space and Social Science,* Macmillan, London, 146–73.

Griffiths, I.L. (1994) *The Atlas of African Affairs,* 2nd edition, Routledge, London.

Grigg, D. (1985) *The World Food Problem 1950–1980,* Blackwell, Oxford.

Groves, P.A. (1987) The northeast and regional integration, 1800–1860, in Mitchell, R.D. and Groves, P.A. (eds) *North America: The Historical Geography of a Changing Continent,* Hutchinson, London.

Gutkind, E.A. (1964) *Urban Development in Central Europe,* Macmillan, London and New York.

H

Halfacree, K. (1993) Locality and social representation: space, discourse and alternative definitions of the rural, *Journal of Rural Studies,* 9, 23–37.

Halfacree, K. (1996) Out of place in the country: travellers and the 'rural idyll', *Antipode,* 18, 42–72.

Hall, P. (ed.) (1966) *Von Thünen's Isolated State,* Pergamon, Oxford.

Hall, P. (1996) *Cities of Tomorrow,* updated edition, Blackwell, Oxford.

Hall, P. (1998) *Cities in Civilization: Culture, Innovation and Urban Order,* Weidenfeld & Nicolson, London.

Hall, R. (1995) Households, families and fertility, in Hall, R. and White, P. (eds) *Europe's Population: Towards the Next Century,* UCL Press, London.

Hall, S. (1992) The West and the rest: discourse and power, in Hall, S. and Gieben, B. (eds) *Formations of Modernity,* ch. 6, Open University Press, Milton Keynes, 275–331.

Hall, S. (1995) New cultures for old, in Massey, D. and Jess, P. (eds) *A Place in the World? Places, Cultures and Globalization*, Oxford University Press, Oxford, 175–213.

Hall, S. (1996) The meaning of new times, in Morley, D. and Chen, K.-H. (eds) *Stuart Hall: Critical Dialogues in Cultural Studies*, Routledge, London, 223–37.

Hamelink, C. (1983) *Cultural Autonomy in Global Communications*, Longman, New York.

Hamnett, C. (1995) Controlling space: global cities, in Allen, J. and Hamnett, C. (eds) *A Shrinking World?*, Oxford University Press, Oxford, 103–42.

Hannerz, U. (1992) The global ecumene, in Hannerz, U. (ed.) *Cultural Complexity: Studies in the Social Organization of Meaning*, Columbia University Press, New York, 217–67.

Hanrou, H. and Obrist, H.-U. (1999) Cities on the move, in Bradley, F. (ed.) *Cities on the Move: Urban Chaos and Global Change. East Asian Art, Architecture and Film Now*, Hayward Gallery, London, 10–15.

Hara, K. (1999) Rethinking the 'Cold War' in the Asia-Pacific, *Pacific Review*, **12**, 515–36.

Harcourt, W. (ed.) (1999) *Women@Internet: Creating New Cultures in Cyberspace*, Zed, London.

Harris, D.R. (1989) An evolutionary continuum of plant-animal interaction, in Harris, D.R. and Hillman, G.C. (eds) *Foraging and Farming: The Evolution of Plant Exploitation*, Unwin Hyman, London, 11–26.

Harris, D.R. (1996a) Introduction: themes and concepts in the study of early agriculture, in Harris, D.R. (ed.) *The Origins and Spread of Agriculture and Pastoralism in Eurasia*, UCL Press, London, 1–9.

Harris, D.R. (1996b) The origins and spread of agriculture and pastoralism in Eurasia: an overview, in Harris, D.R. (ed.) *The Origins and Spread of Agriculture and Pastoralism in Eurasia*, UCL Press, London, 552–70.

Harrison, S. (1998) *Japanese Technology and Innovation Management*, Edward Elgar, Cheltenham.

Hartwick, E. (1998) Geographies of consumption, a commodity-chain approach, *Environment and Planning D, Society and Space*, **16**, 423–37.

Harvey, D. (1982) *The Limits to Capital*, Blackwell, Oxford.

Harvey, D. (1985a) Money, time, space and the city, in Harvey, D. (ed.) *The Urban Experience*, Johns Hopkins University Press, Baltimore, MD.

Harvey, D. (1985b) *The Urbanization of Capital: Studies in the History and Theory of Capitalist Urbanization*, Blackwell, Oxford.

Harvey, D. (1989) *The Condition of Postmodernity: An Enquiry into the Origins of Cultural Change*, Blackwell, Oxford.

Harvey, D. (1997) Contested cities: social process and spatial form, in Jewson, N. and MacGregor, S. (eds) *Transforming Cities: Contested Governance and New Spatial Divisions*, Routledge, London, 19–27.

Hayter, R. and Watts, H.D. (1983) The geography of enterprise: a re-appraisal, *Progress in Human Geography*, 7, 157–81.

Healey, M.J. and Ilbery, B.W. (1990) *Location and Change: Perspectives on Economic Geography*, Oxford University Press, Oxford.

Healey, P., Davoudi, S., Graham, S. and Madani-Pour, A. (eds) (1995) *Managing Cities: The New Urban Context*, John Wiley, London.

Healey, P., McNamara, P., Elson, M. and Doak, A. (1988) *Land Use Planning and the Mediation of Urban Change*, Cambridge University Press, Cambridge.

Hebdige, D. (1979) *Subculture: The Meaning of Style*, Methuen, London.

Hecht, S. and Cockburn, A. (1989) *The Fate of the Forest: Developers, Destroyers and Defenders of the Amazon*, Verso, London and New York.

Heffernan, M. (1998) *The Meaning of Europe: Geography and Geopolitics*, Arnold, London.

Heilbronner, R.L. (1972) *The Economic Problem*, 3rd edition, Prentice Hall, Englewood Cliffs, NJ.

Held, D., McGrew, D., Goldblatt, D. and Perraton, J. (1999) *Global Transformations: Politics, Economics and Culture*, Polity, Cambridge.

Henry, N.D. and Passmore, A. (1999) Rethinking 'global' city centres: the example of Birmingham, *Soundings: A Journal of Politics and Culture*, **13**, 60–6.

Hepple, L. (1986) The revival of geopolitics, *Political Geography Quarterly*, **5**, 521–36.

Hepple, L. (1992) Metaphor, geopolitical discourse and the military in South America, in Barnes, T.J. and Duncan, J.S. (eds) *Writing Worlds*, Routledge, London, 136–54.

Herb, G.H. (1989) Persuasive cartography in Geopolitik and National Socialism, *Political Geography Quarterly*, **8**, 289–303.

Herb, G.H. (1997) *Under the Map of Germany: Nationalism and Propaganda 1918–1945*, Routledge, London.

Herf, J. (1984) *Reactionary Modernism: Technology, Culture and Politics in Weimar and the Third Reich*, Cambridge University Press, Cambridge.

Herzfeld, M. (1992) *The Social Production of Indifference: Exploring the Symbolic Roots of Western Bureaucracy*, University of Chicago Press, Chicago.

Heske, H. (1986) German geographical research in the Nazi period, *Political Geography Quarterly*, **5**, 267–81.

Heske, H. (1987) Karl Haushofer: his role in German geopolitics and Nazi politics, *Political Geography Quarterly*, **6**, 135–44.

Hettne, B. (1995) *Development Theory and the Three Worlds*, 2nd edition, Longman, London.

Hildyard, N. (1994) The big brother bank, *Geographical*, June, 26–8.

Hildyard, N. (1997) *The World Bank and the State: A Recipe for Change?*, N. Hildyard, Bretton Woods Project, London.

Hindess, B. and Hirst, P. (1975) *Pre-Capitalist Modes of Production*, Routledge & Kegan Paul, London.

Hipscher, P. (1998) Democratic transitions as protest cycles: social movement dynamics in democratising Latin America, in Meyer, D. and Tarrow, S. (eds) *The Social Movement Society*, Rowman & Littlefield, Lanham, MD.

Hirschmann, A.O. (1958) *The Strategy of Economic Development*, Yale University Press, New Haven, CT.

Hirst, P. and Thompson, G. (1996) *Globalisation in Question*, Polity Press, Cambridge.

Hobsbawm, E. (1995) *The Age of Extremes: The Short Twentieth Century, 1914–1991*, Abacus, London.

Hodges, R. (1988) *Primitive and Peasant Markets*, Blackwell, Oxford.

Hoggart, K. (1990) Let's do away with rural, *Journal of Rural Studies*, **6**, 245–57.

Holbrook, B. and Jackson, P. (1996) The social milieux of two North London shopping centres, *Geoforum*, **27**, 193–204.

Holdar, S. (1992) The ideal state and the power of geography: the life-work of Rudolf Kjellen, *Political Geography*, **11**, 307–23.

Homburger, E. (1994) *The Historical Atlas of New York City*, Henry Holt, New York.

Honey, R. (1983) The social costs of space: providing services in rural areas, *Geographical Perspectives*, **51**, 24–37.

hooks, b. (1992) Eating the other, desire and resistance, in hooks, b. (ed.) *Black Looks: Race and Representation*, South End Press, Boston, MA, 21–39.

Horticultural Crops Development Authority (HCDA) (1997) *Performance of the Horticultural Industry*, HCDA, Nairobi.

Hosking, G. (1992) *A History of the Soviet Union, 1917–1991*, final edition, Fontana Press, London.

Howard, E. (1902/1965) *Garden Cities of Tomorrow*, edited by Osborn, F.J., with a preface by Mumford, L., Faber and Faber, London.

Huillet, C. (1991) New management for rural services, *OECD Observer*, **172**, 17–19.

Humphreys, J. (1988) Social provision and service delivery: problems of equity, health and health care in rural Australia, *Geoforum*, **19**, 323–38.

Huntington, S.P. (1993) The clash of civilizations?, *Foreign Affairs*, **72**, 22–49.

Hutnyk, J. (1997) Adorno at Womad: South Asian crossovers and the limits of hybridity-talk, in Werbner, P. and Modood, T. (eds) *Debating Cultural Hybridity: Multi-Cultural Identities and the Politics of Anti-Racism*, Zed, London, 106–36.

Ilbery, B. (ed.) (1998) *The Geography of Rural Change*, Longman, Harlow.

Ilbery, B. and Bowler, I. (1996) Industrialisation and world agriculture, in Douglas, I., Huggett, R. and Robinson, M. (eds) *Companion Encyclopaedia of Geography*, Routledge, London, 228–48.

Ilbery, B. and Bowler, I. (1998) From agricultural productivism to post-productivism, in Ilbery, B. (ed.) *The Geography of Rural Change*, Longman, London, 57–84.

Ilbery, B. and Holloway, L. (1997) Responses to the challenge of productivist agriculture, *Built Environment*, **23**, 184–91.

Ilbery, B. and Kneafsey, M. (1998) Product and place: promoting quality products and services in the lagging rural regions of the European Union, *European Urban and Regional Studies*, **5**, 329–41.

Ilbery, B. and Kneafsey, M. (2000a) Producer constructions of quality in regional speciality food production: a case study from south west England, *Journal of Rural Studies*, **16**, 217–30.

Ilbery, B. and Kneafsey, M. (2000b) Registering regional speciality food and drink products in the United Kingdom: the case of PDOs and PGIs, *Area*, **32** (forthcoming).

Ilbery, B., Clark, D., Berkeley, N. and Goldman, I. (1995) Telematics and rural development, *European Urban and Regional Studies*, **2**, 55–68.

Ilbery, B., Healey, M. and Higginbottom, J. (1997) On and off-farm business diversification by farm households in England, in Ilbery, B., Chiotti, Q. and Rickard, T. (eds) *Agricultural Restructuring and Sustainability: A Geographical Perspective*, CAB International, Wallingford, 135–51.

Independent on Sunday (1998) March of the child soldiers, 6 December, 15.

Info Box (1996) *Info Box. The Catalogue*, Dirk Nishen, Berlin.

Inglehart, R. (1990) *Culture Shift in Advanced Industrial Societies*, Princeton University Press, Princeton, NJ.

International Broadcasting Trust (IBT) (1998) *The Bank, the President and the Pearl of Africa*, IBT/Oxfam, Oxford.

International Data Corporation (1999) Internet Commerce Market Model, March.

International Labour Office (1999) *Key Indicators of the Labour Market 1999 (KILM)*, ILO, Geneva.

International Monetary Fund (1996) *International Capital Markets: Developments, Prospects and Key Policy Issues*, IMF, Washington, DC.

Iyer, P. (1989) *Video Nights in Kathmandu: Reports from the Not-so-far East*, Black Swan, London.

Jackson, P. (1993) Towards a cultural politics of consumption, in Bird, J., Curtis, B., Putnam, T., Robertson, G. and Tickner, L. (eds) *Mapping the Futures: Local Cultures, Global Change*, Routledge, London.

Jackson, P. (1999) Commodity culture, the traffic in things, *Transactions of the Institute of British Geographers*, NS, **24**, 95–108.

Jackson, P. and Thrift, N. (1995) Geographies of consumption, in Miller, D. (ed.) *Acknowledging Consumption*, Routledge, London, 204–37.

Jackson, R. (1994) Demand for lottery products, *Journal of Consumer Affairs*, **28**, 313–25.

Jacobs, J. (1961) *The Death and Life of Great American Cities*, Penguin, Harmondsworth.

Japan Institute for Social and Economic Affairs (1996) *Japan 1997: An International Comparison*, Keizai Koho Center, Tokyo.

Jarosz, L. (1992) Constructing the dark continent: metaphor of geographic representation of Africa, *Geografiska Annaler*, **74:2**, 105–15.

Jessop, B. (1989) *Thatcherism: A Tale of Two Nations*, Blackwell, Oxford.

Jezkova, P. (1995) Changes in textiles: implications for women, in Mitter, S. and Rowbotham, S. (eds) *Women Encounter Technology: Changing Patterns of Employment in the Third World*, Routledge, London, 93–110.

Johnston, D. (1998) Why territorial development matters, *OECD Observer*, **210**, February–March.

Johnston, R.J. (1997) *Geography and Geographers: Anglo–American Human Geography since 1945*, 5th edition, Arnold, London.

Johnston, R.J., Gregory, D., Pratt, G. and Watts, M. (eds) (2000) *The Dictionary of Human Geography*, 4th edition, Blackwell, Oxford.

Johnston, R.J., Pattie, C.J. and Allsopp, J.G. (eds) (1988) *A Nation Dividing? The Electoral Map of Great Britain 1979–1987*, Longman, London.

Jones, G. and Hollier, G. (1997) *Resources, Society and Environmental Management*, Paul Chapman, London.

Jones, H. (1990) *Population Geography*, 2nd edition, Paul Chapman, London.

Jones, H., Caird, W., Berry, W. and Dewhirst, J. (1986) Peripheral counterurbanisation: findings from an integration of census and survey data in northern Scotland, *Regional Studies*, **20**, 15–26.

Joseph, A. and Phillips, D. (1984) *Accessibility and Utilization: Geographical Perspectives on Health Care Delivery*, Harper & Row, New York.

Judge, D., Stoker, G. and Wolman, H. (eds) (1995) *Theories of Urban Politics*, Sage, London.

K

Kabeer, N. (1994) *Reversed Realities: Gender Hierarchies in Development Thought*, Verso, London.

Kahn, J.S (1995) *Culture, Multiculture, Postculture*, Sage, London.

Kapferer, B. (1988) *Legends of People, Myths of State: Violence, Intolerance and Political Culture in Sri Lanka and Australia*, Smithsonian Institution Press, Washington DC.

Kasson, J.F. (1978) *Amusing the Million: Coney Island at the Turn of the Century*, Hill & Wang, New York.

Katz, R.S. and Mair, P. (eds) (1994) *How Parties Organize: Change and Adaptation in Party Organizations in Western Democracies*, Sage, London.

Kaye, H.J. (1984) *The British Marxist Historians*, Polity Press, Cambridge.

Kearns, G. and Philo, C. (eds) (1993) *Selling Places: The City as Cultural Capital, Past and Present*, Pergamon, Oxford.

Keck, M.E. and Sikkink, K. (1998) *Activists Beyond Borders: Advocacy Networks in International Politics*, Cornell University Press, Ithaca, NY.

Keeble, D. (1993) Small firm creation, innovation and growth and the urban-rural shift, in Curran, J. and Storey, D. (eds) *Small Firms in Urban and Rural Locations*, Routledge, London, 54–78.

Kennedy, P. (1988) *The Rise and Fall of the Great Powers*, Random House, New York.

Kershaw, I. (1993) *The Nazi Dictatorship: Problems and Perspectives of Interpretation*, 3rd edition, Arnold, London.

Khan, M. (1997) Aligarh city: urban expansion and encroachment, *Asian Profile*, **25**, 47–51.

Kipling, R. (1912) *Rudyard Kipling's Verse: Definitive Edition*, Hodder & Stoughton, London.

Kitschelt, H. (1989) *The Logic of Party Formation: Structure and Strategy of Belgian and West German Ecology Parties*, Cornell University Press, Ithaca, NY.

Kitschelt, H. (1990) New social movements and the decline of party organization, in Dalton, R.J. and Kuechler, M. (eds) *Challenging the Political Order: New Social and Political Movements in Western Democracies*, Oxford University Press, Oxford.

Klandermans, B., Roefs, M. and Olivier, J. (1998) A movement takes office, in Meyer, D. and Tarrow, S. (eds) *The Social Movement Society*, Rowman & Littlefield, Lanham, MD.

Knights, D. and Odih, P. (1995) It's about time! The significance of gendered time for financial services consumption, *Time and Society*, **4**, 205–31.

Knox, P. and Agnew, J. (1994) *The Geography of the World Economy*, 2nd edition, Edward Arnold, London.

Knox, P. and Agnew, J. (1998) *The Geography of the World Economy*, 3rd edition, Edward Arnold, London.

Knox, P. and Marston, S.A. (1988) *Human Geography: Places and Regions in Global Context*, Prentice Hall, Upper Saddle River, NJ.

Kobayashi, A. and Peake, L. (1994) Unnatural discourse: 'race' and gender in geography, *Gender, Place and Culture*, **1**, 225–44.

Koczberski, L. (1998) Women in development: a critical analysis, *Third World Quarterly*, **19**, 395–409.

Kondratieff, N.D. (1925) The major economic cycles, *Voprosy Konjunktury*, **6**, 28–79; an English translation is to be found in *Lloyds Bank Review* (1978) **129**.

Kondratieff, N.D. (1935) The long waves in economic life, *Review of Economic Statistics*, **17**, 105–15.

Korpi, W. (1983) *The Democratic Class Struggle*, Routledge & Kegan Paul, London.

Korten, D.C. (1998) Your mortal enemy, *Guardian*, 21 October, 4–5.

Koter, M. (1990) The morphological evolution of a nineteenth-century city centre: Łódź, Poland, 1825–1973, in Slater, T.R. (ed.) *The Built Form of Western Cities*, Leicester University Press, Leicester and London.

Krieger, J. (ed.) (1993) *The Oxford Companion to Politics of the World*, Oxford University Press, Oxford.

L

Lacey, R. (1986) *Ford: The Men and the Machine*, Heinemann, London.

Lakha, S. (1994) The New International Division of Labour and the Indian computer software industry, *Modern Asian Studies*, **28**, 381–408.

Lall, S. (1983) The rise of multinationals from the third world, *Third World Quarterly*, **5**, 618–26.

Landes, D.S. (1969) *The Unbound Prometheus*, Cambridge University Press, Cambridge.

Lane, J.-E., McKay, D. and Newton, K. (1991) *Political Data Handbook: OECD Countries*, Oxford University Press, Oxford.

Langton, J. (1984) The industrial revolution and the regional geography of England, *Transactions of the Institute of British Geographers*, New Series **9**, 145–67.

Langton, J. and Morris, R.J. (eds) (1986) *Atlast of Industrializing Britain, 1780–1914*, Methuen, London and New York.

Lash, S. and Urry, J. (1987) *The End of Organized Capitalism*, Polity Press, Cambridge.

Latour, B. (1993) *We Have Never Been Modern*, Harvester Wheatsheaf, Brighton.

Laulajainen, R. (1998) *Financial Geography*, School of Economics and Commercial Law, Göteborg University.

Lawton, R. (1989) Introduction: aspects of the development and role of great cities in the Western world in the nineteenth century, in Lawton, R. (ed.) *The Rise and Fall of Great Cities*, Belhaven, London.

Laxton, P. (1986) Textiles, in Langton, J. and Morris, R.J. (eds) *Atlas of Industrializing Britain, 1780–1914*, Methuen, London and New York, 106–13.

Le Corbusier (1929/1987) *The City of Tomorrow*, Architectural Press, London.

Lee, R.B. (1968) What hunters do for a living, or how to make out on scarce resources, in Lee, R.B. and DeVore, I. (eds) *Man the Hunter*, Aldine Press, Chicago, 30–48.

Lee, R. (1999) Local money: geographies of autonomy and resistance?, in Martin, R.L. (ed.) *Money and the Space Economy*, Wiley, Chichester.

LeGates, R.T. and Stout, F. (eds) (1996) *The City Reader*, Routledge, London.

Le Heron, R. (1993) *Globalised Agriculture: Political Choice*, Pergamon, Oxford.

Lemon, J.T. (1987) Colonial America in the eighteenth century, in Mitchell, R.D. and Groves, P.A. (eds) *North America: The Historical Geography of a Changing Continent*, Hutchinson, London.

Lester, A. (1999) in Graham, B. and Nash, C. (eds) *Modern Historical Geographies*, Prentice Hall, Harlow.

Lewis, G. (1990) Community through exclusion and illusion: the creating of social worlds in an American shopping mall, *Journal of Popular Culture*, **24**, 121–36.

Leyshon, A. (1995a) Geographies of money and finance, *Progress in Human Geography*, **19**, 531–43.

Leyshon, A. (1995b) Annihilating space? The speed up of communications, in Allen, J. and Hamnett, C. (eds) *A Shrinking World?,* Oxford University Press, Oxford.

Leyshon, A. (1996) Dissolving difference? Money, disembedding and the creation of 'global financial space', in Daniels, P.W. and Lever, W.F. (eds) *The Global Economy in Transition*, Longman, London.

Leyshon, A. and Thrift, N. (1997) *Money Space: Geographies of Monetary Transformation*, Routledge, London.

Liebes, T. and Katz, E. (1993) *The Export of Meaning: Cross-Cultural Readings of Dallas,* 2nd edition, Polity Press, Cambridge.

Ling, P.J. (1990) *America and the Automobile: Technology, Reform and Social Change,* Manchester University Press, Manchester.

Linklater, A. (1998) Citizenship and sovereignty in the post-Westphalian European state, in Daniele, A., Held, D. and Köhler, M. (eds) *Re-imagining Political Community: Studies in Cosmopolitan Democracy*, Polity Press, Cambridge.

Livingstone, D. (1992) *The Geographical Tradition*, Blackwell, Oxford.

Loewenson, R. (1992) *Modern Plantation Agriculture*, Zed Books, London.

Logan, J. and Molotch, H. (1987) *Urban Fortunes. The Political Economy of Place*, University of California Press, Berkeley.

Lowe, P., Murdoch, J. and Cox, G. (1995) A civilised retreat? Anti-urbanism, rurality and the making of an Anglo-centric culture, in Healey, P., Cameron, S., Davoudi, S., Graham, S. and Madani-Pour, A. (eds) *Managing Cities: The New Urban Context*, John Wiley, London.

Lowe, P., Murdoch, J., Marsden, T., Munton, R. and Flynn, A. (1993) Regulating the new rural spaces: the uneven development of land, *Journal of Rural Studies*, **9**, 205–22.

Lowe, P., Ray, C., Ward, N., Wood, D. and Woodward, R. (1998) *Participation in Rural Development: A Review of European Experience*, Centre for Rural Economy, University of Newcastle upon Tyne.

Löytönen, M. (1995) The effects of the HIV epidemic on the population of Europe, in Hall, R. and White, P. (eds) *Europe's Population: Towards the Next Century*, UCL Press, London.

Luke, T. (1991) The discourse of development: a genealogy of developing nations and modernity, *Current Perspectives in Social Theory*, **11**, 271–93.

Luttwak, E. (1990) From geopolitics to geoeconomics, *National Interest*, **20**, 17–24.

M

McAdam, D. (1983) *Political Process and the Development of Black Insurgency*, University of Chicago Press, Chicago.

McCann, M. (1986) *Taking Reform Seriously: Perspectives on Public Interest Liberalism*, Cornell University Press, Ithaca, NY.

McClintock, A. (1994) *Imperial Leather: Race, Gender and Sexuality in the Colonial Context*, Routledge, London.

McDowell, L. (1994) The transformation of cultural geography, in Gregory, D., Martin, R. and Smith, G. (eds) *Human Geography. Society, Space and Social Science*, Macmillan, London, 146–73.

McDowell, L. (1997) *Capital Culture: Gender at Work in the City*, Blackwell, Oxford.

McDowell, L. and Sharp, J.P. (eds) (1997) *Space, Gender and Knowledge: Feminist Readings*, Arnold, London.

McEwan, C. (forthcoming) 'Walking the walk'? The state and gender transformations in the 'new' South Africa, *South African Geographical Journal*.

Machel, G. (1996) UN Study on the Impact of Armed Conflict on Children, http://www.unicef.org/graca/women/html.

Mackay, H. (ed.) (1997) *Consumption and Everyday Life*, Sage, London.

Mackinder, H. (1904) The geographical pivot of history, *Geographical Journal*, **23**, 421–44.

Mackinder, H. (1919) *Democratic Ideals and Reality: A Study in the Politics of Reconstruction*, Constable, London.

McLaughlin, B. (1986) The rural deprivation debate: retrospect and prospect, in Lowe, P., Bradley, T. and Wright, S. (eds) *Deprivation and Welfare in Rural Areas*, Geo Books, Norwich, 43–54.

McLaughlin, E. and Muncie, J. (eds) (1999) Walled cities: surveillance, regulation and segregation, in Pile, S., Brook, C. and Mooney, G. (eds) *Unruly Cities? Order/Disorder*, Routledge, London.

McLoughlin, J. (1991) *The Demographic Revolution*, Faber and Faber, London.

McLuhan, M. (1964) *Understanding Media: Extensions of Man*, Routledge & Kegan Paul, London.

McLuhan, M. (1975) The implications of cultural uniformity, in Bigsby, C.W.E. (ed.) *Superculture, American Popular Culture and Europe*, Bowling Green University Popular Press, Bowling Green, Ohio, 43–56.

McMichael, P. (1996) Globalisation: myths and realities, *Rural Sociology*, **61**, 25–55.

Madon, S. (1997) Information-based global economy and socio-economic development: the case of Bangalore, *Information Society*, **13:3**, 227–43.

Maffesoli, M. (1996) *The Time of the Tribes: The Decline of Individualism in Mass Society*, translated by Don Smith, Sage, London.

Mahon, R. (1987) From Fordism to ?: new technology, labour markets and unions, *Economic and Industrial Democracy*, **8**, 5–60.

Maisel, S.L. (ed.) (1994) *The Parties Respond: Changes in American Parties and Campaigns*, Westview, Boulder, CO.

Maisels, C.K. (1993) *The Emergence of Civilization*, Routledge, London.

Maitland, B. (1985) *Shopping Malls, Planning and Design*, Construction Press, London.

Malbon, B. (1997) Clubbing, consumption, identity and the spatial practices of every-night life, in Skelton, T. and Valentine, G. (eds) *Cool Places*, Routledge, London, 266–86.

Malbon, B. (1999) *Clubbing*, Routledge, London.

Manin, R. (1997) *The Principles of Representative Government*, Cambridge University Press, Cambridge.

Mann, M. (1988) *States, War and Capitalism: Studies in Political Sociology*, Blackwell, Oxford.

Mann, M. (1993) *The Sources of Social Power*, volume II: *The Rise of Classes and Nation-States, 1760–1914*, Cambridge University Press, Cambridge.

Marchand, M.H. and Parpart, J.L. (1995) *Feminism, Postmodernism, Development*, Routledge, London.

Marcuse, H. (1964) *One-Dimensional Man*, Abacus, London.

Markovits, A.S. and Gorski, P.S. (1993) *The German Left: Red, Green and Beyond*, Polity, Cambridge.

Marsden, T. (1997) Creating space for food: the distinctiveness of recent agrarian development, in Goodman, D. and Watts, M. (eds) *Globalising Food*, Routledge, London, 169–91.

Marsden, T. and Arce, A. (1995) Constructing quality: emerging food networks in the rural transition, *Environment and Planning A*, **27**, 1261–79.

Marsden, T., Lowe, P. and Whatmore, S. (eds) (1990) *Rural Restructuring: Global Processes and Local Responses*, Wiley, London.

Marsden, T., Murdoch, J., Lowe, P., Munton, R. and Flynn, A. (1993) *Constructing the Countryside*, UCL Press, London.

Marshall, T.H. (1963) Citizenship and social class, in Marshall, T.H., *Sociology at the Crossroads*, Heinemann, London.

Martin, R.L. (1994) Stateless monies, global financial integration and national economic autonomy: the end of Geography?, in Corbridge, S., Martin, R. and Thrift, N. (eds) *Money, Power and Space*, Blackwell, Oxford, 253–78.

Martin, R.L. (1999) The new economic geography of money, in Martin, R.L. (ed.) *Money and the Space Economy*, Wiley, Chichester, 3–27.

Martin, R.L. and Minns, R. (1995) Undermining the financial basis of regions: the spatial structure and implications of the UK pension fund system, *Regional Studies*, **29**, 125–44.

Martínez, M. (1997) *Olivenza y el tratado de Alcances*, Ayuntamiento de Olivenza.

Marx, K. (1867) *Das Kapital*; translated into English as *Capital*, 3 vols, International Publishers, New York (1967) and Penguin, London (1976).

Marx, K. (1973) *Grundrisse*, Penguin, London.

Massey, D. (1981) The geography of industrial change, in Potter, D., Anderson, J., Clarke, J., Coombes, P., Hall, S., Harris, L., Holloway, C. and Walton, T. (eds) *Society and the Social Sciences*, Open University Press, Milton Keynes, 302-13.

Massey, D. (1988) What's happening to UK manufacturing?, in Allen, J. and Massey, D. (eds) *The Economy in Question*, Sage, London, 45–90.

Massey, D. (1993) Power-geometry and a progressive sense of place, in Bird, J., Curtis, B., Putnam, T., Robertson, G. and Tickner, L. (eds) *Mapping the Futures: Local Cultures, Global Change*, Routledge, London, 59–69.

Massey, D. (1994) *Space, Place and Gender*, Polity, Cambridge.

Massey, D. (1995) *Spatial Divisions of Labour: Social Structures and the Geography of Production*, Macmillan, London.

Massey, D. (1995) Imagining the world, in Allen, J. and Massey, D. (eds) *Geographical Worlds*, Oxford University Press, Oxford, 5–52.

Massey, D. (1999) Cities in the world, in Massey, D., Allen, J. and Pile, S. (eds) *City Worlds*, Routledge in association with the Open University, London, 99–175.

Massey, D. and Jess, P. (1995a) The contestation of place, in Massey, D. and Jess, P. (eds) *A Place in the World? Places, Cultures and Globalization*, Oxford University Press, Oxford, 133–74.

Massey, D. and Jess, P. (1995b) Places and cultures in an uneven world, in Massey, D. and Jess, P. (eds) *A Place in the World? Places, Cultures and Globalization*, Oxford University Press, Oxford, 215–39.

Massey, D. and Jess, P. (eds) (1995), *A Place in the World? Places, Cultures and Globalization*, Oxford University Press, Oxford.

Massey, D., Allen, J. and Pile, S. (eds) (1999) *City Worlds*, Routledge, London.

Mather, A.S. and Chapman, K. (1995) *Environmental Resources*, Longman, London.

Mattelart, A. (1979) *Multinational Corporations and the Control of Culture*, Harvester Press, Brighton.

Maxwell, S. and Singer, H. (1979) Food aid to developing countries: a survey, *World Development*, **7**, 225–47.

May, J. (1996) A little taste of something more exotic, the imaginative geographies of everyday life, *Geography*, **81**, 57–64.

Mehrotra, R. (1977) *Harvard Design Magazine*, Winter/Spring, 25–7.

Meinig, D.W. (1986) *The Shaping of America. A Geographical Perspective on 500 Years of History: Atlantic America, 1492–1800*, Yale University Press, New Haven and London.

Meyer, D.R. (1990) The new industrial order, in Conzen, M.P. (ed.) *The Making of the American Landscape*, Unwin Hyman, London and Winchester, MA.

Middleton, N. (1999) *The Global Casino: An Introduction to Environmental Issues*, 2nd edition, Arnold, London.

Miller, D. (1992) The young and the restless in Trinidad: a case study of the local and the global in mass consumption, in Silverstone, R. and Hirsch, E. (eds) *Consuming Technologies: Media and Information in Domestic Spaces*, Routledge, London, 163–82.

Miller, D. (1994) *Modernity, an Ethnographic Approach: Dualism and Mass Consumption in Trinidad*, Berg, Oxford.

Miller, D. (1995) Consumption as the vanguard of history, a polemic by way of an introduction, in Miller, D. (ed.) *Acknowledging Consumption: A Review of New Studies*, Routledge, London, 1–57.

Miller, D. (1997) Coca-Cola, a black sweet drink from Trinidad, in Miller, D. (ed.) *Material Cultures*, UCL Press, London.

Miller, D. (1998) *A Theory of Shopping*, Polity, Cambridge.

Miller, D.L. (1997) *City of the Century: The Epic of Chicago and the Making of America*, Simon & Schuster, New York.

Milner, M. (1998) IMF puts UK jobs at risk, *Guardian*, 17 July, 23.

Mintel (1996) *Nightclubs and Discotheques*, Mintel, London.

Mitchell, J.B. (ed.) (1962) *Great Britain: Geographical Essay*, Cambridge University Press, Cambridge.

Mitchell, R.D. and Groves, P.A. (eds) (1987) *North America: The Historical Geography of a Changing Continent*, Hutchinson, London.

Mitchell, W.C. (1937) *The Backward Art of Spending Money and Other Essays*, McGraw-Hill, New York.

Mohanty, C.T. (1991) Under Western eyes: feminist scholarship and colonial discourses, in Mohanty, C.T., Russo, A. and Torres, L. (eds) *Third World Women and the Politics of Feminism*, Indiana University Press, Bloomington, 51–80.

Momsen, J. (1991) *Women and Development in the Third World*, Routledge, London.

Moore, R.I. (ed.) (1981) *The Hamlyn Historical Atlas*, Hamlyn, London and New York.

Mormont, M. (1990) Who is rural? Or, how to be rural: towards a sociology of the rural, in Marsden, T., Lowe, P. and Whatmore, S. (eds) *Rural Restructuring: Global Processes and their Responses*, David Fulton, London, 21–45.

Morris, C. and Potter, C. (1995) Recruiting the new conservationists, *Journal of Rural Studies*, **11**, 51–63.

Morris, C. and Young, C. (1997) *A Preliminary Analysis of UK Farm Quality Assurance Schemes: 'New Deals' for the Rural Environment*, Anglo-French Conference of Rural Geographers, Nantes.

Mortishead, C. (1999) Bangalore's program for success, *The Times*, 24 March, 29.

Moser, C. (1993) *Gender, Planning and Development: Theory, Practice and Training*, Routledge, London.

Mosse, G.L. (1975) *The Nationalization of the Masses*, H. Fertig, New York.

Mouffe, C. (1993) Citizenship, in Krieger, J. (ed.) *The Oxford Companion to Politics of the World*, Oxford University Press, Oxford, 138–9.

Mountjoy, A.B. (1976) Worlds without end, *Third World Quarterly*, **2**, 753–7.

Müller-Rommel, F. (1990) New politics parties and new social movements in Western Europe, in Dalton, R.J. and Kuechler, M. (eds) *Challenging the Political Order: New Social and Political Movements in Western Democracies*, Oxford University Press, Oxford.

Munslow, B. and Ekoko, F. (1995) Is democracy necessary for sustainable development?, *Democratisation*, **2**, 158–78.

Murdoch, J. (1995) Actor-networks and the evolution of economic forms, *Environment and Planning A*, **27**, 731–57.

Murdoch, J. (1998) *Counterurbanisation and the Countryside: Some Causes and Consequences of Urban to Rural Migration*, Papers in Environmental Planning Research 15, Cardiff University, Cardiff.

Myers, N. (1991) *Population, Resources and the Environment: The Critical Challenges*, United Nations Population Fund, New York.

Myrdal, G. (1957) *Economic Theory and Underdeveloped Areas*, Duckworth, London.

N

Nash, G.B. *et al.* (1990) *The American People: Creating a nation and a Society*, 2nd edition, Harper & Row, New York and London.

National Agricultural Lands Study (1980) *Where Have All the Farmlands Gone?*, NALS, Washington, DC.

Nelson, B. and Chowdhury, N. (eds) (1994) *Women and Politics Worldwide*, Yale University Press, New Haven, CT.

Netherlands Scientific Council for Government Policy (1992) *Ground for Choices: Four Perspectives for the Rural Areas in the European Community*, Report 42, NSCGP, The Hague.

New Internationalist (1999) *The A–Z of World Development*, New Internationalist, London.

Newman, O. (1972) *Defensible Space: People and Design in the Violent City*, Architectural Press, London.

Nostrand, R.L. (1987) The Spanish borderlands, in Mitchell, R.D. and Groves, P.A. (eds) *North America: The Historical Geography of a Changing Continent*, Hutchinson, London.

Nove, A. (1987) *The Soviet Economic System*, 3rd edition, Allen & Unwin, London.

Nye, D.E. (1992) *Electrifying America: Social Meanings of a New Technology*, MIT Press, Cambridge, MA.

O

O'Brien, R. (1992) *Global Financial Integration: The End of Geography*, Royal Institute of International Affairs, Pinter Publishers, London.

O'Connor, A. (1976) Third World or one world, *Area*, **8**, 269–71.

Odell, P.R. (1989) Draining the world of energy, in Johnston, R.J and Taylor, P.J. (eds) *A World in Crisis?*, 2nd edition, Blackwell, Oxford.

Ogborn, M. (1999) Historical geographies of globalisation, *c.*1500–1800, in Graham, B. and Nash, C. (eds) *Modern Historical Geographies*, Prentice Hall, Harlow.

Ogden, P.E. (1985) Counterurbanisation in France: the results of the 1982 population census, *Geography*, **70**, 24–35.

Ogilvy, J. and Schwartz, P. (2000) *China's Futures*, Wiley, London.

Ohmae, K. (1990) *The Borderless World: Power and Strategy in the Interlinked Economy*, Collins, London.

Ohnuki-Tierney, E. (1997) McDonalds in Japan, changing manners and etiquette, in Watson, J.L. (ed.) *Golden Arches East: McDonalds in East Asia*, Stanford University Press, Stanford, CA, 161–82.

O'Loughlin, J. (ed.) (1994) *Dictionary of Geopolitics*, Greenwood Press, Westport, CT.

O'Loughlin, J. and van der Wusten, H. (1990) Political geography and panregions, *Geographical Review*, 80, 1–20.

Olson, M. (1965) *The Logic of Collective Action: Public Goods and the Theory of Groups*, Harvard University Press, Cambridge, MA.

O'Neill, M. and Black, M. (1996) Current quality issues in the Northern Ireland tourism sector, *Total Quality Magazine*, 8, 15–19.

Oppong, J. and Hodgson, M. (1994) Spatial accessibility to health care facilities in Suhum District, Ghana, *Professional Geographer*, 46, 199–209.

Organization for Economic Co-operation and Development (1991) *New Ways of Managing Services in Rural Areas*, OECD, Paris.

Organization for Economic Co-operation and Development (1993) *What Future for our Countryside?*, OECD, Paris.

Organization for Economic Co-operation and Development (1998) *OECD in Figures 1998,* OECD, Paris.

Organization for Economic Co-operation and Development (OECD) (1999) *Environment in the Transition to a Market Economy*, OECD, Paris.

Origo, I. (1988) *The Merchant of Prato*, Penguin Books, Harmondsworth.

O'Riordan, T. (ed.) (2000) *Environmental Science for Environmental Management*, 3rd edition, Prentice Hall, Harlow.

Østergaard, L. (1992) *Gender and Development: A Practical Guide*, Routledge, London.

Ó Tuathail, G. (1993) The effacement of place: US foreign policy and the spatiality of the Gulf crisis, *Antipode*, 25, 4–31.

Ó Tuathail, G. (1994) Critical geopolitics and development theory: intensifying the dialogue, *Transactions of the Institute of British Geographers*, New Series 19, 228–38.

Ó Tuathail, G. (1996) *Critical Geopolitics: The Politics of Writing Global Space*, Routledge, London.

Ó Tuathail, G. (1998) Postmodern geopolitics? The modern geopolitical imagination and beyond, in Ó Tuathail, G. and Dalby, S. (eds) *Rethinking Geopolitics*, Routledge, London, 16–38.

Ó Tuathail, G. and Dalby, S. (eds) (1998) *Rethinking Geopolitics*, Routledge, London.

Ó Tuathail, G., Dalby, S. and Routledge, P. (eds) (1998) *The Geopolitics Reader*, Routledge, London.

Owens, B. (1996) *The Case for Aid: A Manifesto*, Confederation of Aid Organisations, London.

Oxfam (1993) *Africa: Make or Break: Action for Recovery*, Oxfam, Oxford.

Oxfam (1995) *A Case for Reform: Fifty Years of the IMF and World Bank*, Oxfam, London.

Oxford Dictionary of Current English (1985) Penguin, Harmondsworth.

P

Paasi, A. (1996) *Territories, Boundaries and Consciousness: The Changing Geographies of the Finnish-Russian Frontier*, Wiley, Chichester.

Packard, V. (1977) *The Hidden Persuaders*, Penguin, Harmondsworth.

Pahl, J. (1989) *Money and Marriage*, Macmillan, London.

Painter, J. (1995) *Politics, Geography, and 'Political Geography': A Critical Perspective*, Arnold, London.

Parasuraman, S. (1995) Economic marginalisation of peasants and fishermen due to urban expansion, *Pakistan Development Review*, **34**, 121–38.

Park, C. (1994) *Sacred Worlds: An Introduction to Geography and Religion*, Routledge, London.

Parker, G. (1987) French geopolitical thought in the interwar years and the emergence of the European idea, *Political Geography Quarterly*, **6**, 145–50.

Past Worlds: The Times Atlas of Archaeology (1988) Times Books, London.

Paterson, T.G. (1992) *On Every Front. The Making and Unmaking of the Cold War*, revised edition, W.W. Norton, New York and London.

Peach, W.N. and Constantin, J.A. (1972) *Zimmerman's World Resources and Industries*, 3rd edition, Harper & Row, London.

Pearson, R. (1995) Gender perspectives on health and safety in information processing: learning from international experience, in Mitter, S. and Rowbotham, S. (eds) *Women Encounter Technology: Changing Patterns of Employment in the Third World*, Routledge, London, 278–302.

Peet, R. (1989) The destruction of regional cultures, in Johnston, R.J. and Taylor, P.J. (eds) *A World in Crisis?*, 2nd edition, Blackwell, Oxford, 150–72.

Perry, R., Dean, K. and Brown, B. (eds) (1986) *Counterurbanisation: International Case Studies of Socio-Economic Change in Rural Areas*, Geo Books, Norwich.

Philo, C. (1992) Neglected rural geographies: a review, *Journal of Rural Studies*, **8**, 193–207.

Pile, S. (1999a) The heterogeneity of cities, in Pile, S., Brook, C. and Mooney, G. (eds) *Unruly Cities? Order/Disorder*, Routledge, London, 7–52.

Pile, S. (1999b) What is a city?, in Massey, D., Allen, J. and Pile, S. (eds) *City Worlds*, Routledge in association with the Open University, London, 3–52.

Pile, S., Brook, C. and Mooney, G. (eds) (1999) *Unruly Cities? Order/Disorder*, Routledge, London.

Pinch, S. (1985) *Cities and Services: The Geography of Collective Consumption*, Routledge & Kegan Paul, London.

Pinch, S. and Henry, N. (1999) Paul Krugman's geographical economics, industrial clustering and the British motor sport industry, *Regional Studies*, **33**, 815–27.

Piven, F.F. and Cloward, R. (1971) *Regulating the Poor: The Functions of Public Welfare*, Vintage, New York.

Pletsch, C. (1981) The three worlds or the division of social scientific labour 1950–1975, *Comparative Studies in Society and History*, **23**, 565–90.

Polanyi, K. (1968) *Primitive, Archaic and Modern Economies*, Beacon Press, Boston.

Pollard, J.S. (1996) Banking at the margins: a geography of financial exclusion in Los Angeles, *Environment and Planning A*, **28**, 1209–32.

Pollard, J.S. (1999) Globalisation, regulation and the changing organisation of retail banking in the Unites States and Britain, in Martin, R.L. (ed.) *Money and the Space Economy*, Wiley, Chichester, 49–70.

Poplawska, I. and Muthesius, S. (1986) Poland's Manchester: 19th-century industrial and domestic architecture in Łódz´ *Journal of Social and Architectural History*, 148–60.

Porter, G. (1995) Scenes from childhood, in Crush, J. (ed.) *Power of Development*, Routledge, London, 63–86.

Porter, G. (1997) Mobility and inequality in rural Nigeria: the case of off-road communities, *Tijdschrift voor Economische en Sociale Geografie*, **88**, 65–76.

Porter, M.E. (1990) *The Competitive Advantage of Nations*, Free Press, New York.

Porter, P.W. and Sheppard, E.S. (1998) *A World of Difference: Society, Nature, Development*, Guilford Press, New York.

Potter, C. (1998) Conserving nature: agri-environmental policy development and change, in Ilbery, B. (ed.) *The Geography of Rural Change*, Longman, London, 85–105.

Potter, R., Binns, T., Elliott, J.A. and Smith, D. (1999) *Geographies of Development*, Longman, London.

Potter, R.B. (1979) Perception of urban retailing facilities: an analysis of consumer information fields, *Geografiska Annaler*, **61B**, 19–27.

Pounds, N.J.F. (1947) *An Historical and Political Geography of Europe*, Harrap, London.

Pounds, N.J.G. (1990) *An Historical Geography of Europe*, Cambridge University Press, Cambridge.

Powe, N. and Whitby, M. (1994) Economies of settlement size in rural settlement planning, *Town Planning Review*, **65**, 415–34.

PRB (1998) *World Population Data Sheet 1998*, Population Reference Bureau, Washington DC.

Preston, P.W. (1996) *Development Theory: An Introduction*, Blackwell, Oxford.

PriceWaterhouseCoopers (1999) *Technology Forecast 1999*, 10th Anniversary Edition, PriceWaterhouseCoopers, London.

Priestley, J.B. (1937) *English Journey,* cheap edition, William Heinemann, London.

Pritchard, W. (1998) The emerging contours of the third food regime: evidence from Australian dairy and wheat sectors, *Economic Geography*, **74**, 64–74.

Prokhovnik, R. (1998) Public and private citizenship: from gender invisibility to feminist inclusiveness, *Feminist Review*, **60**, 84–104.

Prunier, G. (1995) *The Rwanda Crisis 1959–1994: History of a Genocide*, Hurst & Co., London.

Przeworski, A. (1986) *Capitalism and Social Democracy*, Cambridge University Press, Cambridge.

R

Radcliffe, S.A. (1999) Re-thinking development, in Cloke, P., Crang, P. and Goodwin, M. (eds) *Introducing Human Geographies*, Arnold, London, 84–92.

Radcliffe, S. and Westwood, S. (1996) *Remaking the Nation: Place, Identity and Politics in Latin America*, Routledge, London.

Rahnema, R. (1997) Introduction, in Rahnema, R. and Bawtree, V. (eds) *The Post-Development Reader*, Zed, London.

Rathgeber, E. (1990) WID, WAD, GAD: trends in research and practice, *Journal of Developing Areas*, **24**, 489–502.

Rau, B. (1991) *From Feast to Famine: Official Cures and Grassroots Remedies in Africa's Food Crisis*, Zed, London.

Ravenhall, Y. (1999) *Women's Investing*, Part 1. http://www.fool.co.uk/personalfinance.

Ray, C. (1998) Culture, intellectual property and territorial rural development, *Sociologia Ruralis*, **38**, 1–19.

Redclift, M. (1984) *Sustainable Development: Exploring the Contradictions*, Methuen, London.

Reddish, A. and Rand, M. (1996) The environmental effects of present energy policies, in Blunden, J. and Reddish, A. (eds) *Energy, Resources and Environment*, Open University and Hodder & Stoughton, London, 43–91.

Rees, J. (1985) *Natural Resources: Allocation, Economics and Policy,* 2nd edition, Routledge, London.

Rees, J. (1991) Resources and environment: scarcity and sustainability, in Bennett, R. and Estall, R. (eds) *Global Change and Challenge: Geography in the 1990s*, Routledge, London, 5–26.

Reeves, R. (1999) Do we really need gay florists?, *Observer*, 8 August.

Renfrew, C. (1996) Language families and the spread of farming, in Harris, D.R. (ed.) *The Origins and Spread of Agriculture and Pastoralism in Eurasia*, UCL Press, London, 70–92.

Ribeiro, A. (1935) Últimas conquistas e definação territorial, in Peres, D. (ed.) *História de Portugal*. vol. 2, Portucalense Editoria, Porto, 251–73.

Righter, R. (1991) *Utopia Lost: The United Nations and the World Order,* Twentieth Century Fund Press, London.

Rist, G. (1997) *The History of Development*, Zed, London.

Ritzer, G. (1996) *The McDonaldization of Society: An Investigation into the Changing Character of Contemporary Social Life*, Pine Forge Press, California.

Robert, S. and Randolph, W. (1983) Beyond decentralization: the evolution of population distribution in England and Wales, 1961–81, *Geoforum*, **14**, 75–102.

Robins, K. (1991) Tradition and translation: national culture in its global context, in Corner, J. and Harvey, S. (eds) *Enterprise and Heritage: Crosscurrents of National Culture*, Routledge, London, 21–44.

Robins, K. (1995) The new spaces of global media, in Johnston, R.J., Taylor, P. and Watts, M. (eds) *Geographies of Global Change*, Blackwell, Oxford, 248–62.

Robinson, G. (1997) Greening and globalising: agriculture in 'the new times', in Ilbery, B., Chioti, Q. and Rickard, T. (eds) *Agricultural Restructuring and Sustainability: A Geographical Perspective*, CAB International, Wallingford.

Robinson, G. and Ilbery, B. (1993) Reforming the CAP: beyond McSharry, in Gilg, A. (ed.) *Progress in Rural Policy and Planning*, vol. 3, Belhaven Press, London, 197–207.

Robinson, J. (1999) Divisive cities, in Pile, S., Brook, C. and Mooney, G. (eds) *Unruly Cities? Order/Disorder*, Routledge, London, 149–200.

Rogers, A., Viles, H. and Goudie, A. (eds) (1992) *The Student's Companion to Geography*, Blackwell, Oxford.

Rogers, R. and Gumuchdjian, P. (1997) *Cities for a Small Planet*, Faber and Faber, London.

Rollins, W.H. (1995) Whose landscape? Technology, Fascism and environmentalism on the National Socialist Autobahn, *Annals of the Association of American Geographers*, **85**, 494–520.

Rose, N. and Miller, P. (1992) Political power beyond the state: problematics of government, *British Journal of Sociology*, **43**, 173–205.

Rossler, M. (1989) Applied geography and area research in Nazi society: central place theory and planning, *Environment and Planning D: Society and Space*, **7**, 419–31.

Rostow, W.W. (1960) *The Stages of Economic Growth: A Non-Communist Manifesto*, Cambridge University Press, Cambridge.

Rostow, W.W. (1971) *The Stages of Economic Growth,* 2nd edition, Cambridge University Press, Cambridge.

Rostow, W.W. (1975) *How it All Began: The Origins of the Modern Economy,* Methuen, London.

Rubenstein, J.M. (1992) *The Changing US Auto Industry: A Geographical Analysis*, Routledge, London.

Rubin Hudson, B. (1979) Aesthetic ideology and urban design, *Annals of the Association of American Geographers*, **69**, 339–61.

Rueschemeyer, D., Stephens, E.H. and Stephens, J.D. (1992) *Capitalist Development and Democracy*, Polity Press, Cambridge.

S

Sabel, C. (1989) Flexible specialisation and the re-emergence of regional economies, in Hirst, P and Zeitlin, J. (eds) *Industrial Decline? Industrial Structure and Policy in Britain and her Competitors,* Berg, Oxford.

Sachs, W. (1992) *The Development Dictionary,* Zed, London.

Sack, R. (1992) *Place, Modernity and the Consumers World,* Johns Hopkins University Press, Baltimore, MD.

Sack, R.D. (1986) *Human Territoriality: Its Theory and History,* Cambridge University Press, Cambridge.

Salway, P. (1981) Roman Britain, in *Oxford History of England,* Oxford University Press, Oxford.

Sant, M. and Simons, P. (1993) The conceptual basis of counterurbanisation: critique and development, *Australian Geographical Studies,* **31**, 113–26.

Sarre, P. and Blunden, J. (1995) *An Overcrowded World? Population, Resources and Environment,* Oxford University Press, Oxford.

Sassen, S. (1991) *The Global City: New York, London, Tokyo,* Princeton University Press, Princeton, NJ.

Sassen, S. (1994) *Cities in a World Economy,* Pine Forge Press, London.

Schilder, A. and Boeve, B. (1996) An institutional approach to local economic development, *Regional Development Dialogue,* **17**, 94–117.

Schneider, G. and Schubert, P. (1997) *Wer baut wo? Dies grossen Bauprojekte im Überblick,* Jaron Verlag, Berlin.

Schramm, G. (1993) Rural electrification in LDCs as a tool for economic development, *OPEC Review,* **17**, 501–17.

Schumpeter, J. (1939) *Business Cycles: A Theoretical, Historical and Statistical Analysis of the Capitalist Process,* McGraw-Hill, London.

Scott, A.J. (1988) *New Industrial Space,* Pion Press, London.

Scott, J. and Simpson, P. (eds) (1991) *Sacred Spaces and Profane Spaces: Essays in the Geographics of Judaism, Christianity and Islam,* Greenwood Press, New York.

Sennett, R. (1990) *The Conscience of the Eye: Design and Social Life in Cities,* Faber and Faber, London.

Serow, W.J. (1991) Recent trends and future prospects for urban-rural migration in Europe, *Sociologia Ruralis,* **31**, 269–80.

Serrão, J.V. (1979) *História de Portugal,* vol. 1, Verbo, Lisbon.

Service, E.R. (1971) *Primitive Social Organization,* 2nd edition, Random House, New York.

Seton-Watson, H. (1977) *Nations and States: An Enquiry into the Origins of Nations and the Politics of Nationalism,* Westview Press, Boulder, CO.

Shanin, T. (ed.) (1971) *Peasants and Peasant Societies,* Penguin, Harmondsworth.

Sharp, J. (1999) Critical geopolitics, in Cloke, D., Crang, P. and Goodwin, M. (eds) *Introducing Human Geographies,* Arnold, London, 179–88.

Shaw, A. (1994) Urban growth and land-use conflicts: the case of New Bombay, *Bulletin of Concerned Asian Scholars,* **26**, 33–44.

Shen, J. (1995) Rural development and rural to urban migration in China 1978–1990, *Journal of Rural Studies,* **26**, 395–409.

Shiva, V. (1991) *The Violence of the Green Revolution: Third World Agriculture, Ecology and Politics,* Zed, London.

Shiva, V. (1998) *Biopiracy: The Plunder of Nature and Knowledge,* Green Books, Devon.

Shiva, V. (1999) We will drive Monsanto out of the country, *Asian Age,* 6 March, 9.

Shiva, V. and Moser, I. (1995) *Biopolitics: A Feminist and Ecological Reader on Biotechnology,* Zed, London.

Short, J.R. (1996) *The Urban Order: An Introduction to Cities, Culture and Power,* Blackwell, Oxford.

Shucksmith, M. (1993) Farm household behaviour and the transition to post-productivism, *Journal of Agricultural Economics*, **44**, 466–78.

Shucksmith, M., Chapman, P., Clark, G. and Black, S. (1994) Social welfare in rural Europe, in Copus, A. and Marr, P. (eds) *Rural Realities, Trends and Choices*, Proceedings of the 35th EAAE Seminar, Scottish Agricultural College, Aberdeen, 199–219.

Shurmer-Smith, P. and Hannam, K. (1994) *Worlds of Desire, Realms of Power: A Cultural Geography*, Arnold, London.

Sibley, D. (1995) *Geographies of Exclusion: Society and Difference in the West*, Routledge, London.

Sidaway, J.D. (1994) Geopolitics, geography and 'terrorism' in the Middle East, *Environment and Planning D: Society and Space*, **12**, 357–72.

Sidaway, J.D. (1998) What is in a Gulf? From the 'arc of crisis' to the Gulf War, in Ó Tuathail, G. and Dalby, S. (eds) *Rethinking Geopolitics*, Routledge, London and New York, 224–39.

Sidaway, J.D. (1999) American power and the Portuguese Empire, in Slater, D. and Taylor, P.J. (eds) *The American Century*, Blackwells, Oxford, 195–209.

Sidaway, J.D. (2000) Iberian geopolitics, in Dodds, K., and Atkinson, D. (eds) *Geopolitical Traditions: A Century of Geopolitical Thought*, Routledge, London, 118–49.

Sidaway, J.D. and Pryke, M. (2000) The free and the unfree: 'emerging markets', the Heritage Foundation and the 'index of economic freedom', in Bryson, J.R., Daniels, P.W., Henry, N.D. and Pollard, J.S. (eds) *Knowledge, Space, Economy*, Routledge, London.

Siddiquee, N.A. (1996) Health for all by 2000? An assessment of the delivery of health care services in rural Bangladesh, *South Asia*, **19**, 49–70.

Simmel, G. (1978) *The Philosophy of Money*, Routledge & Kegan Paul, London.

Simmel, G. (1991) Money in modern culture, *Theory, Culture and Society*, **8: 3**, 17–31.

Simmons, I.G. (1996) *Changing the Face of the Earth: Culture, Environment, History*, 2nd edition, Blackwell, Oxford.

Simpson, E.S. (1994) The *Developing World: An Introduction*, 2nd edition, Longman, Harlow.

Sittirak, S. (1998) *The Daughters of Development: Women and the Changing Environment*, Zed, London.

Sjöberg, K. (1993) *The Return of the Ainu: Cultural Mobilization and the Practice of Ethnicity in Japan*, Harwood Academic Publishers, London.

Skelton, T. and Valentine, G. (eds) (1998) *Cool Places: Geographies of Youth Cultures*, Routledge, London.

Skidmore, T.E. and Smith, P.H. (1988) *Modern Latin America*, 2nd edition, Oxford University Press, New York.

Skinner, A. (ed.) (1970) *Adam Smith: The Wealth of Nations*, Penguin, London and New York.

Skocpol, T. (1992) *Protecting Soldiers and Mothers: The Political Origins of the American Welfare State*, Harvard University Press, Cambridge, MA.

Slater, D. (1992) On the borders of social theory: learning from other regions, *Environment and Planning D: Society and Space*, **10**, 307–27.

Slater, D. (1993) The geopolitical imagination and the enframing of development theory, *Transactions of the Institute of British Geographers*, New Series **18**, 419–37.

Slater, D. (1997) *Consumer Culture and Modernity*, Polity Press, Cambridge.

Smith, A. (1776) *An Inquiry into the Nature and Causes of the Wealth of Nations*, edited by Skinner, A. (1974) Penguin, London.

Smith, A. (1977) *The Wealth of Nations*, edited by Andrew Skinner, Penguin, Harmondsworth.

Smith A.D. (1986) *The Ethnic Origin of Nations*, Blackwell, Oxford.

Smith, M.D. (1996) The empire filters back, consumption, production, and the politics of Starbucks coffee, *Urban Geography*, **17**, 502–24.

Smith, N. (1984) Isiah Bowman: political geography and geopolitics, *Political Geography Quarterly*, **3**, 69–76.

Soja, E. (1996) *Thirdspace: Journeys to Los Angeles and Other Real and Imagined Places*, Blackwell, Oxford.

Soja, E. (1997) Six discourses on the postmetropolis, in Westwood, S. and Williams, J. (eds) *Imagining Cities: Scripts, Signs, Memory*, Routledge, London, 19–30.

Sondheimer, S. (ed.) (1991) *Women and the Environment: A Reader*, Zed Books, London.

Speer, A. (1995) *Inside the Third Reich*, Phoenix, London.

Spencer, D. (1995) Counterurbanisation: the local dimension, *Geoforum*, **26**, 153–73.

Spivak, G.C. (1985) Three women's texts and a critique of imperialism, *Critical Inquiry*, **12**: **1**, 243–61.

Spivak, G.C. (1993) *Outside in the Teaching Machine*, Routledge, London.

Staehli, L., Kodras, J. and Flint, C. (eds) (1997) *State Devolution in America*, Sage, Beverly Hills, CA.

Stamp, L.D. and Beaver, S. (1933) *The British Isles: A Geographic and Economic Survey*, Longman, London.

Stamp, L.D. and Beaver, S. (1963) *The British Isles: A Geographic and Economic Survey*, 5th edn, Longman, London.

Stempel, J.D. (1981) *Inside the Iranian Revolution*, Indiana University Press, Bloomington.

Stenning, A. and Bradshaw, M. (1999) Globalization and transformation: the changing geography of the Post-Socialist world, in Bryson, J., Henry, N., Keeble, D. and Martin, R. (eds) *The Economic Geography Reader: Producing and Consuming Global Capitalism*, Wiley, Chichester, 88–96.

Stille, A. (1997) *Excellent Cadavers*, Vintage, London.

Stöhr, W.B. and Taylor, D.R.F. (1981) *Development from Above or Below? The Dialectics of Regional Planning in Developing Countries*, John Wiley, Chichester.

Stoker, G. (1995) Regime theory and urban politics, in Judge, D., Stoker, G. and Wolman, H. (eds) *Theories of Urban Politics*, Sage, London.

Storey, D.J. (1997) *Understanding the Small Business Sector*, Thomson, London.

Storper, M. (1993) Regional worlds of production: learning and innovation in the technology districts of France, Italy and the USA, *Regional Studies*, **27**, 433–55.

Strange, S. (1996) *Casino Capitalism*, Blackwell, Oxford.

Strange, S. (1999) *Mad Money: When Markets Outgrow Governments*, University of Michigan Press, Ann Arbor.

Stutz, F.P. and De Souza, A.R. (1998) *The World Economy: Resources, Location, Trade and Development*, 3rd edition, Prentice-Hall, Upper Saddle River, NJ.

Sugden, D. (1982) *Arctic and Antarctic*, Basil Blackwell, Oxford.

Sugrue, T. (1996) *The Origins of the Urban Crisis. Race and Inequality in Postwar Detroit*, Princeton University Press, Princeton, NJ.

Sylvester, C. (1996) Picturing the Cold War: an art graft/eye graft, *Alternatives*, **21**, 393–418.

Tarrant, J. (1980) The geography of food aid, *Transactions of the Institute of British Geographers*, New Series **5**, 125–40.

Taussig, M. (1997) *The Magic of the State*, Routledge, London.

Taylor, P.J. (1989) *Political Geography*, Longman, London.

Taylor, P.J. (1994) From heartland to hegemony: changing the world in political geography, *Geoforum*, **25**, 403–17.

Taylor, P.J. and Flint, C. (2000) *Political Geography: World-Economy, Nation State and Locality*, Prentice Hall, Harlow.

Thiongo, N.W. (1986) *Decolonising the Mind: The Politics of Language in African Literatures*, James Currey, London.

Thompson, E.P. (1963) *The Making of the English Working Class*, Penguin, London and New York.

Thompson, E.P. (1994) The moral economy of the English crowd in the eighteenth century, in Thompson, E.P. (ed.) *Customs in Common*, Penguin, Harmondsworth.

Thompson, G. (ed.) (1999) *Economic Dynamism in the Asia Pacific Region*, Routledge, London.

Thrift, N. (1996) A hyperactive world, in Johnston, R.J., Taylor, P.J. and Watts, M. (eds) *Geographies of Global Change*, Blackwell, Oxford.

Thrift, N. (1998) Virtual capitalism: the globalisation of reflexive business knowledge, in Carrier, J.G. and Miller, D. (eds) *Virtualism: A New Political Economy*, Berg, Oxford.

Tickamyer, A. and Duncan, C. (1990) Poverty and opportunity structure in rural America, *Annual Review of Sociology*, **16**, 67–86.

Tickell, A. (1996) Making a melodrama out of a crisis: reinterpreting the collapse of Barings Bank, *Environment and Planning D: Society and Space*, **14**, 5–33.

Tilly, C. (1984) Social movements and national politics, in Bright, C. and Harding, S. (eds) *Statemaking and Social Movements: Essays in History and Theory*, University of Michigan Press. Ann Arbor, MI.

Tilly, C. (1985) War-making and state-making as organized crime, in Evans, P., Rueschemeyer, R. and Skocpol, T. (eds) *Bringing the State Back In*, Princeton University Press, Princeton, NJ.

Tilly, C. (1986) *The Contentious French*, Harvard University Press, Cambridge, MA.

Times Atlas of World History (1989) 3rd edition, Guild Publishing, London.

Toennies, F. (1887) *Community and Society* (1957 edn), Michigan State University Press, East Lansing, MI.

Tomlinson, J. (1991) *Cultural Imperialism: A Critical Introduction*, Pinter, London.

Tordoff, W. (1997) *Government and Politics in Africa*, 3rd edition, Macmillan, London.

Touraine, A. (1981) *The Voice and the Eye: An Analysis of Social Movements*, Cambridge University Press, Cambridge.

Tratado de Limites (1866) *Tratado de Limites entre Portugal e Espanha assinado em Lisboa aos 29 de Setembro de 1864*, Imprensa Nacional, Lisbon.

Tunstall, J. (1977) *The Media are American*, Constable, London.

Tyler May, E. (1989) Cold War – Warm Hearth: politics and the family in postwar America, in Fraser, S. and Gerstle, G. (eds) *The Rise and Fall of the New Deal Order, 1930*, Princeton University Press, Princeton, NJ, 153–81.

Tyler, M. and Abbott, P. (1998) Chocs away: weight watching in the contemporary airline industry, *Sociology*, **32**, 433–50.

U

Understanding Cities (1999) DD304, Open University, Milton Keynes.

UNESCO (1994) *Statistical Yearbook*, United Nations Educational, Scientific and Cultural Organization, Paris.

United Nations (UN) (1999) *World Population Prospects: The 1998 Revision*, United Nations Population Division, New York.

United Nations Conference on Trade and Development (UNCTAD) (1998) *World Investment Report 1998*, UNCTAD, Geneva.

United Nations Conference on Trade and Development (1999) *World Investment Report 1999: Foreign Direct Investment and the Challenge of Development*, UNCTAD, Geneva.

United Nations Development Programme (UNDP) (1993) *Human Development Report 1993*, Oxford University Press, Oxford.

United Nations Development Programme (UNDP) (1997) *Human Development Report 1997*, Oxford University Press, Oxford.

United Nations Development Programme (UNDP) (1998) *Human Development Report 1998*, Oxford University Press, Oxford.

United Nations Development Programme (UNDP) (1999) *Human Development Report 1999*, Oxford University Press, Oxford.

United Nations Environment Programme (UNEP) (1999) *Global Environmental Outlook 2000*, UNEP-Earthscan Publications, London.

Unwin, T. (ed.) (1994) *Atlas of World Development*, Wiley, Chichester.

Urry, J. (1990) *The Tourist Gaze: Leisure and Travel in Contemporary Societies*, Sage, London.

US Department of Commerce (1998) *The Emerging Digital Economy*, US Department of Commerce, Washington DC (can be found on the Internet at http://www.commerce.gov/emerging.htm).

V

van de Kaa, D.J. (1987) Europe's second demographic transition, *Population Bulletin*, **42**, 1–57.

Vastoia, A. (1997) Perceived quality and certification: the case of organic fruit, in Schiefer, G. and Helbig, R. (eds) *Quality Management and Process Improvement for Competitive Advantage in Agriculture and Food*, Proceedings of the 49th Seminar of the European Association of Agricultural Economists, Bonn.

Veldman, J. (1984) Proposal for a theoretical basis for the human geography of rural areas, in Clark, G., Groenendijk, J. and Thissen, F. (eds) *The Changing Countryside*, Geo Books, Norwich, 17–26.

Verba, S., Schlozman, K.L. and Brady, H.E. (1995) *Voice and Equality: Civic Voluntarism in American Politics*, Harvard University Press, Cambridge, MA.

Vining, D. and Kontuly, T. (1978) Population dispersal from major metropolitan regions: an international comparison, *International Regional Science Review*, **3**, 49–73.

Vitebsky, P. (1996) The northern minorities, in Smith, G. (ed.) *The Nationalities Question in the Post-Soviet States*, Longman, London, 94–112.

W

Walby, S. (1994) Is citizenship gendered?, *Sociology*, **28**, 379–95.

Walker, A.S. (1999) Women's initiatives and activities worldwide, *IWTC Women's Globalnet #126*, International Women's Tribune Centre, 23 July, http://womensnet.org.za/news.

Walker, R. (1996) Another round of globalization in San Francisco, *Urban Geography*, **17**, 60–94.

Wallace, I. (1985) Towards a geography of agribusiness, *Progress in Human Geography*, **9**, 491–514.

Wallerstein, I. (1979) *The Capitalist World Economy*, Cambridge University Press, Cambridge.

Wallerstein, I. (1980) Imperialism and development, in Bergeson, A. (ed.) *Studies of the Modern World System*, Academic Press, New York.

Wallerstein, I. (1991) *Geopolitics and Geoculture: Essays on the Changing World System*, Cambridge University Press, Cambridge.

Walmsley, D.J., Epps, W.R. and Duncan, C.J. (1995) *The New South Wales North Coast 1986–1991: Who Moved Where, Why and with What Effect?*, Australian Government Publishing Service, Canberra.

Ward, D. (1987) Population growth, migration, and urbanization, 1860–1920, in Mitchell R.D. and Groves, P.A. (eds) *North America, the Historical Geography of a Changing Continent*, Hutchinson, London.

Ward, D. (1989) *Poverty, Ethnicity and the American City, 1840–1925*, Cambridge University Press, Cambridge.

Warf, B. (1999) The hypermobility of capital and the collapse of the Keynesian state, in Martin, R.L. (ed.) *Money and the Space Economy*, Wiley, Chichester.

Watnick, M. (1952–3) The appeal of Communism to the peoples of underdeveloped areas, *Economic Development and Cultural Change*, 1, 22–36.

Watson, J.L. (1997a) Introduction, transnationalism, localization, and fast foods in East Asia, in Watson J.L. (ed.) *Golden Arches East: McDonalds in East Asia*, Stanford University Press, Stanford, CA, 1–38.

Watson, J.L. (1997b) McDonalds in Hong Kong, consumerism, dietary change, and the rise of a children's culture, in Watson, J.L. (ed.) *Golden Arches East: McDonalds in East Asia*, Stanford University Press, Stanford, CA, 77–109.

Watson, S. (1999) City politics, in Pile, S., Brook, C. and Mooney, G. (eds) *Unruly Cities? Order/Disorder*, Routledge, London, 201–45.

Watts, M. (1993a) Development I: power, knowledge, discursive practice, *Progress in Human Geography*, 17, 257–72.

Watts, M. (1993b) The geography of post-colonial Africa: space, place and development in sub-Saharan Africa 1960–1993, *Singapore Journal of Tropical Geography*, 14, 173–90.

Weber, A. (1929) *Alfred Weber's Theory of the Location of Industries*, Chicago University Press, Chicago.

Webster, D. (1988) *Looka Yonder! The Imaginary America of Populist Culture*, Comedia, London.

Weeks, J.R. (1988) The demography of Islamic nations, *Population Bulletin*, 43: 4, 1–54.

Werbner, P. (1997) Introduction: The dialectics of cultural hybridity, in Werbner, P. and Modood, T. (eds) *Debating Cultural Hybridity: Multi-Cultural Identities and the Politics of Anti-Racism*, Zed, London, 1–26.

Werbner, P. and Modood, T. (eds) (1997) *Debating Cultural Hybridity: Multi-Cultural Identities and the Politics of Anti-Racism*, Zed, London.

Whatmore, S. (1995) From farming to agribusiness: the global agro-food system, in Johnston, R., Taylor, P. and Watts, M. (eds) *Geographies of Global Change*, Blackwell, Oxford, 36–49.

Whatmore, S. and Thorne, L. (1997) Nourishing networks: alternative geographies of food, in Goodman, D. and Watts, M. (eds) *Globalising Food*, Routledge, London, 287–304.

Whittaker, D.J. (1997) *United Nations in the Contemporary World*, Routledge, London.

Whittlesey, D. (1936) Major agricultural regions of the Earth, *Annals of the Association of American Geographers*, 26, 199–240.

Wiener, M.J. (1985) *English Culture and the Decline of the Industrial Spirit, 1850–1980*, Penguin, Harmondsworth.

Wilkins, P. (1983) *Women and Engineering in the Plymouth Area: Job Segregation and Training at Company Level*, Equal Opportunities Commission, London.

Williams, C.C. (1996) Local exchange and trading systems: a new source of work and credit for the poor and unemployed?, *Environment and Planning A*, 28, 1395–415.

Williams, C.L. (1993) Introduction, in Williams, C.L. (ed.) *Doing 'Women's Work': Men in Nontraditional Occupations*, Sage, London, 1–9.

Williams, K. (1998) *Get me a Murder a Day! A History of Mass Communication in Britain*, Arnold, London.

Willis, P. (1990) *Uncommon Culture*, Open University Press, Milton Keynes.

Wilson, A. (1992) Technological utopias, world's fairs and theme parks, in Wilson, A. (ed.) *The Culture of Nature: North American Landscape from Disney to the Exxon Valdez*, Blackwell, Oxford, 157–90.

Winckler, V. (1987) Women and work in contemporary Wales, *Contemporary Wales: An Annual Review of Economic and Social Research*, 1, 53–72.

Winichakul, T. (1996) Maps and the formation of the geo-body of Siam, in Tønnesson, S. and Antlöv, H. (eds) *Asian Forms of the Nation*, Curzon Press, Richmond, 67–91.

Wirth, L. (1938) Urbanism as a way of life, *American Journal of Sociology*, 44, 1–24, reprinted in Hart, P.K. and Reiss, A.J. (eds) (1957) *Cities and Society*, Free Press, Chicago.

Wolf, E.R. (1966) *Peasants*, Prentice-Hall, Englewood Cliffs, NJ.

Wolfe-Phillips, L. (1987) Why Third World – origins, definitions and usage, *Third World Quarterly*, 9, 1311–19.

Women and Geography Study Group (1997) *Feminist Geographies: Explorations in Diversity and Difference*, Longman, Harlow.

Wood, G. (1985) *Labelling in Development Policy: Essays in Honour of Bernard Schaffer*, Sage, London.

World Bank (1997) *The State in a Changing World*, World Development Report, World Bank, Washington DC.

World Bank (1998) *World Development Indicators 1998*, World Bank, Washington DC.

World Bank (1998b) *World Bank Atlas 1998*, World Bank, Washington DC.

World Bank (1999) *World Development Indicators*, World Bank, Washington DC.

World Bank (1999) *World Development Indicators on CD-Rom*, World Bank, Washington DC.

World Commission on Environment and Development (WCED) (1987) *Our Common Future*, Oxford University Press, Oxford.

World Trade Organization (WTO) (1997) *Annual Yearbook*, WTO, Geneva.

Wrigley, N. and Lowe, M. (eds) (1996) *Retailing, Consumption and Capital: Towards the New Retail Geography*, Longman, Harlow.

Wuyts, M., Mackintosh, M. and Hewitt, T. (eds) (1992) *Development Policy and Public Action*, Oxford University Press, Oxford.

Y

Yan, Y. (1997) McDonalds in Beijing, the localization of Americana, in Watson, J.L. (ed.) *Golden Arches East: McDonalds in East Asia*, Stanford University Press, Stanford, CA, 39–76.

Young, I.M. (1989) Polity and group difference: a critique of the ideal of universal citizenship, *Ethics*, 9, 250–74.

Young, M. and Wilmott, P. (1957) *Family and Kinship in East London*, Routledge & Kegan Paul, London.

Yuval-Davis, N. (1997) *Gender and Nation*, Sage, London.

Z

Zelinsky, W. (1971) The hypothesis of the mobility transition, *Geographical Review*, 61, 219–49.

Zelizer, V. (1989) The social meaning of money: 'special monies', *American Journal of Sociology*, 95, 342–77.

Zlotnik, H. (1998) International migration 1965–96: an overview, *Population and Development Review*, 24, 429–68.

Zukin, S. (1995) *The Cultures of Cities*, Blackwell, Oxford.

INDEX

Chp 7
Chp 4
Chp 5
Chp 6
Chp 12
Chp 11